MARINE BIOLOGY

CLAM FISHERIES AND AQUACULTURE

MARINE BIOLOGY

Additional books in this series can be found on Nova's website
under the Series tab.

Additional e-books in this series can be found on Nova's website
under the e-book tab.

FISH, FISHING AND FISHERIES

Additional books in this series can be found on Nova's website
under the Series tab.

Additional e-books in this series can be found on Nova's website
under the e-book tab.

MARINE BIOLOGY

CLAM FISHERIES AND AQUACULTURE

FIZ DA COSTA GONZÁLEZ
EDITOR

nova publishers
New York

For permission to use material from this book please contact us:
Telephone 631-231-7269; Fax 631-231-8175
Web Site: http://www.novapublishers.com

Library of Congress Cataloging-in-Publication Data

Clam fisheries and aquaculture / editor, Fiz da Costa Gonzalez.
p. cm.
Includes index.
ISBN 978-1-62257-518-3 (hardcover)
1. Clam fisheries. 2. Clam culture. I. Costa Gonzalez, Fiz da.
SH373.C53 2012
639.9'744--dc23

2012025361

Published by Nova Science Publishers, Inc. † New York

CONTENTS

PREFACE

Mollusca is one of the largest phyla in the animal kingdom, over 50,000 species having been described, thus comprising more than a half of the marine species. Molluscs are scientifically and economically important invertebrates, representing 8% of all captured marine species. One of the six molluscan classes is Bivalvia, which is composed of both marine and freshwater animals enclosed in shells with two valves. Among these, the groups of greatest economic importance are oysters, scallops, mussels and clams. This book deals with clams in the broad sense of the term, since there are several bivalve families falling into the "clam" category. The selection of the families is primarily based on their respective economic importance, in terms of fisheries and aquaculture production. These selected families are Veneridae (venus shells), Mactridae (surf clams), Donacidae (wedge shells), Myidae (softshell clams), Pharidae and Solenidae (razor clams). It must be remarked that, in some chapters or sections within them, species belonging to other families of clams are cited.

When I began to compile and edit this book I reflected on the large number of comprehensive and high-quality books concerned with clam biology and aquaculture available at the moment. However, I came to the conclusion that, though there are several books which are very good sources of information, some of these have not been updated, while others have focused on just a single species. Consequently, the purpose of this volume is both to gather together information on the six families of clams, referring to those aspects of biology that influence fisheries and aquaculture, and to provide information about the current status of clam culture and fisheries in a worldwide context.

Chapter 1 is a brief introduction to clams, and provides information about the importance of their captures and aquaculture production within a worldwide context, their taxonomy, the morphological features which define each family, their basic anatomy and habitat preferences. Chapter 2 gathers together information about clam distribution from open-access databases, and provides distribution maps for each clam family. A review of published data about the reproductive cycles of clams is conducted in Chapter 3, describing the different phases of their gametogenic cycles, how a study of them can be tackled and the different factors affecting clam reproduction. Chapter 4 describes the genetic techniques currently available that serve as tools for species identification, conservation biology and stock management, and clam production. Diseases and parasites affecting clams are covered in Chapter 5, whilst Chapter 6 reviews the neoplasms (or tumors) that affect different species of clams. The microbiota associated with clams in natural beds and hatcheries are dealt with in Chapters 7 and 8, respectively. Chapter 9 deals with clam hatchery and nursery culture in land-based facilities, whilst Chapter 10 analyses the perspectives of intermediate culture and grow-out of clams in a specific region (Galicia, NW Spain), where more than 90% of Spanish molluscs are produced. Chapter 11 describes the organization and management tools used in Galicia for artisanal exploitation of clam beds, as an example of how a region or country can manage the production of shellfish resources. Chapter 12 reviews clam fisheries worldwide, analyzing the main species captured, harvesting methods currently employed and the fishing impacts they produce. Chapter 13 deals with the habitat, fisheries and aquaculture of *Galatea paradoxa*, a freshwater clam, as an example of the diversity of habitats occupied by clams. Finally, Chapter 14 reviews currect knowledge about clams as a tool for ecotoxicological studies, highlighting the ecological significance of this group of species and their usefulness with regard to the assessment of coastal pollution.

The names of all the species included in this book have been checked for verification status in *WORMS* (World Register of Marine Species) (http://www.marinespecies.org), due to the huge variability in the use of different scientific names and the use of synonyms for a single species existing in the literature. Our purpose has been to provide the reader with the species names and classification according to an authoritative, updated source. Moreover, the species were also placed in families according to the *WORMS* classification, in order to achieve uniformity within this book. When the species name checked was found not to be included in the *WORMS* database, it was kept unchanged (i.e. as cited in the original source of data).

My gratitude is due to all the authors of the chapters for devoting their scant time to contribute to this book, for their great perseverance in completing their chapters, and for their invaluable advice. I would also like to thank the reviewers who assisted the authors and editor in enhancing the quality of the content of this book. In addition, I would like to acknowledge the postdoctoral grant provided by the Fundación Juana de Vega, which helped me to bring the editing of this volume to a successful conclusion. Finally, thanks are due to the team at Nova Publishers, and especially to Carra Feagaiga, for their patience in dealing with all my questions.

Fiz da Costa
March, 2012

LIST OF REVIEWERS

All the chapters of this book were anonymously peer-reviewed by the following experts. Their contribution is highly acknowledged.

Daniel Adjei-Boateng
Kwame Nkrumah University of Science and Technology. Kumasi, Ghana

Daniel C. Allen
University of Michigan. Michigan, US

Alejo Carballeira
Universidad de Santiago de Compostela. Santiago de Compostela, Spain

Rudo von Cosel
Muséum National d'Histoire Naturelle. Paris, France

Susana Darriba
Instituto Tecnolóxico para o Control do Medio Mariño de Galicia (INTECMAR). Vilagarcía de Arousa, Spain

Camino Gestal
Instituto de Investigaciones Marinas, Consejo Superior de Investigaciones Científicas (CSIC). Vigo, Spain

Bruno Gómez-Gil
CIAD Mazatlan Unit for Aquaculture and Environmental Management. Mazatlan, Mexico

Salvador Guerrero
Centro de Investigacións Mariñas (CIMA). Vilanova de Arousa, Spain

David Iglesias
Centro de Investigacións Mariñas (CIMA). Vilanova de Arousa, Spain

Sandra Joaquim
Instituto Nacional de Recursos Biológicos, I.P. (IPIMAR). Olhão, Portugal

Alexandra Leitão
Instituto Nacional de Recursos Biológicos, I.P. (IPIMAR). Olhão, Portugal

César Lodeiros
Universidad de Oriente. Cumaná, Venezuela

Inés Martínez-Pita
I.F.A.P.A. Centro "Agua del Pino", Cartaya, Spain

Enrique Morsán
Universidad Nacional del Comahue. San Antonio Oeste, Argentina

José María Orensanz
Centro Nacional Patagónico (CONICET). Puerto Madryn, Argentina

Melita Peharda
Institute of Oceanography and Fisheries. Split, Croatia

Emanuele Ponis
ISPRA National Institute for the Protection and Environmental Research. Chioggia, Italy

Emilia Quesada
Universidad de Granada. Granada, Spain

In: Clam Fisheries and Aquaculture
Editor: Fiz da Costa González
ISBN: 978-1-62257-518-3

Chapter 1

INTRODUCTION TO THE BIOLOGY OF CLAMS

F. da Costa*

Centro de Cultivos Marinos de Ribadeo-CIMA,
Muelle de Porcillán, Ribadeo (Lugo), Spain

1.1. INTRODUCTION

This chapter aims to provide the reader of this book with basic information about general aspects of clam biology, such as systematics, morphological features of each family, anatomy and habitat, which are not covered in depth in the rest of the chapters. Its purpose is both to serve as a starting point for the more specialized information contained in the rest of the book and to fill the gaps of chapters that are lacking in this volume. This book deals with clams in a broad sense and the following families were selected to be included on the basis of their economic importance: Veneridae (venus shells), Mactridae (surf clams), Donacidae (wedge shells), Myidae (softshell clams), Pharidae and Solenidae (razor clams).

Clam production and aquaculture are important economic resources in a worldwide context. Despite the fact that FAO statistics do not always reflect the real situation of a single species or country, they provide a good overview of the worldwide context of captures and aquaculture. Moreover, in some of the families listed below, the production data does not reach species level. In other cases, e.g. *Siliqua patula* Dixon, 1789, recreational fishery is much more important than its commercial counterpart (Roach et al., 2011). Table 1 shows the most important species of each family in terms of captures and aquaculture production. Within these six families of clams, those which have a greater volume of landings from capture fisheries are Veneridae and Mactridae, according to FAO statistics. Regarding aquaculture data, Veneridae head the field in terms of both production and value, and the Manila clam *Venerupis* (=*Ruditapes*) *philippinarum* A. Adams and Reeve, 1850 is the species which contributes most, by far, within Veneridae.

* E-mail address: fiz.da.costa.gonzalez@ifremer.fr; present address: Ifremer, Laboratoire de Physiologie des Invertébrés Marins, Station Expérimentale d'Argenton, Presqu'île du Vivier, Landunvez, France

Table 1. Volume of landings from capture fisheries and aquaculture operations and value of cultured clam´s families in 2009. Production in tonnes and value in thousands of US dollars. Source: Production and aquaculture database 1950-2009 of the Fishstat software (FAO, 2010)

		Captures	Aquaculture	
	Species	Production	Production	Value
DONACIDAE	*Donax* spp.	930	0	0
	Total	**930**	**0**	**0**
MACTRIDAE	*Spisula solidissima*	122,899	0	0
	Mactromeris (=Spisula) polynyma	25,594	0	0
	Total	**153,070**	**4**	**5**
MYIDAE	*Mya arenaria*	9,979	701	484
	Total	**9,979**	**701**	**484**
SOLENOIDEA	*Solen* spp.[1]	2,323	3	10
	Siliqua patula	244	0	0
	Sinonovacula constricta[2]	0	683,806	615,425
	Total	**4,375**	**683,809**	**615,435**
VENERIDAE	*Venerupis (=Ruditapes) philippinarum*	53,997	3,248,013	3,034,814
	Chamelea gallina	46,111	0	0
	Mercenaria mercenaria	5,021	27,004	67,508
	Meretrix spp.	18,880	0	0
	Meretrix lusoria	1,396	51,884	76,930
	Paphia spp.	14,891	0	0
	Leukoma (=Protothaca) thaca	15,739	0	0
	Tawera gayi	7,725	0	0
	Callista chione	2,239	0	0
	Venerupis decussata (=Ruditapes decussatus)	1,445	3,516	31,364
	Venerupis corrugata (=V. pullastra)	2,159	164	2,514
	Total	**183,917**	**3,332,502**	**3,222,047**

[1]FAO do not differentiate at species level for razor clams within Europe. [2]FAO include this species in family Solecurtidae.

1.1.1. Family Donacidae

Despite the importance of Donacidae or wedge shell species for artisanal and small-scale fishing, FAO statistics do not differentiate between species. *Donax trunculus* Linnaeus, 1758 represents an important fishery in Portugal, France, Spain and Italy, and may account for most of the recorded catches of FAO in these countries for genus *Donax*.

1.1.2. Family Mactridae

Mactridae include genera such as *Anatina* Schumacher, 1817, *Lutraria* Lamarck, 1799, *Mactra* Linnaeus, 1767, *Mactrinula* Gray, 1853, *Meropesta* Iredale, 1929, and *Spisula* Gray, 1837. *Mactra quadrangularis* (=*M. veneriformis*) Reeve, 1854 is a popular low-cost seafood abundant in Chinese coastal areas, especially in the coastal shoals of Jiangsu province (Wang et al., 2011). Also, *Mactra chinensis* Philippi, 1846 is a commercially important bivalve in China (Li et al., 2011). *Spisula solidissima* Dillwyn, 1817 (Atlantic surfclam) is a large and commercially important bivalve distributed along the Gulf of Maine, where populations are limited to the north shore of Massachusetts, the south-eastern portion of Maine, and Georges Bank. In the United States, Surf clams represented approximately 25% of total harvested molluscs, with landings valued at $38 million in 2007.

S. solidissima ranked second in total landings behind *Placopecten magellanicus* Gmelin, 1791 (Shumway et al., 1994; Hare et al., 2010). Most of the landings have been reported from waters off New Jersey since 1980, but landings per unit effort have recently decreased in that region, while they have increased off Long Island (Hare et al., 2010; Marzec et al., 2010). Regarding FAO statistics, *S. solidissima* is the species which accounts for the higher percentage of captures. It has to be noted that no *Mactra* species are recorded in FAO statistics (Table 1), although some works have mentioned *Mactra* species as important commercial species (Li et al., 2011; Wang et al., 2011).

1.1.3. Family Myidae

Mya contains most species of this family. *M. arenaria* Linnaeus, 1758, *M. baxteri* Coan and Scott, 1997, *M. pseudoarenaria* Schlesch, 1931 and *M. truncata* Linnaeus, 1758, are among the well-researched species. *M. arenaria* fishing and culture is an important socioeconomic activity in many small communities in the USA and Canada; however, production has generally declined, mainly as a result of overfishing, environmental degradation and diseases (Beal and Vencile, 2001; Beal, 2006). It is the only species within *Mya* recorded in FAO statistics (Table 1).

1.1.4. Superfamily Solenoidea (Families Pharidae and Solenidae)

In some sections of the chapter, information for families Pharidae and Solenidae is placed together in the superfamily Solenoidea, due to the similarities between the two families in the aspects dealt with in those parts of the chapter. The razor clams included in the superfamily Solenoidea include important commercial species worldwide, such as *Siliqua patula*, *Ensis directus* Conrad, 1843 and *Sinonovacula constricta* Lamarck, 1818. Research into these species is rendered more difficult by sampling difficulty, their fast deep burrowing ability, and the low volume of worldwide production compared with other bivalves (Cosel, 1990; Guerra Diaz et al., 2011). Regarding captures in wild beds, the main species are *S. patula* and *E. directus*, while *S. patula* fishery has transformed from commercial fishery to recreational fishery in many states within the USA (Roach et al., 2011). *E. directus* accounted for half of the captures of razor clams within the USA in 2009. *E. directus* was accidentally introduced into Europe at the end of the 1970's (Cosel et al., 1982; Essink, 1985, 1986) and has now spread to many countries in Europe (Cosel, 2009), accounting for 57% of the landings of razor clams in Europe in 2009, and reaching high densities in The Netherlands. Within Europe, FAO statistics do not differentiate at species level and all captures are included under the denomination of *Solen* spp. The aquaculture production of razor clams is exclusively concentrated on the culture of *S. constricta* in China, primarily in the southern Zhejiang and Fujian provinces, where seeds are usually collected from the wild in September-November (Su, 2006 in Yan et al., 2009).

1.1.5. Family Veneridae

Venus shells or Veneridae include the most important representatives of clam species in the world. For example, *Venerupis philippinarum* or Manila clam, which has been introduced to various parts of the world since the beginning of the twentieth century, is by far the most commonly cultured clam species. The hard clam *Mercenaria mercenaria* Linnaeus, 1758 is an important recreational and commercial species harvested in the United States (Kraeuter and Castagna, 2001). FAO statistics recorded a higher aquaculture production than fishery of wild stocks in 2009. *Meretrix lusoria* Röding, 1798 is a commercially important bivalve in Korea, Japan and China (Chung, 2007), being the second most commonly cultured clam species in the world in 2009 (FAO, 2010). Despite the fact that other species belonging to genus *Meretrix*, such as *M. meretrix* Linnaeus, 1758, are an important commercial species in coastal areas of South and Southeast Asia, including China, Korea, Japan and India (Ho and Zheng, 1994), it is not reported in FAO statistics. Other species which account for a lower capture production are also valuable as aquaculture resources, such as *Venerupis decussata* (=*Ruditapes decussatus*) Linnaeus, 1758 and *Venerupis corrugata* (=*V. pullastra*) Gmelin, 1791.

1.2. TAXONOMY

The current taxonomic position of the six clam families is provided following the classification provided by the World Register of Marine Species (WORMS) (http://www.marinespecies.org). If not it will be noted in the text. These families of clams are included within Class Bivalvia Linnaeus, 1758; Subclass Heterodonta Neumayr, 1884 and Infraclass Euheterodonta.

1.2.1. Family Donacidae

The *Donax* genus was first established by Linnaeus in 1758 and the family Donacidae was created by Fleming (1828). It was assigned to Tellinacea by Olsson (1931), Vokes (1980) and Ward and Blackwelder (1987). Later, it was placed in Mactroidea by Ward (1992) and moved to Tellinoidea by Coan et al. (2000). Nevesskaja (2009) included this family in Donacoidea (bean clams or wedge clams). This family includes ten genera (Bouchet, 2011). The current taxonomic position of this family is shown in Table 2.

Table 2. Current systematic position of Donacidae

Order Veneroida GRAY, 1854
Superfamily Tellinoidea BLAINVILLE, 1824
Family Donacidae FLEMING, 1828

1.2.2. Family Mactridae

The type genus of trough shells or duck clams is *Mactra* Linnaeus, 1767. Mactridae was included in the suborder Mactracea of the order Teleodesmacean by Dall (1889), and several authors assigned Mactridae to this suborder (Glenn, 1904; Olsson, 1931; Vokes, 1980; Ward and Blackwelder, 1987). The current classification can be seen in Table 3 which includes 5 subfamilies.

Table 3. Current systematic position of Mactridae

Order Veneroida GRAY, 1854
Superfamily Mactroidea LAMARCK, 1809
Family Mactridae LAMARCK, 1809[1]
Subfamily Mactrinae LAMARCK, 1809
Subfamily Lutrariinae GRAY, 1853
Subfamily Kymatoxinae STENZEL and KRAUSE, 1957
Subfamily Tanysiphoninae SCARLATO and STAROBOGATOV, 1971
Subfamily Zenatiinae DALL, 1895

[1]Subfamily content following Bouchet et al. (2010).

1.2.3. Family Myidae

The Superfamily Myoidea has three families, Corbulidae, Erodonidae and Myidae. Some important genera such as *Distugonia* Iredale, 1936, *Paramya* Conrad, 1860, *Sphenia* Turton, 1822, *Tugonella* Jousseaume, 1891, *Tugonia* Gray, 1842, and *Mya* Linnaeus, 1758 are placed in the family Myidae (softshell clams). The order Myacea was established by Stoliczka (1870) as an order including the families Myidae, Mactridae, Anatinidae, Saxicavidae, Glauconomyidae and Solenidae. Later, it was assigned to the suborder Myacea by Fischer (1887). Newell (1965) emended it to order Myina, including the superfamilies Myoidea, Gastrochaenoidea and Hiatelloidea. Its currents taxonomic position is shown in Table 4.

1.2.4. Superfamily Solenoidea (Families Pharidae and Solenidae)

The genus *Solen* was first established by Linnaeus (1758) and the family Solenidae (razor clams) was proposed by Lamarck in 1809. Later, some of the species previously assigned to *Solen* by Linnaeus were included in the newly established genera *Cultellus* and *Ensis* (Schumacher, 1817).

Table 4. Current systematic position of Myidae

Order Myoida STOLICZKA, 1870
Superfamily Myoidea LAMARCK, 1809
Family Myidae LAMARCK, 1809[1]
Subfamily Myinae LAMARCK, 1809
Subfamily Cryptomyinae HABE, 1777
Subfamily Spheniinae F.R. BERNARD, 1983

[1]Subfamily content following Bouchet et al. (2010).

Table 5. Systematic arrangement of the Solenoidea by different authors

H. Adams and A. Adams, 1856:
Family Solenidae
Subfamily Soleninae
Subfamily Pharinae

Tryon, 1884:
Superfamily Solenaceae
Family Solenidae
Subfamily Soleninae
Subfamily Pharellinae

Ghosh, 1920:
Family Solenidae
Subfamily Soleninae
Subfamily Novaculininae
Subfamily Solecurtinae

Thiele, 1935:
Family Solenidae
Subfamily Soleninae
Subfamily Glaucomyidae

Vokes (1967):
Superfamily Solenoidea
Family Pharellidae
Family Solenidae

Table 6. Current systematic position of Pharidae and Solenidae

Order uncertain
Superfamily Solenoidea LAMARCK, 1809
Family Pharidae H. ADAMS and A. ADAMS, 1856[1]
Subfamily Pharinae H. ADAMS and A. ADAMS, 1856
Subfamily Siliquinae BRONN, 1862
Subfamily Pharellinae STOLICZKA, 1870
Subfamily Novaculininae GHOSH, 1920
Subfamily Cultellinae DAVIES, 1935
Family Solenidae LAMARCK, 1809

[1]Subfamily content following Cosel (1990, 1993) with the adjustments of Bouchet et al. (2010).

The systematic arrangement of this family by different authors is shown in Table 5. The Superfamily Solenoidea (Lamark, 1809) was placed in the Heterodonta subclass (Neumayr, 1884), and Order Veneroida (H. Adams and A. Adams, 1856) (Bieler and Mikkelsen, 2006; CLEMAM, 2011). However, molecular studies placed the Solenoidea in Euheterodonta (Giribet and Distel, 2003; Taylor et al., 2007). Table 6 shows the current classification.

1.2.5. Family Veneridae

The Veneridae (venerids or venus clams) is the most species rich bivalve family, with over 800 extant species in approximately 170 genera (Mikkelsen et al., 2006). It was first established by Rafinesque in 1815 and it was included in the order Veneracea (H. Adams and A. Adams, 1856). The early classifications (e.g. Adams and Adams, 1857; Chenu, 1862; Gill, 1871; Tryon, 1884; Fischer, 1887) grouped Veneridae with Petricolidae and Glauconomidae (Mikkelsen et al., 2006). The systematic arrangement of this family by different authors is shown in Table 7. A phylogeny based on morphology and molecules was produced by Mikkelsen et al. (2006) who proposed a subfamily classification that is accepted in the Nomenclator of Bivalve Families of Bouchet et al. (2010) (Table 8).

Table 7. Systematic arrangement of the Veneridae by different authors

Deshayes, 1853:
Family Veneridae
- Subfamily or tribe Dosiniana
- Subfamily or tribe Meretriciana
- Subfamily or tribe Venusina
- Subfamily or tribe Tapesina

Fischer, 1887:
Family Veneridae
- Tribe Meretricinse
- Tribe Venerinse
- Tribe Tapetinae

Dall, 1902:
Family Veneridae
- Subfamily Dosininae
- Subfamily Meretricinae
- Subfamily Venerinae
- Subfamily Gemminae

Jukes-Browne, 1914:
Family Veneridae
- Subfamily Dosininae-Meretricinae
- Subfamily Venerinae
- Subfamily Gemminae

Keen, 1969:
Family Veneridae
- Subfamily Chioninae
- Subfamily Circinae
- Subfamily Clementinae
- Subfamily Cyclinae
- Subfamily Dosiniinae
- Subfamily Gemminae
- Subfamily Meretricinae
- Subfamily Pitarinae
- Subfamily Samaranginae
- Subfamily Sunettinae
- Subfamily Tapetinae
- Subfamily Venerinae

Habe 1977:
Followed Keen (1969) classification and divided Subfamily Pitarinae in:
- Subfamily Lioconchinae
- Subfamily Callistinae

Table 8. Current systematic position of Veneridae

Order Veneroida GRAY, 1854
Superfamily Veneroidea RAFINESQUE, 1815
Family Veneridae RAFINESQUE, 1815[1]
Subfamily Venerinae RAFINESQUE, 1815
Subfamily Callocardiinae DALL, 1895
Subfamily Chioninae FRIZZELL, 1936
Subfamily Clementiinae FRIZZELL, 1936
Subfamily Dosiniinae DESHAYES, 1853
Subfamily Gemminae DALL, 1895
Subfamily Gouldiinae STEWART, 1930
Subfamily Lioconchinae HABE, 1777
Subfamily Meretricinae GARY, 1847
Subfamily Samarangiinae KEEN, 1969
Subfamily Petricolinae d'ORBIGNY, 1840
Subfamily Sunettinae STOLOCZKA, 1870
Subfamily Tapetinae GRAY, 1851
Subfamily Turtoniinae W. CLARK, 1855

[1]Subfamily content following Milkkensen et al. (2006).

1.3. MORPHOLOGY

The main morphological features of the six families studied here are equivalve shells (Table 9, Figures 1-6). However, shells vary from equilateral or inequilateral in Mactridae to inequilateral in the others. Hinge teeth are absent in Myidae, whilst the other families have cardinal and lateral teeth (Table 9). They all exhibited a pallial sinus.

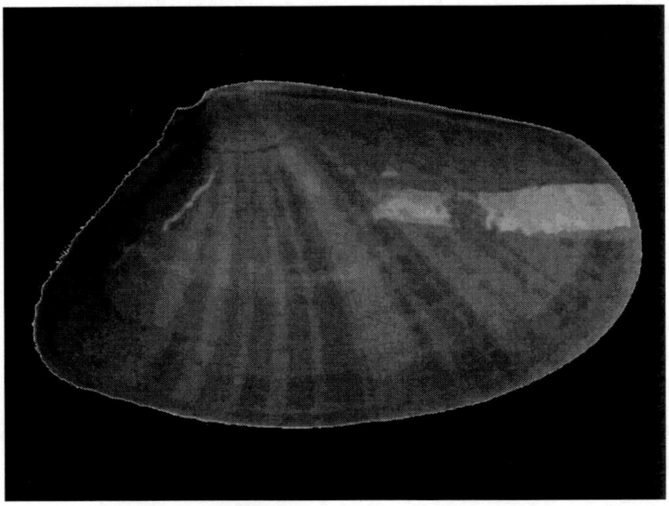

Figure 1. *Donax trunculus.* Family Donacidae.

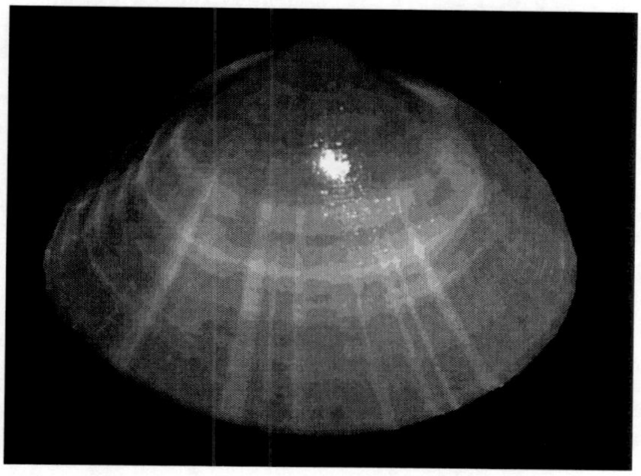

Figure 2. *Mactra stultorum* Linnaeus, 1758. Family Mactridae. Photo credit: Hans Hillewaert. Reproduced with permission.

Figure 3. *Mya truncata.* Familiy Myidae. Photo credit: Claude Nozères. Reproduced with permission.

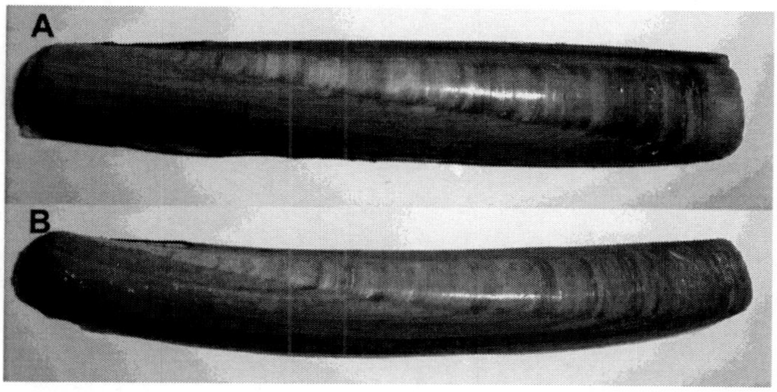

Figure 4. A. *Ensis siliqua* Linnaeus, 1758. B. *Ensis magnus* (=*E. arcuatus*) Schumacher, 1817. Family Pharidae.

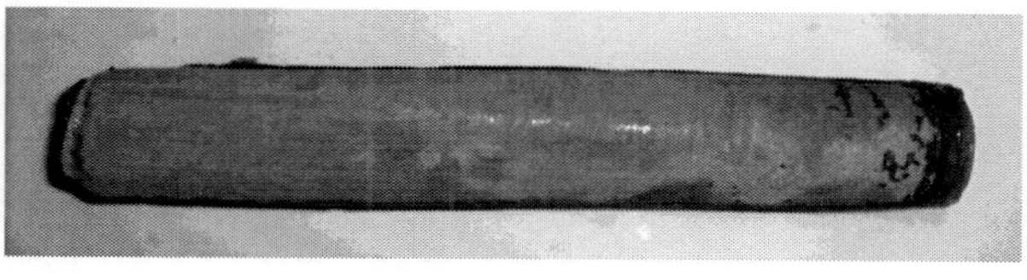

Figure 5. *Solen marginatus* Pulteney, 1799. Family Solenidae.

Table 2. Morphological characteristic of six studied families

Family	Shells	Hinge teeth and ligament	Other characteristics	Siphons	Foot
Donacidae	Equivalve or nearly so, usually solid, compressed, trigonal to wedge-shaped, not gaping. Inequilateral, with a shorter and somewhat obliquely truncated posterior end. Umbones opisthogyrate. Posterodorsal slope is often differently sculptured than the rest of the shell. Interior of shell porcelaneous. Margin crenulated or smooth.	Two small cardinal teeth in each valve, the strongest commonly bifid, and more or less developed lateral teeth. Ligament external, in a groove behind umbones.	Two adductor muscle scars, about equal in size. Pallial sinus deep, largely confluent ventrally with the pallial line. Cruciform muscle scars obscure. Internal margins crenulated or smooth. Cruciform muscles present. Mantle margins wide open antero-ventrally.	Quite short, naked, separated from their base, with six lobes on top.	Strong, laterally compressed.
Mactridae	Equivalve shell, subequilateral to inequilateral, ovate or trigonal to transversely elongated, closed to somewhat gaping posteriorly. Umbones prosogyrate, more or less prominent. Outer surface of the shell smooth or, mainly, concentrically sculptured, often with an obvious periostracum. Interior of shell porcelaneous.	Each valve has two cardinal teeth and smooth or striated, more or less developed, lateral teeth. Cardinal teeth of the left valve forming an inverted V-shaped process. Delicate additional cardinal lamellae often present in either valve. External ligament short and not prominent, just behind the umbones; internal ligament well developed, set in each valve in a deep trigonal pit of the hinge plate and pointing towards the umbo.	Two, often sub-equal, adductor muscle scars. Pallial line with a well-developed sinus. Internal margins usually smooth. Mantle margins smooth, more or less cuticularly united or fused ventrally, with a large pedal opening anteriorly and an additional aperture beneath the inhalant siphon.	United, generally rather short, naked or sheathed with an expansion of the periostracum, papillate on top.	Large and compressed, heeled, without a byssus.
Myidae	Shell thick to thin and fragile, equivalve and inequilateral. Fragile, widely oval to irregular oblong with a posterior gape, glossy, often whitish with smooth sculpture or irregular commarginal growth lines with a thin periostracum.	Absent, but possesses a chondrophore to support the resilium. Left valve with a platform for the internal ligament. Right valve with a pit under the umbo to accommodate the corresponding ligament from the left valve.	Pallial sinus sometimes present. Pallial line undulating and interrupted. Dimyarian-type adductor muscle scars almost equal.	One long, fused siphon.	Compressed and grooved, with a byssal gland and a byssus to anchor to the substrate.

Family	Shells	Hinge teeth and ligament	Other characteristics	Siphons	Foot
Pharidae	Equivalve and inequilateral. Extremely thin and fragile to more or less solid shells, very short and oval to very long and slender, more or less compressed laterally, gaping anteriorly and posteriorly. Valves straight to considerably curved with dorsal margin concave and ventral margin convex, or oval to jackknife-shaped with both dorsal and ventral margin or only ventral margin more or less convex. Beaks terminal, subterminal or subcentral. Ends blunt with rounded corners or more or less rounded to semicircular and often tapering. Interior with or without reinforcements in front of and behind the hinge.	Left valve with two vertical cardinal and two more or less subhorizontal cardinal teeth behind it. Right valve with one vertical cardinal and one subhorizontal cardinal. Two middle teeth in the left valve can be partly or totally merged to form one strongly bicuspid or bifid tooth. External ligament.	Anterior adductor scar short and nearly circular to very long and narrow. Posterior adductor scar relatively small and united to the posterior pallial line or separate from it. Pallial sinus from short to very short, rounded, square, trapeze-shaped or irregular triangular.	Very short to rather long, separate or fused, retractible or not, with or without surrounding tentacles.	Compressed laterally, at the end obliquely truncated with a flattened anterior surface and a surrounding keel.
Solenidae	Very thin and fragile to strong and solid shells, from moderately elongated to very long and slender, rectangular with straight dorsal and ventral margins, slightly curved with concave dorsal and convex ventral margin, or knife-shaped, with both ventral and dorsal margin or only the ventral margin slightly convex.	Hinge uniform, each valve with just one cardinal tooth. Lateral teeth are lacking. External ligament.	Anterior adductor scar elongated, from very short and oval to very long, narrow and slender.		
Solenidae	Shells more or less inflated laterally, in cross-section oval-tubular, gaping anteriorly and posteriorly, with square, blunt or more or less rounded ends with vertical or positively or negatively oblique margins. Exterior with or without a more or less pronounced furrow parallel to the anterior margin or with only a slight depression. Beaks terminal or just subterminal.		Posterior adductor scar only somewhat elongated, oval to triangular, united with the upper part of the posterior pallial line, just touching it or well separated. Pallial sinus from short to very short, trapezoid to triangular or nearly square. Periostracum from thin to quite thick and folding over the margin of the valves.	Siphonal tubes completely fused, with transverse constrictions, extensible beyond the posterior end of the valves and dropped by autotomy when the animal is disturbed.	Cylindrical and club-shaped foot which is dilated towards its end and which ends more or less conically with a blunt tip.

Table 2. (Continued)

Family	Shells	Hinge teeth and ligament	Other characteristics	Siphons	Foot
Veneridae	Shell mostly solid, equivalve or subequivalve, obliquely rounded, or ovate to subtrigonal in outline and usually not gaping; inequilateral, with generally prominent, prosogyrate umbones, at or in front of the midline of the shell. Lunule and/or escutcheon usually present. Sculpture only commarginal, or with a radial component. Periostracum most of the time inconspicuous. Interior of shell porcelaneous.	Ligament external, behind the umbones, often inserted in a deep groove. Hinge with 3 usually radially disposed cardinal teeth in each valve (1 or more of which may be grooved or bifid), anterior lateral teeth sometimes present.	Two more or less equal adductor muscle scars, the posterior sometimes slightly larger. Pallial sinus usually present, in some species indistinct. Internal margins smooth to denticulate. Mantle broadly open ventrally.	Siphons short to long, naked, fused or separate, with simple tentacles on tips and inside the inhalant opening to strain out large particles.	Large and rather short, hatchet-shaped, rarely byssate in the adult.

Figure 6. Clam species within Veneridae family. A. *Venerupis corrugata* (=*V. pullastra*). B. *Venerupis* (=*Ruditapes*) *philippinarum*. C. *Venerupis decussata* (=*Ruditapes decussatus*).

1.4. Anatomy

The general characteristics of bivalve mollusc anatomy are compiled in some books such as Gosling (2002) and Grizel and Auffret (2003). There are also book chapters dealing with specific families or species dealt with in this volume (Eble, 2001; Darriba and López, 2011). In this section the general features of clams' anatomy are presented; however, more detailed information can be found in the aforementioned manuals and in specific publications on each family. Photographs of the general appearance of the internal features of the shell and the soft tissue anatomy of Veneridae (Figure 7), Donacidae (Figure 8) and Pharidae (Figure 9) are shown in this section to illustrate the families discussed in this volume. The umbo or hinge area, where the valves are joined together, is the dorsal part of the animal, while the region opposite is the ventral margin. The foot is located in the anterior-ventral position and the siphons are in the posterior area.

1.4.1. Shell

The shell consists of 3 layers, one on top of the other, mainly composed of calcite and segregated by the mantle. These are: the periostracum (the horny outermost layer consisting of conchiolin), the prismatic layer that forms most of the shell, and the nacreous layer (the inner layer). Muscle scars can be seen on the inner surface of valves (Figures 7-9). Alongside the shell, without reaching its ends, is located the pallial scar (dorsal, ventral, anterior and posterior), which outlines the area where the mantle is attached to the valves. At the posterior end, this pallial scar varies in shape and size according to the family, with the open part towards the end of the shell, called the pallial sinus.

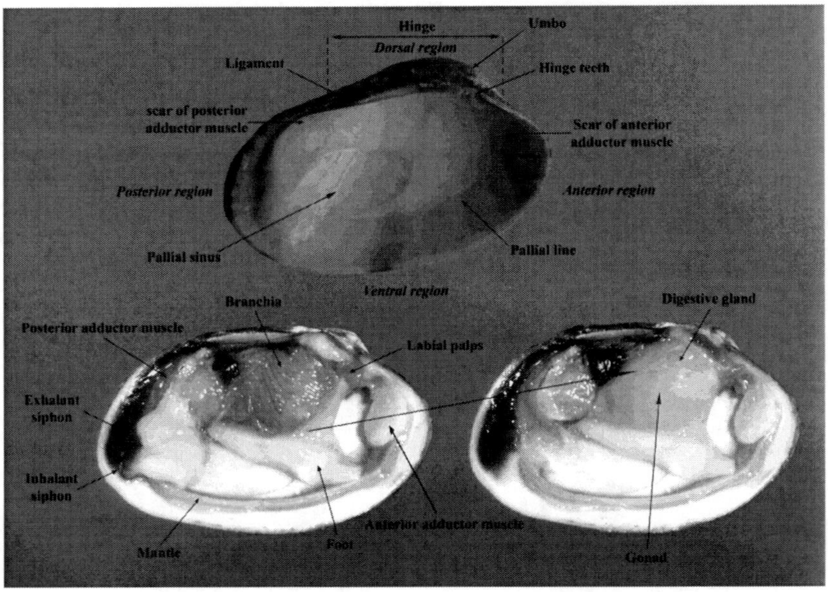

Figure 7. Internal features of the shell valves and soft tissue anatomy of *Venerupis corrugata* (Veneridae). Photo credit: Ana Cerviño Otero. Reproduced with permission and modified.

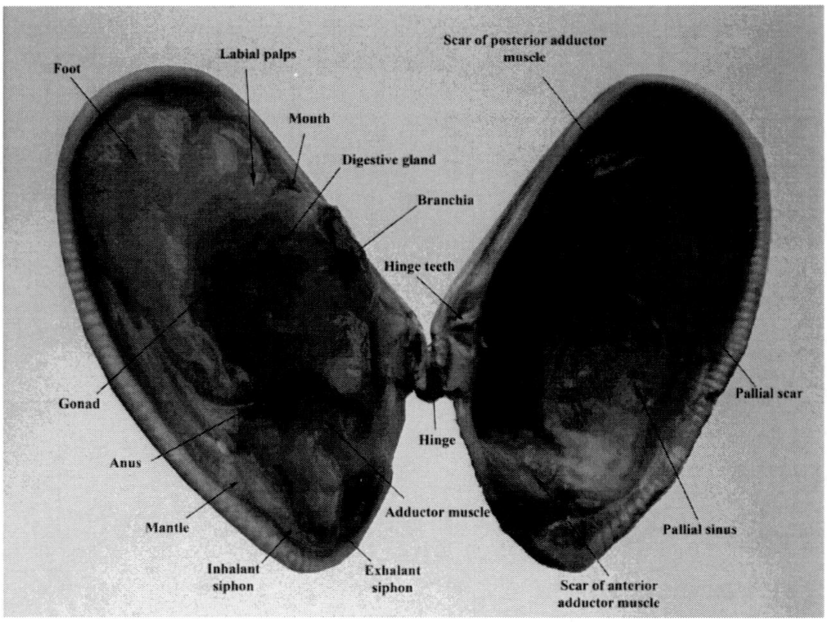

Figure 8. Internal features of the shell valves and soft tissue anatomy of *Donax trunculus* (Donacidae). Photo credit: Andrea Louzán Pérez. Reproduced with permission and modified.

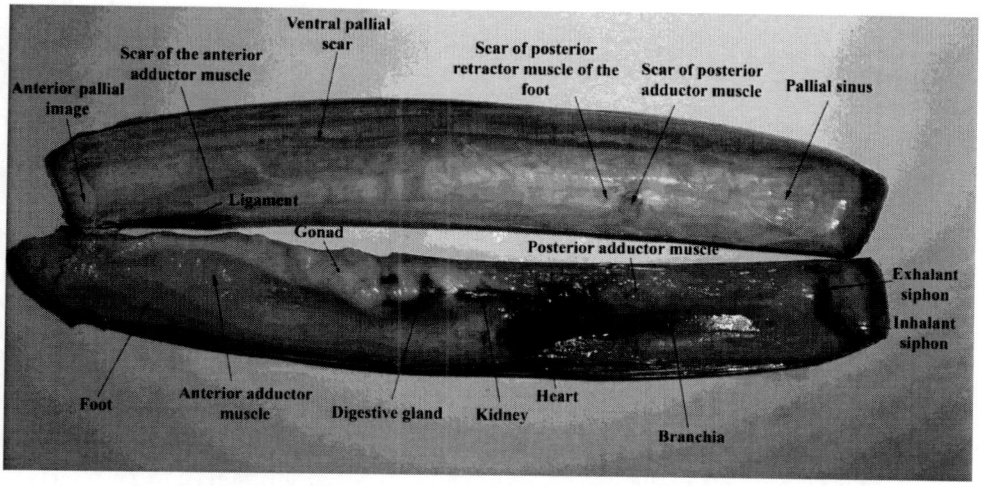

Figure 9. Internal features of the shell valves and soft tissue anatomy of *Ensis magnus* (Pharidae).

1.4.2. Mantle

The mantle is formed by a tissue sheet at the inside of the valves, which is composed of an internal and external epithelium, with connective tissue connecting both layers, which is crossed by numerous blood sinuses, nerves and muscles that act in mantle edge contraction and retraction. The two halves of the mantle are attached to the shell from the hinge ventral to the pallial line, but are free at their edges. Mantle edges are generally divided into three folds: the outer fold secretes the two outer layers of the shell, the middle fold is sensory and the inner fold is muscular and controls water movements. In Veneridae four folds are to be found (Eble, 2001), whilst Pharidae and Solenidae lack these folds (Darriba and López, 2011). The mantle protects the soft parts of the animal, creating between them and the visceral mass a space called the pallial cavity. In Veneridae mantle lobes are free in the ventral region, following the shape of the shell, and in the posterior region they are fused, forming the inhalant and the exhalant siphons. By contrast, in Pharidae and Solenidae the mantle edges are sealed, leaving only three apertures, which communicate with the outer area: the inhalant siphon, exhalant siphon and anterior aperture for the foot. At least in some *Ensis* species (Pharidae), a fourth aperture is also found in the middle of the ventral part, though its function is unknown.

1.4.3. Foot and Muscular System

In clams there are two adductor muscles, anterior and posterior, both formed by portions of the muscle called the “catch muscle”, which holds the valves in position when they have been fully or partially closed, and the “quick muscle”, which contracts to shut the valves. These muscles close the valves, acting in opposition to the ligament and resilium, which maintain the valves open when the muscles relax.

The foot of a clam is a well-developed organ that is used to burrow into the substrate and anchor the animal into position. Razor clams (Pharidae and Solenidae) have long, thin shells and a large foot that can occupy up to half of the mantle cavity, thus allowing them to burrow fast and quite deep. The nerves of the foot originate in the pedal ganglion, while its muscles are distributed in all directions and arranged in layers. It is formed by an epithelium in the outer part. In clams such as *Mercenaria mercenaria*, a layer of circular muscle bundles lie just peripheral to the centrally located, longitudinal muscles (Eble, 2001), whereas in razor clams two layers of diagonal muscles can be found in a total of six layers, as described by Graham (1931). Unlike the majority of Venerids, *Nutricola tantilla* Gould, 1853 exhibits byssal threads, which are used for attaching to different substrates, both as an adult and a juvenile (Narchi, 1970).

1.4.4. Circulatory System

Bivalves have a simple circulatory system, which is rather difficult to trace. It is an open one, with the hemolymph being transported to the tissues through "lakes" and accumulating in blood sinuses. The heart lies in the mid-dorsal region, close to the hinge line of the shell. It is located in the pericardial cavity or pericardium, and consists of a single, muscular ventricle and two thin-walled auricles. The circulatory system is very important for the burrowing mechanism.

1.4.5. Gills

Gills or ctenidia are used in part for respiration and partly for filtering food from the water. For the latter function, gills are responsible for selecting particles that enter through the inhalant siphon. These particles are carried to the labial palps, where selection takes place, and are then transferred to the mouth.

Two pairs of gill lamellae (inner and outer) are located on each side of the body, composed of an ascending and a descending lamella. Each pair of lamellae is joined by interlamellar bridges. Lamellae are pleated, with each plica being composed of a variable number of gill filaments, which can be of two types: ordinary and primary filaments. The former are found at the crest of the fold and have ciliated cells in the distal area, whilst the latter lack cilia and are joined by inter-filamentous bridges.

1.4.6. Digestive System

The digestive system in clams consists of the mouth, oesophagus, stomach, digestive gland, intestine and anus. It begins in the mouth, which opens externally, and where the food arrives from the labial palps, carried there by the ciliary current from the gills.

A short oesophagus leads from the mouth to the stomach, which is an irregularly-shaped sac divided into three compartments: oesophagus portion, cardiac portion and pyloric portion. A crystalline style secreted by the style sac extends well into the stomach, assists in mixing the food in the stomach and releases enzymes that help in digestion.

The stomach is completely surrounded by the digestive gland, where intracellular digestion takes place. An opening from the stomach leads to the intestine, which has a variable series of loops in the visceral mass, ending in the rectum and the anus.

1.4.7. Nervous System

The nervous system consists of three pairs of ganglia (cerebral, pedal and visceral ganglia), emitting nerves that ramify and reach different parts of the body.

1.4.8. Excretory System

The excretory system in clams follows the general pattern observed in bivalves and is formed by a pair of kidneys and pericardial glands. However, the body surface, and mainly gills, may also emit excretion products (Bayne et al., 1976). The kidney consists of a network of nephridial tubules with an epithelium of columnar cells, and excretes the urine into the mantle chamber.

1.4.9. Reproductive System

The gonad forms a mass that surrounds the intestinal loop in the visceral region in Veneridae, Donacidae, Myidae and Mactridae, whilst in Solenidae the gonad surrounds the intestinal loops and invades the inner part of the foot in the period of maturity.

In Pharidae, the gonad is found not only in the same parts as in Solenidae but is also located on the anterior adductor muscle covering the digestive gland.

A pair of gonoducts branching into several minor channels that end in a network of follicles or alveoli are responsible for discharging the gametes. Gonadal follicles are responsible for the formation of gametes in males (spermatogenesis) and females (ovogenesis), with a series of different cells, depending on the phase of the process which will lead to the production of spermatozoids and mature oocytes. More comprehensive information about reproduction is found in Chapter 3.

1.5. Life Cycle

Most of the clams studied in this volume are gonochoric, i.e. sexes are separated in different individuals, although some cases of hermaphroditism have been described in strictly gonochoric species of Veneridae (Ponurovsky and Yakovlev, 1992; Delgado and Pérez-Camacho, 2002) and Pharidae (Valli and Giglio, 1980; Darriba et al., 2005). Other species, such as *Mercenaria mercenaria*, are protandrous hermaphrodites (although some individuals mature directly into females, whereas others are simultaneous hermaphrodites (Loosanoff, 1937), or *Chamelea gallina* Linnaeus, 1758, in which the primary gonad is hermaphrodite and protandry occurs (Ansell, 1961). The general pattern for clams is to have external fertilization and planktonic larvae. Nonetheless, four genera have been reported as larviparous: *Gemma*, *Parastarte*, *Psephidea* and *Transenella* (Narchi, 1971; Kabat, 1985; Commito et al., 1995; Geraghty et al., 2008). In these larviparous species the early stages of larval development occur in the inhalant chamber of the mantle cavity of the female. Spawned eggs are passed through the gills and retained in the mantle chamber. Sperm is taken in through the inhalant siphon. The length of time larvae are held in the mantle chamber varies according to species and all these brooding clams release crawling juveniles (Kabat, 1985; Commito et al., 1995; Geraghty et al., 2008). Nonetheless, in most clam species, gametes are discharged into the open environment, where fertilization occurs. The general pattern of the life cycle in clams is shown in Figure 10.

Larval development in clams is characterized by different stages. Once the female gametes are fertilized, the embryo initiates cleavage, the next stage being a pyriform trochophore, which is able to swim. The next larval form is the straight-hinged larva or D-larva stage, which has two valves, a complete digestive system and an organ called the velum that is peculiar to bivalve larvae.

When the larva swims through the water column the velum collects phytoplankton upon which the larva feeds. Larvae continue to swim, feed and grow, while the umbo, which is a protuberance of the shell near the hinge, is developing. When larvae approach maturity, a foot develops and gill rudiments become evident, but the velum still remains in this pediveliger stage.

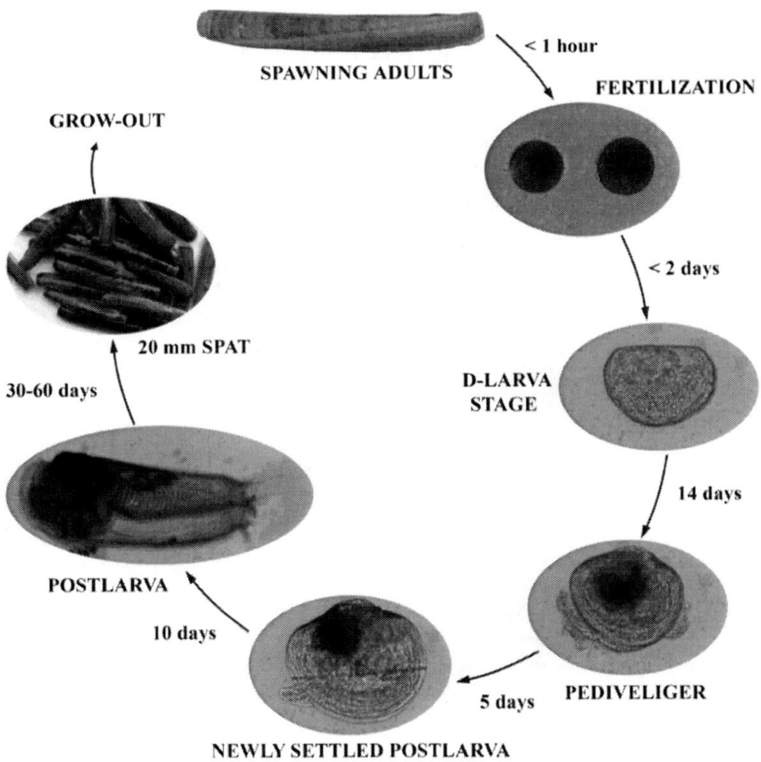

Figure 10. Representation of the life cycle of *Ensis magnus* (Pharidae). The duration of the period between the various stages may differ for other species of clams.

The velum will be lost and branchia will develop, and thus the larvae will settle and metamorphose into newly settled postlarvae. Metamorphosis is a critical time in the development of bivalves, during which the animal changes from a swimming, planktonic existence to a sedentary benthic one. Postlarvae start to become similar in appearance to adults and bury in the substrate thanks to the foot, and at the same time faster growth is observed. The time spent to reach commercial size varies greatly between species, with several factors influencing it, such as location, subtidal and intertidal beds, environmental conditions, genetic variability, etc.

1.6. HABITAT

1.6.1. Family Donacidae

The superfamily Tellinoidea is one of the most diverse bivalve superfamilies, and most of the species inhabit soft sediment in shallow water. However, they can live in a variety of ecosystems, from littoral sand to deep mud areas. Most species are active burrowers within sediments (Bosch et al., 1995). Members of the Donacidae family are important in food chains of sandy coastal areas and have been used as bio-indicators, since they are sensitive to environmental changes and coastal industrial activities (Bosch et al., 1995). The tidal migratory behaviour of the members of this genus has been studied widely (Donn et al., 1986). Donacidae are suspension-feeders on phytoplankton (Mouëza and Chessel, 1976) and suspended particulate organic matter (Wade, 1964). This basically determines the vertical distribution of this species, which normally inhabits the wash zones of the beaches, where hydrodynamism favours the presence of suspended particles and impedes rapid sedimentation of the organic matter (Zeichen et al., 2002). For example, *D. trunculus* inhabits the high-energy environment of exposed sandy beaches, where it forms extensive, dense beds and occurs down to 6 m deep (Gaspar et al., 1999).

Wedge shells live in the intertidal zone of low-profile sandy beaches, but also occur subtidally, such as *D. variabilis* Say, 1822 (Cobb et al., 2011). *D. vittatus* da Costa, 1778 is found in the lower part of the intertidal and in greater densities in the shallow sublittoral, whilst on the Atlantic coast of France, it is mainly confined between low water and 5-6 m in depth (Ansell and Boyou, 1979). This family also includes freshwater bivalves, such as *Galatea paradoxa* Born, 1778, which is restricted to the lower reaches of a few rivers in West Africa (Etim and Brey, 1994).

1.6.2. Family Mactridae

Representatives of the Mactridae are infaunal bivalves which usually occur in relatively shallow water, to a depth of approximately 50 m, in medium to coarse substrata, in sheltered areas (Coan et al., 2000; Hare et al., 2010). They are active burrowers in mobile sand substrates. The most widely-researched genera, *Spisula* and *Mactra*, belong to Mactrinae. *Spisula solidissima* (Atlantic surfclam) inhabits sandy substrates, with high concentrations in the turbulent waters of oceanic beaches, just beyond the breaker zone. Species of the genus

Mactrinula, which also belongs to Mactrinae, are reported from the tropical western Pacific, and occur predominantly in either silt or mud from the low intertidal area to depths of 250 m (Habe, 1977; Jiang et al., 1986; Bernard et al., 1993).

1.6.3. Family Myidae

Most species of Myidae are shallow-water burrowers in a variety of soft sediments (Bosch et al., 1995), such as mud or sand, or huddle in rocky crevices or old bore holes (Lamprell et al., 1998). Some myids are facultative anaerobes and are able to go without oxygen for many days. Myidae remain at the same level of sediment depth and can only stretch out their siphon or draw it in. They cannot dig again into the sediment when taken out and thus become victims of predators.

Mya arenaria and *Mya truncata* are two large species which burrow deeply into sandy-muddy coasts. *M. arenaria* is native to the Atlantic coast of North America (Connell et al., 2007), but was introduced into the North Sea around 1600 B.C. and, in the last few decades, into the sandy-muddy upper intertidal zones of the Tagus estuary, in central Portugal (Conde et al., 2010). *M. truncata* (Arctic clam) lives in mixed muddy and sandy substrates on all British coasts (Amaro et al., 2003; Birkely et al., 2003; Camus et al., 2003; Yuan et al., 2011).

1.6.4. Superfamily Solenoidea (Families Pharidea and Solenidae)

Economically important edible razor clams (Pharidae and Solenidae) live in soft bottom habitats from the lower intertidal zone down to 60-110 m. They constitute a major component of infaunal soft-bottom communities. Razor shells live buried vertically in a tube which they form, in which they ascend to keep the edge of the siphons at sediment surface level, and in which they descend to flee predators. If a Solenoidean is dug out from its habitat in the sediment, it can dig itself in again immediately and form a new vertical hole, in which it can ascend and descend. Most species are found from the lower part of the intertidal zones to 20-30 m depth. Henderson and Richardson (1994) suggested a movement of juveniles further down, where they are safer from wave action and currents. They live in soft sandy sediment, mostly fine sand, fine muddy sand or silt. *Ensis siliqua* tend to inhabit beaches more exposed to swell than *Ensis magnus*, which occupy substrate with thicker grains (Holme, 1954). *Solen marginatus* prefers substrates with very fine grain, internal areas and those protected from swell, with medium-low salinities (Darriba and Fernández-Tajes, 2011). A few species tolerate lower salinities, such as *Solen annandalei* Preston, 1915 and *Solen kempi* Preston, 1915 in the mouth of Chilka Lake (Northeast India) (Cosel, 1990). *Sinonovacula constricta* tolerates wide temperature and salinity ranges and prefers substrata with a muddy top layer and fine sand bottom (Yan et al., 2009).

1.6.5. Family Veneridae

Members of the Veneridae are ecologically and economically important, and famous as food sources in most parts of the world. They are common members of macro-benthos in intertidal and inshore areas, and inhabit a wide range of substrates. However, most are important fauna of sand and consolidated sediments offshore. Veneridae's adaptation to different environmental conditions leads to high diversity and abundance. For example, *Chamelea gallina* is an infaunal filter-feeder that lives in well-sorted fine sand in shallow waters (0-10 m depth) (Moschino et al., 2008), whilst Tawera mawsoni Hedley, 1916 lives in sandy substrates from 12 to 450 m depth (Luckens, 1990).

Paphia undulata Born, 1778 inhabits muddy bottoms (del Norte-Campos et al., 2010), whereas other clams, such as Venus verrucosa Linnaeus, 1758 live in poorly sorted sand, sometimes with coralline rhodoliths, and in channels between beds of Posidonia oceanica Delile, 1813 down to a depth of about 30 m (Arneri et al., 1998).

ACKNOWLEDGMENTS

Fiz da Costa is grateful to the staff of Centro de Cultivos Marinos de Ribadeo-CIMA (Xunta de Galicia) and he also acknowledges the Fundación Juana de Vega (Spain) for the postdoctoral fellowship at IFREMER.

REFERENCES

Adams, H. and Adams, A. (1853-1858). *The Genera of Recent Mollusca: Arranged According to Their Organization.* Van Voorst, London. 2 vols. Of text. Bivalves in vol. 2. 1 vol of plates. Published in parts.

Amaro, T., Duineveld, G., Bergman, M. and Witbaard, R. (2003). Growth variations in the bivalve *Mya truncata*: A tool to trace changes in the Frisian Front macrofauna (southern North Sea)? *Helgol. Mar. Res.* 57, 132-138.

Ansell, A.D. (1961). The development of the primary gonad in *Venus striatula* (Da Costa). *J. Molluscan Stud.* 34, 243-247.

Ansell, A.D. and Bodoy, A. (1979). Comparison of events in the seasonal cycle for *Donax vittatus* and *Donax trunculus* in European waters. In: E. Naylor, R.G. Hartnoll (Eds.), *Cyclic phenomena in marine plants and animals.* pp. 191-198, Pergamon Press, Oxford.

Arneri, E., Giannetti, G. and Antolini, B. (1998). Age determination and growth of *Venus verrucosa* L. (Bivalvia: Veneridae) in the southern Adriatic and the Aegean Sea. *Fish. Res.* 38, 193-198.

Bayne, B.L., Windows, J. and Thompson, R.J. (1976). Physiology II. In: B.L. Bayne (Ed.), *Marine Mussels: Their Ecology and Physiology.* pp. 121-206, Cambridge University Press, Cambridge.

Beal, B.F. (2006). Biotic and abiotic factors influencing growth and survival of wild and cultured individuals of the softshell clam (*Mya arenaria* L.) in eastern Maine. *J. Shellfish Res.* 25, 461-474.

Beal, B.F. and Vencile, K.W. (2001). Short-term effects of commercial clam (*Mya arenaria* L.) and worm (*Glycera dibranchiata* Ehlers) harvesting on survival and growth of juveniles of the soft-shell clam. *J. Shellfish Res.* 20, 1145-1157.

Bernard, F.R., Cai, Y.Y. and Morton, B. (1993). *Catalogue of the living marine bivalve molluscs of China.* Hong Kong University Press, Hong Kong.

Bieler, R. and Mikkelsen, P.M. (2006). Bivalvia-a look at the Branches. *Zool. J. Linn. Soc.-Lond.* 148, 223-235.

Birkely, S.R., Grahl-Nielsen, O. and Gulliksen, B. (2003). Temporal variations and anatomical distributions of fatty acids in the bivalve *Mya truncata*, L. 1758, from Isfjorden, Spitsbergen. *Polar Biol.* 26, 83-92.

Bosch, D., Dance, P., Moolenbeek, R. and Oliver, G. (1995). *Sea shells of eastern Arabia.* Motative Publishing, Dubai.

Bouchet, P., Rocroi, J.-P., Bieler, R., Carter, J.G. and Coan, E.V. (2010). Nomenclator of Bivalve Families with a Classification of Bivalve Families. *Malacologia.* 52, 1-184.

Bouchet, P. (2011). Donacidae. Accessed through: World Register of Marine Species at http://www.marinespecies.org/aphia.php?p=taxdetailsandid=236 on 12th September 2011.

Camus, L., Birkely, S.R., Jones, M.B., Børseth, J.F., Grøsvik, B.E., Gulliksen, B. and Depledge, M.H. (2003). Biomarker responses and PAH uptake in *Mya truncata* following exposure to oil-contaminated sediment in an Arctic fjord (Svalbard). *Science Sci. Total Environ.* 308, 221-234.

Chenu, J.C. (1862). *Manuel de Conchyliologie et de Paleontologie Conchyliologique, II.* Masson, Paris.

Chung, E.-Y. (2007). Oogenesis and sexual maturation in *Meretrix lusoria* (Röding 1798) (Bivalvia: Veneridae) in Western Korea. *J. Shellfish Res.* 26, 71-80.

CLEMAM (2011). *Taxonomic Database on European MArine Mollusca.* Accessed at http://www.somali.asso.fr/clemam on 10th December 2011.

Coan, E.V., Valentich Scott, P. and Bernard, F.R. (2000). Bivalve sheashells of western North America. Marine bivalve mollusks from Arctic Alaska to Baja California. *Santa Barbara Museum Natural History Monographs.* 2, viii.

Cobb, J.C., Stephenson, S.P. and Arnold, W.S. (2011). Reproductive cycle and recruitment patterns of the coquina clam *Donax variabilis* Say along the central gulf coast of Florida. *J. Shellfish Res.* 30, 25-34.

Commito, J.A., Currier, C.A., Kane, L.R., Reinsel, K.A. and Ulm, I.M. (1995). Dispersal dynamics of the bivalve *Gemma Gemma* in a patchy environment. *Ecol. Monogr.* 65, 1-20.

Conde, A., Novais, J. and Domínguez, J. (2010). Southern limit of distribution of the soft-shell clam *Mya arenaria* on the Atlantic East Coast. *Biol. Invasions.* 12, 429-432.

Connell, L.B., MacQuarrie, S.P., Twarog, B.M., Iszard, M. and Bricelj, V.M. (2007). Population differences in nerve resistance to paralytic shellfish toxins in softshell clam, *Mya arenaria*, associated with sodium channel mutations. *Mar. Biol.* 150, 1227-1236.

Cosel, R. von., Dorjes, J. and Mühlenhardt-Siegel, U. (1982). Die Amerikanische Schwertmuschel *Ensis directus* (Conrad) in der Deutschen Bucht. I. Zoogeographie und Taxonomie in Vergleich mit einheimischen Schwertmuschel-Arten. *Senckenbergiana Maritime.* 14, 143-173.

Cosel, R. von (1990). An introduction to the razor shells (Bivalvia: Solenacea). In: B. Morton (Ed.), *The Bivalvia Proceedings of a Memorial Symposium in Honour of Sir Charles Maurice Yonge (1899-1986).* Hong Kong University Press, Hong Kong. Viii + 355 pp.

Cosel, R. von (1993). The razor shells of the eastern Atlantic. Part I: Solenidae and Pharidae I (Bivalvia: Solenacea). *Archiv für Molluskenkunde.* 122, 207-231.

Cosel, R. von (2009). The razor shells of the eastern Atlantic, part 2. Pharidae II: the genus *Ensis* Schumacher, 1817 (Bivalvia, Solenoidea). *Basteria.* 73, 9-26.

Dall, W.H. (1889). On the hinge of pelecypods and its development, with an attempt toward a better subdivision of the group. *Am. J. Sci.* 38, 445-462.

Dall, W.H. (1902). Synopsis of the family Veneridae and of the North American Recent species. *Proc. U.S. Nat. Mus.* 26, 335-412.

Darriba, S., San Juan, F. and Guerra, A. (2005). Gametogenic cycle of *Ensis siliqua* (Linnaeus, 1758) in the Ría de Corcubión, Northwest Spain. *J. Molluscan Stud.* 71, 47-51.

Darriba, S. and Fernández-Tajes, J. (2011). Chapter 1: Systematics and distribution. In: A. Guerra, C. Lodeiros, M.B. Gaspar, F. da Costa (Eds.), *Razor clams: biology, aquaculture and fisheries.* pp. 29-44, Xunta de Galicia, Consellería do Mar, Santiago de Compostela, Spain.

Darriba, S. and López, C. (2011). Chapter 2: Anatomy. In: A. Guerra, C. Lodeiros, M.B. Gaspar, F. da Costa (Eds.), *Razor clams: biology, aquaculture and fisheries.* pp. 45-64, Xunta de Galicia, Consellería do Mar, Santiago de Compostela, Spain.

Delgado, M. and Pérez-Camacho, A. (2002). Hermaphroditism in *Ruditapes decussatus* (L.) (Bivalvia) from the Galician coast (Spain). *Sci. Mar.* 66, 183-185.

del Norte-Campos, A., Nabuab, F.M., Palla, R.Q. and Burlas, M.R.M. (2010). The early development of the short-necked clam, *Paphia undulata* (Born 1778) (Mollusca, Pelecypoda: Veneridae) in the laboratory. *Science Diliman.* 22, 13-20.

Deshayes, G.P. (1853). Observations sur les animaux de quelques genres de mollusques acéphalés (*Chamostrea*, *Glauconome*, *Circe* and *Capsa*). *Proc. Zool. Soc. Lond.* 21, 167-173.

Donn Jr., T.E., Clarke, D.J., McLachlan, A. and Toit, P.D. (1986). Distribution and abundance of *Donax serra* Röding (Bivalvia: Donacidae) as related to beach morphology. I. Semilunar migrations. *J. Exp. Mar. Biol. Ecol.* 102, 121-131.

Eble, A.F. (2001). Chapter 4. Anatomy and histology of *Mercenaria mercenaria.* In: J.N. Kraeuter, M. Castagna (Eds.), *Biology of the Hard Clam.* pp. 117-220, Elsevier, Amsterdam.

Essink, K. (1985). On the occurrence of the American jack-knife clam *Ensis directus* (Conrad, 1843) (Bivalvia, Cultellidae) in the Dutch Wadden Sea. *Basteria.* 49, 73-80.

Essink, K. (1986). Note on the distribution of the American jack-knife clam *Ensis directus* (Conrad, 1843) in N.W. Europe (Bivalvia, Cultellidae). *Basteria.* 50, 33-34.

Etim, L. and Brey, T. (1994). Growth, productivity, and significance for fishery of the bivalve *Egeria radiata* (Donacidae) in the Cross River, Nigeria. *Arch. Fish. Mar. Res.* 42, 63-75.

FAO (2010) Fishstat - FAO Fisheries Department, Fishery Information, Data and Statistics Unit.

Fischer, P. (1887). *Manuel de Conchyliologie et de Paleontologie Conchyliologique ou Historie Naturelle des Mollusques Vivants et Fossiles suivi d'un Appendice sur les Brachiopodes par D. P. Oehlert, fasc. 11.* Librairie F Savy, Paris. 1009-1369.

Fleming, J. (1828). *A history of British animals.* Bell and Bradfute, Edinburgh.

Gaspar, M.B., Ferreira, R. and Monteiro, C.C. (1999). Growth and reproductive cycle of *Donax trunculus* L., (Mollusca: Bivalvia) off Faro, southern Portugal. *Fish. Res.* 41, 309-316.

Geraghty, J., Russell, M.P. and Dollahon, N. (2008). A quantitative assessment of spermatozoan morphology in *Nutricola confusa* and *Nutricola tantilla* (Bivalvia: Veneridae). *The Veliger.* 50, 263-268.

Ghosh, E. (1920). Taxonomic studies on the soft parts of the Solenidae. *Rec. Indian Mus., Calcutta.* 19, 49-78.

Gill, T. (1871). *Arrangement of the families of mollusks.* Smithsonian Miscellaneous Collections.

Giribet, G. and Distel, D.L. (2003). Bivalve phylogeny and molecular data. In: C. Lydeard, D. Lindberg (Eds.), *Molecular Systematics and Phylogeography of Mollusks.* pp. 45-90, Smithsonian Institution Press, Washington, DC.

Glenn, L.C. (1904). Pelecypoda. *Maryland Geological Survey.* Volume Miocene, 274-401.

Gosling, E. (2002). *Bivalve Molluscs: Biology, Ecology and Culture.* Fishing News Books. Blackwell Publishing, UK, 443 pp.

Graham, A. (1931). On the morphology, feeding mechanisms and digestion of *Ensis siliqua* (Schumacher). *Trans. R. Soc. Edin.* 56, 725-751.

Grizel, H. and Auffret, M. (2003). *An atlas of histology and cytology of marine bivalve molluscs.* Editions Quae, 201 pp.

Guerra, A., Lodeiros, C., Gaspar, M.B. and da Costa, F. (Eds.) (2011). *Razor clams: biology, aquaculture and fisheries.* Xunta de Galicia, Consellería do Mar, Santiago de Compostela, Spain, 428 pp.

Habe, T. (1977). *Systematics of Mollusca in Japan, Bivalvia and Scaphopoda.* Zukan-No-Hokuryukan, Tokyo, 244-275 pp.

Hare, M.P., Weinberg, J., Peterfalvy, O. and Davidson, M. (2010). The "southern" surfclam (*Spisula solidissima similis*) found north of its reported range: A commercially harvested population in long island sound, New York. *J. Shellfish Res.* 29, 799-807.

Henderson, S.M. and Richardson, C.A. 1994. A comparison of the age, growth rate and burrowing behaviour of the razor clams, *Ensis siliqua* and *Ensis ensis*. *J. Mar. Biol. Assoc. UK.* 74, 939-954.

Ho, J.S. and Zheng, G.X. (1994). *Ostrincola koe* (Copepoda, Myicolidae) and mass mortality of cultured hard clam (*Meretrix meretrix*) in China. *Hydrobiologia.* 284, 169-173.

Holme, N.A. (1954). The ecology of british species of *Ensis. J. Mar. Biol. Ass. UK.* 33, 145-172.

Jiang, J.X., Huang, L.Q. and Meng, F. (1986). Distributional characteristics of benthic communities on the continental shelf of Donghai Sea. *Taiwan Haixiu.* 5, 70-77.

Jukes-Browne, A.J. (1914). A synopsis of the family Veneridae. Parts I and II. *Proc. Malacol. Soc. Lond.* 11, 58-94.

Kabat, A. (1985). The allometry of brooding in *Transenella tantilla* (Gould) (Mollusca: Bivalvia). *J. Exp. Mar. Biol. Ecol.* 91, 271-279.

Keen, A.M. (1969). Superfamily Veneracea. In: L.R. Cox, N.D. Newell, D.W. Boyd, C.C. Branson, R. Casey, A. Chavan, A.H. Coogan, C. Dechaseaux, C.A. Fleming, F. Haas, L.G. Hertlein, E.G. Kauffman, A.M. Keen, A. Larocque, A.L. McAlester, R.C. Moore, C.P. Nuttall, B.F. Perkins, H.S. Puri, L.A. Smith, T. Soot-Ryen, H.B. Stenzel, E.R. Trueman, R.D. Turner, J. Weir (Eds.), *Treatise on Invertebrate Paleontology. Part N, Vol. 2 Mollusca, 6: Bivalvia*, pp. 670-690. Geological Society of America, and University of Kansas Press, Lawrence, Kansas.

Kraeuter, J.N. and Castagna, M. (Eds.) (2001). *Biology of the Hard Clam*. Elsevier, Amsterdam, 751 pp.

Lamarck, J.B. (1809). *Philosophie zoologique*. Vol. 1. Dentu, Paris. Xxv + 428 pp.

Lamprell, K., Healy, J.M. and Dyne, G.R. (1998). *Superfamily Myoidea*. In: P.L. Beesley, G.J.B. Ross, A. Wells (Eds). *Mollusca, Pt. A*. pp. 363-366.

Li, Q., Yang, L., Ke, Q. and Kong, L. (2011). Gametogenic cycle and biochemical composition of the clam *Mactra chinensis* (Mollusca: Bivalvia): Implications for aquaculture and wild stock management. *Mar. Biol. Res.* 7, 407-415.

Linnaeus, C. (1758). *Systema naturae*. Ed. 10, vol. 1. L. Salvvi, Holmiae. iii + 824 pp.

Loosanoff, V.L. (1937). Development of the primary gonad and sexual phases in *Venus mercenaria* Linnaeus. *Biol. Bull.* 72, 389-405.

Luckens, P.A. (1990). Distribution, size-frequency, and growth-ring analyses of *Tawera mawsoni* (Bivalvia: Veneridae) at Macquarie Island. *New Zeal. J. Mar. Fresh. Res.* 24, 59-73.

Marzec, R.J., Kim, Y. and Powell, E.N. (2010). Geographical trends in weight and condition index of surfclams (*Spisula solidissima*) in the Mid-Atlantic Bight. *J. Shellfish Res.* 29, 117-128.

Mikkelsen, P.M., Bieler, R., Kappner, I. and Rawlings, T.A. (2006). Phylogeny of Veneroidea (Mollusca: Bivalvia) based on morphology and molecules. *Zool. J. Linn. Soc.-Lond.* 148, 439-521.

Moschino, V., Chicharo, L.M.Z. and Marin, M.G. (2008). Effects of hydraulic dredging on the physiological responses of the target species *Chamelea gallina* (Mollusca: Bivalvia): laboratory experiments and field surveys. *Sci. Mar.* 72, 493-501.

Mouëza, M. and Chessel, D. (1976). Contribution à l'étude de la biologie de *Donax trunculus* L. dans l'Algerois: analyse statistique de la dispersion le long d'un plage en baie de Bou-Ismael. *J. Exp. Mar. Biol. Ecol.* 21, 211-221.

Narchi, W. (1970). The presence of byssus in adult *Transennella tantilla* (Gould) Veneridae-Bivalvia. *Wasmann J. Biol.* 28, 233-236.

Narchi, W. (1971). Structure and Adaptation in *Transennella Tantilla* (Gould) and *Gemma Gemma* (Totten) (Bivalvia: Veneridae). *Bull. Mar. Sci.* 21, 866-885.

Neumayr, M. (1884). Zur morphologie des bivalves schlosses. *Sitzungsberichte der Mathematisch-Naturwissenschaflinchen Klasse der Kaiserlichen Akademie der Wissenschanften.* 88, 395-419.

Nevesskaja, L. (2009). Principles of systematics and the system of bivalves. *Paleont. J.* 43, 1-11.

Newell, N.D. (1965). Classification of Bivalvia. *American Museum Novitates.* 2206, 1-25.

Olsson, A.A. (1931). Contributions to the Tertiary Paleontology of Northern Peru: Part 4, The Peruvian Oligocene. *Bull. Am. Paleontol.* 17, 97-264.

Ponurovsky, S.K. and Yakovlev, Y. (1992). The reproductive biology of the Japanese littleneck, *Tapes philippinarum* (Adams and Reeve, 1850) (Bivalvia: Veneridae) *J. Shellfish Res.* 11, 265-277.

Rafinisque, C.S. (1815). *Analyse de la nature ou tableau de l'univers et des corps organisés.* Palerme, 223 pp.

Roach, S., MacDonald, D. and Kenchington, E. (2011). North American jackknife and razor clam fisheries. In: A. Guerra, C. Lodeiros, M.B. Gaspar, F. da Costa (Eds.), *Razor clams: biology, aquaculture and fisheries.* pp. 369-379, Xunta de Galicia, Consellería do Mar, Santiago de Compostela, Spain.

Schumacher, C.F. (1817). *Essai d'un nouveau systéme des hatitations des vers testacés.* Copenhagen, 287 pp.

Shumway, S.E., Sherman, S.A., Cembella, A.D. and Selvin, R. (1994). Accumulation of paralytic shellfish toxins by surfclams, *Spisula solidissima* (Dillwyn, 1897) in the Gulf of Maine: Seasonal changes, distribution between tissues, and notes on feeding habits. *Nat. Toxins.* 2, 236-251.

Stoliczka, F. (1870). Cretaceous fauna of southern India, 3 Paleontologica Indica, being figures and descriptions of the organic remains produced during the progress of the Geological Survey of India. *Memoirs of the Geological Survey of India, 5. Paleontologica Indica* Ser. 6, 3, xxii + 538 pp., 550 + 534 pls.

Su, Y. (2006). Seed production techniques for the clam *Sinonovacula constricta* in natural sea region. *Fisheries Science and Technology Information.* 33, 53-55 (in Chinese).

Taylor, J.D., Williams, S.T., Glover, E.A. and Dyal, P. (2007). A molecular phylogeny of heterodont bivalves (Mollusca: Bivalvia: Heterodonta): new analyses of 18S and 28S rRNA genes. *Zool. Scr.* 36, 587-606.

Thiele, J. (1935). *Handbuch der systematischen Weichtierkunde. Vol. 2. Scaphopoda. Bivalvia. Cephalopoda.* Gustav Fischer Verlag Jena.

Tryon, G.W. (1884). Structural and systematic conchology: An introduction to the study of Mollusca. *Academy of Sciences of Philadelphia, Philadelphia.* 3, 128-134.

Valli, G. and Gioglio, I. (1980). Periodo riproduttivo de alcuni aspetti della biometria di *Ensis minor* (Chenu) nel Golfo de Trieste. *Quad. Lab. Tecnol. Pesca.* 2, 285-306.

Vokes, H.E. (1967). *Genera of the Bivalvia: A systematic and bibliographic catalogue (revised and updated).* Paleontological Research Institucion, Ithaca, New York.

Vokes, H.E. (1980). *Genera of the Bivalvia: a systematic and bibliographic catalogue.* Paleontological Research Institution, Ithaca.

Wade, B.A. (1964). Notes on the ecology of *Donax denticulatus* (Linné). *Proc. Gulf. Carib. Fish. Inst. (17 Ann. Sess.)*, pp. 36-41.

Wang, L.C., Zhang, K., Di, L.Q., Liu, R. and Wu, H. (2011). Isolation and structural elucidation of novel homogenous polysaccharide from *Mactra veneriformis. Carbohydr. Polym.* 86, 982-987.

Ward, L.W. and Blackwelder, B.W. (1987). Late Pliocene and Early Pleistocene Mollusca from the James City and Chowan River Formations at Lee Creek Mine. *Smithsonian Contributions to Paleobiology.* 61, 113-283.

Ward, L.W. (1992). *Molluscan biostratigraphy of the Miocene Middle Atlantic Coastal Plain of North America.* Virginia Museum of Natural History 2.

Yan, H., Li, Q., Liu, W., Yu, R. and Kong, L. (2009). Seasonal changes in reproductive activity and biochemical composition of the razor clam *Sinonovacula constricta* (Lamarck 1818). *Mar. Biol. Res.* 6, 78-88.

Yuan, L., Sun, L., Wei, G., Long, N., Xie, Z. and Wang, Y. (2011). 9,400 yr B.P.: The mortality of mollusk shell (*Mya truncata*) at high Arctic is associated with a sudden cooling event. *Environ. Earth Sci.* 63, 1385-1393.

Zeichen, M.M., Agnesi, S., Mariani, A., Maccaroni, A. and Ardizzone, G.D. (2002). Biology and Population Dynamics of *Donax trunculus* L. (Bivalvia: Donacidae) in the South Adriatic Coast (Italy). *Estuar. Coast. Shelf Sci.* 54, 971-982.

In: Clam Fisheries and Aquaculture
Editor: Fiz da Costa González

ISBN: 978-1-62257-518-3

Chapter 2

ASPECTS OF GLOBAL DISTRIBUTION OF SIX MARINE BIVALVE MOLLUSC FAMILIES

H. Saeedi*[*] and M. J. Costello
Leigh Marine Laboratory, University of Auckland, New Zealand

ABSTRACT

We compared the global distribution of six families of bivalves based on their economic importance in terms of fisheries and aquaculture production: Veneridae (venus shells), Mactridae (surf clams), Donacidae (wedge shells), Myidae (softshell clams), Pharidae, and Solenidae (razor clams). All distribution data were gathered from the open-access databases GBIF (Global Biodiversity Information Facility) and OBIS (Ocean Biogeographic Information System). Species nomenclature and synonyms were reconciled using WoRMS (World Register of Marine Species). Geographic coordinates related to species records noted as fossils, that lacked a geogographic precision, and where precision was >100 km were excluded from the analysis. Comparison with WoRMS indicated that about half of the known species and 64% of genera had data in GBIF and OBIS combined.

All distribution records were from shallow coastal areas, and a sampling bias in Europe was evident. There were no records of any of the families in Antarctica, and only the Mactridae and Veneridae occurred in New Zealand. The GBIF and OBIS data tended to indicate wider distribution ranges than found in a survey of the literature. However, in several cases this reflected species introduced outside their native range. A significant amount of species distribution data was easily accessible from GBIF and OBIS for about half the described species of these bivalve families. However, the metadata that describes the datasets in GBIF and OBIS merited improvement, and considerable cleaning of the data was necessary before use.

Studies on biogeography need to consider the effect of species introductions outside their native range on their analyses. Despite these limitations, the analysis found distinct biogeographic patterns at a family level that merit further research into the evolutionary origins and dispersal patterns of the six families.

[*] E-mail address: hsae002@aucklanduni.ac.nz; hanieh.saeedi@gmail.com

1. INTRODUCTION

Molluscs are one of the largest phyla. Of the 230,000 described marine species, almost 52,000 are marine molluscs (Bouchet, 2006). In 2009, molluscs contributed 8% (6 million tonnes) of global fisheries and 38% (13 million tonnes) of the global mariculture (FAO, 2011). Most of this harvest was of bivalve molluscs, including mussels and clams. Here, we review the global distribution of six clam families due to their economic importance in terms of the world fisheries and aquaculture: namely Veneridae (venus shells), Mactridae (surf clams), Donacidae (wedge shells), Myidae (softshell clams), Pharidae, and Solenidae (razor clams). Not all species distribution data are included in the considerable and diverse literature, and gathering such data is very time consuming. However, the publication of such data through the online and open-access Global Biodiversity Information Facility (GBIF) and Ocean Biogeographic Information System (OBIS) makes it easily accessible. A concern in using this data is that its fitness for use may be compromised by misapplication of species names, misspellings of names, occurrence of synonyms, and errors in geo-referencing (Costello et al., 2007; Robertson, 2008). We used the World Register of Marine Species (Appeltans et al., 2011) to validate the taxonomy, and compared the relative occurrence of species and their distribution data in GBIF, OBIS and the literature.

2. METHODS

All data on species distributions were gathered from the Global Biodiversity Information Facility (GBIF) and Ocean Biogeographic Information System (OBIS, Costello et al., 2007) (Table 1).

Each resource integrates numerous datasets from different sources, and OBIS provides datasets to GBIF to publish. Although our preference would be to cite each dataset, citations were not available for many datasets and many had poor metadata to indicate their content and sampling methods. We cross matched datasets between OBIS and GBIF to avoid duplication. The data included notes that indicated the geographic precision of the location and if the record was a fossil. All data that was fossil, that mapped onto land, and where the coordinates either had no precision or if precision was more than 100 km, were excluded.

All species' names were verified in WoRMS (World Register of Marine Species) (Appeltans et al., 2011; Bouchet, 2011a,b; Gofas, 2011a-e), synonyms and misspellings were reconciled, and accepted species were used to create distribution maps. Nomenclature follows WoRMS and taxonomic authorities are given with species names in Table 2. We compared the distribution of a sample of 77 species that were more commercially and scientifically important with the 692 species found in GBIF and OBIS. ArcGIS version 10 was used to create all distribution maps. Space limitations prevented the inclusion of maps for genera and species.

Table 1. Availability of distribution data on the species in GBIF and OBIS compared to the species listed in WoRMS

	Genera		Species		% in GBIF and OBIS	
Family	**WoRMS**	**GBIF and OBIS**	**WoRMS**	**GBIF and OBIS**	**Genera**	**Species**
DONACIDAE Fleming, 1828	10	3	102	56	30	55
MACTRIDAE Lamarck, 1809	33	25	200	111	76	56
MYIDAE Lamarck, 1809	7	5	33	16	71	48
PHARIDAE H. Adams and A. Adams, 1856	16	7	73	33	44	45
SOLENIDAE Lamarck, 1809	2	2	68	43	100	63
VENERIDAE Rafinesque, 1815	112	72	863	433	64	50
Total	180	114	1339	692	64	53

3. Results

From all datasets for six families, distribution records were found for 692 species. Comparison with the species listed in WoRMS indicated that our dataset included half of all known species, and 64% of genera, for these families (Table 1).

Our GBIF and OBIS dataset had 17 species for which we did not find distribution data in the literature (Tables 2 and 3). However, 30 species lacked data although we found some in the literature. The dataset indicated a wider geographic distribution (range) for 16 species than was evident from the literature, whereas the reverse was true for only 1 species. Similar species ranges were found in the dataset and literature for around 40% of the species (Table 3). All distribution records were from shallow coasts (Figures 1-6). Some families have been reported from all continental coasts and large islands, suggesting that global sampling is adequate to represent Family level distribution.

No species of any of the families have been reported from Antarctica. Otherwise the Mactridae and Veneridae were cosmopolitan. The Donacidae, Myidae, Phardidae and Solenidae were absent from New Zealand and the Pacific islands. The Myidae has the most restricted distribution, only occurring in the Americas and north-east Atlantic.

Table 2. Comparison of species distribution data available in the GBIF and OBIS dataset with that in the literature. *Species introduced outside their native range

Species	References	Location in literature	Difference	Location from GBIF and OBIS	Introduced range from GBIF/OBIS
DONACIDAE					
Donax cuneata (=*D. cuneatus*) Linnaeus, 1758	Bosch et al. (1995)	South-eastern Persian Gulf		No data	
Donax deltoides Lamarck, 1818	McLachlan et al. (1996)	South coast of Australia from Eyre Peninsula to Kingston, through Tasmania to Fraser in south-eastern Queensland	=	South Australia	
Donax fossor Say, 1822		No data		East North America	
Donax hanleyanus Philippi, 1847	Penchaszadeh and Olivier (1975); Defeo and de Alava (1995); Herrmann et al. (2009)	Western Atlantic sandy beaches, South America, from tropical (17 S Caravelas, Brazil) to temperate regions (37 S Punta Mogotes in Mar del Plata, province of Buenos Aires)	>	East North America	
Donax incarnatus Gmelin, 1791		No data		Asia	
Donax scalpellum Gray, 1825	Bosch et al. (1995)	South-eastern Persian Gulf, Masirah, Gulf of Oman and Southern Oman		No data	
Donax semistriatus Poli, 1795	Manca Zeichen et al. (2002)	Mediterranean and the Black Sea		No data	
Donax serra Röding, 1798	Donn et al. (1986); Dugan and McLachlan (1999); Laudien (2002)	Northern boundary of Namibia to South Africa's Eastern Cape		No data	
Donax texasianus Philippi, 1847		No data		South coast of North America	
Donax trunculus Linnaeus, 1758	Ramon et al. (1995); Deval (2009)	Mediterranean Sea, the Black Sea, and from Senegal to the northern Atlantic coast of France		No data	
Donax variabilis Say, 1822	Jones et al. (2004)	Southeastern United States which has been occurred from Virginia to southern Florida around the Gulf coast of Texas	<	Coasts of Asia, North and South America	
Donax variegatus (Gmelin, 1791)	Ramon et al. (1995); Deval (2009)	Mediterranean Sea, the Black Sea, and from Senegal to the northern Atlantic coast of France		No data	
Donax venustus Poli, 1795	Manca Zeichen et al. (2002)	Mediterranean and the Black Sea as well as Atlantic coasts, south of the Gibraltar Strait to Cap Blanc		No data	

Species	References	Location in literature	Difference	Location from GBIF and OBIS	Introduced range from GBIF/OBIS
MACTRIDAE					
Mactra achatina Holten, 1802		No data		Western Asia (e.g. Kagoshima, Japan and Gulf of Suez)	
Mactra aequisulcata G.B. Sowerby III, 1894	Bosch et al. (1995)	Offshore Gulf of Oman		No data	
Mactra chinensis Philippi, 1846		No data		Eastern coast of China	
Mactra lilacea Lamarck, 1818*	Bosch et al. (1995)	Eastern Arabia	=	Western Asia (e.g. Al Bahr al Ahmar and Janub Sina, Egypt)	
Mactrinula tryphera Melvill, 1899	Bosch et al. (1995)	North and southeast Persian Gulf and the Gulf of Oman		No data	
Meropesta nicobarica Gmelin, 1791		North and southeast Persian Gulf and the Gulf of Oman, northwest Persian Gulf and Masirah	<	Eastern Africa, eastern and South Asia (e.g. Kushi, Andaman and Nicobar Islands, India), and Australia (e.g. Tin Can Bay, Gulf of Carpentaria, Hervey Bay)	
Spisula solidissima (Dillwyn, 1817)*	Shumway et al. (1994); Hare et al. (2010)	Cape Hatteras, North Carolina to the Gulf of St. Lawrence, Canada	<	Eastern North America (e.g. Block Island, Clearwater Beach, Pinellas County, and Florida, the United State)	Europe (e.g. Netherlands)
MYIDAE					
Cryptomya californica Conrad, 1837	Yonge (1951)	Along the coast of California		No data	
Mya arenaria Linnaeus, 1758*	Høpner-Petersen (1999); Beal et al. (2001); Hunt (2004); Connell et al. (2007); Conde et al. (2010)	Estuaries and sheltered bays in many parts of North America and Europe; soft-bottoms of Maine in the USA, north-eastern of the United State; Kattegat near Frederikshavn in Denmark	<	Europe, eastern and western North America, eastern Asia (e.g. Fukura, Awaji, Hiroshima, Kisarazu, Chiba-Ken, and Aichi and Mikawa, Japan)	Netherlands
Mya eideri Høpner-Petersen, 1999	Høpner-Petersen (1999)	Greenland from Thule in northwest Greenland down along the coast of West Greenland to the East Greenland fjords and also from Alaska		No data	
Mya neoovata Høpner-Petersen, 1999	Høpner-Petersen (1999)	Inner parts of Ikka Fjord, Denmark		No data	
Mya pseudoarenaria Schlesch, 1931		Svalbard, Vest Spitsbergen in Europe and Eld Inlet, Mud Bay, Tuktoyaktuk Harbour, Kodiak Bay, South and southeast off St. Lawrence Island in North America		No data	

Table 2. (Continued)

Species	References	Location in literature	Difference	Location from GBIF and OBIS	Introduced range from GBIF/OBIS
Mya truncata Linnaeus, 1758	Amaro et al. (2003); Birkely et al. (2003); Camus et al. (2003); Yuan et al. (2011)	British coasts, from northeast Atlantic (including the North Sea and the coast of Norway) to Biscay and around Arctic fjord, Svalbard (748–818N, 108–358E) between the Barents Sea and Greenland Sea; southern North Sea, Frisian Front	<	Along Europe (e.g. western Norway), North America (e.g. North Carolina), and Asia (e.g. Russia)	
Paramya africana Cosel, 1995		Africa (Tropical West Africa)		No data	
Paramya subovata Conrad, 1845		Destin, eastern North America		No data	
Platyodon cancellatus Conrad, 1845	Yonge (1951)	Along the coast of California	=	Western North America (e.g. San Diego and Skidegate Inlet)	
Sphenia antillensis Dall and Simpson, 1901 (=*Sphenia fragilis* H. Adams and A. Adams, 1854	Narchi and Domaneschi (1993)	Southern Brazilian coasts	=	Along eastern South America (Engenho Dagua, Brazil)	
Sphenia binghami Turton, 1822	Yonge (1951)	Mediterranean, along the Atlantic coasts of Spain and France and around the British Isles as far north as Scarborough on the east and Skye on the west	=	Europe	
Tugonella decurata A. Adams, 1851	Bosch et al. (1995)	Southern Oman		No data	
Tugonia anatina Gmelin, 1791		Gambia, western Africa		No data	
Tugonia inopinata Iredale, 1936		Hervey Bay, Elliot River, and Keppel Bay in eastern Australia		No data	
Tugonia nobilis H. Adams and A. Adams, 1856	Bosch et al. (1995)	Masirah, Indian Ocean such as Aden in Yemen and Masirah in Oman		No data	

Species	References	Location in literature	Difference	Location from GBIF and OBIS	Introduced range from GBIF/OBIS
PHARIDAE					
Ensis arcuatus Jeffreys, 1865 (=*E. magnus* Schumacher, 1817)	Cosel (1990); Hayward and Ryland (1998); Varela et al. (2007); Guerra et al. (2011)	From the Portuguese coasts to the Scandinavian Peninsula and also Ireland and the United Kingdom	≤	Europe	
Ensis directus Conrad, 1843*	Beukema and Dekker (1995)	Along the North Sea coasts from northern Denmark to northern France	<	Western and eastern North America	Europe
Ensis ensis Linnaeus, 1758	Cosel (1990); Guerra et al. (2011)	From the Portuguese coasts to the Scandinavian Peninsula and also Ireland and the United Kingdom	≤	Europe	
Ensis macha Molina, 1782	Cosel, 1990; Barón et al. (2004); Guerra et al. (2011)	Along southern end of South America	<	North America, South America	
Ensis minor Chenu, 1843	Cosel (1990, 1993); Guerra et al. (2011)	Mediterranean Sea basin, Atlantic from Scotland southward to northern Morocco	≤	Eastern North America, Europe	
Ensis siliqua Linnaeus, 1758	Holme (1954); Cosel (1990); Henderson and Richardson (1994); Guerra et al. (2011)	From the Portuguese coasts to the Scandinavian Peninsula and also Ireland and the United Kingdom	≤	Europe	
SOLENIDAE					
Solen capensis Fischer, 1881	Hodgson et al. (1983); Hodgson and Fielden (1986)	Bushmans River Estuary, South Africa	=	South Africa	
Solen corneus Lamarck, 1818	Cosel (2002)	Java (Leschenault)		No data	
Solen crosnieri Cosel, 1989	Cosel (1989)	West coast of Madagascar from Tuléar to Nosy Be		No data	
Solen dactylus Cosel, 1989	Cosel (1989); Saeedi et al. (2009)	Northern parts of the Persian Gulf (Bandar Abbas, Iran) and eastward along the coast of Pakistan to Kathiawar State, and India (Gulf of Kutch)		No data	
Solen darwinensis Cosel, 2002	Cosel (2002)	Northern Queensland, north and westward to the tropical northern part of west Australia		No data	
Solen fonesii Dunker, 1862				Keppel Bay, Statute Point, Arnhem Land, Maningrida, and Crocodile Research Station, Australia	
Solen grandis Dunker, 1862	Xu and Song (2008); Guerra et al. (2011)	Andaman Sea, Trung Satun Province; Shandong Peninsula, coastal Regions of the Yellow Sea, China	=	Eastern Asia	

Table 2. (Continued)

Species	References	Location in literature	Difference	Location from GBIF and OBIS	Introduced range from GBIF/OBIS
Solen kajiyamai Habe, 1964		No data		Gulf of Carpentaria, beaches around Mapoon, Australia	
Solen kikuchii Cosel, 2002	Cosel (2002)	Isahaya Bay, Ariake Inland Sea, west Kyushu in Japan		No data	
Solen linearis Spengler, 1794	Cosel (2002)	Nicobar Island	=	Nicobar Island	
Solen madagascariensis Cosel, 1989	Cosel (1989)	West coast of Madagascar and the coast of central East Africa (Kenya)		No data	
Solen malaccensis Dunker, 1862	Cosel (2002); Guerra et al. (2011)	Malacca; Samut Songkhram, Thailand		No data	
Solen marginatus Pulteney, 1799	Remacha-Triviño and Anadón (2006)	Along the west Mediterranean Sea and Atlantic Ocean from Britain to Mauritania comprising all of Spain's Coasts	=	Western and southern parts of Europe	
Solen pseudolinearis Cosel, 2002	Cosel (2002)	Southern parts of the Chinese coast from Beibu Gulf (Gulf of Tonkin) northward to Xiamen (Amoy)		No data	
Solen sarawakensis Cosel, 2002	Cosel (2002)	Buntal Beach, Kuching, and Sarawak in Malaysia	=	Eastern Asia (e.g. North of Kuching, Buntal Beach, Malaysia)	
Solen sicarius Gould, 1850		No data		Western North America (e.g. west Moresby Island, Canada)	
Solen soleneae Cosel, 2002	Cosel (2002)	Malaysia eastward to the South Chinese coast		No data	
Solen strictus Gould, 1861	Hong and Lee (1990); Kim et al. (2004)	Cheju-do and Simpo tideland, Outfal Estuary of Kumkang River, Korea	=	Eastern Asia (e.g. Honmoku, Tokyo Bay, Kushima, Ariake Bay, Aichi, Mikawa, Japan)	
Solen tehuelchus Hanley, 1842	Penchaszadeh et al. (2006)	Mar Del Plata, Northern Argentina	≤	Eastern South America (e.g. Desembocadura Río Negro, Argentina and Santana's Archipelago, Portugal)	
Solen thachi Cosel, 2002	Cosel (2002)	Long Hai, Vietnam	=	Khanh Hoa, Nha Trang, Vietnam	
Solen thailandicus Cosel, 2002	Cosel (2002)	Gulf of Thailand, from Phetchaburi Province, west coast of Bight in Bangkok, East and south-eastern ward to the west coast of Vietnam		No data	
Solena oblique (Spengler, 1794)		No data		Northern South America and southern parts of North America (e.g. Manaure, Colombia and Veracruz, Mexico)	

Species	References	Location in literature	Difference	Location from GBIF and OBIS	Introduced range from GBIF/OBIS
VENERIDAE					
Amiantis purpurata Lamarck, 1818	Morsan (2007)	Along Patagonia (Argentina)	<	Eastern parts of South America and southwest Asia	
Amiantis umbonella Lamarck, 1818 (was not in WORMS)	Bosch et al. (1995); Saeedi et al. (2010)	Northern parts of the Persian Gulf (e.g. Bandar Abbas, Iran)	<	Eastern to southern Asia and both North and South America	
Dosinia mactracea Broderip, 1835		No data		Anakiwa Marlborough Sounds, New Zealand	
Gafrarium Australe G.B. Sowerby II, 1851		No data		North to east Australia	
Gafrarium dispar Holten, 1802		No data		South Asia (e.g. Parangipettai coast, Tamil Nadu, India)	
Gafrarium pectinatum Linnaeus, 1758*		No data		South Asia (e.g. Parangipettai coast, Tamil Nadu, India)	
Mercenaria mercenaria Linnaeus, 1758*	Pritchard (2004)	North America, the United States	=	North America	
Ruditapes decussatus Linnaeus, 1758 (=*Venerupis decussata* Linnaeus, 1758*	Gomez-León et al. (2007); Gharbi et al. (2010)	Along the Mediterranean and its adjacent Atlantic waters from the North Sea to the coast of Senegal and west coast of Galicia (northwest Spain); along the coastline of Tunisia except the Cape Bon and Gulf of Hammamet	<	Eastern and southern Europe, Galicia	Asia
Venerupis philippinarum Adams and Reeve, 1850*	Ponurovskii (2008); Mao et al. (2011)	Philippines, the South China, Yellow Sea, Japan, and Okhotsk Seas to the shoals near the South Kurils	<	Eastern and southern Asia	Europe
Tivela compressa G.B. Sowerby II, 1851		No data		Africa and Europe	
Tivela polita Sowerby		No data		Africa and Europe	
Tivela stultorum Mawe, 1823		No data		Western parts of North America (e.g. Point Mugu, the United States) and East Asia	

Table 3. Comparison of the available species distribution data between the literature, and GBIF and OBIS combined. *Includes species introduced outside its native range. Where species range information is included in both sources, the percent that are similar is given.

	No. species	Not in literature	Not in GBIF or OBIS	> range in literature	> range in GBIF and OBIS combined	Range same in literature, GBIF and OBIS	
						No. species	%
DONACIDAE	13	3	7	1	1	1	33
MACTRIDAE	9	3	3	0	2*	1	33
MYIDAE	15	0	10	0	2*	3	60
PHARIDAE	6	0	0	0	6*	0	0
SOLENIDAE	22	3	10	0	1	7	87
VENERIDAE	12	7	0	0	4*	1	20
Total	77	17	30	1	16	13	39

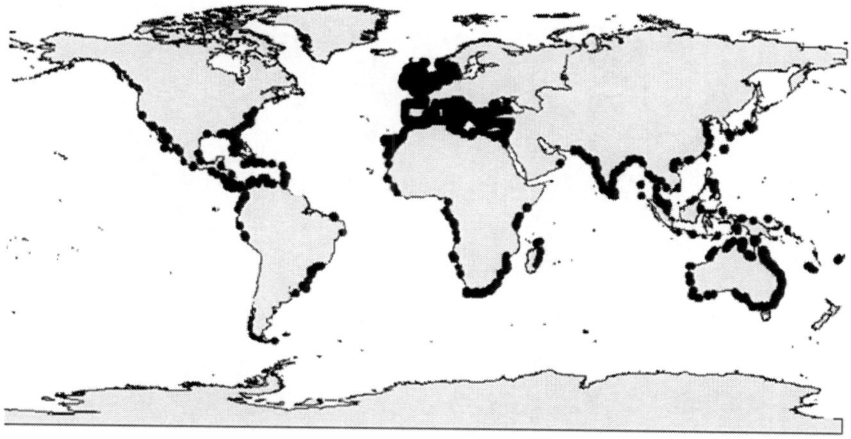

Figure 1. Distribution of Family Donacidae.

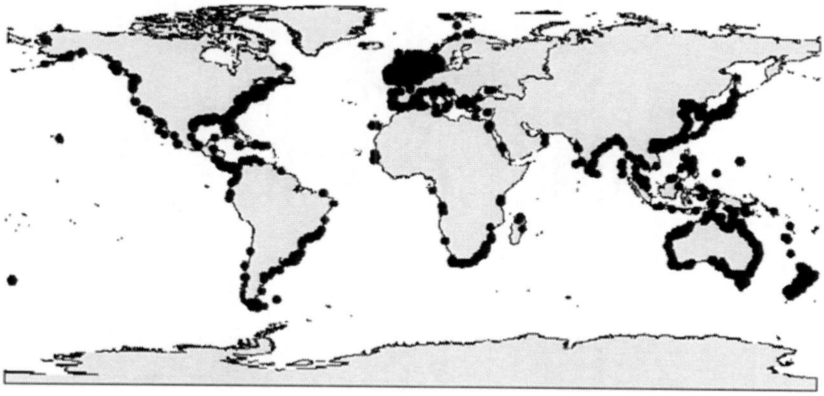

Figure 2. Distribution of Family Mactridae.

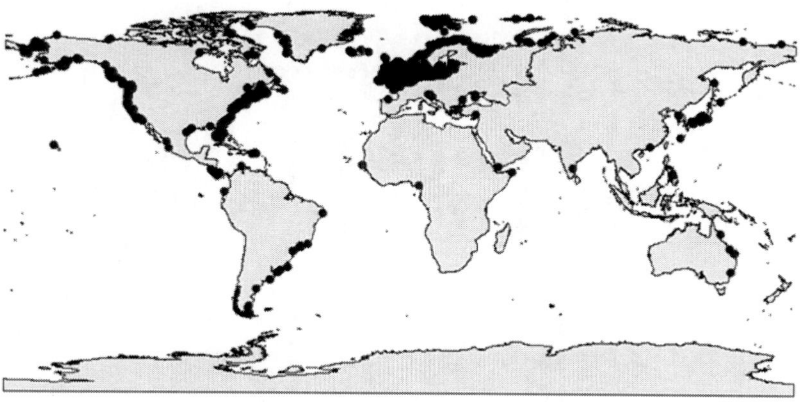

Figure 3. Distribution of Family Myidae.

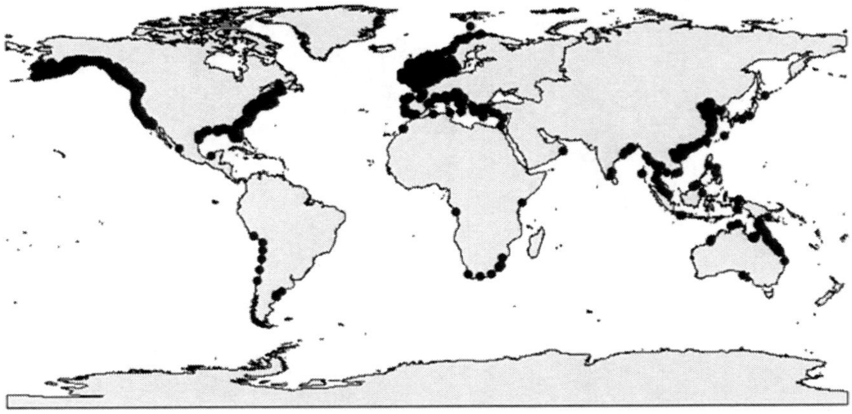

Figure 4. Distribution of Family Pharidae.

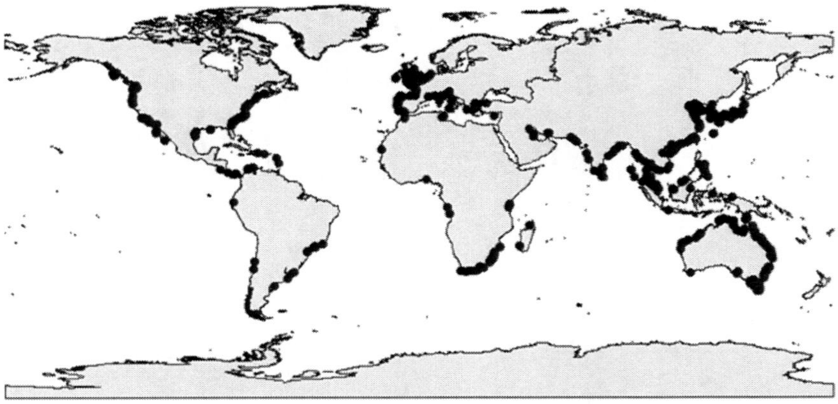

Figure 5. Distribution of Family Solenidae.

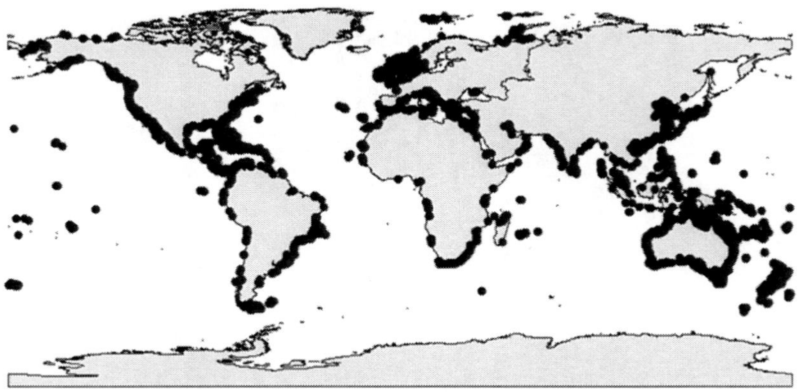

Figure 6. Distribution of Family Veneridae.

4. DISCUSSION

On average, 64% of genera and 53% of species reported in WoRMS had distribution data in GBIF and OBIS combined (Table 1). That these open-access resources contain some data on over half of the species is a significant achievement. They also had data additional to that in the literature (Table 2). Of the species with occurrence data in the literature and GBIF and OBIS, about 40% recorded similar geographic distributions, while the databases indicated a wider distribution than in the literature for 53% of these species. In some cases, this may have been because species had been introduced outside of their native range. For example, at least 5 species from 4 families had a wider distribution in our dataset because it included their introduced range, namely *Ruditapes decussatus* (=*Venerupis decussata*), *V. philippinarum, Ensis directus*, *Spisula solidissima* and *Mya arenaria* (Table 2). One species, *Mercenaria mercenaria* was only reported from its native range in the GBIF and OBIS dataset (Table 2). However, there is no indicator for users of GBIF or OBIS to know which species have different native and introduced ranges.

This could be achieved by having an indicator for species known to live outside their native range so a reader could click on the indicator and be taken to a web page that indicates the species native and introduced ranges. Alternatively, known introduced species could be excluded from the analysis. In other cases, the databases may have included geo-referencing or other errors that led to inaccuracies, as Robertson (2008) found for marine fish species.

It was evident that considerable useful data published in OBIS and GBIF lack adequate metadata, including informative titles for datasets. As this was the most visible aspect of the published dataset it did not provide confidence in the quality of the actual data whose quality were more difficult to assess. In addition to potential errors in the sources of the data, errors may occur in the application of species names, geo-referencing, not recognising species introductions, and in data processing (e.g. omission or addition of plus and minus signs to latitude or longitude coordinates).

We agree with Robertson (2008) that for OBIS and GBIF to fulfil their aspirations greater scrutiny of metadata and data by editors and peer's is needed. Quality assurance indicators to indicate which datasets have passed various steps in quality control (e.g. taxonomy, mapping) would be helpful to readers (Costello and Vanden Berghe, 2006; Costello, 2009). Nevertheless, while GBIF and OBIS contain data on only half of all known clam species studied here, they generally provided more distribution data than available from the literature alone. Thus these databases significantly add to data available from the literature, as well as making it easily accessible in a standardised format.

We consider the fact that most records have been reported from Europe reflects sampling and publishing effort in the region. The presence of only Veneridae and Mactridae in New Zealand suggests these families evolved prior to the separation of New Zealand from Australia. Although several other bivalve families are known from Antarctica (Kirkwood and Burton, 1988; Arnaud et al., 2001; Cattaneo-Vietti et al., 2000), the absence of any species from these families studied here may suggest that none of the families had evolved before the break up of Gondwanaland, an absence of suitable coastal sandy habitat due to ice cover and/or ice scour, the lack of sampling in their near-shore habitats due to permanent ice cover, and/or past extinctions during glaciations. A study of the evolutionary history or these familes and fossil data may help elucidate these questions.

ACKNOWLEDGMENTS

HS was supported by New Zealand International Doctoral Research Scholarship (NZIDRS) and University of Auckland Doctoral Scholarship. We thank the referees for helpful comments that improved this paper.

REFERENCES

Amaro, T., Duineveld, G., Bergman, M. and Witbaard, R. (2003). Growth variations in the bivalve *Mya truncata*: A tool to trace changes in the Frisian Front macrofauna (southern North Sea)? *Helgol. Mar. Res.* 57, 132-138.

Appeltans, W., Bouchet, P., Boxshall, G.A., Fauchald, K., Gordon, D.P., Hoeksema, B.W., Poore, G.C.B., van Soest, R.W.M., Stöhr, S., Walter, T.C. and Costello, M.J. (Eds.) (2011). World Register of Marine Species. Accessed at http://www.marinespecies.org on 2011-12-10.

Arnaud, P.M., Troncoso, J.S. and Ramos, A. (2001). Species diversity and assemblages of macrobenthic Mollusca from the South Shetland Islands and Bransfield Strait (Antarctica). *Polar Biol.* 24, 105-112.

Barón, P.J., Real, L.E., Ciocco, N.F. and Ré, M.E. (2004). Morphometry, growth and reproduction of an Atlantic population of the razor clam *Ensis macha* (Molina, 1782). *Sci. Mar.* 68, 211-217.

Beal, B.F., Parker, M.R. and Vencile, K.W. (2001). Seasonal effects of intraspecific density and predator exclusion along a shore-level gradient on survival and growth of juveniles of the soft-shell clam, *Mya arenaria* L., in Maine, USA. *J. Exp. Mar. Biol. Ecol.* 264, 133-169.

Beukema, J. J. and Dekker, R. (1995). Dynamics and growth of a recent invader into European coastal waters: The American razor clam, *Ensis directus*. *J. Mar. Biol. Assoc. U. K.* 75, 351-362.

Birkely, S.R., Grahl-Nielsen, O. and Gulliksen, B. (2003). Temporal variations and anatomical distributions of fatty acids in the bivalve *Mya truncata*, L. 1758, from Isfjorden, Spitsbergen. *Polar Biol.* 26, 83-92.

Bosch, D., Dance, P., Moolenbeek, R. and Oliver, G. (1995) *Sea shells of eastern Arabia*, Motative Publishing: Dubai.

Bouchet, P. (2006). The magnitude of marine biodiversity. In: C.M. Duarte (Ed.), *The Exploration of Marine Biodiversity. Scientific and Technological Challenges.* pp. 33-64, Fundación BBVA.

Bouchet, P. (2011a). Donacidae. Accessed through: World Register of Marine Species at http://www.marinespecies.org/aphia.php?p=taxdetailsandid=236 on 2011-09-12.

Bouchet, P. (2011b). Veneridae. Accessed through: World Register of Marine Species at http://www.marinespecies.org/aphia.php?p=taxdetailsandid=243 on 2011-09-13.

Camus, L., Birkely, S.R., Jones, M.B., Børseth, J.F., Grøsvik, B.E., Gulliksen, B. and Depledge, M.H. (2003). Biomarker responses and PAH uptake in *Mya truncata* following exposure to oil-contaminated sediment in an Arctic fjord (Svalbard). *Science Sci. Total Environ.* 308, 221-234.

Cattaneo-Vietti, R., Chiantore, M., Schiaparelli, S. and Albertelli, G. (2000). Shallow- and deep-water mollusc distribution at Terra Nova Bay (Ross Sea, Antarctica). *Polar Biol.* 23, 173-182.

Conde, A., Novais, J. and Domínguez, J. (2010). Southern limit of distribution of the soft-shell clam *Mya arenaria* on the Atlantic East Coast. *Biol. Invasions* 12, 429-432.

Connell, L.B., MacQuarrie, S.P., Twarog, B.M., Iszard, M. and Bricelj, V.M. (2007). Population differences in nerve resistance to paralytic shellfish toxins in softshell clam, *Mya arenaria*, associated with sodium channel mutations. *Mar. Biol.* 150, 1227-1236.

Cosel, R.v. (1989). Three new species of *Solen* (Bivalvia: Solenidae) from the Indian Ocean, with remarks on the Solenidae of Madagascar. *J. Conchol.* 33, 189-208.

Cosel, R.v. (1990). An introduction to the razor shells (Bivalvia: Solenacea). The Bivalvia-*Proceedings of a Memorial Symposium in Honour of Sir Charles Maurice Yonge, Edinburgh 1986.* Hong Kong, Hong Kong University Press.

Cosel, R.v. (1993). The razor shells of the eastern Atlantic.Part 1: Solenidae and Pharidae I (Bivalvia: Solenacea). *Archiv für Molluskenkunde.* 122, 207-321.

Cosel, R.v. (2002). Seven new species of *Solen* (Bivalvia: Solenidae) from the tropical Western Pacific, with remarks on other species. In: M. Falkner, K. Groh, C.D. Speight (Eds.), *Collectanea Malacologica - Festschrift für Gerhard Falkner.* 301-341; col. pls. 31-38, Conchbooks, Hackenheim/Frierich-Held-Gesellschaft, München.

Costello, M.J. (2009) Motivation of online data publication. *Biosci.* 59, 418-427.

Costello, M.J. and Vanden Berghe, E. (2006). "Ocean Biodiversity Informatics" enabling a new era in marine biology research and management. *Mar. Ecol. Prog. Ser.* 316, 203-214.

Costello, M.J., Stocks, K., Zhang, Y., Grassle, J.F. and Fautin, D.G. (2007). About the Ocean Biogeographic Information System. Retrieved from http://hdl.handle.net/2292/5236

Defeo, O. and de Alava, A. (1995) Effects of human activities on longterm trends in sandy beach populations: the wedge clam *Donax hanleyanus* in Uruguay. *Mar. Ecol. Prog. Ser.* 123, 73-82.

Deval, M.C. (2009). Growth and reproduction of the wedge clam (*Donax trunculus*) in the Sea of Marmara, Turkey. *J. Appl. Icthyol.* 25, 551-558.

Donn Jr, T. E., Clarke, D. J., McLachlan, A. and Toit, P.D. (1986). Distribution and abundance of *Donax serra* Röding (Bivalvia: Donacidae) as related to beach morphology. I. Semilunar migrations. *J. Exp. Mar. Biol. Ecol.* 102, 121-131.

Dugan, J.E. and McLachlan, A. (1999). An assessment of longshore movement in *Donax serra* Roding (Bivalvia: Donacidae) on an exposed sandy beach. *J. Exp. Mar. Biol. Ecol.* 234, 111-124.

Food and Agriculture Organisation (2011). *Capture fisheries.* In: FAO Fisheries and Aquaculture Department, Rome. Accessed on 20 October 2011 at http://www.fao.org/fishery/capture/en.

Gharbi, A., Chatti, N., Said, K. and van Wormhoudt, A. (2010). Genetic variation and population structure of the carpet shell clam *Ruditapes decussatus* along the Tunisian coast inferred from mtDNA and ITS1 sequence analysis. *Biologia.* 65, 688-696.

Global Biodiversity Information Facility. (2011). Accessed http://data.gbif.org/datasets/resource/1542, on 2011-09-18).

Gofas, S. (2011a). Mactroidea. Accessed through: World Register of Marine Species athttp://www.marinespecies.org/aphia.php?p=taxdetailsandid=14634 on 2011-09-13.

Gofas, S. (2011b). Myoida. Accessed through: World Register of Marine Species at http://www.marinespecies.org/aphia.php?p=taxdetailsandid=245 on 2011-09-03.

Gofas, S. (2011c). Pharidae. Accessed through: World Register of Marine Species at http://www.marinespecies.org/aphia.php?p=taxdetailsandid=23091 on 2011-09-13.

Gofas, S. (2011d). Solenoidea. Accessed through: World Register of Marine Species at http://www.marinespecies.org/aphia.php?p=taxdetailsandid=14635 on 2011-12-03.

Gofas, S. (2011e). Veneroidea. Accessed through: World Register of Marine Species at http://www.marinespecies.org/aphia.php?p=taxdetailsandid=14638 on 2011-09-13.

Gomez-León, J., Aranguren, R., Gestal, C., Novoa, B. and Figueras, A. (2007). Temporal distribution of potentially pathogenic agents detected on carpet-shell clam, *Ruditapes decussatus* cultured in Galicia (NW Spain). *Aquat. Living Resour.* 20, 185-189.

Guerra, A., Lodeiros, C.; Gaspar, M.B. and da Costa, F. (2011). *Razor clams: biology, aquaculture and fisheries*. Xunta de Galicia, Consellería do Mar, Santiago de Compostela, Spain.

Hare, M.P., Weinberg, J., Peterfalvy, O. and Davidson, M. (2010). The "southern" surf clam (*Spisula Solidissima Similis*) found north of its reported range: A commercially harvested population in long island sound, New York. *J. Shellfish Res.* 29, 799-807.

Hayward, P.J. and Ryland, J.S. (1998). *Handbook of the marine fauna of North–West Europe*. Oxford: Oxford University Press.

Henderson, S.M. and Richardson, C.A. (1994). A comparison of the age, growth rate and burrowing behaviour of the razor clams, *Ensis siliqua* and *Ensis ensis*. *J. Mar. Biol. Assoc. U.K.* 74, 939-954.

Herrmann, M., Carstensen, D., Fischer, S., Laudien, J., Penchaszadeh, P.E. and Arntz, W.E. (2009). Population structure, growth, and production of the wedge clam *Donax hanleyanus* (Bivalvia: Donacidae) from Northern Argentinean Beaches. *J. Shellfish Res.* 28, 511-526.

Hodgson, A.N. and Fielden, L.J. (1986). The ultrastructure of ciliated cells from the siphon of *Solen capensis* (Mollusca, bivalvia). *Molluscan Stud.* 52, 161-168.

Hodgson, A.N., Cross, R.H.M. and Fielden, L.J. (1983). The structure and distribution of stacked plates in the siphon wall of *Donax sordidus* and *Solen capensis* (Mollusca: bivalvia). *Micron (1969)*. 14, 141-145.

Holme, N.A. (1954). The ecology of British species of *Ensis*. *J. Mar. Biol. Assoc. U.K.* 33, 145-172.

Hong, S.S. and Lee, J.J. (1990). Histological studies on the gametogenesis and the reproductive cycle of razor clam, *Solen strictus* Gould, in Cheju-do. *Bull. Mar. Res. Inst. Cheju Natl. Univ.* 14, 39-59.

Høpner-Petersen, G. (1999). Five recent *Mya* species, including three new species and their fossil connections. *Polar Biol.* 22, 322-328.

Hunt, H.L. (2004). Effects of epibenthic predators in flow: Transport and mortality of juveniles of the soft shell clam *Mya arenaria*. *Mar. Ecol. Prog. Ser.* 279, 151-160.

Jones, D.S., Quitmyer, I.R. and Andrus, C.F.T. (2004). Seasonal shell growth and longevity in *Donax variabilis* from northeastern Florida: Evidence from oxygen isotopes. *J. Shellfish Res.* 23, 707-714.

Kim, Y.G., Yu, J.E., Chung, E.Y. and Chung, P.R. (2004). *Acanthoparyphium tyosenense* (Digenea: Echinostomatidae): experimental confirmation of the cercaria and its complete life history in Korea. *J. Parasitol.* 90, 97-102.

Kirkwood, J.M. and Burton, H.R. (1988). Macrobenthic species assemblages in Ellis Fjord, Vestfold Hills, Antarctica. *Mar. Biol.* 97, 445-457.

Laudien, J. (2002). *Population dynamics and ecology of the surf clam Donax serra (Bivalvia, Donacidae) inhabiting beaches of the Benguela upwelling system.* Berichte zur Polar- und Meeresforschung. 99 pp.

Manca Zeichen, M., Agnesi, S., Mariani, A., Maccaroni, A. and Ardizzone, G.D. (2002). Biology and population dynamics of *Donax trunculus* L. (Bivalvia: Donacidae) in the South Adriatic coast (Italy). *Estuar. Coast. Shelf. Sci.* 54, 971-982.

Mao, Y., Gao, T., Yanagimoto, T. and Xiao, Y. (2011). Molecular phylogeography of *Ruditapes philippinarum* in the Northwestern Pacific Ocean based on COI gene. *J. Exp. Mar. Biol. Ecol.* 407, 171-181.

McLachlan, A., Dugan, J.E., Defeo, O., Ansell, A.D., Hubbard, D.M., Jaramillo, E. and Penchaszadeh, P.E. (1996). Beach clam fisheries. Oceanogr. *Mar. Biol. Annu. Rev.* 34, 163-232.

Morsan, E. (2007). Spatial pattern, harvesting and management of the artisanal fishery for purple clam (*Amiantis purpurata*) in Patagonia (Argentina). *Ocean. Coast. Manage.* 50, 481-497.

Narchi, W. and Domaneschi, O. (1993). The functional anatomy *of Sphenia antillensis* Dall and Simpson, 1901 (Bivalvia: Myidae). *J. Molluscan Stud.* 59, 195-210.

Penchaszadeh, P.E., Arrighetti, F., Cledãn, M., Livore, J.P., Botto, F. and Iribarne, O.O. (2006). Bivalve contribution to shallow sandy bottom food web off Mar Del Plata (Argentina): Inference from stomach contents and stable isotope analysis. *J. Shellfish Res.* 25, 51-54.

Penchaszadeh, P.E. and Olivier, S.R. (1975). Ecologiáde una poblacióndе ''berberecho'' (*Donax hanleyanus*) en Villa Gesell, Argentina. *Malacologia.* 15, 133-146.

Ponurovskii, S.K. (2008). Population structure and growth of the Japanese littleneck clam *Ruditapes philippinarum* in Amursky Bay. Sea of Japan. *Russ. J. Mar. Biol.* 34, 329-332.

Pritchard, E.S. (2004). *Fisheries of the United States 2003.* National Marine Fisheries Service, Office of Science and Technology. Silver Spring, Md.: Fisheries Statistics.

Ramon, I., Abello, P. and Richardson, C.A. (1995). Population structure and growth of *Donax trunculus* (Bivalvia: Donacidae) in the western Mediterranean. *Mar. Biol.* 121, 665-671.

Remacha-Triviño, A. and Anadón, N. (2006). Reproductive cycle of the razor clam *Solen marginatus* (Pulteney 1799) in Spain: A comparative study in three different locations. *J. Shellfish Res.* 25, 869-876.

Robertson, D.R. (2008). Global biogeographical data bases on marine fishes: caveat emptor. *Divers. Distrib.* 14, 891-892. doi:10.1111/j.1472-4642.2008.00519.x.

Saeedi, H., Ardalan, A.A., Kamrani, E. and Kiabi, B.H. (2010). Reproduction, growth and production of *Amiantis umbonella* (Bivalvia: Veneridae) on northern coast of the Persian Gulf, Bandar Abbas, Iran. *J. Mar. Biol. Assoc. U.K.* 90, 711-718.

Saeedi, H., Raad, S. P., Ardalan, A.A., Kamrani, E. and Kiabi, B.H. (2009). Growth and reproduction of *Solen dactylus* (Bivalvia: Solenidae) on northern coast of the Persian Gulf (Iran). *J. Mar. Biol. Assoc. U.K.* 89, 1635-1642.

Shumway, S.E., Sherman, S.A., Cembella, A.D. and Selvin, R. (1994). Accumulation of paralytic shellfish toxins by surfclams, *Spisula solidissima* (Dillwyn, 1897) in the Gulf of Maine: Seasonal changes, distribution between tissues, and notes on feeding habits. *Nat. Toxins* 2, 236-251.

Varela, M.A., González-Tizón, A., Francisco-Candeira, M. and Martínez-Lage, A. (2007). Isolation and characterization of polymorphic microsatellite loci in the razor clam *Ensis siliqua*: Primer note. *Mol. Ecol. Notes*. 7, 221-222.

Xu, K. and Song, W. (2008). Two trichodinid ectoparasites from marine molluscs in the Yellow Sea, off China, with the description of *Trichodina caecellae* n. sp. (Protozoa: Ciliophora: Peritrichia). *Syst. Parasitol.* 69, 1-11.

Yonge, C.M. (1951). Observations on *Sphenia binghami* Turton. *J. Mar. Biol. Assoc. U.K.* 30, 387-392.

Yuan, L., Sun, L., Wei, G., Long, N., Xie, Z. and Wang, Y. (2011). 9,400 yr B.P.: The mortality of mollusk shell (*Mya truncata*) at high Arctic is associated with a sudden cooling event. *Environ. Earth Sci.* 63, 1385-1393.

In: Clam Fisheries and Aquaculture
Editor: Fiz da Costa González
ISBN: 978-1-62257-518-3

Chapter 3

CLAM REPRODUCTION

F. da Costa[1,2*], J. A. Aranda-Burgos[1], A. Cerviño-Otero[1], A. Fernández-Pardo[1], A. Louzán[1], S. Nóvoa[1], J. Ojea[1] and D. Martínez-Patiño[1]

[1]Centro de Cultivos Marinos de Ribadeo-CIMA, Ribadeo (Lugo), Spain
[2]Ifremer, Laboratoire de Physiologie des Invertébrés Marins, Station Expérimentale d'Argenton, Landunvez, France

ABSTRACT

The reproductive cycle of bivalves has been widely investigated, mainly focusing in commercial species, due to its importance in aquaculture development and fishery management. In this chapter, we review the investigations dealing with the reproduction of six species of clams of commercial interest in Spain carried out by our research group and we discuss it with published data about other species of clams in the world.

These clam species belong to the families Veneridae (*Venerupis philippinarum*, *Venerupis decussata* and *Venerupis corrugata* (=*V. pullastra*)), Donacidae (*Donax trunculus*), Pharidae (*Ensis siliqua*) and Solenidae (*Solen marginatus*). The different ways of assessing the reproductive cycles in clams are analyzed and information is provided on the studies using these methods. Since reproduction in bivalves is influenced by exogenous (mainly temperature and food availability) and endogenous factors (of endocrine and neurological type), we discuss the investigations dealing with the effect of both factors in clams. Moreover, we tackle the study of biochemical composition during the reproductive process because marine bivalves present energy storage and utilization cycles closely linked with gametogenic cycles.

* Corresponding author: F. da Costa. Centro de Cultivos Marinos de Ribadeo-CIMA, Muelle de Porcillán, s/n, 27700, Ribadeo (Lugo), Spain. Present address: Ifremer, Laboratoire de Physiologie des Invertébrés Marins, Station Expérimentale d'Argenton, Presqu'île du Vivier, 29840, Landunvez, France. E-mail address: fiz.da.costa.gonzalez@ifremer.fr; mptea@cimacoron.org.

J. A. Aranda-Burgos, A. Cerviño-Otero, A. Fernández-Pardo, A. Louzán, S. Nóvoa, J. Ojea, D. Martínez-Patiño: Centro de Cultivos Marinos de Ribadeo-CIMA, Muelle de Porcillán, s/n, 27700, Ribadeo (Lugo), Spain.

1. INTRODUCTION

The study of reproductive cycles is an essential part in the knowledge of the biology of a species and its determination provide the information needed to tackle fisheries management and culture development of each species. The reproductive cycle is defined by Seed (1976) as "the entire cycle of events from activation of the gonad, through gametogenesis to spawning and subsequent recession of the gonad", differentiating the reproductive period from a rest period. Generally, bivalves exhibit an annual reproductive cycle which includes a sequence of events that start with gonad activation, followed by a stage of gametogenic development, the maturity, the gamete release or spawning event, and then, a vegetative or rest period. These stages are linked to the seasonal environmental changes, thus producing a characteristic model for each species which could differ depending on population location. The reproductive activity is determined by the interaction among exogenous factors, mainly temperature and food availability, and endogenous factors of endocrine and neurological type. All the factors, to a greater or lesser extent, have an influence on gametogenesis depending on the species, and thus, intra-specific variations in gametogenic cycles could be observed due to the variation of endogenous and exogenous factors in different locations (Barber and Blake, 1991). The stages of the cycle among individuals could be synchronic or asynchronic (Sastry, 1979). Gamete formation is called spermatogenesis in males and ovogenesis in females and it takes place in gonadal follicles. The characteristic cells of each stage of the reproductive process are formed inside the follicles which finally lead to the formation of spermatozoids in males and ripe oocytes in females that will be released to seawater. Follicular asynchrony could be observed within an individual, thus coexisting different developmental stages within gonad follicles. In general, evolution of a biochemical substrate is closely linked to the degree of sexual maturity of the bivalves, and consequently is related to energy supply from ingested food or from previously stored reserves (Sastry, 1979). Reproductive strategy for each species can be classified as either opportunistic or conservative based on the relationship between gonad development and the cycles of storage and depletion of nutrients (Bayne, 1976). In the former, gametogenic development and sexual maturing is closely coupled with the accumulation of nutrients and thus with food abundance in the environment. In the latter category, gametogenesis takes place at the expense of previously stored reserves of the organism. Consequently, the aim of this chapter is reviewing the clams´ reproduction and the different factors affecting the gametogenic cycles in clams. We will focus on the clams´ species studied in our laboratory (*Venerupis philippinarum*, *Venerupis decussata*, *Venerupis corrugata*, *Donax trunculus*, *Ensis siliqua* and *Solen marginatus*).

We bear in mind that reviewing the whole scientific literature in clam reproduction currently available in the world is unfathomable. Moreover, despite the great importance of the culture of *Mercenaria mercenaria*, only specific studies on this species are included in this chapter due to the detailed review on *M. mercenaria* reproduction by Eversole (2001).

2. Gametogenic Cycles

2.1. Means of Assessing

Investigations on reproductive cycle of bivalves have been carried out using several methods that can be classified in direct and indirect methods. The former are based on gonadal tissue analysis and they are more reliable. Indirect methods are based on detecting larvae in plankton or estimating the subsequent postlarval settlement or seed recruitment, which could provide information about the spawning period.

Macroscopic observation of the gonad has been a method widely used (Mazé and Laborda, 1990; Aracena et al., 2003; Remacha-Triviño and Anadón, 2006) due to its easiness. It consists on the observation of the relative size, shape and color of the gonads. However, the information provided is scarce and difficult to interpret.

Frotis or the examination of fresh gonad smears microscopically provides more direct information of the development and viability of gametes. The drawback of this method is that only provides a measure of gamete functionality and that it is qualitative.

A simple but effective mean of assessing gametogenic cycles in clams is to determine mean gonad weight on a regular basis throughout the year. Dry weights are preferable to wet weights as water content may vary seasonally. Moreover, as gonad size is highly influenced by the age of the individuals, it is necessary to sample individuals of a discrete size range or determine the gonad weight of a "standard" clam (e.g. Ojea et al., 2004). Gonad weights alone provide useful information, however, the elaboration of gonad condition index gives a more reliable information about the gametogenic cycles. Condition indexes are mathematical relations between biometric variables easy to determine. There are several condition indexes used by different investigators in clam reproduction, which can be divided in two categories: the ones that use the flesh and the ones that involves gonad weight. Among the former category it can be found the following methods: flesh dry weight/valve dry weight (Robert et al., 1993; Laruelle et al., 1994); ash-free dry weight/dry shell weight (Gaspar et al., 1999; Joaquim et al., 2011), flesh dry weight/total dry weight (Morvan and Ansell, 1988; Hamida et al., 2004b) and visceral mass dry weight/total dry weight (Morvan and Ansell, 1988). The main disadvantage of total fresh weight to study the gametogenic cycle is that variation in other organs different that the gonad could mask gonad weight evolution throughout the reproductive cycle. Among the second group, the gonad condition indexes, are: dry weight of the gonad-visceral mass/dry weight of shell (Ojea et al., 2004) and gonad fresh weight/valve dry weight (Darriba et al., 2004; 2005a).

Despite the aforementioned methods are simple, fast and inexpensive, do not provide anatomical details of the gonad maturation. Histological preparation of gonad tissue is time consuming and costly, but allows us gathering detailed information about the reproductive cycle at a structural level. The commonest method to analyze histological preparations is a qualitative method consisting of ascribing each preparation to a previously established gametogenic scale. In the next section a gametogenic scale in *V. corrugata* is described as an example.

There are other methods based on the histological observation of the gonads that are quantitative, which are complementary to the qualitative study of the gonads with gametogenic scales. One method is gonad index, which consist in assigning numerical values to each developmental stage (i.e. each of the stages of a gametogenic scale) allowing a calculation of a histological index for each sample. As an example, Joaquim et al. (2011) calculated the following one for *V. corrugata* (Gonad index (GI)=($\sum$ individuals in each stage x stage ranking)/total individuals sampled each month). Other way of quantifying the gonad by histology involves measuring mean oocytes diameter, which clearly reflects the gametogenic cycle. Oocytes gradually increase in size as they mature, reaching their maximum size prior to spawning. Then, mean oocyte diameter decreases sharply after spawning due to the release of the larger and mature female gametes. It has been used to study the reproductive cycle in *V. decussata* and *V. philippinarum* (Laruelle et al., 1994; Xie and Burnell, 1994; Hamida et al., 2004b; Meneghetti et al., 2004). Closely linked with the estimation of oocyte diameter are stereological techniques, which allow quantifying not only all gamete stages volume fraction (i.e. developing, mature and resorbing gametes) but also other cytological features of the gonad, such as: connective tissue and lumen space.

The reproductive pattern of a species can also be determined estimating larval and spat abundances. Thus, peaks of spawning of wild population can be determined using the duration of planktonic larvae in water column and growth of settlers in wild population.

2.2. Gametogenic Scale

There are several gametogenic scales for bivalves in the scientific literature. Herein, we present as an example the one described by Cerviño-Otero (2011) for the clam *V. corrugata*, which is based on some qualitative scales for bivalves (Lubet, 1959; Holland and Chew, 1974; Wilson and Seed, 1974), although slight variations were introduced to adapt it for this species after the observation of the histological preparations. A stage of this gametogenic scale was assigned to each individual. Since it was frequently observed follicular asynchrony in the same individual, it was assigned the gametogenic stage corresponding to the majority of the gonad. Rest stage (stage 0) is characterized for the abundance of connective tissue and the absence of follicles and gametes in other species (Figure 1). Rest stage is very short in *V. corrugata*, being possible sex determination of the individuals all year round by the presence of residual gametes remaining in the gonads and the onset of the formation of vesicular cells. Sex of the clams could not be determined only in 0.73% of the sampled population.

Gonadal development starts in stage I, start of gametogenesis, and it is characterized by the presence of fully developed follicles which are full of vesicular cells. Interfollicular muscle fibers were observed in the gonad.

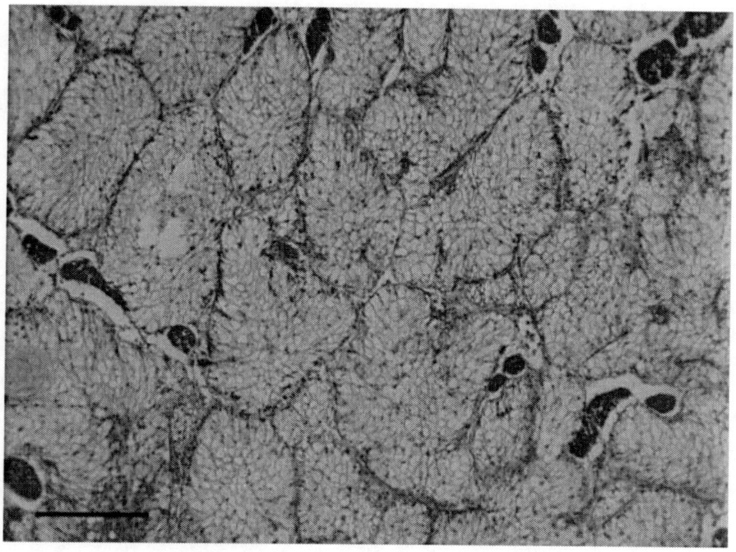

Figure 1. Photomicrograph showing an indeterminate individual (stage 0) of *Venerupis corrugata*. Scale bar: 100 μm.

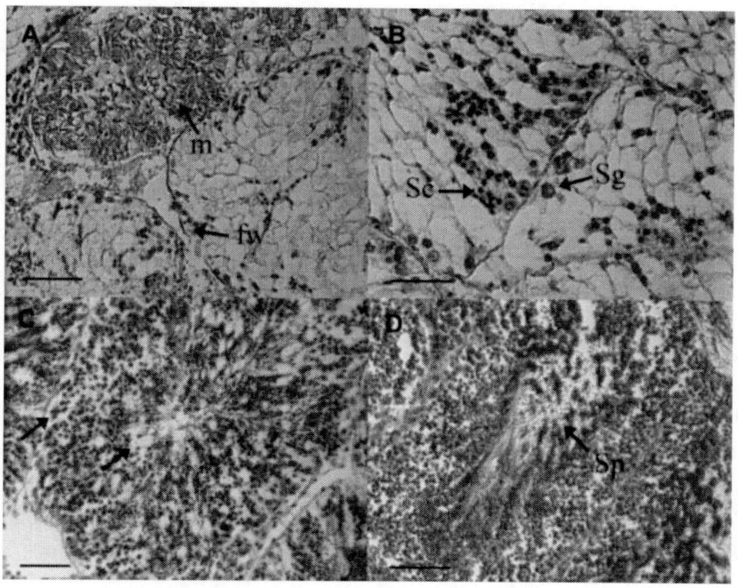

Figure 2. Photomicrographs showing stages in the development of male gonad of *Venerupis corrugata*. A. Stage I (start of gametogenesis); m: muscle; fw: follicle wall. B. Stage I; Sc: spermatocyte; Sg: spermatogonia. C. Stage II (advanced gametogenesis); between arrows are all cells in the germinal line. D. Stage II; Sp: spermatozoa. Scale bar: 50 μm in A, C and D; 25 μm in B.

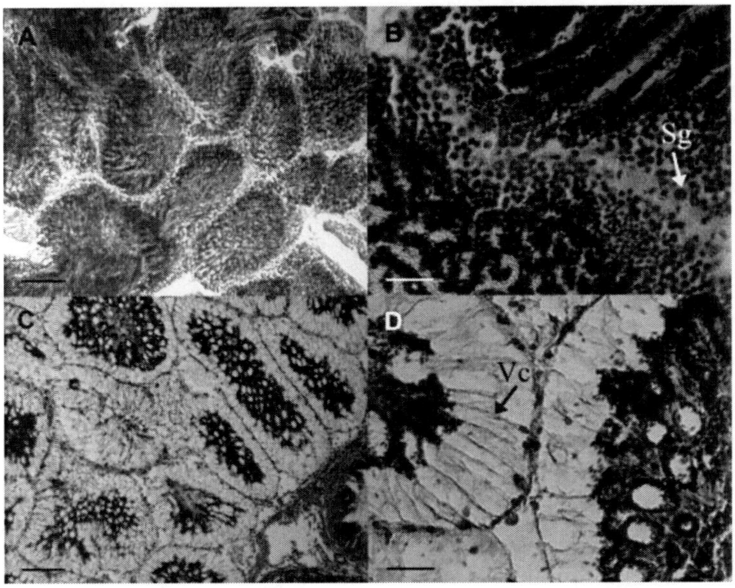

Figure 3. Photomicrographs showing stages in the development of male gonad of *Venerupis corrugata*. A. Stage III (ripe and spawning). B. Stage III; Sg: spermatogonia. C. Stage IV (restoration). D. Detail of stage IV showing a vesicular cell (Vc). Scale bar: 100 μm in A, C and D; 25 μm in B.

In males, plenty of spermatogonias are found close to follicular walls and few spermatocytes are observed in the interior part of the follicle (Figure 2A-B). In females, ovogonias attached to follicular walls increased and few oocytes started to store vitellum (Figure 4A). Non-atresic residual oocytes are commonly found in this stage.

In stage II (advanced gametogenesis) vesicular cell decreased, albeit some remain stuck to follicular walls. In this stage males had germinal cells in all phases of spermatogenesis (protogonias, spermatogonias, spermatocytes, spermatids and spermatozoids) (Figure 2C-D).

In females, previtellogenic and pedunculated oocytes were frequently observed and in a lesser extent ripe oocytes are present (Figure 4B-C).

In stage III (ripe and spawning) follicles were fully occupied by spermatozoids in males (Figure 3A-B). Free spermatozoids ready to be released and few lines of spermatogonias were found in some of the follicles. Female follicles were full of ripe oocytes and empty spaces of the released oocytes were found (Figure 4D-E). Similar to the observed pattern in males, ovogonias attached to follicle walls were observed. Few vesicular cells were present close to follicle walls.

Stage IV or restoration stage was characterized by the presence of newly-formed vesicular cells. In males, spermatozoids were found in the center of the follicle surrounded by few lines of vesicular cells (Figure 3C-D). Gonias increased and neither spermatogonia nor spermatocytes were found. In females, non-atresic residual oocytes surrounded by newly-formed vesicular cells were observed. There were not any previtellogenic oocytes present in the female follicles (Figure 4F). Sometimes, small regions of conjunctive tissue can be found organizing themselves in follicles (Figure 5).

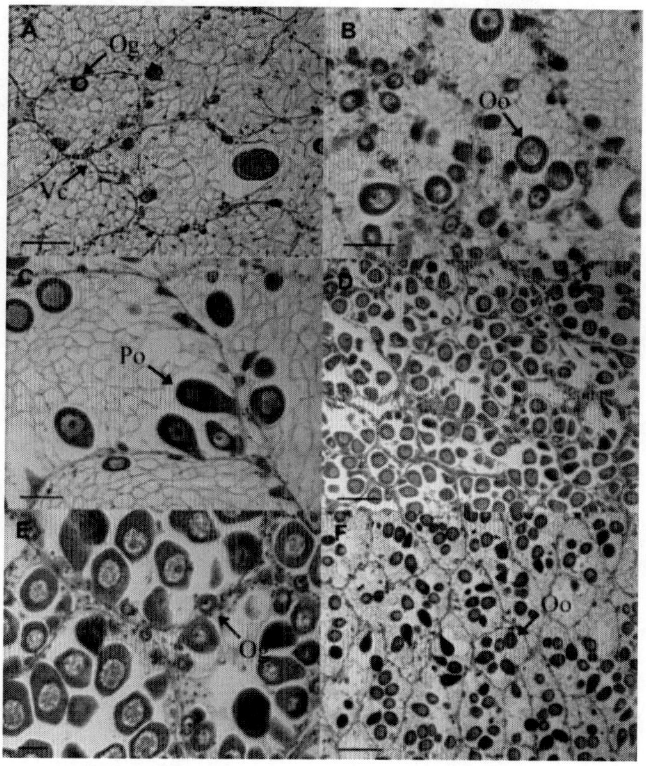

Figure 4. Photomicrographs showing stages in the development of female gonad of *Venerupis corrugata*. A. Stage I; Og: ovogonia; Vc: vesicular cell. B. Stage II (advanced gametogenesis); Oo: oocytes. C. Detail of stage II showing a pedunculated oocyte (Po) stuck to the follicle wall. D. Stage III. E. Detail of stage III; Og: ovogonia. F. Stage IV (restoration); Oo: oocytes. Scale bar: 100 μm in C, D and F; 50 μm in A and B; 25 μm in E.

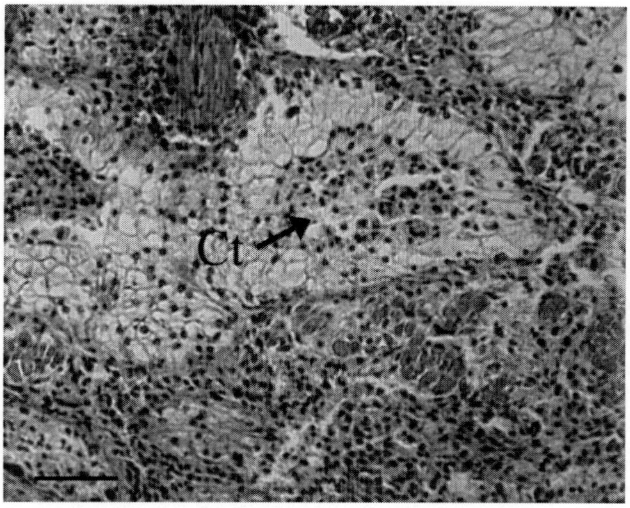

Figure 5. Photomicrograph of a gonad region invaded by connective tissue. Scale bar: 100 μm.

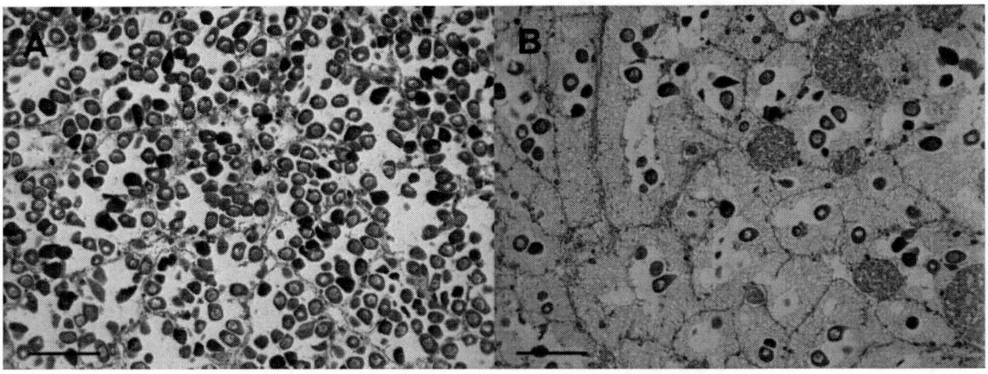

Figure 6. Photomicrographs of *Venerupis corrugata* gonad in the same female showing different stages of gametogenic development. A. Follicles in ripe and spawning stage. B. Follicles in restoration stage. Scale bar: 100 μm.

Follicular asynchrony within an individual is frequently found in restoration stage (stage IV), being observed mainly during summer and autumn because in this season is when individuals go through ripeness to a new gametogenic cycle (Figure 6A-B). In this season are present follicles in ripe stage, follicles that start to form vesicular cells, and even few follicles in restoration can be found.

Table 1. Spawning periods for various clam species

Species	Country	Location	Spawning period	Reference
VENERIDAE				
Venerupis philippinarum	France	Ile Tudy, Brittany	July-August	Beninger and Lucas (1984)
		Arcachon Bay	Autumn	Robert et al. (1993)
		Bay of Brest Morbihan Gulf	May, July, mid-August-mid-September Early June, early July, early September, early October	Laruelle et al. (1994)
		Arcachon Bay	Autumn, varies within locations	Dang et al. (2010)
	Ireland	Cork	September	Xie and Burnell (1994)
		Dungloe Bay and Drumcliff Bay	May-September	Drummond et al. (2006)
	Spain	Ría de Vigo	June-November	Rodríguez-Moscoso et al. (1992)
		Ría de Camariñas	May-October	Ojea et al. (2005)
	Italy	River Po Delta	June-early autumn	Sbrenna and Campioni (1994)
		Lagoon of Venice	June-September	Marin et al. (2003)

Species	Country	Location	Spawning period	Reference
Venerupis philippinarum		Lagoon of Venice	June-September	Meneghetti et al. (2004)
	USA	Hood Canal, Washington	June-September	Holland and Chew (1974)
		Hood Canal, Washington	May-September	Nosho and Chew (1972)
	Canada	British Columbia	June to autumn	Bourne (1982)
	Japan	Musaka Soal	Late spring, early to late fall	Ohba (1959)
		Ariake Sound	Fall to early winter	Ishii et al. (2001)
		Tokyo Bay	Spring to early summer	Ishii et al. (2001)
		Tokyo Bay	May-October	Toba et al. (2007)
		Matsukawa-ura	Late June and early August, late September and early October	Kanazawa and Sato (2008)
	China	Jiaozhou Bay	May-June	Ren et al. (2008)
	Korea	Jinju Bay	Summer-October	Kang et al. (2007)
	Russia	Sea of Japan, 5 locations	July-August	Ponurovsky and Yakovlev (1992)
Venerupis decussata	France	Ile Tudy, Brittany	July-August	Beninger and Lucas (1984)
		Bay of Brest Etel Ria	July-October June to mid-July, late August, late September	Laruelle et al. (1994)
	Spain	Ría de Vigo	June-July	Figueras (1957)
		Ría de Arousa	July-August	Villalba et al. (1993)
		Ría de Arousa	April-August, August/September	Rodríguez-Moscoso and Arnaiz (1998)
		Urdaibai Estuary	April-mid-October	Urrutia et al. (1999)
		Lagunas de Baldaio	June-August	Ojea et al. (2004)
	Ireland	Cork	August-September	Xie and Burnell (1994)
	France	Lagoon of Thau	June, August	Borsa and Millet (1992)
	Italy	Venice Lagoon	August-September	Breber (1980)
	Morocco	Oualidia Lagoon Moulay Bousselham Lagoon	May-June, August-September May-June, August-September	Shafee and Daoudi (1991)
	Tunisia	Gabès Gulf	June-December (irregularly)	Hamida et al. (2004b)
		Gabès Gulf	August-September	Ketata et al. (2007)
	Turkey	Sufa Lagoon	July-October	Serdar and Lök (2009)
Venerupis corrugata	Spain	Ría de Vigo	January, June	Figueras (1957)
		Ría de Arousa	March-May	Pérez-Camacho (1980)
		Ría de Vigo	May-June, autumn and winter	Villalba et al. (1993)
		Ría de Arousa and Ría de Camariñas	February-July	Cerviño-Otero (2011)
	Portugal	Ria de Aveiro	May-November	Maia et al. (2006)
		Ria de Aveiro	May-November	Joaquim et al. (2011)

Table 1. (Continued)

Species	Country	Location	Spawning period	Reference
Polititapes virgineus	France	Bay of St. Malo	Late May, July-September	Morvan and Ansell (1988)
	Spain	Ría de Vigo	May-June, August and winter	Villalba et al. (1993)
		Málaga	April	Yamuza-Clavijo et al. (2010)
Ameghinomya antiqua (=*Venus antiqua*)	Chile	Dichato Bay	March	Urban (1996)
Venus verrucosa	Spain	Málaga	March-April, May-August	Tirado et al. (2003)
Chamelea gallina	Spain	Andalusia	January-September	Rodríguez de la Rúa et al. (2003)
C. gallina (=*Venus striatula*)	Portugal	Vilamoura	April-August	Gaspar and Monteiro (1998)
Gafrarium pectinatum (=*G. tumidum*)	India	Chinnapalam, Pamban	November (peak), April (minor spawning)	Jagadis and Rajagopal (2007)
Callista chione	Portugal	Arrábida	January-March, April-May, August-October	Moura et al. (2008)
	Greece	Northern Euboikos Gulf	May, December-January	Metaxatos (2004)
Marcia opima	India	Tuticorin Bay Ashtamudi estuary	May-July, September-December March-May, September-December	Suja and Muthiah (2007)
Cyclina sinensis	China	Yellow River delta	August	Yan et al. (2010)
Meretrix lusoria	Korea	Simpo	June-September	Chung (2007)
	Japan	Ariake Sound and Tokyo Bay	May-October	Nakamura et al. (2010)
DONACIDAE				
Donax trunculus	France	Camarque coast	June-July, August-September	Ansell and Bodoy (1979)
		Ile d'Oleron	August	Ansell and Lagardère (1980)
	Spain	Ría del Barquero	May-July	Mazé and Laborda (1990)
		Cullera	June-September	Ramón et al. (1995)
	Portugal	Faro	March, May-August	Gaspar et al. (1999)
	Morocco	Mehdia	Spring-July	Bayed (1990)
	Italy	Tuscanina coast	July-August	Voliani et al. (1997)
		Central Tirreno Sea	July-August	La Valle (2005)
	Algeria	-	January-April	Mouëza and Frenkiel-Renault (1973)
	Turkey	Sea of Marmara	April-July	Deval (2009)
	Israel	Haifa Bay	July-September	Neuberger-Cywiak et al. (1989)

Species	Country	Location	Spawning period	Reference
D. semistriatus	Spain	Málaga	April, June-November	Tirado and Salas (1999)
D. vittatus	Scotland, UK	Firth of Forth	April-May, August-September	Ansell and Bodoy (1979)
D. variabilis	USA	Central Gulf Coast of Florida	January-May	Cobb et al. (2011)
D. venusta	Spain	Málaga	February, April-October	Tirado and Salas (1999)
PHARIDAE				
Ensis magnus (=*E. arcuatus*)	Spain	Ría de Vigo	From December-January to May-June	Darriba et al. (2004)
	Ireland	County Galway	April	Fahy et al. (2001)
E. siliqua	Spain	Ría del Barquero	May-June	Martínez (2002)
		Ría de Corcubión	May-June	Darriba et al. (2005a)
	Ireland	Gormanstown	mid-May-early August	Fahy and Gaffney (2001)
	Portugal	Vilamoura	May-June	Gaspar and Monteiro (1998)
E. minor	Italy	Gulf of Trieste	April-May	Del Piero et al. (1980)
		Gulf of Trieste	April-May	Valli and Gioglio (1980)
		Gulf of Manfredonia	March-April	Casavola et al. (1985)
		Tuscany, Latium and Campania	April-May	Costa et al. (1987)
E. macha	Chile	Dichato Bay	April	Urban (1996)
		Tubul and Corral Ancud	spring-summer summer	Avellanal et al. (2002)
		Golfo de Arauco	November-February	Aracena et al. (2003)
	Argentina	San Matias Gulf	September-November, May-June	Barón et al. (2004)
E. directus	The Netherlands	Wadden Sea	April-May, August-September	Cardoso et al. (2009)
Sinonovacula constricta	China	Yellow River delta	September-November	Yan et al. (2009)
SOLENIDAE				
Solen marginatus	Spain	Ría de Ortigueira	May-June	Martínez (2002)
		Eo Estuary Santander Bay Terrón Estuary	May-June June-August May-July	Remacha-Triviño and Anadón (2006)
	Portugal	Ría de Aveiro	May-September (2002) and May-July (2003)	Maia et al. (2006)
	Tunisia	Gabès Gulf	May-October	Hamida et al. (2010)
S. dactylus	Iran	Bandar Abbas	January	Saeedi et al. (2009)
Siliqua patula	Canada	British Columbia	July-September	Bourne (1979)

3. Variations in Gametogenic Cycles

Gametogenic cycles in bivalves are affected by the geographical location and consequently by the environmental factors (for more information see section Exogenous regulation in this chapter). For example, various authors have indicated the existence of one single period of gamete release during summer in *Venerupis decussata* in Spain (Figueras, 1957; Pérez-Camacho, 1980; Villalba et al., 1993), Italy (Breber, 1980) and Ireland (Xie and Burnell, 1994).

Other authors reported two major periods of spawning, in the spring and again in the summer in France (Borsa and Millet, 1992; Laruelle et al., 1994) and Morocco (Shafee and Daoudi, 1991) (Table 1).

Similarly, Joaquim et al. (2011) reported an extended single spawning event in *Venerupis corrugata* in Portugal that began in late winter and ended in the early summer, whilst Cerviño-Otero (2011) in Spain also indicated that in spite of the fact that this species exhibits a wide period of gamete release from February to June, the presence of ripe gametes throughout the year may indicate that new recruitments could occur all year round.

In *V. philippinarum* it has been observed variation in the number of annual spawning within Europe. For example, Robert et al. (1993) reported a single spawning period in France, other authors identified two spawning periods as Beninger and Lucas (1984) in France and Rodríguez-Moscoso et al. (1992) in Spain or three spawning periods as Laruelle et al. (1994) in France.

Therefore, in the light of the studies reviewed local environmental conditions seem to affect the number of spawning periods. In the razor clams spawning patterns varies from one single spawning event in *Ensis siliqua* and *Solen marginatus* in European locations (Gaspar and Monteiro, 1998; Fahy and Gaffney, 2001; Martínez, 2002; Darriba et al., 2005a; Remacha-Triviño and Anadón, 2006), two spawning events as reported for Argentinean populations of *E. macha* (Barón et al., 2004) and several spawning events during few months in *E. magnus* in Spanish beds (Darriba et al., 2004).

4. Biochemical Composition

Marine bivalves exhibit seasonal cycles of energy storage and depletion which are closely linked to gametogenic cycle. Seasonal variations of the metabolic activities are the result of the interactions between available food, environmental conditions, growth and reproductive activity (Gabbott, 1983). In temperate climates food availability exhibited seasonal cycles. In some species, gamete production is coupled with high abundances of food and consequently, they show an opportunistic strategy (Lubet, 1986).

In other species, nutrients are stored in different tissues and gamete production takes place when low availability of food is found in the wild beds, thus these species exhibited a conservative strategy (Sastry, 1979). Examples of conservative species are *Venerupis decussata* (Ojea et al., 2004) and *Venerupis corrugata* (Pérez-Camacho, 1980; Joaquim et al., 2011). However, other authors reported for the same species an intermediate strategy between opportunistic and conservative lifestyles. Aníbal et al. (2011) concluded that *V. decussata* exhibited this intermediate strategy since both stored and recently assimilated nutrients are used for gametogenesis. Similarly, *V. corrugata* stored reserves in specialized tissues (vesicular cells), which allows a certain independence between gametogenesis and environmental conditions (Cerviño-Otero, 2011). Glycogen and lipids are stored during summer and autumn (conservative strategy), gametogenesis starts in winter, thus allowing the first spawning release when seawater temperature increases. From spring onwards, the first phytoplanktonic blooms are observed and gamete production continues (opportunistic strategy).

Vesicular cells are specific cells involved in storage of reserves and that are located in the interior of the gonadal follicles in venerids, such as *V. philippinarum*, *V. decussata* and *V. corrugata* (Rodríguez-Moscoso and Arnaiz, 1998; Drummond et al., 2006; Cerviño-Otero, 2011). Medhioub and Lubet (1988) stated that intra-gonadic vesicular cells are rich in glycogen and progressively release glucose and its derivates during the gametogenic cycle. Moreover, the development of vesicular cells is generally opposite to gamete development and ripening, since during rest stage the volume of vesicular cells is highest and as gonad development proceeds showed a declining trend (Rodríguez-Moscoso and Arnaiz, 1998). Contrary to that, the absence of vesicular cells in the razor clam *Ensis magnus* suggested that mobilization of nutrients from other tissues was necessary to provide energy for gametogenesis (Darriba et al., 2005b).

In *E. magnus* the digestive gland appears to act as an important reserve storage site for lipids whilst muscle tissues store glycogen. Among the major biochemical substrates, glycogen plays a central role in energetic and metabolic supply of gametogenesis in many bivalves (Giese, 1967) and gonad development may involve the metabolic conversion of glycogen to lipid (Gabbott, 1975).

It can be an energy source allocated for growth and at the same time stored in specific cells as an energy reserve during the vitellogenic process (Marin et al., 2003). Rodríguez et al. (1993) concluded that glycogen is the main energy reserve for gametogenesis in *V. philiphinarum* and *V. decussata* from the Ría de Muros and Noya (Galicia, Spain). Glycogen content decreased after the resting phase and reached minimum values during the spawning period and then recovered after spawning (Ojea et al., 2004; Kang et al., 2007; Cerviño-Otero, 2011).

Endogenous reserves laid down in the eggs during vitellogenesis are important for providing energy during embryogenesis before larvae feed exogenously. Among them, lipid content of eggs has been considered one of the main energy sources for bivalve larvae (Holland, 1978; Gallager and Mann, 1986; Whyte et al., 1990, 1991, 1992). Seasonal variation of its concentration is closely linked to the reproductive cycles in bivalves.

Lipid seasonal variations are inversely related to glycogen contents in *V. philiphinarum* and *V. decussata* (Beninger and Lucas, 1984; Robert et al., 1993; Marin et al., 2003; Ojea et al., 2004), whilst in *V. corrugata* a positive relationship was observed between total lipids and glycogen (Cerviño-Otero, 2011; Joaquim et al., 2011). In general, lipid content increases before mass spawning takes place, and then markedly decreases.

Proteins are the predominant respiratory substrate during gonad maturation in *V. philippinarum* (Adachi, 1979; Beninger and Lucas, 1984; Marin et al., 2003). Moreover, protein constitutes the major biochemical component of gametes in this species (Beninger and Lucas, 1984).

Table 2. Size and age at first maturity in clams´species

Species	Country	Location	Minimum length (mm) with ripe gametes	Age (years) first maturity	Length (mm) maturity 50	Age (years) maturity 50	Reference
VENERIDAE							
Venerupis philippinarum	USA	Hood Canal, Washington	5-10	1	-	-	Holland and Chew (1974)
Venerupis corrugata	Portugal	Ría de Aveiro	-	-	22	-	Maia et al. (2006)
Callista chione	Greece	Euboikos Gulf	12	2	-	-	Metaxatos (2004)
Chamelea gallina	Spain	Andalusia	-	-	16	1	Rodríguez-Rúa et al. (2003)
C. gallina (=*Venus striatula*)	Portugal	Vilamoura	-	1	-	-	Gaspar and Monteiro (1998)
Meretrix lusoria	Korea	Simpo	30-35	-	40-45	2	Chung (2007)
	Japan	Tokyo Bay	19	-	-	-	Nakamura et al. (2010)
Gafrarium pectinatum (=*G. tumidum*)	India	Chinnapalam, Pamban	-	-	22.3	-	Jagadis and Rajagopal (2007)
DONACIDAE							
Donax trunculus	Portugal	Faro	13-21	1	-	-	Gaspar et al. (1999)
	Spain	Ría del Barquero	12	-	28	> 1	Martínez-Patiño et al. (2003)
	Italy	Apulia	-	-	18.39 (females)	-	Zeichen et al. (2002)
		Central Tirreno Sea	13	-	14	-	La Valle (2005)
	Algeria	-	10-12	-	16	-	Mouëza and Frenkiel-Renault (1973)
	Turkey	Sea of Marmara	-	-	19.1	-	Deval (2009)
Ensis siliqua	Portugal	Vilamoura	-	1	-	-	Gaspar and Monteiro (1998)
E. magnus (=*E. arcuatus*)	Ireland	County Galway	2-3	-	-	-	Fahy et al. (2001)
SOLENIDAE							
Solen marginatus	Portugal	Ría de Aveiro	-	-	47.7 (females) 44.7 (males)	-	Maia et al. (2006)
Solen dactylus	Iran	Bandar Abbas	-	-	46.5	1-1.5	Saeedi et al. (2009)

Despite Marin et al. (2003) observed no regular seasonal trend in protein content in *V. philippinarum*, it has been suggested an increase during maturation perhaps related to the increasing accumulation in gametes and protein content depletion during spawning. Contrary to that, Cerviño-Otero (2011) reported that proteins exhibited an opposite pattern to lipids and glycogen in *V. corrugata*, showing higher values when individuals are in ripe and spawning stage.

5. Size and Age at First Maturity

Some studies dealing with reproductive processes in clams has focus partially on first maturity (see Table 2). Size at first maturity determination is important for the establishment of minimum size of capture, which is one of the parameters which are needed for fisheries management. Despite most authors investigating first maturity in bivalves have studied length at first maturity other authors have focused on age. Gaspar et al. (1999) pinpointed that first maturity is a function of age, not size in *Donax trunculus*, since they found that individuals reached maturity during the first year of life.

6. Exogenous Regulation

Temperature and food abundance has been reported as the main environmental parameters affecting the reproductive process in bivalves (Ruiz et al., 1992; Pazos et al., 1997; Darriba et al., 2004; Dridi et al., 2007; Enríquez-Díaz et al., 2009). Other environmental variables affecting gametogenic activity in bivalves are salinity, light (photoperiod), lunar phase or tides and variations in immersion times (i.e. subtidal vs intertidal locations).

Temperature could affect directly the metabolic rate of bivalves or indirectly the availability of food (Yan et al., 2009). It is influential on the onset of both gametogenesis and spawning. In temperate climates the most common pattern among clams is that the start of gametogenesis occurred when sea water temperature begin to increase and reaches a certain level (Ohba, 1959; Holland and Chew, 1974; Xie and Burnell, 1994; Dang et al., 2010). Some of the studies determining this temperature has been performed in the laboratory and are summarized in chapter 9 of this book. Moreover, temperature is closely linked to geographic locations (see table 1) influencing timing of the cycle, timing and duration of spawning and number of spawnings per year (Robert et al., 1993; Laruelle et al., 1994; Xie and Burnell, 1994). For example, variation in the number of annual spawning is evident in *Venerupis philippinarum* and *V. decussata* within Europe as aforementioned.

Opposite to the general pattern observed for clams in temperate climates, gametogenesis initiation in the razor clam *Ensis magnus* takes place at low sea water temperatures and the last spawning event is observed when the surface temperature increased (Darriba et al., 2004). In tropical locations, such as where *Solen dactylus* inhabits, gametogenic cycle started when the sea-surface temperature decreased below a certain threshold (27ºC) and clam spawned at 20ºC in winter (Saeedi et al., 2009).

In addition to temperature, food availability also plays an important role in the reproductive cycle of bivalves. Despite phytoplankton is a major food source for bivalves is not the only one, as bivalves also feed on dissolved organic matter (DOM), particulate organic matter (POM), benthic microalgae and macroalgal detritus (Manahan, 1990; Page and Lastra, 2003). Chlorophyll a concentration is considered to be an indicator of the amount of phytoplankton (food).

In temperate climates, as temperature and food supply increase the rate of gonad development in many clam species speeds up (Ojea et al., 2004; Yan et al., 2009, 2010). In many bivalve species spawning is coupled with phytoplankton blooms, ensuring that the amount of food in the water column is sufficient when larvae are released and thus maximizing larval success. Contrary to that, the first spawning of the season in the razor clam *E. magnus* was observed when the amount of food was at the lowest level of the year and razor clams continued releasing larvae during the winter until the first phytoplankton bloom occurred (Darriba et al., 2004). Consequently, the strategy of *E. magnus* consists of releasing the larvae when seawater temperature and food availability are low, thus reducing competition with other species.

Salinity also influences the reproductive cycle of bivalves. Darriba and Miranda (2005) concluded that salinity decreases interrupted gonadal development during months when in normal years successive spawning and restorations would take place in the razor clam *E. magnus*. Whilst, other authors reported that an increase in temperature and salinity favors the spawning in *Marcia opima* (Suja and Muthiah, 2007).

Other environmental variable which affected timing and duration of reproductive cycles in bivalves in a same area is immersion time (i.e. subtidal vs intertidal locations). Cerviño-Otero (2011) found a lesser degree of maturity in *Venerupis corrugata* inhabiting intertidal beds compared to clams of a subtidal bed in the locality. Similarly, Walker and Hefferman (1994) found differences in the gonadal development of *Mercenaria mercenaria* sowed at four different tidal levels, observing more immature individuals in the intertidal clams. Eversole et al. (1980) observed in the same species a higher condition index in subtidal clams. Nonetheless, it is not clear yet whether clams respond to temperature, food supply, exposure time to food or any other physical factor caused by tidal exposure (Eversole, 2001).

7. Endogenous Regulation

Neuropeptides, sex steroids and eicosanoids are essential for the regulation of reproductive processes in invertebrates (Morishita et al., 2010). However, to our knowledge there are no studies in clam species on the effect of neuropeptides in reproduction. Although, a comprehensive review of the regulatory activity of neuropeptides in molluscs, with a section dealing with bivalves, was published by Morishita et al. (2010). Conversely, there are investigations reporting the effect of sex steroids in bivalves, mainly focusing in scallops and oysters (see review by Croll and Wang (2007)), whereas studies in clams are scanty. Sex steroids can be divided in groups: estrogens, androgens and progestins (Croll and Wang, 2007). Among the former, the more common ones are: 17β-estradiol, estrone and estriol. Among the androgens found in molluscs are testosterone, 11-keto-testosterone, 5α-dihydrotestosterone, 3α-androstanediol, androsterone, dehydroepiandrosterone (DHEA), and androstenedione. The latter group includes pregnenolone, 17α-hydroxypregnenolone, progesterone and 17α-hydroxyprogesterone. Estrogens are more abundant in females whilst androgens are more abundant in males, thus suggesting a specific role in reproduction. The main steroid-producing organs have been reported to be the gonad and the digestive gland (Gauthier-Clerc et al., 2006; Negrato et al., 2008). Moreover, temperature could act as a stimulus for steroid hormone production (Siah et al., 2003).

Yan et al. (2011) pointed out that oestradiol-17β and testosterone contents may be related to sex in *Sinovacula constricta*, since oestradiol-17β was more abundant in females, whereas testosterone was more abundant in males during gametogenesis. However, sex differences in steroid content have not been found in other clam species, such as *Mya arenaria* (Gauthier-Clerc et al., 2006) and *Venerupis decussata* (Ketata et al., 2007). In *V. philippinarum* no differences in steroids related to sex were observed except for oestradiol-17β in ripe and spawning stages (Negrato et al., 2008).

A slight increase in oestradiol-17β levels prior to gametogenesis in *S. constricta* led Yan et al. (2011) to think that it may be associated with sex determination. Oestradiol-17β in females may play an important role in regulating vitellogenesis in *M. arenaria* (Gauthier-Clerc et al., 2006), *V. decussata* (Ketata et al., 2007) and *S. constricta* (Yan et al., 2011) since its values slightly increase prior to gametogenesis. It is also involved in the regulation of spawning through its mediation on the production of cathecholamines and prostaglandins (neurohormones) implicated in this process (Osada and Nomura, 1989, 1990). High levels of oestradiol-17β has been observed in *M. arenaria* and *S. constricta* during the spawning followed by decreased levels after spawning, thus supporting its involvement in spawning process (Gauthier-Clerc et al., 2006; Yan et al., 2011).

The same role attributed to oestradiol-17β in *S. constricta* during the initiation of the gametogenetic cycle has been suggested to testosterone (Yan et al., 2011). The high levels of testosterone during ripening period suggest its role in the regulation of spermatogenesis in *Mulinia lateralis* (Moss, 1989) and *V. decussata* (Ketata et al., 2007). Moreover, testosterone is involved in spawning in *M. lateralis* (Moss, 1989), *M. arenaria* (Gauthier-Clerc et al., 2006) and *V. decussata* (Ketata et al., 2007).

Progesterone levels increase as maturation takes place, reaching the highest values at the end of gametogenesis (Siah et al., 2002; Ketata et al., 2007; Negrato et al., 2008). The high levels of progesterone observed during the resting period may suggest that progesterone stock have to be renewed prior to the initiation of a new gametogenic cycle in *M. arenaria* (Siah et al., 2003). In addition to that, progesterone has a role in spawning in several clam species (Moss, 1989; Gauthier-Clerc et al., 2006; Ketata et al., 2007).

In *Spisula solidissima* oocytes are shed at the germinal vesicle stage and fertilization is directly responsible for reinitiation of the meiotic divisions (Dubé, 1988), whilst in other species, such as *V. philippinarum*, fertilization takes place only after they undergo germinal vesicle breakdown (Guerrier et al., 1993). In these cases, resumption of meiosis undergoes a two step process since release of the prophase block is followed by a second arrest in metaphase 1 (Gobet et al., 1994). Serotonin (5-hydroxytryptamine, 5-HT) is a neurotransmitter/neurohormone mediating various functions in bivalves, among which exerts the nervous control of the sexual function. Its levels in neural and gonadic tissue of bivalves show that it increases during gametogenesis, followed by a decrease after spawning (Martínez and Rivera, 1994). 5-HT induces spawning when injected directly into the body of some clam species, such as *S. solidissima* and *S. sachalinensis* (Gibbons and Castagna, 1984; Hirai et al.,

1988), *V. philippinarum* (Campioni and Sbrenna, 1995), *Mercenaria mercenaria* (Gibbons and Castagna, 1984, 1985), *Katelysia scalarina* and *V. largillierti* (Kent et al., 1998, 1999), *Tivela stultorum* (Alvarado-Álvarez et al., 1996), *Venus verrucosa* (Siniscalchi et al., 2004) and *M. arenaria* (Garnerot et al., 2006). Moreover, it also promotes *in vitro* oocyte maturation and meiosis reinitiation in clams when added to the suspending medium in *S. solidissima* (Kadam and Koide, 1989), *V. philippinarum* (Osanai and Kuraishi, 1988; Gobet et al., 1994; Durocher and Guerrier, 1996), *V. decussata* (Hamida et al., 2004a), *T. stultorum* (Alvarado-Álvarez et al., 1996), *V. verrucosa* (Siniscalchi et al., 2004) and *M. arenaria* (Garnerot et al., 2006). These reproductive responses in *V. verrucosa* are mediated by 5-HT, which is released in a pool of neurons in the visceral ganglion (Siniscalchi et al., 2004). Then, 5-HT is transmitted through serotonergic fibers that come from a branching of the cerebro-visceral connectives connecting with the gonads in the follicle walls of both sexes.

Acknowledgments

We are grateful to the staff of Centro de Cultivos Marinos de Ribadeo. The studies of the Centro de Cultivos Marinos de Ribadeo reviewed in this chapter were supported in part by funds for marine investigation from the Xunta de Galicia and by the grants ALMEJAS (2005-2007), ALMEJAS (2008-2010), and HATCHERIES from the Junta Asesora Nacional de Cultivos Marinos (JACUMAR), and grant PGIDITO6RMA50801PR (ALBA) from Xunta de Galicia. Fiz da Costa was partly supported by a Fundación Juana de Vega postdoctoral fellowship at IFREMER.

References

Adachi, K. (1979). Seasonal changes of the protein level in the adductor muscle of the clam, *Tapes philippinarum* (Adams and Reeve, 1850) with reference to the reproductive seasons. *Comp. Biochem. Physiol. A.* 64, 85-89.

Alvarado-Álvarez, R., Gould, M.C. and Stephano, J.L. (1996). Spawning, in vitro maturation, and changes in oocyte electrophysiology induced by serotonin in *Tivela stultorum*. *Biol. Bull.* 190, 322-328.

Aníbal, J., Esteves, E. and Rocha, C. (2011). Seasonal variations in gross biochemical composition, percent edibility, and condition index of the clam *Ruditapes decussatus* cultivated in the Ria Formosa (South Portugal). *J. Shellfish Res.* 30, 17-23.

Ansell, A.D. and Bodoy, A. (1979). Comparison of events in the seasonal cycle for *Donax vittatus* and *Donax trunculus* in European waters. In: E. Naylor, R.G. Hartnoll (Eds.), *Cyclic phenomena in marine plants and animals*. pp. 191-198, Pergamon Press, Oxford.

Ansell, A.D. and Lagardère, F. (1980). Observations on the biology of *Donax trunculus* and *D. vittatus* at Ile d'Oleron (French Atlantic coast). *Mar. Biol.* 57, 287-300.

Aracena, O.L., Lépez, I.M., Sánchez, J., Carmona, A.M., Medina, L. and Saavedra, A. (2003). On two new macroscopic indexes to evaluate the reproductive cycle of *Ensis macha* (Molina, 1782). *J. Shellfish Res.* 22, 675-680.

Avellanal, M.H., Jaramillo, E., Clasing, E., Quijón, P. and Conteras, H. (2002). Reproductive cycle of the bivalves *Ensis macha* (Molina, 1782) (Solenidae), *Tagelus dombeii* (Lamarck, 1818) (Solenicurtidae), and *Mulinia edulis* (King, 1831) (Mactridae) in Southern Chile. *The Veliger.* 45, 33-44.

Barber, B.J. and Blake, N.B. (1991). Reproductive Physiology. In: S.E. Shumway (Ed.), *Scallops: Biology, Ecology and Aquaculture, Developments in Aquaculture and Fisheries Science.* pp. 357-416, Elsevier, Amsterdam.

Barón, P.J., Real, L.E. and Ciocco, N.F. and Ré, M.E. (2004). Morphometry, growth and reproduction of an Atlantic population of the razor clam *Ensis macha* (Molina, 1782). *Sci. Mar.* 68, 211-217.

Bayed, A. (1990). Reproduction de *Donax trunculus* L. sur la côte Atlantique Marocaine. *Cah. Biol. Mar.* 31, 159-169.

Bayne, B.L. (1976). Aspects of reproduction in bivalve molluscs. In: M. Wiley (Ed.), *Estuarine processes. 1. Uses, stresses and adaptation to the estuary.* pp. 432-448, Academic Press, New York.

Beninger, P.G. and Lucas, A. (1984). Seasonal variations in condition, reproductive activity, and gross biochemical composition of two species of adult clam reared in a common habitat: *Tapes decussatus* L. (Jeffreys) and *Tapes philippinarum* (Adams and Reeve). *J. Exp. Mar. Biol. Ecol.* 79, 19-37.

Borsa, P. and Millet, B. (1992). Recruitment of the clam *Ruditapes decussatus* in the Lagoon of Thau, mediterranean. *Estuar. Coast. Shelf Sci.* 35, 289-300.

Bourne, N. (1979). Razor clam, *Siliqua patula* Dixon, breeding and recruitment at Masset, British Columbia. *Proc. Nat. Shellfish. Assoc.* 69, 21-29.

Bourne, N. (1982). Distribution, reproduction, and growth of Manila clam, *Tapes philippinarum*, in British Columbia. *J. Shellfish Res.* 2, 47-54.

Breber, P. (1980). Annual gonadal cycle in the carpet-shell clam *Venerupis decussata* in Venice Lagoon, Italy. *Proc. Nat. Shellfish. Assoc.* 70, 31-35.

Campioni, D. and Sbrenna, G. (1995). Induzione dello spawning in *Tapes philippinarum* (Bivalvia: Veneroida) e *Scapharca inaequivalvis* (Bivalvia: Arcidae) mediante somministrazione di serotonina (5-idrossitriptamina). *Biol. Mar. Mediterr.* 2, 307-309.

Cardoso, J., Witte, J.I. and van der Veer, H.W. (2009). Reproductive investment of the American razor clam *Ensis americanus* in the Dutch Wadden Sea. *J. Sea Res.* 62, 295-298.

Casavola, N., Rizzi, E. and Marano, G.S. and Sacacino, C. (1985). Ciclo reproduttivo e biometria di *Ensis minor* (Chenu) (Bivalvia: Solenidae) nel Golfo di Manfredonia. *Oebalia.* 11, 439-449.

Cerviño-Otero, A. (2011). Ciclo reproductivo, cultivo en criadero y en el medio natural de la almeja babosa *Venerupis pullastra* (Montagu, 1803). *Ph.D Thesis. University of Santiago de Compostela.* 288 pp.

Chung, E.-Y. (2007). Oogenesis and sexual maturation in *Meretrix lusoria* (Röding, 1798) (Bivalvia: Veneridae) in Western Korea. *J. Shellfish Res.* 26, 71-80.

Cobb, J.C., Stephenson, S.P. and Arnold, W.S. (2011). Reproductive cycle and recruitment patterns of the coquina clam *Donax Variabilis* Say along the central gulf coast of Florida. *J. Shellfish Res.* 30, 25-34.

Costa, C., Bianchini, M., Ceccarelli, P., Orecchia, P., Rambaldi, E. and Volterra, L. (1987). Indagine sui molluschi bivalvi di interesse commerciale (Telline, Cannolicchi e

Vongole) delle coste della Toscana, del lazio e della campania. 1985-1987. *Quaderni Ist. Idrobiol. Acquacolt. Brunelli.* 7, 3-58.

Croll, R.P. and Wang, C. (2007). Possible roles of sex steroids in the control of reproduction in bivalve molluscs. *Aquaculture.* 272, 76-86.

Dang, C., de Montaudouin, X., Gam, M., Paroissin, C., Bru, N. and Caill-Milly, N. (2010). The Manila clam population in Arcachon Bay (SW France): Can it be kept sustainable? *J. Sea Res.* 63, 108-118.

Darriba, S., San Juan, F. and Guerra, A. (2004). Reproductive cycle of the razor clam *Ensis arcuatus* (Jeffreys, 1865) in northwest Spain and its relation to environmental conditions. *J. Exp. Mar. Biol. Ecol.* 311, 101-115.

Darriba, S. and Miranda, M. (2005). Impacto del descenso de salinidad en la reproducción de la navaja (*Ensis arcuatus*). In: M. Rey-Méndez, J. Fernández-Casal, M. Izquierdo, A. Guerra (Eds.), *VIII Foro dos Recursos Mariños e da Acuicultura das Rías Galegas*. pp. 239-242, O Grove, Spain.

Darriba, S., San Juan, F. and Guerra, A. (2005a). Gametogenic cycle of *Ensis siliqua* (Linnaeus, 1758) in the Ría de Corcubión, Northwest Spain. *J. Molluscan Stud.* 71, 47-51.

Darriba, S., San Juan, F. and Guerra, A. (2005b). Energy storage and utilization in relation to the reproductive cycle in the razor clam *Ensis arcuatus* (Jeffreys, 1865). *ICES J. Mar. Sci.* 62, 996-896.

Del Piero, D., Orel, G., Valli, G., Vio, E. and Froglia, C. (1980). Aspetti e problemi della biologia e della pesca di *Ensis minor* (Chenu) nel golfo de Trieste. *Nova Thalassia.* 4, 173-191.

Deval, M.C. (2009). Growth and reproduction of the wedge clam (*Donax trunculus*) in the Sea of Marmara, Turkey. *J. Appl. Ichthyol.* 25, 551-558.

Dridi, S., Romdhane, M.S. and Elcafsi, M. (2007). Seasonal variation in weight and biochemical composition of the Pacific oyster, *Crassostrea gigas* in relation to the gametogenic cycle and environmental conditions of the Bizert lagoon, Tunisia. *Aquaculture.* 263, 238-248.

Drummond, L., Mulcahy, M. and Culloty, S. (2006). The reproductive biology of the Manila clam, *Ruditapes philippinarum*, from the North-West of Ireland. *Aquaculture.* 254, 326-340.

Dubé, F. (1988). The relationships between early ionic events, the pattern of protein synthesis, and oocyte activation in the surf clam, *Spisula solidissima. Dev. Biol.* 126, 233-241.

Durocher, Y. and Guerrier, P. (1996). Activation of an 85 kDa ribosomal S6 kinase during serotonin-induced oocyte maturation. *Int. J. Dev. Biol.* 40, 557-566.

Enríquez-Díaz, M., Pouvreau, S., Chávez-Villalba, J. and Le Pennec, M. (2009). Gametogenesis, reproductive investment, and spawning behavior of the Pacific giant oyster *Crassostrea gigas*: evidence of an environment-dependent strategy. *Aquacult. Int.* 17, 491-506.

Eversole, A.G. (2001). Chapter 5. Reproduction in *Mercenaria mercenaria*. In: J. Knauer, M. Castagna (Eds.), *Biology of the Hard Clam*. pp. 221-260, Elsevier, Amsterdam.

Eversole, A.G., Michener, W.K. and Eldridge, P.J. (1980). Reproductive cycle of *Mercenaria mercenaria* in a South Carolina estuary. *Proc. Nat. Shell. Ass.* 70, 20-30.

Fahy, E. and Gaffney, J. (2001). Growth statistics of an exploited razor clam (*Ensis siliqua*) bed at Gormanstown, Co Meath, Ireland. *Hydrobiologia.* 465, 139-151.

Fahy, E., Norman, M., Browne, R., Roantree, V., Pfeiffer, N., Stokes, D., Carroll, J. and Hannaffy, O. (2001). Distribution, population structure, growth and reproduction of the razor clam *Ensis arcuatus* Jeffreys (Solenaceae) in Coastal waters of western Ireland. *Irish Fish. Invest.* 10. 24 pp.

Figueras, A. (1957). Moluscos de las playas de la ría de Vigo: II. Crecimiento y reproducción. *Investig. Pesq.* 7, 49- 97.

Gabbott, P.A. (1975). Storage cycles in marine bivalve molluscs: a hypothesis concerning the relation between glycogen and gametogenesis. In: H. Barnes (Ed.), *Proceedings of the Ninth European Marine Biology Symposium.* pp. 191-211, Aberdeen Univ. Press, Aberdeen, Scotland.

Gabbott, P.A. (1983). Developmental and seasonal metabolic activities in marine mollusca. In: P. W. Hochachka (Ed.), *The Mollusca. Vol. 2. Environmental biochemistry and physiology*, pp. 165-269. Academic Press, New York.

Gallager, S.M. and Mann, R. (1986). Growth and survival of larvae of *Mercenaria mercenaria* (L.) and *Crassostrea virginica* (Gmelin) relative to broodstock conditioning and lipid content of eggs. *Aquaculture.* 56, 105-121.

Garnerot, F., Pellerin, J., Blaise, C. and Mathieu, M. (2006). Immunohistochemical localization of serotonin (5-hydroxytryptamine) in the gonad and digestive gland of *Mya arenaria* (Mollusca: Bivalvia). *Gen. Comp. Endocr.* 149, 278-284.

Gaspar, M.B. and Monteiro, C.C. (1998). Reproductive cycles of the razor clam *Ensis siliqua* and the clam *Venus striatula* off Vilamoura, Southern Portugal. *J. Mar. Biol. Assoc. U.K.* 78, 1247-1258.

Gaspar, M.B., Ferreira, R. and Monteiro, C.C. (1999). Growth and reproductive cycle of *Donax trunculus* L., (Mollusca: Bivalvia) off Faro, southern Portugal. *Fish. Res.* 41, 309-316.

Gauthier-Clerc, S., Pellerin, J. and Amiard, J.C. (2006). Estradiol-17b and testosterone concentrations in male and female *Mya arenaria* (Mollusca bivalvia) during the reproductive cycle. *Gen. Comp. Endocr.* 145, 133-139.

Gibbons, M.C. and Castagna, M. (1984). Serotonin as an inducer of spawning in six bivalve species. *Aquaculture.* 40, 189-191.

Gibbons, M.C. and Castagna, M. (1985). Response of the hard clam *Mercenaria mercenaria* (Linné) to induction of spawning by serotonin. *J. Shellfish Res.* 5, 65-67.

Giese, A.C. (1967). Some methods for study of the biochemical constitution of marine invertebrates. *Oceanogr. Mar. Ann. Rev.* 5, 159-186.

Gobet, I., Durocher, Y., Leclerc, C., Moreau, M. and Guerrier, P. (1994). Reception and transduction of the serotonin signal responsible for meiosis reinitiation in oocytes of the Japanese clam *Ruditapes philippinarum*. *Dev. Biol.* 164, 540-549.

Guerrier, P., Leclerc-David, C. and Moreau, M. (1993). Evidence for the involvement of internal calcium stores during serotonin-induced meiosis reinitiation in oocytes of the bivalve mollusc *Ruditapes philippinarum*. *Dev. Biol.* 159, 474-484.

Hamida, L., Medhioub, M.-N., Cochard, J.C. and Pennec, M.L. (2004a). Evaluation of the effects of serotonin (5-HT) on oocyte competence in *Ruditapes decussatus* (Bivalvia, Veneridae). *Aquaculture.* 239, 413-420.

Hamida, L., Medhiouband, M.N., Cochard, J.C., Romdhane, M.S. and Le Pennec, M. (2004b). A comparative study of the reproductive cycle of *Ruditapes decussatus* under natural (South of Tunisia) and controlled conditions (hatchery). *Cah. Biol. Mar.* 45, 291-303.

Hamida, L., Ayache, N., Haouas, Z. and Romdhane, M.S. (2010). Oocyte cohort analysis: criteria for an evaluation of the reproductive cycle in *Solen marginatus* (Pennánt, 1777), (Bivalvia: Solenacea) in Southern Tunisia. *J. Shellfish Res.* 29, 129-134.

Hirai, S., Kishimoto, T., Kadam, A.L., Kanatani, H. and Koide, S.S. (1988). Induction of spawning and oocyte maturation by 5-hydroxytryptamine in the surf clam. *J. Exp. Zool.* 245, 318-321.

Holland, D.A. and Chew, K.K. (1974). Reproductive cycle of the manila clam (*Venerupis japonica*), from Hood Canal, Washington. *Proc. Natl. Shellfish Assoc.* 64, 53-58.

Holland, D.L. (1978). Lipid reserves and energy metabolism in the larvae of benthic marine invertebrates. In: D.C. Malius, J.R. Sargent (Eds.), *Biochemical and biophysical perspectives in marine biology, Vol 4.* pp. 85-123, Academis Press, London.

Ishii, R., Sekiguchi, H., Nakahara, Y. and Jinnai, Y. (2001). Larval recruitment of the manila clam *Ruditapes philippinarum* in Ariake Sound, southern Japan. *Fish. Sci.* 67, 579-591.

Jagadis, I. and Rajagopal, S. (2007). Reproductive biology of Venus clam *Gafrarium tumidum* (Roding, 1798) from Southeast coast of India. *Aquac. Res.* 38, 1117-1122.

Joaquim, S., Matias, D., Matias, A.M., Moura, P., Arnold, W.S., Chícharo, L. and Gaspar, M.B. (2011). Reproductive activity and biochemical composition of the pullet carpet shell *Venerupis senegalensis* (Gmelin, 1791) (Mollusca: Bivalvia) from Ria de Aveiro (northwestern coast of Portugal). *Sci. Mar.* 75, 217-226.

Kadam, A.L. and Koide, S.S. (1989). Characterization of a factor with oocyte maturation inducing activity in *Spisula. Biol. Bull.* 176, 8-13.

Kanazawa, T. and Sato, S. (2008). Environmental and physiological controls on shell microgrowth pattern of *Ruditapes philippinarum* (Bivalvia: Veneridae) from Japan. *J. Molluscan Stud.* 74, 89-95.

Kang, C.K., Kang, Y.S., Choy, E.J., Kim, D.S., Shim, B.T. and Lee, P.Y. (2007). Condition, reproductive activity, and gross biochemical composition of the Manila clam, *Tapes philippinarum* in natural and newly created sandy habitats of the southern coast of Korea. *J. Shellfish Res.* 26, 401-412.

Kent, G.N., Maguire, G.B., Pugh, R. and Duthie, I. (1999). Spawning, settlement, and growth of the New Zealand Venerid *Ruditapes largillierti* (Philippi 1849) in culture. *New Zeal. J. Mar. Fresh. Res.* 33, 55-62.

Kent, G.N., Maguire, G.B., John, M., Cropp, M. and Frankish, K. (1998). Broodstock conditioning, spawning induction, and larval rearing of the stepped venerid, *Katelysia scalarina* (Lamarck, 1818). *J. Shellfish Res.* 17, 1065-1070.

Ketata, I., Guermazi, F., Rebai, T. and Hamza-Chaffai, A. (2007). Variation of steroid concentrations during the reproductive cycle of the clam *Ruditapes decussatus*: A one year study in the gulf of Gabès area. *Comp. Biochem. Phys. A.* 147, 424-431.

La Valle, P. (2005). *Donax trunculus* (Bivalvia: Donacidae) quale indicatore biologico degli equilibri costieri e del bilancio sedimentario. *Ph.D Thesis. Universitá degli Studi di Roma "La Sapienza".* 233 pp.

Laruelle, F., Guillou, J. and Paulet, Y.M. (1994). Reproductive pattern of the clams, *Ruditapes decussatus* and *R. philippinarum* on intertidal flats in Brittany. *J. Mar. Biol. Assoc. U.K.* 74, 351-366.

Lubet, P. (1959). Recherches sur le cycle sexuel et l'emission des gamètes chez les Mytilidae et les Pectinidae (Moll. Bivalves). *Revue Trav. Inst (scient. tech). Pêch. Marit.* 23, 387-548.

Lubet, P. (1986). The reproductive strategies of marine bivalve molluscs. In: M. Porchet, J.C. Andries, A. Dhainaut (Eds.), *Advances in Invertebrate Reproduction 4*. pp. 401-408, Elsevier, Amsterdam.

Maia, F., Sobral, M.P. and Gaspar, M. (2006). Ciclo reproductivo e primeira maturaçao de *Solen marginatus* e *Venerupis pullastra* na Ría de Aveiro. *Relatórios científicos e técnicos do IPIMAR.* 30, 1-35.

Manahan, D.T. (1990). Adaptations by invertebrate larvae for nutrients acquisition from seawater. *Am. Zool.* 30, 147-160.

Marin, M.G., Moschino, V., Deppieri, M. and Lucchetta, L. (2003). Variations in gross biochemical composition, energy value and condition index of *T. philippinarum* from the Lagoon of Venice. *Aquaculture.* 219, 859-871.

Martínez-Patiño, D., Nóvoa, S., Ojea, J. and Rodríguez-Moscoso, E. (2003). Determinación de la talla de primera maduración sexual de la coquina, *Donax trunculus* (Linné, 1758) (Mollusca:Bivalvia), en un banco natural de la Ría del Barquero (NO de Galicia). *IX Congreso Nacional de Acuicultura.* Cádiz, Spain.

Martínez, D. (2002). Estudio de los Solénidos, *Solen marginatus* (Pennánt, 1777) y *Ensis siliqua* (Linné, 1758), de los bancos naturales de la Ría de Ortigueira y Ría del Barquero: ciclo gametogénico, composición bioquímica y cultivo larvario. *Ph.D Thesis. Universidad de Santiago de Compostela.* 240 pp.

Martínez, G. and Rivera, A. (1994). Role of monoamines in the reproductive process of *Argopecten pupuratus*. *Invertebr. Reprod. Dev.* 25, 167-174.

Mazé, R.A. and Laborda, A.J. (1990). Cambios estacionales de una población de *Donax trunculus* (Linnaeus, 1758) (Pelecypoda: Donacidae) en la ría de El Barquero (Lugo, N.O. de España). *Sci. Mar.* 54, 131-138.

Medhioub, N.M. and Lubet, P.E. (1988). Recherches cytologiques sur l'environnement cellulaire (tissu de reserve) des gonades de la Palourde (*Ruditapes philippinarum* Adams et Reeve), Mollusque bivalve. *Ann. Sci. Nat. Zool. (Paris)* 9, 87-102.

Meneghetti, F., Moschino, V. and Da Ros, L. (2004). Gametogenic cycle and variations in oocyte size of *Tapes philippinarum* from the Lagoon of Venice. *Aquaculture.* 240, 473-488.

Metaxatos, A. (2004). Population dynamics of the venerid bivalve *Callista chione* (L.) in a coastal area of the eastern Mediterranean. *J. Sea Res.* 52, 293-305.

Morishita, F., Furukawa, Y., Matsushima, O. and Minakata, H. (2010). Regulatory actions of neuropeptides and peptide hormones on the reproduction of molluscs. *Can. J. Zool.* 88, 825-845.

Morvan, C. and Ansell, A.D. (1988). Stereological methods applied to reproductive cycle of *Tapes rhomboides*. *Mar. Biol.* 97, 355-364.

Moss, S.M. (1989). Effects of exogenous androgens on growth, biochemical composition, and reproduction of the coot clam (*Mulinia lateralis*). *Pac. Sci.* 43, 200.

Mouëza, M. and Frenkiel-Renault, L. (1973). Contribution à l'etude de la biologie de *Donax trunculus* L. (Mollusques Lamellibranches) dans l'algerois: la reproduction. *Cah. Biol. Mar.* 14, 261-283.

Moura, P., Gaspar, M.B. and Monteiro, C.C. (2008). Gametogenic cycle of the smooth clam *Callista chione* on the south-western coast of Portugal. *J. Mar. Biol. Assoc. U.K.* 88, 161-167.

Nakamura, Y., Nakano, T., Yurimoto, T., Maeno, Y., Koizumi, T. and Tamaki, A. (2010). Reproductive cycle of the venerid clam *Meretrix lusoria* in Ariake Sound and Tokyo Bay, Japan. *Fish. Sci.* 76, 931-941.

Negrato, E., Marin, M.G., Bertotto, D., Mattozo, V., Poltronieri, C. and Simontacchi, C. (2008). Sex steroids in *Tapes philippinarum* (Adams and Revee 1850) during gametogenic cycle: preliminary results. *Fresenius Environ. Bull.* 17, 1466-1470.

Neuberger-Cywiak, L., Achituv, Y. and Mizrahi, L. (1989). The ecology of *Donax trunculus* Linnaeus and *Donax semistriatus* Poli from the Mediterranean coast of Israel. *J. Exp. Mar. Biol. Ecol.* 134, 203-220.

Nosho, T.Y. and Chew, K.K. (1972). The setting and growth of the Manila clam, *Venerupsis japonica* (Deshayes), in Hood Canal, Washington. *Proc. Nat. Shell. Ass.* 62, 50-58.

Ohba, S. (1959). Ecological studies in the natural population of a clam, *Tapes japonica*, with special reference to seasonal variations in the size and structure of the population and to individual growth. *Biol. J. Okayama. Univ.* 5, 13-42.

Ojea, J., Pazos, A. J., Martínez, D., Novoa, S., Sánchez, J.L. and Abad, M. (2004). Seasonal variation in weight and biochemical composition of the tissues of *Ruditapes decussatus* in relation to the gametogenic cycle. *Aquaculture.* 238, 451-468.

Ojea, J., Martínez, D., Novoa, S. and Cerviño-Otero, A. (2005). Ciclo gametogénico de una población de almeja japonesa *Ruditapes philippinarum* (Adams and Reeve, 1850) en la ría de Camariñas (noroeste de España) y relación con la composición bioquímica mayoritaria. *Bol. Inst. Esp. Oceanogr.* 21, 337-342.

Osada, M. and Nomura, T. (1989). Estrogen effect on the seasonal levels of catecholamines in the scallop *Patinopecten yessoensis*. *Comp. Biochem. Phys. C.* 93, 349-353.

Osada, M. and Nomura, T. (1990). The levels of prostaglandins associated with the reproductive cycle of the scallop, *Patinopecten yessoensis*. *Prostaglandins.* 40, 229-239.

Osanai, K. and Kuraishi, R. (1988). Response of oocytes to meiosis-inducing agents in pelecypods. *Bull. Mar. Biol. Stn. Asamushi, Tohoku Univ.* 18, 45-56.

Page, H.M. and Lastra, M. (2003). Diet of intertidal bivalves in the Ría de Arosa (NW Spain): evidence from stable C and N isotope analysis. *Mar. Biol.* 143, 519-532.

Pazos, A.J., Román, G., Acosta, C.P., Abad, M. and Sánchez, J.L. (1997). Seasonal changes in condition and biochemical composition of the scallop *Pecten maximus* L. from suspended culture in the Ría de Arousa (Galicia, N.W. Spain) in relation to environmental conditions. *J. Exp. Mar. Biol. Ecol.* 211, 169-193.

Pérez-Camacho, A. (1980). Biología de *Venerupis pullastra* (Montagu, 1803) y *Venerupis decussata* (Linné, 1767) (Mollusca: Bivalvia) con especial referencia a factores determinantes de la producción. *Bol. Inst. Esp. Oceanogr.* 281, 353-358.

Ponurovsky, S.K. and Yakovlev, Y. (1992). The reproductive biology of the Japanese littleneck, *Tapes philippinarum* (Adams and Reeve, 1850) (Bivalvia: Veneridae). *J. Shellfish Res.* 11, 265-277.

Ramón, M., Abellá, P. and Richardson, C.A. (1995). Population structure and growth of *Donax trunculus* (Bivalvia: Donacidae) in the western Mediterranean. *Mar. Biol.* 121, 665-671.

Remacha-Triviño, A. and Anadón, N. (2006). Reproductive cycle of the razor clam *Solen marginatus* (Pulteney, 1799) in Spain: a comparative study in three different locations. *J. Shellfish Res.* 25, 869-876.

Ren, Y., Xu, B., Guo, Y., Yang, M. and Yang, J. (2008). Growth, mortality and reproduction of the transplanted Manila clam (*Ruditapes philippinarum* Adams and Reeve 1850) in Jiaozhou Bay. *Aquac. Res.* 39, 1759-1768.

Robert, R., Trut, G. and Laborde, J.L. (1993). Growth, reproduction and gross biochemical composition of the Manila clam *Ruditapes philippinarum* in the Bay of Arcachon, France. *Mar. Biol.* 116, 291-299.

Rodríguez-Moscoso, E. and Arnaiz, R. (1998). Gametogenesis and energy storage in a population of the grooved carpet-shell clam, *Tapes decussatus* (Linné, 1787), in northwest Spain. *Aquaculture.* 162, 125-139.

Rodríguez-Moscoso, E., Pazo, J.P., García, A. and Fernández-Cortés, F. (1992). Reproductive cycle of Manila clam, *Ruditapes philippinarum* (Adams and Reeve, 1850) in Ría de Vigo (NW Spain). *Sci. Mar.* 56, 61-67.

Rodríguez-Rúa, A., Prado, M.A., Romero, Z. and Bruzón, M. (2003). The gametogenic cycle of *Scrobicularia plana* (da Costa, 1778) (Mollusc: Bivalve) in Guadalquivir estuary (Cádiz, SW Spain). *Aquaculture.* 217, 157-166.

Rodríguez de la Rúa, A., Prado, M.A. and Bruzón, M.A. (2003). Estudio del ciclo reproductor de *Chamelea gallina* (L., 1758) (Mollusca: Bivalvia) en tres poblaciones del litoral andaluz. *Bol. Inst. Esp. Oceanogr.* 19, 57-63.

Rodríguez, S., Quintana, R., Lamas, L., Ayensa, G., Velasco, F.J. and Pascual, C. (1993). Etude comparative du cycle gamétogénique et composition biochímique de *Tapes decussatus* et *Ruditapes philippinarum* dans la Ría de Muros y Noya. In: G. Bernabe, P. Kestemont (Eds.), *Bordeaux Aquaculture 92. Production, Environment and Quality. Special Publication-European Aquaculture Society*, pp. 503- 511, Bordeaux.

Ruiz, C., Abad, M., Sedano, F., García-Martín, L.O. and Sánchez-López, J.L. (1992). Influence of seasonal environmental changes on the gamete production and biochemical composition of *Crassostrea gigas* (Thunberg) in suspended culture in El Grove, Galicia, Spain. *J. Exp. Mar. Biol. Ecol.* 155, 249-262.

Saeedi, H., Raad, S.P., Ardalan, A.A., Kamrani, E. and Kiabi, B.H. (2009). Growth and reproduction of *Solen dactylus* (Bivalvia: Solenidae) on northern coast of the Persian Gulf (Iran). *J. Mar. Biol. Assoc. U.K.* 89, 1635-1642.

Sastry, A.N. (1979). Pelecypoda (excluding Ostreidae). In: A.C. Giese, J.S. Pearse (Eds.), *Reproduction of marine invertebrates, Vol 5. Molluscs: Pelecypods and lesser classes.* pp. 113-292, Academic Press, New York.

Sbrenna, G. and Campioni, D. (1994). Gametogenic and spawning patterns of Manila clams *Tapes philippinarum* (Bivalvia: Veneroida) in two lagoons of the River Po Delta, Italy. *J. Shellfish Res.* 13, 37-46.

Seed, R. (1976). Ecology. In: B.L. Bayne (Ed.), *Marine Mussels: Their Ecology and Physiology*. pp. 13-65, Cambridge University Press, Cambridge.

Serdar, S. and Lök, A. (2009). Gametogenic cycle and biochemical composition of the transplanted carpet shell clam *Tapes decussatus*, Linnaeus 1758 in Sufa (Homa) Lagoon, Izmir, Turkey. *Aquaculture*. 293, 81-88.

Shafee, M.S. and Daoudi, M. (1991). Gametogenesis and spawning in the carpet-shell clam, *Ruditapes decussatus* (L.) (Mollusca: Bivalvia), from the Atlantic coast of Morocco. *Aquac. Res.* 22, 203-216.

Siah, A., Pellerin, J., Benosman, A., Gagné, J.P. and Amiard, J.C. (2002). Seasonal gonad progesterone pattern in the soft-shell clam *Mya arenaria*. *Comp. Biochem. Phys. A.* 132, 499-511.

Siah, A., Pellerin, J., Saint-Louis, R., Amiard, J.C. and Pelletier, E. (2003). TBT effects on steroid hormones in (*Mya arenaria*). *30th Annual Aquatic Toxicology Workshop.* Ottawa, Ontario.

Siniscalchi, A., Cavallini, S., Sonetti, D., Sbrenna, G., Capuano, S., Barbin, L., Turolla, E. and Rossi, R. (2004). Serotonergic neurotransmission in the bivalve *Venus verrucosa* (Veneridae): a neurochemical and immunohistochemical study of the visceral ganglion and gonads. *Mar. Biol.* 144, 1205-1212.

Suja, N. and Muthiah, P. (2007). The reproductive biology of the baby clam, *Marcia opima*, from two geographically separated areas of India. *Aquaculture*. 273, 700-710.

Tirado, C. and Salas, C. (1999). Reproduction of *Donax venustus* Poli 1795, *Donax semistriatus* Poli 1795 and Intermediate Morphotypes (Bivalvia: Donacidae) in the Littoral of Málaga (Southern Spain). *Mar. Ecol.* 20, 111-130.

Tirado, C., Salas, C. and Márquez, I. (2003). Reproduction of *Venus verrucosa* L., 1758 (Bivalvia: Veneridae) in the littoral of Málaga (southern Spain). *Fish. Res.* 63, 437-445.

Toba, M., Yamakawa, H., Kobayashi, Y., Sugiura, Y., Honma, K. and Yamada, H. (2007). Observations on the maintenance mechanisms of metapopulations, with special reference to the early reproductive process of the manila clam *Ruditapes philippinarum* (Adams and Reeve) in Tokyo Bay. *J. Shellfish Res.* 26, 121-130.

Urban, H.J. (1996). Population dynamics of the bivalves *Venus antiqua*, *Tagelus dombeii*, and *Ensis macha* from Chile at 36°S. *J. Shellfish Res.* 15, 719-727.

Urrutia, M.B., Ibarrola, I., Iglesias, J.I.P. and Navarro, E. (1999). Energetics of growth and reproduction in a high-tidal population of the clam *Ruditapes decussatus* from Urdaibai Estuary (Basque Country, N. Spain). *J. Sea Res.* 42, 35-48.

Valli, G. and Gioglio, I. (1980). Periodo riproduttivo de alcuni aspetti della biometria di *Ensis minor* (Chenu) nel Golfo de Trieste. *Quad. Lab. Tecnol. Pesca.* 2, 285-306.

Villalba, A., Carballal, M.J. and López, M.C. (1993). Estudio del ciclo gonadal de tres especies de almeja, *Ruditapes decussatus*, *Venerupis pullastra* y *Venerupis rhomboides* de las rías gallegas. *Actas del IV Congreso nacional de Acuicultura*, pp. 341-346.

Voliani, A., Auteri, R., Baino, R. and Silvestri, R. (1997). Insediamento nel substrato ed accescimento di *Donax trunculus* L. sul litorale toscano. *Biol. Mar. Mediterr.* 4, 458-460.

Walker, R.L. and Hefferman, P.B. (1994). Temporal and spatial effects of tidal exposure on the gametogenic cycle of the northern quahog, *Mercenaria mercenaria* (Linnaeus, 1758), in coastal Georgia. *J. Shellfish Res.* 13, 479-486.

Whyte, J.N.C., Bourne, N. and Ginther, N.G. (1990). Biochemical and energy changes during embryogenesis in the rock scallop *Crassadoma gigantea*. *Mar. Biol.* 106, 239-244.

Whyte, J.N.C., Bourne, N. and Ginther, N.G. (1991). Depletion of nutrient reserves during embryogenesis in the scallop *Patinopecten yessoensis* (Jay). *J. Exp. Mar. Biol. Ecol.* 149, 67-79.

Whyte, J.N.C., Bourne, N., Ginther, N.G. and Hodgson, C.A. (1992). Compositional changes in the larva to juvenile development of the scallop *Crassadoma gigantea* (Gray). *J. Exp. Mar. Biol. Ecol.* 163, 13-29.

Wilson, J. and Seed, R. (1974). Reproduction in *Mytilus edulis* L. (Mollusca: Bivalvia) in Carlingford Lough, Northern Ireland. *Irish Fish. Invest.* 15, 1-30.

Xie, Q. and Burnell, G. M. (1994). A comparative study of the gametogenic cycles of the clams *Tapes philippinarum* (Adams and Reeve, 1850) and *Tapes decussatus* (Linnaeus) on the south coast of Ireland. *J. Shellfish Res.* 13, 467-472.

Yamuza-Clavijo, M.P., Rodríguez-Rúa, A. and Bruzón-Gallego, M.A. (2010). Ciclo gametogénico de *Venerupis rhomboides* Pennant, 1777 en el litoral Andaluz (sur España). *Hidrobiológica.* 20, 195-202.

Yan, H., Li, Q., Yu, R. and Kong, L. (2010). Seasonal variations in biochemical composition and reproductive activity of Venus Clam *Cyclina sinensis* (Gmelin) from the Yellow River Delta in Northern China in relation to environmental factors. *J. Shellfish Res.* 29, 91-99.

Yan, H., Li, Q., Liu, W., Yu, R. and Kong, L. (2009). Seasonal changes in reproductive activity and biochemical composition of the razor clam *Sinonovacula constricta* (Lamarck 1818). *Mar. Biol. Res.* 6, 78-88.

Yan, H., Li, Q., Liu, W., Ke, Q., Yu, R. and Kong, L. (2011). Seasonal changes of oestradiol-17β and testosterone concentrations in the gonad of the razor clam *Sinonovacula constricta* (Lamarck, 1818). *J. Molluscan Stud.* 77, 116-122.

Zeichen, M.M., Agnesi, S., Mariani, A., Maccaroni, A. and Ardizzone, G.D. (2002). Biology and Population Dynamics of *Donax trunculus* L. (Bivalvia: Donacidae) in the South Adriatic Coast (Italy). *Estuar. Coast. Shelf Sci.* 54, 971-982.

In: Clam Fisheries and Aquaculture
Editor: Fiz da Costa González

ISBN: 978-1-62257-518-3

Chapter 4

Genetic Studies on Commercially Important Species of Veneridae

A. Arias-Pérez[1]; A. Insua[1]; R. Freire[1]; J. Méndez[1] and J. Fernández-Tajes[1,2]*

[1] Departamento de Biología Celular y Molecular, Área de Genética, Facultad de Ciencias, Universidade da Coruña, A Zapateira s/n, A Coruña, Spain

[2] Rheumatology Division, Genomic Group, INIBIC - Hospital Universitario A Coruña, As Xubias s/n, A Coruña, Spain

Abstract

Veneridae is a diverse bivalve family comprising members that represent key components in fisheries and aquaculture. Nowadays, there is unanimity in assuming that conservation, management and production of clam resources requires the genetic characterization of species and populations. In this chapter, we treated genetic aspects of main commercial species of the Veneridae family, involving cytogenetics, ploidy manipulation, population genetics, species differentiation and genomics. First of all, it is provided a perspective of cytogenetic data obtained by classical and more recent techniques, followed by the trials achieved for chromosome manipulation including methods for induction and detection of polyploidy and the major consequences of induced polyploidy. Then, molecular markers, the tools employed to carry out population genetic studies, are described, as well as their use to define how genetic variation is distributed within and among populations, and how it is influenced by evolutionary forces. Finally, a description of genomic resources for these species is outlined together with the molecular methods available for species identification of economically important species.

* E-mail address: jfernandezt@udc.es

1. INTRODUCTION

Genetic characterization of marine species and populations provides essential information to know the biodiversity, which is indispensable for its conservation especially when they are commercially exploited. In the family Veneridae over 800 species have been described (Mikkelsen et al., 2006), and about 20 species are harvested from natural beds and/or cultivated around the world to increase the production and cover the market demand. A good example is *Venerupis philippinarum*, a native clam of the Indo-Pacific coastal seas, introduced accidentally or deliberately as hatchery broodstock in many countries. In 2009, more than three million tonnes of this clam were produced, and 98% came from aquaculture, according to FAO statistics.

This chapter covers different aspects of Veneridae clams related to the major advances in cytogenetics, population genetics and genomics of main commercial species of the family to support the improvement of resource conservation and management, aquaculture and product commercialization. In the field of cytogenetics, current knowledge of karyotypes, chromosome banding and physical mapping of DNA sequences by fluorescence in situ hybridization (FISH) is reviewed. Research about chromosome manipulation is also addressed. This includes methods for induction and detection of polyploidy and the major consequences of induced polyploidy. Studies of population genetics are also reported, including the molecular markers used. Assessment of the genetic variation and differentiation of populations as well as the possible influence of factors such us evolutionary forces or geographical barriers in the results obtained are mentioned. The implications of genetic population data for conservation, management, hatchery or restocking programs are also taken into account. Given that clams species differs on the economic value and species identification based on morphological traits is often difficult, molecular markers and techniques employed for species identification are compiled. Moreover, it was outlined the effort that is being carried out for increasing the genomic resources (e.g. EST libraries) in this group of bivalves in the last years.

2. CHROMOSOMES AND PLOIDY MANIPULATION

Chromosomes are the structures that carry the genetic material. During cell division they are compactly coiled and are visible under optical microscope after appropriate staining. The chromosome number is usually constant within species but may vary between species. Somatic cells are diploid (2N), chromosomes occurring in pairs of homologous, and undergo mitotic division where the two chromatids of each duplicated chromosome separate into the daughter cells. The egg and sperm are haploid (N), containing one set of chromosomes, as result of the meiotic divisions (I and II) of germ cells; meiosis I is a reduction division where homologous chromosomes are separated into two cells and meiosis II is similar to a mitotic division with separation of chromatids into the daughter cells. Polyploid individuals contain one or more additional sets of chromosomes and may originate from alterations of meiotic or mitotic divisions or reproductive contact among species. Natural polyploids are rare in animals but the level of ploidy can be artificially modified during the egg meiotic divisions.

Chromosomes can be studied in cells in mitotic or meiotic division but most studies involve cells arrested in the mitotic metaphase stage by means of colchicine. Subsequent treatments including hypotonic shock, fixation and basic staining (e.g. Giemsa staining) allow chromosome counting and characterization of chromosome morphology, which is determined by the size and centromere position. Chromosomes whose centromeres are located in median, submedian, subterminal, and terminal regions are designated as metacentric, submetacentric, subtelocentric and telocentric, respectively. The representation of the chromosomes of an individual arranged in homologous pairs constitutes the karyotype. The use of differential staining techniques makes visualization of finer details of the chromosomes possible such as location of the nucleolus organizer regions (NORs) or distribution and composition of the heterochromatin. A higher level of resolution can be reached when the technique of fluorescence in situ hybridization (FISH) is applied. This is a more recent molecular cytogenetic technique that indicates the chromosomal location of specific DNA sequences.

Chromosomal analysis plays an important role in different fields of bivalve biology. Chromosome counting inform about aneuploidy (loss or gain of individual chromosomes) and polyploidy events. Accurate distinction of homologous pairs provides information on fixed chromosomal rearrangements and polymorphisms. Identification of functional and structural entities of the chromosomes reflects the gene and genome organization. Comparative analysis enables the establishment of relationships between taxa and helps to understand the species evolution since speciation is frequently promoted or followed by karyotype rearrangements (White, 1978; King, 1993). Also contributes to the study of the hybridization process by means of the characterization of parental chromosomes and the chromosome disturbances that can be associated to this process as well as the estimation of its efficiency. Finally, genetic improvement programs based on genetic maps are also facilitated by the assignment of linkage groups to specific chromosomes and the knowledge of chromosomal location of molecular markers.

Cytogenetic data of Veneridae clams are given in Table 1. This includes data compiled by Nakamura (1985), Insua (1993), Thiriot-Quiévreux (2002), Leitão & Chaves (2008) and recent papers. Early data comes from chromosome spreads obtained mostly from gonadic tissue by means of the squashing method but colchicine treatment and hypotonick shock were combined with the air drying technique (Thiriot-Quiévreux & Ayraud, 1982) and mitotic metaphase spreads from embryos, larvae or juvenile gills become the choice material for most cytogenetic studies.

Most Veneridae clams have a chromosome number of 2N=38 which correspond with the most frequent number within the Bivalvia class (Thiriot-Quiévreux, 2002). In the cases where the chromosome number of a species is reported by different studies data are usually consistent. This is the expected given that in bivalves intraspecific variation in the chromosome number is very unusual. In *Chamelea gallina* Rasotto et al. (1981) report N=15 but the authors highlight that this haploid number should be kept in reserve due to difficulties in the bivalent count. Therefore, the most probable chromosome number of this species is also 2N=38, as was reported by Corni & Trentini (1986). *Meretrix lusoria* with 2N=50 represents an exception to the present-day picture of 2N=38. Although available data indicate that the chromosome number in Veneridae clams is practically constant it can not be discarded that a more significant variation in chromosome number among species exists, taking into account that the family comprises over 800 species (Mikkelsen et al., 2006). Comparing with other families involving commercial species, the modal diploid chromosome number of Veneridae clams is identical to that of scallops, cockles, and razor clams but different from mussels (2N=28) and oysters (2N=20). Wang & Guo (2004) suggested that al least one genome duplication event has occurred during the evolution of bivalves and that species with a haploid number of 19 are likely tetraploid derived from a diploid ancestor 2N=20 similar to the extant species of Ostreidae. This hypothesis, based on chromosomal and DNA content data in scallops, provides a new vision of bivalve evolution to be tested in further studies involving more species.

When bivalve chromosomes are examined, aneuploid complements are usually detected. This phenomenon was extensively studied in the oyster *Crassostrea gigas* (Leitão et al., 2001a) and a negative relationship between aneuploidy and growth rate was demonstrate; fast-growing individuals always display a lower percentage of aneuploid somatic cells than slow-growing ones. Moreover, evidences for an association of the aneuploidy phenomenon with genetic (Leitão et al., 2001b) and environmental (Bouilly et al., 2003) factors have been reported. In Veneridae clams, the aneuploidy in somatic cells was studied in *Venerupis decussata* where the values reported were substantially higher than those observed in other bivalves and a negative correlation between the level of aneuploidy and length was also demonstrated (Teixeira de Sousa et al., 2011). This means that to maximize growth, aneuploidy should be taken into account.

Karyotypes of Veneridae clams are characterized by the fact that most if not all of the homologous pairs are biarmed chromosomes (metacentric or submetacentric). However, a variety of karyotypes is observed (Table 1) indicating that chromosomal changes have taken place during the Veneridae evolution; although it seems that Robertsonian changes (fusion and fision of chromosomes) did not play an essential role taking into account that chromosome number is nearly constant. When different authors reported the karyotype of a species, as is the case of *Circe scripta*, *V. decussata* and *V. philippinarum* (see Table 1 for references), discrepancies regarding composition are observed. This may represent the occurrence of polymorphism among populations but more probably result from the analysis of chromosomes that differ in the condensation degree and imprecise centromere location when chromosomes are measured.

Banding or FISH techniques, indispensable for the accurate cytogenetic characterization of species, were scarcely applied in Veneridae clams (Table 1) but the few studies carried out represent a significant progress in the field of bivalve cytogenetics. It should be noted that the combination of surface spreading of synaptonemal complexes with FISH (SC-FISH) constitute a powerful tool to analyse meiotic chromosomes and was used for the first time in bivalves in *Dosinia. exoleta* (Hurtado & Pasantes, 2005). When applied in *V. philippinarum*, *V. decussata* and their putative hybrids, SC-FISH provided cytological evidence of hybridization between the two species (Hurtado et al., 2011).

Table 1. Cytogenetic data of Veneridae clams

Species	2n/n	Karyotype	Banding Techniques	FISH	References
Chione cancellata	38/**19**				Menzel (1968)
Circe scripta	38	19 m			Ieyama (1980)
	38	6 m, 6 sm, 4 st, 3 t			Ebied & Aly (2004)
Dosinia exoleta	38/**19**	11 m, 8 sm	**FB (DAPI/PI); Ag-NORs**	**mrDNA; TS**	Hurtado & Pasantes (2005)
Irus mitis	38/**19**	14 m, 5 sm			Ieyama (1980)
Mercenaria campechiensis	38/**19**				Menzel & Menzel (1965)
Mercenaria mercenaria	38/**19**				Menzel & Menzel (1965)
	38			TS	Wang & Guo (2001)
	38	11 m, 8 sm		mrDNA	Wang & Guo (2007)
	38	15 m, 4 sm			Lin et al. (2008)
Meretrix lusoria	50	10 m, 14 sm, 1st			Tsai et al. (1996)
Paphia vernicosa	38/**19**	19 m/sm			Ieyama (1980)
Pitaria chione	19				Rasotto et al. (1981)
Venerupis decussata (=*Ruditapes decussatus*)	38/**19**				Gerard (1978); Rasotto et al. (1981)
	38	6m, 3 sm, 10 st			Borsa & Thiriot-Quiévreux 1990
	38	6 m, 5 sm, 3 st, 5 t			Ebied & Aly (2004)
			REB		Leitão et al. (2006)
				mrDNA; 5S rDNA	Hurtado et al. (2011)
Venerupis philippinarum (=*Tapes philippinarum*; =*Ruditapes philippinarum*)	38				Gerard (1978); Corni & Trentini (1990)
	38	10 m, 8 sm, 1 st			Ieyama (1985)
	38	9 m, 10 sm			Borsa & Thiriot-Quiévreux (1990)
			Ag-NORs	Satellite DNA	Passamonti et al. (1998)
				mrDNA; 5S rDNA	Hurtado et al. (2011)
Saxidomus gigantea	38				Rasotto et al. (1981)

Table 1. (Continued)

Species	2n/n	Karyotype	Banding Techniques	FISH	References
Saxidomus nutalli	38				Rasotto et al. (1981)
Polititapes aureus	38/**19**	19 m/sm			Corni & Trentini (1990)
(=*Ruditapes aureus*;	38	8 m, 9 sm, 1 st, 1 t			Borsa & Thiriot-Quiévreux (1990)
=*Venerupis aurea*)			FB (DAPI/CMA/PI)	mrDNA; 5S rDNA; CHG; TS	Carrilho et al. (2011)
Venerupis corrugata (=*Venerupis pullastra*)	38	3 m, 8 sm, 8 st			Insua & Thiriot-Quiévreux (1992)
Polititapes virgineus (=*Tapes rhomboides*;	38	4 m, 8 sm, 4 st, 3 t			Insua & Thiriot-Quiévreux (1992)
=*Venerupis rhomboides*)			FB (DAPI/CMA/PI)	mrDNA; 5S rDNA; CHG; TS	Carrilho et al. (2011)
Chamalea gallina	15				Rasotto et al. (1981)
	38/**19**	15 m/sm, 4 st			Corni & Trentini (1986)
Venus verrucosa	**19**				Rasotto et al. (1981)
	38	7 m, 4 sm, 4 st, 4 t			Ebied & Aly (2004)

m: metacentric; sm: submetacentric; st: subtelocentric; t. telocentric; FB: fluorochrome banding; DAPI: 4',6-diamidino-2-phenylindole; CMA: chromomycin A3; PI: propidium iodide; REB: restriction enzyme banding; mrDNA: major ribosomal DNA; TS: telomeric sequences; CHG: core histone genes. Boldfaced items indicate that meiotic chromosomes were analysed.

Silver staining of NORs, which reveals NORs that were transcriptionally active at the precedent interphase, has been frequently used in other bivalve species (Thiriot-Quiévreux, 2002; Leitão & Chaves, 2008). Nevertheless, in Veneridae it was only applied in *D. exoleta* and *V. philippinarum*. In contrast, chromosomal location of major ribosomal DNA (mrDNA), that contains the gene family coding for 18S, 5.8S and 28S rRNA and occupies NOR sites, was identified in all species where FISH was employed. When fluorochrome staining was applied (*D. exoleta*, *Polititapes. virgineus*, and *P. aureus*) differential staining at mrDNA sites was observed indicating that these correspond to GC rich regions. Taking together the data from silver staining and FISH, all species display a chromosome pair with NORs/mrDNA sites except *Mercenaria. mercenaria* that displays two pairs. The position on the bearing chromosomes is highly divergent among species, being on short or long arms and on interstitial, sub-terminal or terminal positions, which reflects that considerable chromosomal changes have occurred during the evolution of Veneridae. However, FISH signals of telomeric DNA were always found on telomeric regions in all examined species instead of interstitial sites (at least some of them) as expected if chromosomal rearrangements took place through evolution. Discrepancies between both types of data may indicate that chromosomal changes have occurred but telomeric regions were not involved.

Distinction of all homologous chromosomes is usually achieved with classical techniques inducing longitudinal banding patters (G or R banding) but these protocols do not work well in bivalves as do in other organisms such as mammals. To overcome this difficulty, a molecular cytogenetic technique based on *in situ* digestion with restriction endonucleases can be used. Treatment with the restriction endonuclease *Hae*III produce interstitial, centromeric and telomeric bands in *V. decussata* chromosomes, and the banding pattern obtained allowed the identification of the nineteen chromosome pairs (Leitão et al., 2006). On the other hand, advances on distinction of homologous chromosomes were also obtained by FISH using different probes. In *P. aureus* and *P. virgineus* physical location of mrDNA, 5S rDNA containing the gene family coding for 5S rRNA, and core histone genes by double-color FISH and re-hybridization experiments showed that all genes tested are on different chromosome pairs, which allowed to distinguish five and four pairs, respectively (Carrilho et al., 2011).

Techniques inducing C banding were employed in some bivalve species to determine the chromosomal distribution of constitutive heterochromatin (Leitão & Chaves, 2008) but not in Veneridae clams. However, a family of satellite DNA, defined by tandemly repeated non-coding DNA sequences, and probably involved in heterochromatin constitution and other functions, was mapped by FISH mainly on pericentromeric position of most chromosomes of *V. philippinarum* (Passamonti et al., 1998).

Most bivalve species display external fertilization once eggs and sperm are released into the water. At the moment of fertilisation, sperm is haploid but meiosis is not completed in eggs. These are arrested at prophase or metaphase of meiosis I and the cell division re-initiation take place when it is activated by the entry of spermatozoon (Colas & Dubé, 1998). In contrast to male meiosis, female meiosis produces unequal size cells: one egg and two/three polar bodies (smaller cells). Because female meiosis is not completed, ploidy manipulation is possible.

Ploidy can be manipulated by physical or chemical treatments of fertilised eggs allowing chromosomal replication but preventing cell division (Beaumont & Fairbrother, 1991; Gosling, 2003; Dunham, 2004; Beaumont, 2006; Piferrer et al., 2009). Physical treatments include pressure and temperature shocks that seem to disturb the meiotic spindle. Chemical treatment involves cytochalasin B (CB), a toxic fungus metabolite that prevents the formation of the cleavage furrow by interfering with the formation of contractile microfilaments, or 6-dimethylaminopurine (6-DMAP), a compound less toxic than CB. Any treatment applied at the appropriate time during meiosis I or meiosis II suppresses cell division, prevents the extrusion of a polar body and produces triploids, while suppression of the first mitotic cleavage produces tetraploids.

Because triploids have two chromosomes from the mother and one from the father they should have an increase in overall heterozygosity with respect to diploids. Moreover, a higher overall heterozigosity is expected from triploids induced at meiosis I (MI triploids) than from those induced at meiosis II (MII triploids). Nevertheless, the actual heterozigosity increase at a locus also depends on the frequency of crossing-over (recombination) between the centromere and the locus (Beaumont & Fairbrother, 1991). In absence of recombination, MI triploids retain maternal heterozigosity and MII triploids produce homozygous eggs. However, when recombination happens, the heterozigosity in MI triploids is reduced and MII triploids produce heterozygous eggs. This means that high levels of recombination will reduce the differences in overall heterozygosity between MI and MII triploids (Beaumont & Fairbrother, 1991).

Several methods have been described to assess the ploidy level in molluscs following treatment for inducing triploids or tetraploids (Beaumont & Fairbrother, 1991; Gosling, 2003; Dunham, 2004; Piferrer et al., 2009). Those applied to Veneridae clams include chromosome count, measurement of the cell nucleus, image analysis and flow cytometry. Chromosome count constitutes a low cost direct method that can be carried out at any stage of development but is time consuming, depends on cell division rate, which may be low especially in adult tissues, and requires animal killing. Although this was the method of choice in early works, more recent reports made use of faster non-dependent cell division methods. Based on the assumption that the nucleus of the triploid is 1.5 greater in volume compared with that of the diploid due to the extra DNA content, the diameter measurement of suitably stained nuclei makes possible to distinguish between diploids and triploids (Child & Watkins, 1994). Diameters of diploid and triploid nuclei would show the expected ratio 1:1.145. This method is usually applied to nuclei of gill tissue but nuclei of haemolymph cells can also be tested, avoiding the animal killing. Image analysis relies on the use of a microscope and a computer, both linked to a camera, to determine the optical density (OD) of stained nuclei and DNA indices (DI) computed by comparison with a mean OD derived from a diploid control (Gerard et al., 1994a). Expected values of DI for a diploid, triploid and tetraploid sample will be 1.0, 1.5 and 2, respectively. This method can be employed at different stages of the life cycle. Flow cytometry is a technology that measures and then analyzes multiple physical characteristics including relative fluorescence intensity of single cells, as they flow in a fluid stream through a beam of light. Determination of the ploidy level is based on the fact that the intensity of fluorescence emitted by stained cells with nucleic acids specific fluorescent dyes (propidium iodide or DAPI) is proportional to the DNA content. Known diploid samples are used as standards to estimate the relative DNA content. The modal value of triploid or tetraploid DNA fluorescence value should be 1.5 or 2-fold that of the diploid value (Allen, 1983). The assay uses fresh or frozen cells from different tissues including haemolymph cells and allows the analysis of several hundred individuals per day.

For purposes of commercial culture, triploids possess advantages over diploids (Beaumont & Fairbrother, 1991; Piferrer et al., 2009). Triploids are expected to be sterile on the premise that the occurrence of three sets of homologous chromosomes disturbs the synapse and segregation of chromosomes during meiosis. In sterile triploids, energy usually used for gametogenesis could be diverted into somatic growth, resulting in enhanced growth rates. Potential problems associated with sexual maturation could be also evaded. The use of sterile triploids could limit the impact of enhancement programs with genetically differentiated stocks and restrict the proliferation of non-native species preventing potential hybridization and competition with native species. On the other hand, triploidy condition could theoretically redound to fast growth due to two additional reasons. First, the extra set of chromosomes in triploids causes an increase in cell size; assuming a constant number of cell divisions, triploids would reach a larger size than diploids. Second, based on the observation of a positive correlation between heterozygosity and growth in bivalves (Zouros & Mallet, 1989), triploids could grow faster than diploids due to their higher overall heterozygosity. The interest in producing tetraploids relies on the fact that the effectiveness of triploid induction methods is less than 100%. Since tetraploids should produce diploid gametes, crosses between tetraploids and diploids are expected to produce 100% triploid offspring.

Triploidy induction in Veneridae clams was achieved on *V. philippinarum*, *M. mercenaria*, *V. decussata* and *Tapes dorsatus* with most of the efforts focussed on the first species (Table 2). In most cases, it was carried out using CB; only Gosling & Nolan (1989) used thermal shock as induction method in *V. philippinarum*. Since CB is a hydrophobic compound, it must be dissolved in dimethyl sulphoxide (DMSO) before use. CB was usually administrated to eggs at 0.5 mg/l during 15 min in *V. philippinarum* but in the other species the concentration was 1 mg/l and the duration of treatment involved 15 or 20 min. Depending on administration time after fertilization, MI or MII triploids can be obtained. In the case of *V. philippinarum* meiosis I is generally complete by 20 min and meiosis II by 35 min at a temperature of 23ºC (Beumont & Contaris, 1988). However, MII triploids were preferably induced and solely Beumont & Contaris (1988) in *V. philippinarum*, and Eversole et al. (1996) and El-Wazzan & Scapa (2009) in *M. mercenaria* reported MI triploids. As detection method, chromosome count represents the most frequent choice in early works and when triploidy assessment was undertaken in embryos. For triploidy assessment in later stages of life cycle (larvae, juveniles or adults) both nucleus diameter measurement, as well as, flow cytometry were employed. Although these two methods can be applied to haemolymph cells, which are obtained by a non-destructive method, they were always performed using larval cells or gill cells, except in the work carried out by Child & Watkins (1994) in *V. philippinarum*.

As occurs in other bivalves, none of the tested treatments routinely produce all-triploid populations in Veneridae clams. The triploid proportion reported varies between trials and depends on the time of assessment. The highest record (94-95%) was reached in *V. decussata* embryos (Gerard et al., 1994b). In *V. philippinarum* the best percentage of triploid embryos is comprised between 73 (Laing & Utting, 1994) and 82% (Beaumont & Contaris, 1988). Lower values (50%) were observed in one case but probably a proportion of the fertilised eggs had completed the meiosis before treatment (Gosling & Nolan, 1989).

Survival of CB treated eggs is usually lower than that of controls until larval stage (Dufy & Diter, 1990; Utting & Doyou, 1992; Gerard et al., 1994b; Utting & Child, 1994) but no differences are detected later (Utting & Doyou, 1992; Laing & Utting, 1994; Utting & Child, 1994; Nell et al., 1995; Utting et al., 1996; Eversole et al., 1996) although a significant decline in ploidy status, from an initial level of 76% to 1%, was reported in *V. philippinarum*

(Shpigel & Spencer, 1996). On the other hand, the prospect of enhanced growth in triploids is not evident in the reported trials. Growth of CB treated larvae and juveniles of *V. philippinarum* was typically similar to non treated controls (Dufy & Diter, 1990; Utting & Doyou, 1992; Laing & Utting, 1994; Utting & Child, 1994). Afterwards, in adult clams, Ekaratne & Davenport (1993) observed no significant differences neither in dry body weight nor in shell breadth between mature diploids and triploids. Also, Shpigel & Spencer (1996) reported no significant differences in growth rates, condition indices, carbohydrate and lipid content between diploid and triploid clams. However, Utting et al. (1996) found that triploid clams were heavier and had a higher condition index and carbohydrate content than diploids of the same age. Therefore, it seems that performance of triploid adults of *V. philippinarum* might depend on the environmental conditions. In the case of *M. mercenaria*, growth of triploids was slower or similar to that of diploids until 27 months of age, with MI triploids displaying a slower growth than MII triploids, however at 47 months of age, triploids were significantly larger than diploids (Eversole et al., 1996; El-Wazzan et al., 2009). For triploids of other species (*V. decussata* and *T. dorsatus*), no differences were observed in the growth rate of diploid and triploid larvae (Gerard et al., 1994b; Nell et al., 1995) but data from adult specimens are not available.

Triploidy in Veneridae clams does not necessarily produce total sterility but rather a significant reduction of the breeding potential. Studies carried out in *V. philippinarum* by Utting et al. (1996) demonstrated that triploid clams produced gametes and that the spawning occurred. Nevertheless, these authors report that triploid populations produces only 12.5% of the number of eggs produced by diploid populations, due to a significant reduction of the number of spawners and eggs released in triploids, and that the ratio of male to female spawners (1:44) differs from the expected (1:1). Although it is unknown whether triploid parents produced viable progeny, when eggs from triploids were fertilized with sperm from diploids, viable larvae were obtained but fewer than diploid-diploid fertilizations (Utting et al., 1996). On the contrary, *M. mercenaria* triploids failed to respond to spawning stimuli and exhibited signs of abnormal and severely retarded gametogenesis which evidence that they should be considered sterile (Eversole et al., 1996).

The fact that the tested methods for inducing triploidy are not 100% effective and that triploid-diploid fertilizations may produce viable progeny indicate that triploids of *V. philippinarum* may be used to restrict but not to avoid their proliferation in the wild when this species is introduced in non native places for cultivation purposes. Despite this, the use of triploids can contribute significantly to reduce the impact of the culture of this species, especially in areas where *V. philippinarum* can hybridize with native species such as *V. decussata* in the European coasts (Hurtado et al., 2011).

Tetraploidy was induced in *V. philippinarum* and *M. mercenaria* (Table 2) with variable results. CB treatments applied by Diter & Duffy (1990) in *V. philippinarum* and the heat shock applied by Yang & Guo (2006) in *M. mercenaria* were effective in producing tetraploid embryos, the highest percentage reaching at least 80%, but none survived to juvenile stage. This was attributed to an abnormal development probably due to a deficiency in cytoplasm or cells, since the cleavage of an egg with a large tetraploid nucleus would lead to either a reduction in the number of blastomeres or blastomeres with inadequate amounts of cytoplasm, and also to an imbalance in gene expression (Yang & Guo, 2006). However, Allen et al. (1994) reported the incidental production of a few tetraploid viable to the adult stage

Table 2. Methods for induction and assessment of triploids and tetraploids in Veneridae clams

Ploidy/Species	Induction method			Assesment method	References
	Treatment	Time[1]	Duration		
Triploids					
Venerupis philippinarum	CB: 0.5 ml/l	0-15 min (MI), 15-30 min (MII)	15 min	CC	Beaumont & Contaris (1988)
	T: 32°C	20 min	10 min	CC	Gosling &Nolan (1989)
	CB: 0.5 mg/l	20 min	15 min	CC	
	CB: 1mg/l	20-35 min	15 min	CC	Dufy & Diter (1990)
	CB: 0.5 mg/l	15 min	15 min	CC	Utting & Doyou (1992)
	CB: ns	ns	ns	ND	Ekaratne & Davenport (1993)
	CB: 0.5 ml/l	15-20 min	15 min	CC; ND; FC	Utting & Child (1994)
	CB: 0.5 ml/l	15-20 min	15-20 min	CC; ND	Laing & Utting (1994)
	CB: 0.5 ml/l	15-20 min	15 min	ND	Utting et al. (1996)
	CB: 0.5 ml/l	20 min	15 min	CC; ND	Shpigel & Spencer (1996)
Mercenaria mercenaria	CB: 1mg/l	5 min (MI); 10 min (MII)	20 min	FC	Eversole et al. (1996)
	CB: 1mg/l	ns (MI) ns (MII)	15 min	FC	El-Wazzan & Scarpa (2009)
Venerupis decussata	CB: 1mg/l	15 min	20 min	CC; IA	Gerard et al. (1994b)
Tapes dorsatus	CB: 1 mg/l	45% FPB	15 min	FC	Nell et al. (1995)
Tetraploids					
Venerupis philippinarum	CB: 1 mg/l	0-10 and 45 min	15 min	CC	Diter & Dufy (1990)
	CB: 0.5 ml/l	20 min	15 min	ND; FC	Allen et al. (1994)
Mercenaria mercenaria	T: 35 and 38°C	60-70% BPB	10-20 min	FC	Yang & Guo (2006)

[1]time after fertilization; CB: cytochalasin B; T: temperature chock; FPB: when ~45 of eggs had developed the first polar body; BPB: when 60-70% had released both polar bodies; ns: not specified; CC: chromosome count; ND: nucleus diameter; IA: image analysis; FC: flow cytometry.

when eggs of *V. philippinarum* were treated to induce MII triploids. These incidental tetraploids were considered an artefact of asynchronous development of eggs at the time of treatment and could have arisen from the inhibition of the first mitotic division, the inhibition of the first polar body alone, the inhibition of both polar bodies, or a combination of all of this. Regardless of the mechanism involved in the induction of tetraploidy, the results of Allen et al. (1994) provide evidences that tetraploidy can be tolerated in *V. philippinarum*.

3. GENETIC ANALYSIS OF POPULATIONS

Population analysis requires the use of genetic markers. A genetic marker is a heritable character with polymorphism. The first group of genetic markers employed was morphologic markers. This kind of markers is easy to analyse. Nevertheless, subjective observations and environmental influences can invalidate their use.

Molecular markers are genetic markers based in macromolecules. Protein markers arrived thanks to the development of starch gel electrophoresis and they imply an extraordinary progress in genetic variability and population structure analysis. Isoenzymes show specificity for the same substrate but they show different chemical properties, such as differences in charge and or molecular weight. When an electric current is applied to a protein solution, different isoenzymes migrate along the gel and are separated according to their different chemical properties. The use of specific substrates, which interact with a group of isoenzymes, allow the detection of all alleles present in the sample. So, because both alleles can be scored, most of isoenzymes are codominant markers. Isoenzyme studies have been developed in most of the animals and plants analysed, and nowadays they are still the basis of a lot of surveys.

DNA molecular markers allow access to a broad range of variability, due to the fact that both coding and no coding sequences can be analysed. They are now the most popular markers for genetic variation analysis. First DNA markers were based on hybridization techniques. DNA hybridization as a tool for detecting specific sequences after electrophoresis separation was first described by Southern (1975), and since then, it was one of the most popular techniques in molecular biology. DNA is transferred to a nylon membrane, and it is hybridized with probes to detect specific fragments of DNA. Hybridization techniques are relatively easy to apply, but they need large quantities of high quality DNA. This problem can be settled with the use of amplification of DNA with polymerase chain reaction (PCR). This technique was developed by Saiki et al. (1985) and allows exponential *in vitro* amplification of specific sequences of DNA from a small quantity of DNA target thanks to two complementary primers flanking the sequence of interest. These primers allow the synthesis of DNA in presence of a thermostable polymerase, dNTPs and other components like buffer and cofactors. A series of repetitive cycles consist of denaturalization, annealing and synthesis of new DNA chains, result in the exponential accumulation of a specific DNA fragment. So, after a few PCR cycles, a million of DNA copies can be achieved.

Molecular markers can be obtained from both nuclear and mitochondrial DNA. For population studies, mitochondrial DNA is sometimes preferred over nuclear DNA because of its higher rate mutation and for the maternal inheritance, which produces only haploid genotypes. Nevertheless, doubly uniparental inheritance (DUI) for mitochondrial DNA detected in some species of bivalves (Skibinski et al., 1994; Zouros et al., 1994) may cause problems in the interpretation of results.

Examination of DNA variability can be done in specific or anonymous sequences. In general, DNA molecular markers can be classified according to the technique used.

3.1. Molecular Markers Based on Hybridization Techniques

Restriction Fragment Length Polymorphisms (RFLPs)

In RFLPs, DNA hybridization is combined with restriction enzyme digestion. Restriction enzymes are DNAse enzymes that cut DNA in specific sites where a recognition sequence is. Variation in the sequence of this recognition sequence produces differences in the RFLP pattern observed. These differences can be analysed with DNA hybridization by Southern-blot. RFLP based on hybridization are restricted to the existence of probes. The use of PCR can settle this problem. In this case, PCR fragments are cut with restriction enzymes and the result is visualized using electrophoresis agarose gel.

Variable Number of Tandem Repeats (VNTRs)

Digested DNA is transferred to a membrane and is hybridized with minisatellite probes. Fragment length varies according to the number of tandem repeats of the minisatellite.

3.2. Molecular Markers Based on Arbitrary or Semi Arbitrary PCR Amplification

In this case, no information about DNA sequence is necessary and multilocus dominant markers are generated.

Random Amplified Polymorphic DNA (RAPD)

When no information is known about genomic DNA of a species, RAPD markers are a good option for genetic analysis. This marker explores the existence of inverted repetitions in DNA. Genomic DNA is amplified with a single random primer of 10 nt long with low annealing temperatures. Whenever two inverted repetitions are close enough, amplification will be produced. The final product is a collection of several fragments of different sizes. These are separated by agarose gel electrophoresis and bands can be sized and scored. Presence or absence of specific bands can be determined in each individual. These markers are considered biallelic and dominant markers. Despite RAPDs are easy and fast markers, the main problem of this use is their unreliability.

Inter Simple Sequence Repeats (ISSR

The principle of this kind of markers is the same that for RAPD markers. In this case, semi arbitrary primers anchored in microsatellite sequences are used.

Amplified Fragment Length Polymorphism (AFLP)

These markers combine RFLP and RAPD, by using restriction enzymes and random primer amplification. First of all, genomic DNA is cut by two restriction enzymes to produce different size fragments. Second, two adapters are anchored to the fragments produced. Third, amplification of fragments is done thanks to complementary primers to adapters.

3.3. Molecular Markers Based on Specific PCR Amplification

When specific primers are used, monolocus codominant markers are analysed, but some knowledge about DNA sequence is necessary.

Sequence Characterized Amplified Region (SCAR)

This kind of markers implies sequencing of discrete bands from RAPD markers and designing of primers for the specific amplification of these bands. So, dominant multilocus markers become codominant monolocus markers.

Microsatellite Markers

Microsatellites are tandemly repeated genomic sequences, which are present in both prokaryotic and eukaryotic species. They have high rates of mutation and they vary in their number of repetitions. Because their high polymorphism and because they can be easily amplified with polymerase chain reaction, they become one of the most popular molecular markers in population genetics. Nevertheless, microsatellite markers are difficult to develop in some species, and interspecific analyses are not possible in most of the cases.

Most of population studies in clams are based in enzyme loci, but molecular DNA markers are being used more and more frequently, as in other bivalve species. Below we describe the population genetics works in main commercial clam species, which concentrate the majority of this kind of studies.

3.4. Mercenaria mercenaria

Pesch (1972) analyzed variation at lactate dehydrogenase (LDH) in four geographical sites, one from Canada (Bideford River, Prince Edward Island) and three from the USA (Boothbay Harbor, Maine; Narragansett Bay, Rhode Island; Wadmalaw Island, South Carolina). In a posterior work (Pesch 1974), he expanded the study with data from three additional protein loci. Two loci (LDH and tretazolium oxidase) showed a North-South cline, another did not show notable differences between samples and the fourth was almost monomorphic (only four individuals of 324 were heterozygous). Surprisingly, all the individuals from the Canadian sample were heterozygous for the same alleles of the LDH *loci*. Pesch (1972, 1974) suggested that the cline, the lack of homozygous individuals for the LDH loci in Canada, and the heterozygote excesses observed for LDH, could be related to selective pressures. Specifically, LDH plays an important role in the respiratory metabolism. Individuals from Northern localities are subjected to more extreme ambient conditions and during coolest months they have to close their valves to support temperatures below 4 ºC, having low activity and periods of anaerobiosis. On the contrary, individuals from southern localities do not suffer from such low temperatures and have aerobic respiration all the year. Alternatively, a genomic duplication could have affected the LDH locus in the Canadian sample (Pesch, 1972) or, as suggested by Hilbish (2004), two genes in this species could code this enzyme or there is single gene intercorvension.

A work of Humphrey (1981) analyzed the genetic variation of six wild samples of *M. mercenaria* in the coast of Georgia (USA) with eleven enzyme loci. This work was posteriorly expanded with five additional sampling sites (Humphrey & Crenshaw, 1989), from Massachusetts to Florida, which covered most of the natural distribution of this species. Nevertheless, it is important to note that the two samples taken in Florida (Tampa Bay and Port Saint Joe) were assumed to be composed by the congeneric species *M. campechiensis*. Significant heterozygote deficits were observed for four enzymes in Humphrey's (1981) work. This is a common observation in studies of marine bivalves, first reported for allozymes (see Gaffney, 1994) but also common in DNA markers such as microsatellites (e.g. Launey et al., 2002; Kenchington et al., 2006). To explain this phenomenon both biological (inbreeding, Wahlund effect, or selection) and technical explanations (null alleles) have been proposed. Both studies showed low levels of population differentiation, being the differences in allelic frequencies observed among the Georgia samples similar to those observed for the whole geographic range. Accordingly, the values of Nei's Standard Genetic Distance (Nei & Roychoudhury, 1974) were low (range: 0.005-0.020).

Dillon & Manzy (1987) compared the genetic variation at seven polymorphic enzyme loci for two nursery stocks and two wild samples taken at the sites that served as sources for the founding parents (Martha's vineyard, Massachusetts, and Hog Island, near Wachapreague, Virginia; USA). The Mendelian inheritance for five of these loci had previously been demonstrated by Adamkewicz et al. (1984). The seven loci showed genotypic frequencies mostly fitting to Hardy-Weinberg expectations for each of the four samples. The allelic frequencies showed significant differences at several loci between hatchery stocks and wild samples. In addition the hatchery samples seemed to have lost some rare alleles, although the heterozigosity did not show evident changes. This is not a completely unexpected observation, since the allelic diversity is more sensitive to the effects of population bottlenecks (Allendorf, 1986). On the contrary, the wild samples did not show significant differences.

The allozymic study of Slattery et al (1991) examined the existence of genetic differences between samples taken from two close adjacent habitats, one with vegetation and other without, from three localities along the east coast of the US (Nauset Marsh, Cape Cod, Massachusetts; Marshedler Island, Little Egg Harbor, New Jersey; Shackleford Banks, Middle Marsh, Cape Lookout, North Carolina). Of the fifteen loci initially screened nine could be used in the genetic diversity analysis. Significant heterozygote deficits were reported for several enzymes in all six samples. A hierarchical analysis of genetic diversity (Nei, 1973) was carried out and the total genetic variance was partitioned among three levels: within local habitat samples, between habitat samples within geographic localities and between localities. Most of the genetic variance (98%) corresponded to within local habitat samples. The variance between localities although low (1.7%), was significant. On the contrary, the variance between habitats within localities was not significant. Similarly, comparisons of allelic frequencies between habitats did not show evidences of differentiation after Bonferroni correction.

The populations in the Northern limit of distribution range of *M. mercenaria* were examined in more detail by Dillon & Manzy (1992) using seven enzyme loci. They used the data obtained in a previous work (Dillon & Manzi, 1987; sample from Massachusetts) and analyzed two samples from Canada (Prince Edwards Island and Sam Orr's Pond, St. Andrews, New Brunswick) and another one from the US (Brunswick, Maine). Genotypic frequencies showed generally good agreement with HWE expectations and Canadian samples showed lower heterozygosity values than US ones. This contrast with the results of Pesch (1972, 1974), which observed a heterozygote excess for LDH enzyme. Nevertheless, as stated above the basis of the LDH polymorphism is not clear and the number of loci used by Pesch (1972, 1974) was low. In addition, most of the loci showed allelic frequencies significantly different between Massachusetts and Maine samples. US samples showed significant differences at two loci (mannose-6-phosphate isomerase and superoxide dismutase). A hierarchical F-statistics analysis indicated that most of the differentiation was attributable to differences between Canadian and US samples. Nevertheless, the authors point out that there should be a gene flow strong enough to prevent a substantial differentiation due to genetic drift. The differences between Maine and Massachusetts are probable caused by isolation by distance.

The genetic effects of aquaculture practices carried out in the Folly River (near Charleston, South Carolina, USA) over natural populations were examined by Metzner-Roop (1994). She analyzed the frequencies of the alleles of the glucose phosphate isomerase, some of which had higher frequencies in the aquaculture stock, and the *notata* shell-color polymorphism, that is infrequent in the natural populations. Four samples were obtained in the area where aquaculture practices were carried out and a sample taken in the same area by Dillon & Manzi (1989) was used as control. The results did not show changes in the allele frequencies or in the *notata* coloration incidence attributable to the aquaculture stocks. Nevertheless, one test for linkage disequilibrium was significant in one of the samples. Although, the authors suggests that may be attributable to type I statistical error. Therefore, the aquaculture practices do not seem to have affected significantly the composition of natural populations.

The population structure of *M. mercenaria*, deduced from allozyme studies suggests the existence of genetic homogeneity over most of its geographic range. The differentiation observed between northernmost and southern localites could be related to bottleneck events or episodes of extinction/recolonization.

The first work that analyzed the genetic diversity and population structure of *M. mercenaria* using DNA-based markers is that of Baker et al. (2008). They sequenced a fragment of the mitochondrial cytochrome c oxidase subunit I (COI) in ten samples taken throughout its range, but specially around the Florida peninsula, extending from Charlottetown (Prince Edwards Island, Canada) to Cedar Key (Florida, USA). The sampling locations were selected to study the effect of putative biogeographic boundaries. Eighty-five haplotypes were detected in the 297 individuals analyzed for 528 bp of COI. The relationships among haplotypes were typical of intraspecific phylogenies, with a low number of differences between haplotypes and numerous unresolved relationships.

The value of Fu's statistic suggested the existence of deviation from neutrality or a recent population expansion, if near-neutrality is assumed. Diversity indices, haplotype diversity and nucleotide diversity, showed increasing values from north to south, although not in a linear fashion. This is probably caused by a population range expansion. Pairwise differentiation between samples, estimated as average corrected values of nucleotide diversity (π) between populations, was low, ranging from -0.016 to 0.069. Similarly, an analysis of molecular variance indicated the existence of low, but significant, population structure. The Φ_{ST} (and F_{ST} analogue) between pairs of localities was significant for 16 of 45 comparisons.

Wang et al. (2010) reported the development of 29 polymorphic microsatellite markers for *M. mercenaria*, but these markers were not used to assess the population structure of this species. During the isolation of markers the authors found that a high proportion of primers amplified more fragments than expected, suggesting the existence of duplicated microsatellite loci in the genome of this species.

3.5. Venerupis decussata

Jarne et al. (1988) analysed genetic variation in eight enzymatic loci from five localities covering a broad range of distribution of this species. They collected samples from Mediterranean (Spain, France and Tunisia), from lake Timsah (Egypt) and from Atlantic Ocean (Portugal). The five localities showed the same major alleles, with exclusive alleles at low (0.036) or very low (0.015-0.019) frecuencies. Expected heterozigosity values were similar (0.229-0.282). Absolute genetic distances (Gregorius, 1984) ranged from 0.069 between Ebro (Spain) and Bizerte (Tunisia) to 0.133 between Thau (France) and Faro (Portugal), and none of the values were significantly distinct from zero. High gene flow was assumed to explain the absence of differentiation.

One of the most habitual problem concerning to population genetics is the absence of Hardy-Weinberg equilibrium in the majority of the localities. Borsa et al. (1991) analysed this question in seven sites along a French locality, Etang de Thau. One or several cohorts were identified according to the distribution of shell lengths in every site sampled. Analysis was done with seven enzymatic loci. Significant positive F_{IS} values were found (heterozygote deficiency) in some loci. No significant F_{IS} values were found in samples with only one cohort. No F_{ST} value differed significantly from zero. The analysis of a multi-cohort sample showed heterozygote deficiency and significant differences between cohorts in one of the loci (Ldh1). Based on this, the authors suggest than a temporal Wahlund effect is acting in *V. decussata*, with several cohorts that differ in allele frequency mixed.

Borsa et al. (1994) carried out an analysis of several localities from Mediterranean and Timsah Lake. They studied 10 enzymatic systems and they calculated several Theta values: among cohorts from a same sample; among different sites from the same lake (Etang de Thau, France); among close localities; and among all localities sampled. Only this last comparison was significantly different from zero. Theta per locus values were significant in regional and global comparisons, but no correlation between geographic and genetic distance was found. Nevertheless, localities are too far to perform a reliability Mantel test.

Passamonti et al. (1997) analysed the genetic variation in 12 enzymatic loci in three samples from Adriatic Sea (Italy) and one from Tunisia. Although the main purpose of this survey was the genetic relationship among several Tapetinae species, intra and interpopulation genetic variation was analyzed too. Genetic distances (Nei, 1972) varied from 0.030 between two of the Italian localities to 0.128 between the Tunisian locality and one of the Italian localities. These values were consistently low. In contrast, global F_{ST} value was significantly different from zero, but authors related this with differences in cohorts and the presence of a temporal Wahlund effect. So, no genetic differentiation among Mediterranean localities was found.

In Azores archipielago, in São Jorge island (Lagoa de Santo Cristo), there is a well-established population of *V. decussata.* It seems likely that this population was introduced by man, due to the fact that the most closely populations from Iberian Peninsule and Morocco are separated from Azores by more than 1500 km. This distance is unlikely to be solved by larval transport, since they remain planktonic from 10 days (Borsa et al., 1991) and the maximum distance they can afford is 100 km. Because of this, Jordaens et al. (2000) carried out an enzymatic analysis of seven loci in three localities of *V. decussata*, one from Lagoa de Santo Cristo, one from North Atlantic coast of Spain and one from Mediterranean coast of France. They observed a decreased genetic variability in Azorean locality, with low values of allelic diversity and expected heterozygosity. The loss of genetic variability was observed in bivalves with hatchery stocks established from a few number of individuals. Moreover isolation of Azorean locality can produce a decrease of genetic variation due to genetic drift and inbreeding. Overexplotation of this population can reinforce these effects. Despite the isolation of population from Santo Cristo, the authors found low values for genetic distances among localities, and there is only a slight differentiation from continental samples.

DNA markers were employed in three surveys of genetic population structure in *V. decussata*. The first study was the development of molecular markers based on Exon Primer Intron Crossing (EPIC) PCR. In this method, primers anchored in conserved exon sequences were used to amplify more variable intron sequences. This kind of PCR can produce two types of polymorphisms: intron length polymorphism and intron RFLP polymorphism. Cordero et al. (2008) used two approximations for the development of EPIC-PCR markers. First of all they tried published intron markers for other species. Second, they used databases for searching new intron markers. With these two approaches they developed molecular markers for two clam species, *V. decussata* and *V. philippinarum*. In the grooved carpet shell, two intron markers were analysed, one intron length marker and one intron RFLP marker. They used these two markers for the analysis of genetic variability in two populations of *V. decussata*, one from Northwest Atlantic coast of Spain and one from Mediterranean coast of Italy. Both markers were polymorphic in the two populations. TBP marker, with intron length polymorphism, showed 2 and 3 alleles in Spanish and Italian localities respectively. Allele frequencies were strikingly different and F_{ST} values were highly significant. By contrast, SRP54 marker, with intron RFLP polymorphism, showed four and three alleles in Spanish and Italian localities, and no significant F_{ST} value was found. These contrasting results need the analysis of more populations and more markers to give an accurate explanation. Nevertheless, the authors suggest the existence of stabilizing selection at SRP54 or diversifying selection at TBP to explain the contrasting patterns observed at the two loci.

Gharbi et al. applied both DNA sequencing (2010) and allozymes (2011) for the analysis of genetic variation and population structure in several localities of Tunisia. In their first survey, they sequenced the first COI, and the internal transcribed spacer region between 18S and 5.8S ribosomal genes (ITS1, nuclear DNA). Haplotype and nucleotide diversity in COI were lower than those reported in other bivalves, and they were also lower than in ITS1. Because mitochondrial DNA has an effective population size approximately one-quarter that of nuclear DNA (Hurst & Jiggins, 2005) it can detect more easily population events like bottleneck. In fact, network obtained by Gharbi et al. (2010) results in a star-shaped genealogy, which is commonly interpreted as a result of recent population expansion followed by a population bottleneck (Slatkin & Hudson, 1991). The overexplotation of the populations analysed here can also explain the low genetic diversity obtained. Nevertheless, these results contrast with those obtained in allozyme study where the average heterozygosity obtained with 15 allozymes was higher than the average reported in other marine invertebrates (Berger, 1983; Buroker, 1983; Gallardo et al., 1998). The different nature of both markers could explain the discrepancies.

Molecular variance analysis (AMOVA) of DNA sequences failed to detect differentiation among the pooled groups on either side of the Siculo-Tunisian strait. In contrast, hierarchical F_{ST} values in allozyme analysis were significant among eastern and western localities. Other genetic studies show this strait as an important genetic barrier (eg. Nikula & Vainola, 2003). DNA markers, especially mitochondrial markers reflect more recent events than allozyme markers, which can show evidence of past historical barriers to dispersal. In fact, the connection between the eastern and western Mediterranean Sea has been closed several times during glacial periods in the region of the Siculo-Tunisian strait, but nowadays there is water circulation currents that could have permitted sufficient gene flow, preventing the accumulation of substantial genetic differentiation. So, allozyme markers can show ancient barriers while DNA markers can show recent gene flow.

The last study about genetic polymorphism and population structure in *V. decussata* employs RAPD markers to assess the variability of two Portuguese populations (Pereira et al., 2011). The analysis of 245 reproducible bands showed that both localities have very high levels of genetic polymorphism and no genetic differentiation was detected. High levels of gene flow and restocking can explain these results.

3.6. Venerupis philippinarum

This species is indigenous to the Western Pacific Ocean, but it is cultivated in a broad range of localities, including Asian an European countries. So, several studies concerned the genetic variation in cultivated populations. This is the case of Moraga survey (1986), who analysed 11 allozyme loci in a French cultivated population. Observed heterozygosity was 0.26, a similar value than those found in natural populations of *V. decussata*. The authors concluded that although more studies with the same loci in natural populations of *V. philippinarum* was needed, it seems that there is not genetic variation loss in the cultivated population.

Passamonti et al. (1997) analysed the genetic variation in 12 enzymatic loci in four samples from Adriatic Sea (Italy). Observed heterozygosity values were similar to those found by the authors in *V. decussata* localities. No significant departure of Hardy-Weinberg equilibrium was found with a p=0.02. F statistics showed significant values for F_{IS} and F_{IT} and no significance for F_{ST}.

Park et al. (2002) used two DNA regions, the mitochondrial COI gene and the nuclear ribosomal ITS2 region to determine the genetic variability of *V. philippinarum* from three geographical regions in Korea. They obtained sequences from seven individuals of the three sampled localities. Although this is not a population study, since the number of localities and individuals analysed are too low, they found few differences in the nucleotide sequence of the two regions between the three localities. They concluded that more polymorphic markers, like microsatellites, should be used for a more detailed analysis of relationships between the populations.

Liu et al. (2007) analysed genetic variation and population structure of four Chinese localities of *V. philippinarum* employing AFLP markers. The analysis of 264 AFLP loci in 195 individuals revealed high values of genetic diversity, and these values were similar among localities. In AMOVA test, 15.38% of variability was found between localities, and this value was highly significant. F_{ST} revealed differentiation among the four populations too. The individual-based distance matrix was used to construct dendogram by UPGMA cluster analysis to explore population structure. In general, clusters obtained reflected geographical location of individuals with a few exceptions. Individuals from one of the locations grouped more dispersedly than the others. In this location clam seeds were transplanted from other areas. So, it seems that human activities would have some influence in population structure of *V. philippinarum*.

Vargas et al. (2008) examined the allozyme genetic variability in four Japanese and one Chinese localities of *V. philippinarum*. In two of the Japanese localities from Ariake Sea, foreign clam seeds were released in aquaculture ground separated from natural grounds, which are sampled in this study. They analysed one Korean locality of the close species *R. bruguieri* too, identified by morphological characters. The analysis of eight loci indicated a high genetic variability and homozygote excess in almost all loci. F_{ST} pairwise values were between 0.0015 and 0.0122 in the comparisons among *V. philippinarum* samples. Those localities close to aquaculture grounds showed lower F_{ST} values in comparisons with Chinese sample than those with no transplanted seeds. In comparisons with *R. bruguieri* and *V. philippinarum* samples, F_{ST} values ranged from 0.0314 to 0.0634. Pairwise F_{ST} distances were used to calculate a UPGMA tree to evaluate relationships among samples. *R. bruguieri* sample was clearly isolated in a different cluster from the other samples. In *V. philippinarum* samples from Ariake Sea clustered together with Chinese sample, while the other two samples were placed in another cluster. Based on allozyme allele frequencies, authors calculated the mixture proportions in Ariake Sea samples, using as background samples the Chinese locality and one of the Japanese localities out of Ariake Sea. They found high proportions of Chinese individuals (0.41 and 0.49) although no foreign seed was released in the areas that were sampled. The authors concluded that interbreeding between areas with culture and no culture occurred. Nevertheless, taking account the high similarity of allele frequencies in the samples analysed, the precision of mixing calculation was assumed to be low. Moreover, the allele frequencies of the Japanese localities before the release of Chinese seed was unknown, and the authors have to use as background a 60 km far locality. So, genetic invasion of stocked Chinese individuals into wild populations of Ariake Sea remains unclear due to the low precision of the estimates. High polymorphic DNA markers could improve the precision of the analysis.

Cordero et al. (2008) used three intron markers for the analysis of two *V. philippinarum* localites, one in Japan, and the other in Europe, probably founded with Japanese individuals. The number of alleles ranged from 2 (H3-iA locus homologous to mouse histone 3.3A gene) to 5 (in the TFP marker). Almost all alleles of the three markers were present in both samples. Only one allele in SRP54 marker was present in Japanese sample and absent in European sample, although it frequency was low. F_{ST} for this marker was significantive, but no for the other two markers. The authors concluded that it seems that introduction of Japanese individuals didn't produce loss of genetic variability in European sample, but more markers and localities should be analysed to asses this question.

Microsatellite markers were developed for *V. philippinarum* (Yasuda et al., 2007; An et al., 2009), and eight of them amplified in *R. variegate* too (An et al., 2009). Chiesa et al. (2011) employed markers developed by Yasuda et al. (2007) together with mitochondrial 16S sequences to evaluate the genetic structure in six localities of Northern Adriatic Sea. Spite the low number of samples used, they suggest more than one events of introduction of *V. philippinarum* in Adriatic Sea.

3.7. Cyclina sinensis

Zhao et al. (2007) examined population structure between four samples of *C. sinensis* from China using AFLP markers. The value of θ_B, a Bayesian analog of F_{ST} (Holsinger et al, 2002) was 0.205, suggesting the existence of differentiation. Pairwise θ_B and Nei's (1978) genetics distance, which ranged from 0.081 to 0.158 (average distance=0.117), indicated that the samples could be grouped in two clusters. One cluster including the northernmost sites (Lvshun in Liaoning province and Lianyungang in Jiangsu Province) and the other including Yueqing (Zhejiang Province) and Dongxing (Guangxi Zhuang autonomous region) sites. In addition, a principal coordinate analysis placed almost every individual close to other individuals of their sampling site. The authors suggest that currents along coastal areas in China could be one of the main causes limiting the gene flow.

A posterior work of Zhao et al. (2009) included two additional sites (Maoming, Guangdong province; and Tianjin; Tianjin Municipality) and examined morphological and allozymic variation. Nei's (1978) genetic distance calculated with nine polymorphic loci ranged from 0.002 to 0.388, showing low values (<=0.010) when comparing samples north or south to the Yangtze river, The exception was Dongxing that displayed relatively high genetic distance (>=0.119) with close samples. The overal (F_{ST}=0.0778, P <0.05) and pairwise F_{ST} (0.0004 to 0.1697) described a similar trend. In addition, the morphometric analysis indicated a clear separation between northern and southern samples. The observed genetic divergence suggests that the samples could not represent only one species, since populations of the same species usually have lower values (see for example Thorpe 1983, Backeljau et al. 1994). Additional studies are required to clarify the taxonomical status of this species.

Recently, Feng et al. (2010) presented 20 microsatellite markers for *C. sinensis*. Nevertheless, they only evaluated the diversity of microsatellites in individuals from the locality of Qingdao (Shandong, China) in the Northwest of the East China Sea.

3.8. Chamelea gallina

The genetic variaton of two morphotypes of *Chamelea gallina* (L) was studied with seven polymorphic enzime loci by Backeljau et al. (1994). The sampling of these morphotypes, *C. gallina sensu lato* and *C. striatula*, included four points of the Ría Formosa (southern Portugal), where they occur simpatrically. In one of the sampled points *C. striatula* was not found. None loci could distinguish between morphotypes unequivocally, although the locus malate dehydrogenase (MDH) allowed a good separation between them. Strong heterozygote deficits were reported for both morphotypes. Nei's (1978) unbiased genetic distances between morphotypes (0.771-1.509) were remarkably higher than distances within morphotypes (*C. gallina* s.l.: 0.000-0.251; *C. striatula*: 0.103-0.373). These values are typical of intraspecific comparisons so despite living in the same areas introgresion seems to be low or absent between *C. gallina* and *C. striatula*. Therefore the subspecies status between these morphotypes it is probably not appropriate.

3.9. Venerupis corrugata

The pullet carpet shell *V. corrugata* is commercially exploited in several countries of Europe. The overfishing of this species in some areas has left to a high density decrease of populations. This is the case in Ria Formosa (Portugal), where it seems necessary to restore stocks to reproductive viability. In order to know the genetic variability of this population and to assess the most appropriate founder population to develop an aquaculture program, Joaquim et al. (2010) used RAPD markers to evaluate two Portuguese populations, Ria Formosa and Ria de Aveiro. Both localities are separated 500 km. The analysis of 222 reproducible bands showed high and similar values of genetic variation and Shanon's information index in both localities. Nei distance value revealed small genetic distance between the two populations, and F_{ST} index indicated absence of genetic differentiation. The high variability in Ria Formosa suggests that this locality has enough genetic plasticity to overcome changes in environmental conditions despite the overfishing. The low genetic differentiation suggests a high genetic flow between both localities. Nevertheless, due to the high variability found in Ria Formosa, the authors suggest that the best way to improve the population is to use Formosa as broodstock for aquaculture purposes, although Aveiro could be a second option in the case that Ria Formosa broodstock was insufficient.

Microsatellite markers for *V. corrugata* were developed by Pereira et al. (2010) but they only tested their polymorphism in one population. These markers will be a valuable tool for future population genetic studies in this clam.

3.10. Ameghinomya antiqua

Gallardo et al. (1998) used enzyme loci for the analysis of polymorphism and genetic structure of *A. antiqua* in Chilean coast. They analysed four close localities (Fátima, Martín Balmaceda, Yaldad and Guapiquilán) and one further (San Juan) for Southern Chile. Eleven loci were polymorphic and the level of heterogeneity varied between 6.8% to 10.3%. Heterozygote deficiency was found in some loci in almost all localities. Nei's (1978) distance and Rogers's (1972) similarity index were used for clustering in a UPGMA tree. Results with both measures coincided, with San Juan locality as the most divergent sample. Wright's (1978) F indexes showed intrapopulation heterozygote deficiency and global F_{ST} revealed genetic differences. This value can be explained by the presence of San Juan locality. In fact, the highest pairwise F_{ST} values were obtained in all comparisons involved San Juan sample. Estimates of number of migrants (Nm) showed high variability, from only less than two migrants per generation between San Juan and the other samples to 118 individuals per generation between the two close locations of Yaldad and Guapiquilán.

3.11. Leukoma staminea

Parker et al. (2003) examined the allozyme polymorphism of *Leukoma staminea* at Puget Sound (Washington, US), a fjord-like estuary composed by several basins. Twenty-five enzymes were assayed but only five allozyme loci were employed in the three samples obtained from the estuary. Significant differences were reported for up to four of the five loci in the samples. Differentiation tests based on allelic and genotypic frequencies showed significant differences among all sites. This differentiation was also reflected by the F_{ST} and genetic distance values. The results obtained for this species were remarkably different to those obtained for other bivalve species, *Macoma balthica*, sampled in the same work and area. These differences could be related to the ecological niches that these species occupy and/or larval behaviour. Although this work demonstrates the existence of genetic differentiation in this species a wider study covering the range of distribution is needed.

3.12. Meretrix lusoria

In a work where the genetic relationships among several species of the *Meretrix* genus were investigated using allozyme loci (Yamakawa 2008), several samples of *M. lusoria* were analyzed. Based on the shell morphology and coloration, samples of canonical (Mutsu Bay and Ariake Sea, Japan) and putative (Korea and Taiwan) *M. lusoria* individuals were recollected from four locations. As expected, the Japanese samples displayed between them a high and low Nei's (1978) unbiased measure of genetic identity (0.970) and genetic distance (0.003), respectively. On the contrary, the Taiwanese sample differed considerably from the other samples (maximum Genetic identity of 0.741 and minimum Genetic distance of 0.299). Lastly, the Korean sample values indicated that this location was closer to the Japanese ones. The results suggest that the Taiwanese sample could represent other *Meretrix* species with a morphology and coloration similar to *M. lusoria*. Therefore, further studies are required to characterize the different species of the genus *Meretrix* and additional sampling is needed to study the genetic variation and population structure in *M. lusoria*.

3.13. *Polititapes aureus*

Passamonti et al. (1997) analysed genetic variability in three Italian localities of *P. aureus* by using 12 enzyme loci. Observed and expected heterozygosity values were higher than those observed in the other species included in this study (*V. decussata*, *V. philippinarum* and *Paphia undulata*). Like in this species, heterozygote deficiency was found in two of the three localities analysed. F_{ST} values were significantly distinct from zero.

4. Clams Identification

Bivalve species identification has traditionally rely on morphological criteria. This requires the presence of the valves, which is subjectively evaluated by the observer, and may often present great plasticity due to environmental influences. Moreover, the loss of these species-specific features due to manufacturing processes (i.e. boiling and packaging) is often loss, as it is the case for commercialised species. This makes species identification sometimes difficult or even impossible.

Given these troubleshooting with the morphological criteria, focus has turned lately towards molecular marker-based species identification. As in population genetic studies, the first molecular markers developed for commercial species identification were proteins. Allozyme analysis and species-specific antigens offered a quick, easy and reliable technique for polymorphism studies. However, this kind of markers presents some inconvenient properties, such as differential expression between tissues or developmental stages. Furthermore, protein studies are almost impossible in the case of manufactured products, since most of them are denatured during commercial preparation.

DNA-based molecular markers are in this sense much more effective than proteins. Firstly, genomic direct analysis is not affected by tissue or developmental stage. Secondly, DNA remains intact or at least unchanged after manufacture procedures. However, DNA integrity is sometimes an essential starting point for analysis, and will most of the times determine the technique to use. Individual conservation methods will also influence DNA quality: DNA is not degraded at low temperatures and is very stable (Cerda & Koppen, 1998), but this relationship quickly changes with temperature increases. Heating produces DNA fragmentation in segments of around 500 base pairs (bp) or less (Chikuni et al., 1990; Unseld et al., 1995).

Both nuclear as well as mitochondrial genomes may be used for the development of molecular markers in species identification. The mitochondrial genome is present in abundance compared with the nuclear portion. Besides, it being circular offers a greater resistance to degradation (Borgo et al., 1996). This implies that PCR amplification from mitochondrial DNA is probably easier to obtain in manufactured samples than its nuclear respective. Also, mitochondrial genes present an evolution rate that is optimal for species identification (Wan et al., 2004). On the other hand, nuclear DNA, although more easily disrupted, presents a much larger sequence, and the number of potential loci in a single analysis is much greater than that for mitochondrial DNA.

4.1. Molecular Methods for Clam Identification

Clam genetic studies related to species identification are scarce. The reasons for this may be related to a lower plasticity in this bivalve group, and their global lower commercial importance, in terms of capture, when compared to mussels or oysters.

The first of the works developed for the identification of the three "major" European clam species: *V. decussata*, *V. corrugata*) and *V. phillippinarum*, was published by Fernández et al. (2000). In this work, the sequence analysis of α-actin and the building of a restriction map allowed these authors to identify species-specific restriction sites. Electrophoretic analyses of amplification product digested with *Mae*III and *Rsa*I produced differential patterns for the three species. In 2001, the same authors (Fernández et al., 2001) also described species identification by restriction patterns in the internal spacer transcript ITS. PCR amplification with *Mytilus*-based ITS primers produced a 1195 bp fragment in *V. decussata*, 1074 bp for *V. corrugata* and 1188 bp in *V. phillippinarum*. PCR product digestion with *Hinf*I and *Rsa*I endonucleases, followed by electrophoresis, showed a restriction profile that allowed for discrimination amongst the studied species. In 2002a, RFLP analysis of the mitochondrial 16S region allowed for the identification of the four commercial species *V. decussata*, *V. corrugata*, *V. phillippinarum* and *P. virgineus* (=*Venerupis rhomboides*). rDNA 16S amplification and a later digestion with *Bsm*AI and *Bsr*I identified specific restriction patterns for each of the species (Fernández et al., 2002a). Later, Fernández et al. (2002b) the same group designed a protocol for the differentiation of *V. decussata* and *V. corrugata* based on PCR-SSCP amplification of α-actin.

5. Genomic Resources in Clams

Despite the importance of genetic resources in aquaculture, the use of genomic methodologies in mollusks, and particularly in clams, is still at an early stage. Nevertheless, there has been in the recent years a considerable effort to increase the amount of genomic data, mainly thanks to the development of next generation sequencing methodologies. This generation of genomic data in aquaculture is mostly focused on obtaining linkage maps with the aim to detect the genes that are responsible for variation in productive traits, and on the study of cellular and genetic response in relation to pathogens.

There are two main strategies to obtain genomic data: whole genome sequencing and sequencing of the expressed genome. The former has been used to reveal the genome of *Aplysia californica* (California sea hare) from the Gastropoda class, with the data being now publicly available in the UCSC genome browser (http://genome.ucsc.edu/cgi-bin/hgGateway). In the case of the Bivalvia, the sequencing of *Crassostrea gigas* is currently being carried out by the *Oyster Genome Consortium*, formed by 11 countries (Hedgecock et al., 2005). Nevertheless, genome size, the lack of reference sequences, the economic cost, and the number of repetitive regions make this whole genome sequencing approach difficult, and nowadays, expressed genome sequencing is the methodology preferred. This latter approach involves the sequencing of only a fraction of the genome, which is being expressed (i.e the coding one). This fraction involves around 5% of the total genomic DNA load, but includes the information needed for protein production. In order to obtain this coding fraction, the messenger RNA is extracted and converted to DNA (cDNA). Depending on the cDNA library construction method and sequencing technology we can obtain a whole mRNA or just a fragment. Although whole transcript sequencing is *a priori* desirable, the economic and technical costs make the sequencing of small fragments a better strategy for most researchers. The sequences obtained by this method are named ESTs (Expressed Sequence Tags), and are widely used for transcriptomic studies. One advantage of EST is that a gene identity might be assigned and thus the researcher captures a fragment of "what is going on" in the bivalve cells.

As it has been mentioned before, the availability of genomic data for the Veneridae family is still scarce. One single search in the NCBI Entrez Taxonomy Browser in December of 2011 retrieves 15321 entries, whereas other bivalve families with a higher economic profile, such as Mytilidae or Ostreidae, produce 80866 and 225336 entries, respectively. Although these data could be regarded as poor, a considerable effort has been carried out from the scientific community in order to reach these numbers in the last years. In this context, an important initiative has been carried out in the context of the *European Network of Marine Genomics* with the common goal of increasing the amount of ESTs from commercial marine mollusk species, and particularly in the less studied but ecologically and commercially important groups, such as the mussel and clam genera (Tanguy et al., 2007). Around 2000 ESTs were obtained by these means for the clam *V. decussata*.

Although the number of studies using cDNA libraries in species from the Veneridae family is growing exponentially, the two main objectives for undertaking these projects are still the same: to obtain molecular markers, (e.g Single Nucleotide Polymorphisms (SNPs) or microsatellites) for population analysis and linkage mapping, and to study the transcriptomic profile in response to different conditions. There are several examples of these purposes in the bibliography, such as SNP search associated with productive traits (Wang et al., 2011a,b) or the obtaining of immune genes related to exposure to different pathogens (Adhya et al., 2010; Li et al., 2011; Moreira et al., 2012) or also for instance environmental monitoring (e.g. Milan et al., 2011)

While it is exciting to watch some of the major milestones in aquaculture genomics being achieved, mainly in fish and mollusk economically important species, great challenges still lie ahead. These include the continued lack of genomic information for many aquaculture species, such as those being the main character of this book.

REFERENCES

Adamkewicz, L., Taub, S.R. & Wall, J.R. (1984). Genetics of the clam *Mercenaria mercenaria*. I. Mendelian inheritance of allozyme variation. *Biochem. Genet.* 22, 215-219.

Adhya, M., Chol, K.S., Yu, Y. & Cho, M. (2010). Expression and localization of MCsialec, a sialic acid-specific lectin in the marine bivalve Manila clam, *Ruditapes philippinarum*. *J. Fish Dis*. 33, 889-899.

Allen, S. (1983). Flow cytometry: assaying experimental polyploid fish and shellfish. *Aquaculture*. 33, 317-328.

Allen, S., Shpigel, M., Utting, S. & Spencer, B. (1994). Incidental production of tetraploid Manila clams, *Tapes philippinarum* (Adams and Reeve). *Aquaculture*. 128, 13-19.

Allendorf, F.W. (1986). Genetic drift and the loss of alleles versus heterozygosity. *Zoo Biol.* 5, 181-190.

An, H.S., Kim, E.M. & Park, J.Y. (2009). Isolation and characterization of microsatellite markers for the clam *Ruditapes philippinarum* and cross-species amplification with the clam *Ruditapes variegate*. *Conserv. Genet.* 10, 1821-1823.

Backeljau, T., Bouchet, P., Gofas, S. & Debruyn, L. (1994). Genetic variation, systematics and distribution of the Venerid clam *Chamelea gallina*. *J. Mar. Biol. Assoc. U.K.* 74, 211-223.

Baker, P., Austin, J.D., Bowen, B.W. & Baker, S.M. (2008). Range-wide population structure and history of the northern quahog (*Merceneria merceneria*) inferred from mitochondrial DNA sequence data. *ICES J. Mar. Sci.* 65, 155-163.

Beaumont, A. & Contaris, M. (1988). Production of triploid embryos of *Tapes semidecussatus* by the use of cytochalasin B. *Aquaculture*. 73, 37-42.

Beaumont, A. (2006). Genetics. In: S.E. Shumway, G.J. Parsons (Eds.), *Scallops: biology, ecology and aquaculture*. pp. 543-594, Elsevier, Amsterdam.

Beaumont, A.R. & Fairbrother, J.E. (1991). Ploidy manipulation in molluscan shellfish: a review. *J. Shellfish Res.* 10, 1-18.

Berger, E.M. (1983). Population genetics of marine gastropods and bivalves. *Mollusca.* 6, 563-596.

Borgo, R., Souty-Grosset, C., Bouchon, D. & Gomot, L. (1996). PCR-RFLP analysis of mitochondrial DNA for identification of snail meat species. *J. Food. Sci.* 61, 1-4.

Borsa, P. & Thiriot-Quiévreux C. (1990). Karyological and allozymic characterization of *Ruditapes philippinarum*, *R. aureus* and *R. decussatus* (Bivalvia, Veneridae). *Aquaculture.* 90, 209-227.

Borsa, P., Jarne, P., Belkhir, K. & Bonhomme, F. (1994). Genetic-structure of the palourde *Ruditapes decussatus* L in the Mediterranean. In: A.R. Beaumont (Ed.), *Genetics and evolution of aquatic organisms.* pp. 103-113, Chapman and Hall, London.

Borsa, P., Zainuri, M. & Delay, B. (1991). Heterozygote deficiency and population structure in the bivalve *Ruditapes decussatus. Heredity.* 66, 1-8.

Bouilly, K., Leitão A., McCombie, H. & Lapègue S. (2003). Impact of atrazine on aneuploidy in Pacific oysters, *Crassostrea gigas. Environ. Toxicol. Chem.* 22, 219-223.

Buroker, N.E. (1983). Population genetics of the American oyster *Crassostrea virginica* along the Atlantic coast and the Gulf of Mexico. *Mar. Biol.* 75, 99-112.

Carrilho, J., Perez-Garcia, C., Leitão, A., Malheiro, I. & Pasantes, J.J. (2011). Cytogenetic characterization and mapping of rDNAs, core histone genes and telomeric sequences in *Venerupis aurea* and *Tapes rhomboides* (Bivalvia: Veneridae). *Genetica.* 139, 823-831.

Cerda, H., & Koppen, G. (1998). DNA degradation in chilled fresh chicken studied with the neutral comet assay. *Zeitschrift für Lebensmittel Untersuchung und Forschung.* 207, 22-25.

Chiesa, S., Marzano, F.N., Minervini, G., De Lucrezia, D., Baccarani, G., Bordignon, G., Poli, I., Ravagnan, G. & Argese, E. (2011). The invasive Manila clam *Ruditapes philippinarum* (Adams and Reeve, 1850) in Northern Adriatic Sea: Population genetics assessed by an integrated molecular approach. *Fish. Res.* 110, 259-267.

Chikuni, K., Ozutsumi, K., Koishikawa, T. & Kato, S. (1990). Species identification of cooked meats by DNA hybridization assay. *Meat Sci.* 27, 119-128.

Child, A. & Watkins, H. (1994). A simple method to identify triploid molluscan bivalves by the measurement of cell nucleus diameter. *Aquaculture.* 125, 199-204.

Colas, P. & Dube, F. (1998). Meiotic maturation in mollusc oocytes. *Semin. Cell Dev. Biol.* 9, 539-548.

Cordero, D., Pena, J.B. & Saavedra, C. (2008). Polymorphisms at three introns in the Manila clam (*Ruditapes philippinarum*) and the grooved carpet-shell clam (*R. decussatus*). *J. Shellfish Res.* 27, 301-306.

Corni, M. & Trentini, M. (1986). A chromosomic study of *Chamelea gallina* (L) (Bivalvia, Veneridae). *Boll. Zool.* 53, 23-24.

Corni, M. & Trentini, M. (1990). The chromosomes of *Venerupis aurea* and *Ruditapes philippinarum* of the northern Adriatic Sea (Bivalvia, Heterodonta, Veneridae). *Venus: Jap. J. Malacol.* 49, 258-261.

de Sousa, J.T., Matias, D., Joaquim, S., Ben-Hamadou, R. & Leitão A. (2011). Growth variation in bivalves: new insights into growth, physiology and somatic aneuploidy in the carpet shell *Ruditapes decussatus. J. Exp. Mar. Biol. Ecol.* 406, 46-53.

Dillon, R.T. & Manzi, J.J. (1987). Hard clam, *Mercenaria mercenaria*, broodstocks: genetic drift and loss of rare alleles without reduction in heterozygosity. *Aquaculture.* 60, 99-105.

Dillon, R.T. & Manzi, J.J. (1989). Genetics and shell morphology in a hybrid zone between the hard clams *Mercenaria mercenaria* and *Mercenaria campechiensis. Mar. Biol.* 100, 217-222.

Dillon, R.T. & Manzi, J.J. (1992). Population genetics of the hard clam, *Mercenaria mercenaria*, at the northern limit of its range. *Can. J. Fish. Aquat. Sci.* 49, 2574-2578.

Diter, A. & Dufy, C. (1990). Polyploidy in the Manila clam, *Ruditapes philippinarum*. II. Chemical induction of tetraploid embyos. *Aquat. Living Resour.* 3, 107-112.

Dufy, C. & Diter, A. (1990). Polyploidy in the Manila clam, *Ruditapes philippinarum* I. Chemical induction and larval performances of triploids. *Aquat. Living Resour.* 3, 55-60.

Dunham, R.A. (2004). *Aquaculture and fisheries biotechnology: genetic approaches.* CABI Publishing, Wallingford.

Ebied, A.-B.M. & Aly, F.M. (2004). Cytogenetic studies on metaphase chromosomes of six bivalve species of families Mytilidae and Veneridae (Nucinelloidea, Mollusca). *Cytologia.* 69, 261-273.

Ekaratne, S. & Davenport, J. (1993). The relationships between the gametogenetic status of triploids or diploids of Manila clams, *Tapes philippinarum*, and their oxygen uptake and gill particle transport. *Aquaculture.* 117, 335-349.

El-Wazzan, E. & Scarpa, J. (2009). Comparative growth of triploid and diploid juvenile hard clams *Mercenaria mercenaria notata* under controlled laboratory conditions. *Aquaculture.* 289, 236-243.

Eversole, A., Kempton, C., Hadley, N. & Buzzi, W. (1996). Comparison of growth, survival, and reproductive success of diploid and triploid *Mercenaria mercenaria. J. Shellfish Res.* 15, 689-694.

Feng, Y., Li, Q. & Kong, L. (2010). Twenty microsatellite DNA markers for the Venus clam (*Cyclina sinensis* Gmelin). *Conserv. Genet.* 11, 1189-1192.

Fernandez, A., Garcia, T., Asensio, L., Rodríguez, M., Gonzalez, I., Hernandez, P. & Martin, R. (2001). PCR-RFLP analysis of the internal transcribed spacer (ITS) region for identification of 3 clam species. *J. Food Sci.* 66, 657-661.

Fernandez, A., Garcia, T., Asensio, L., Rodríguez, M., Gonzalez, I., Lobo, E., Hernandez, P. & Martin, R. (2002). Genetic differentiation between the clam species *Ruditapes decussatus* (grooved carpet shell) and *Venerupis pullastra* (pullet carpet shell) by PCR-SSCP analysis. *J. Sci. Food Agric.* 82, 881-885.

Fernández, A., Garcia, T., Asensio, L., Rodríguez, M.A., Gonzalez, I., Céspedes, A., Hernandez, P.E. & Martin, R. (2000). Identification of the clam species *Ruditapes decussatus* (Grooved carpet shell), *Venerupis pullastra* (Pullet carpet shell), and *Ruditapes philippinarum* (Japanese carpet shell) by PCR-RFLP. *J. Agric. Food Chem.* 48, 3336-3341.

Fernández, A., Garcia, T., Gonzalez, I., Asensio, L., Rodríguez, M.A., Hernandez, P.E. & Martin, R. (2002). Polymerase chain reaction-restriction fragment length polymorphism analysis of a 16S rRNA gene fragment for authentication of four clam species. *J. Food Prot.* 65, 692-695.

Gaffney, P.M. (1994). Heterosis and heterozygote deficiencies in marine bivalves: more light? In: A.R. Beaumont (Ed.), *Genetics and evolution of aquatic organisms.* pp. 146-153, Chapman & Hall, London.

Gallardo, M.H., Penaloza, L. & Clasing, E. (1998). Gene flow and allozymic population structure in the clam *Venus antiqua* (King of Broderip), (Bivalvia, Veneriidae) from Southern Chile. *J. Exp. Mar. Biol. Ecol.* 230, 193-205.

Gerard, A. (1978). Étude des garnitures chromosomiques de deux Veneridae: *Ruditapes decussatus* (L.) et *Ruditapes philippinarum* (Adams et Reeve). *Haliotis.* 9, 69-71.

Gerard, A., Naciri, Y., Peignon, J.M., Ledu, C., Phelipot, P., Noiret, C., Peudenier, I. & Grizel, H. (1994a). Image analysis: a new method for estimating triploidy in commercial bivalves. *Aquacult. Fish. Manage.* 25, 697-708.

Gerard, A., Naciri, Y., Noiret, C., Ledu, C., Peignon, J.M. & Phelipot, P. (1994b) Induced triploidy in the European clam, *Ruditapes decussatus* (L.), and performance of triploid larvae. *Aquacult. Fish. Manage.* 25, 769-779.

Gharbi, A., Chatti, N., Said, K. & Van Wormhoudt, A. (2010). Genetic variation and population structure of the carpet shell clam *Ruditapes decussatus* along the Tunisian coast inferred from mtDNA and ITS1 sequence analysis. *Biologia.* 65, 688-696.

Gharbi, A., Zitari-Chatti, R., Van Wormhoudt, A., Dhraief, M.N., Denis, F., Said, K. & Chatti, N. (2011). Allozyme variation and population genetic structure in the carpet shell clam *Ruditapes decussatus* across the Siculo-Tunisian Strait. *Biochem. Genet.* 49, 788-805.

Gosling, E. & Nolan, A. (1989). Triploidy induction by thermal shock in the Manila clam, *Tapes semidecussatus*. *Aquaculture.* 78, 223-228.

Gosling, E. (2003). *Bivalve molluscs: biology, ecology and culture*. Fishing News Books, Oxford.

Gregorius, H. (1984). A Unique Genetic Distance. *Biometrical J.* 26, 13-18.

Hedgecock, D., Gaffney, P.M., Goulletouer, P., Guo, X., Reece, K. & Wark, G.W. (2005). The case for sequencing the Pacific oyster genome. *J. Shellfish Res.* 24, 429-441.

Hilbish, T.J. (2001). Genetics of hard clams, *Mercenaria mercenaria*. In: J.N. Kraeuter, M. Castagna (Eds.), *Biology of the hard clam*. pp. 261-280, Elsevier Science, Amsterdam.

Holsinger, K.E., Lewis, P.O. & Dey, D.K. (2002). A Bayesian approach to inferring population structure from dominant markers. *Mol. Ecol.* 11, 1157-1164.

Humphrey, C.M. (1981). Ecological genetics of the hard clams (*Mercenaria mercenaria* Linné and *Mercenaria campechiensis* Gmelin): electrophoretic estimates of enzyme variation and the use of shell morphology as a species indicator. *PhD. thesis, University of Georgia*. pp. 93.

Humphrey, C.M. & Crenshaw Jr., J.W., (1989). Clam genetics. In: J.J. Manzi, M. Castagna (Eds.), *Clam Mariculture in North America.* pp. 323-356, Elsevier, Amsterdam.

Hurst, G.D. & Jiggins, F.M. (2005). Problems with mitochondrial DNA as a marker in population, phylogeographic and phylogenetic studies: the effects of inherited symbionts. *Proc. Biol. Sci.* 272, 1525-1534.

Hurtado, N. & Pasantes, J. (2005). Surface spreading of synaptonemal complexes in the clam *Dosinia exoleta* (Mollusca, Bivalvia). *Chromosome Res.* 13, 575-580.

Hurtado, N.S., Pérez-García, C., Morán, P. & Pasantes, J.J. (2011). Genetic and cytological evidence of hybridization between native *Ruditapes decussatus* and introduced *Ruditapes philippinarum* (Mollusca, Bivalvia, Veneridae) in NW Spain. *Aquaculture.* 311, 123-128.

Ieyama, H. (1980). Studies on the chromosomes in three species of the Veneridae (Bivalvia, Heterodonta). *Venus: Jap. J. Malacol.* 39, 49-55.

Ieyama, H. (1985). Karyotype in *Ruditapes philippinarum* (Bivalvia: Veneridae). *Special publication of the Mukaishima Marine Biological Station*. pp. 83-85.

Insua, A. & Thiriot-Quiévreux, C. (1992). Karyotypes of *Cerastoderma edule*, *Venerupis pullastra* and *Venerupis rhomboides* (Bivalvia, Veneroida). *Aquat. Living Resour.* 5, 1-8.

Insua, A. (1993). Estudio cariológico en moluscos bivalvos. *Ph.D Thesis, University of A Coruña.*

Jarne, P., Berrebi, P. & Guelorget, O. (1988). Genetic and morphometric variability of 5 populations of the clam *Ruditapes decussatus* (Mollusca, Bivalvia). *Oceanol. Acta.* 11, 401-407.

Joaquim, S., Pereira, J., Leitao, A., Matias, D., Chaves, R., Guedes-Pinto, H., Chicharo, L. & Gaspar, M. (2010). Genetic diversity of two Portuguese populations of the pullet carpet shell *Venerupis senegalensis*, based on RAPD markers: contribution to a sustainable restocking program. *Helgol. Mar Res.* 64, 289-295.

Jordaens, K., De Wolf, H., Willems, T., Van Dongen, S., Brito, C., Martins, A.M.F. & Backeljau, T. (2000). Loss of genetic variation in a strongly isolated Azorean population of the edible clam, *Tapes decussatus. J. Shellfish Res.* 19, 29-34.

Kenchington, E.L., Patwary, M.U., Zouros, E. & Bird, C.J. (2006). Genetic differentiation in relation to marine landscape in a broadcast-spawning bivalve mollusc (*Placopecten magellanicus*). *Mol. Ecol.* 15, 1781-1796.

King, M. (1993). *Species evolution: the role of chromosome change*. Cambridge University Press, Cambridge.

Laing, I. & Utting, S. (1994). The physiology and biochemistry of diploid and triploid Manila clam (*Tapes philippinarum* Adams and Reeve) larvae and juveniles. *J. Exp. Mar. Biol. Ecol.* 184, 159-169.

Launey, S., Ledu, C., Boudry, P., Bonhomme, F. & Naciri-Graven, Y. (2002). Geographic structure in the European flat oyster (*Ostrea edulis* L.) as revealed by Microsatellite polymorphism. *J. Hered.* 93, 331-351.

Leitão, A. & Chaves, R. (2008). Banding for chromosomal identification in Bivalves: a 20-years history. Dyn. Biochem., Process Biotech. Mol. Biol. 2(Special Issue 1), 1-20.

Leitão, A., Boudry P. & Thiriot-Quiévreux, C. (2001a). Negative correlation between aneuploidy and growth in the Pacific oyster, *Crassostrea gigas*: ten years of evidence. *Aquaculture.* 193, 39-48.

Leitão, A., Boudry, P., McCombie, H., Gerard, A. & Thiriot-Quiévreux C. (2001b). Experimental evidence for a genetic basis to differences in aneuploidy in the Pacific oyster (*Crassostrea gigas*). *Aquat. Living Resour.* 14, 233-237.

Leitão, A., Chaves, R., Matias, D., Joaquim, S., Ruano, F. & Guedes-Pinto H. (2006). Restriction enzyme digestion chromosome banding on two commercially important venerid bivalve species: *Ruditapes decussatus*and *Cerastoderma edule*. *J. Shellfish Res.* 25, 857-863.

Li, C., Li, L., Liu, F., Ning, X., Chen, A., Zhang, L., Wu, H. & Zhao, J. (2011). Alternation of *Venerupis philippinarum* Hsp40 gene expression in response to pathogen challenge and heavy metal exposure. *Fish Shellfish Inmunol.* 30, 447-450.doi:10.1016/j.fsi.2010.10.023.

Lin, Z.H., Lu, Z.M., Chai, X.L., Fang, J. & Zhang, J.M. (2008). Karyotypes of diploid and triploid *Mercenaria mercenaria* (Linnaeus). *J. Shelfish Res.* 27, 297-300.

Liu, X., Bao, Z., Hu, J., Wang, S., Zhan, A., Liu, H., Fang, J. & Wang, R. (2007). AFLP analysis revealed differences in genetic diversity of four natural populations of Manila clam (*Ruditapes philippinarum*) in China. *Acta Oceanol. Sin.* 26, 150-158.

Menzel, R.W. & Menzel, M.Y. (1965). Chromosomes of two species of quahog clams and their hybrids. *Biol. Bull.* 129, 181-188.

Menzel, R.W. (1968). Chromosome number in nine families of marine pelecypod mollusks. *Nautilus.* 82, 45-58.

Metzner-Roop, K.L. (1994). The effect of aquaculture on the genetics of natural-populations of the hard clam, *Mercenaria mercenaria* (L). *J. Shellfish Res.* 13, 487-491.

Mikkelsen, P.M., Bieler, R., Kappner, I. & Rawlings, T.A. (2006). Phylogeny of Veneroidea (Mollusca: Bivalvia) based on morphology and molecules. *Zool. J. Linn. Soc.* 148, 439-521.

Moraga, D. (1986). Genetic polymorphism of cultivated populations of the Manila clam, *Tapes philippinarum. Comptes Rendus Acad. Sci. Ser. III-Sci. Vie-Life Sci.* 302, 621-624.

Moreira, R., Balseiro, P., Romero, A., Dios, S., Posada, D., Novoa, B., Figueras, A. (2012). Gene expression analysis of clams *Ruditapes philippinarum* and *Ruditapes decussatus* following infection yields insights into pathogen resistance and immunituy. *Dev. Comp. Immunol.* 36, 140-149..

Nakamura, H.K. (1985). A review of molluscan cytogenetic information based on CISMOCH-computerized index system for molluscan chromosomes. Bivalvia, Polyplacophora and Cephalopoda. *Venus: Jap. J. Malacol.* 44, 193-225.

Nei, M. & Roychoudhury, A.K. (1974). Sampling variances of heterozygosity and genetic distance. *Genetics.* 76, 379-390.

Nei, M. (1972). Genetic distance between populations. *Am. Nat.* 106, 283-292.

Nei, M. (1973). Analysis of gene diversity in subdivided populations. *Proc. Natl. Acad. Sci. U. S. A.* 70, 3321-3323.

Nei, M. (1978). Estimation of average heterozygosity and genetic distance from a small number of individuals. *Genetics.* 89, 583-590.

Nell, J., O'Connor, W., Hand, R. & McAdam, S. (1995). Hatchery production of diploid and triploid clams, *Tapes dorsatus* (Lamarck 1818): a potential new species for aquaculture. *Aquaculture.* 130, 389-394.

Nikula, R. & Väinölä, R. (2003). Phylogeography of *Cerastoderma glaucum* (Bivalvia: Cardiidae) across Europe: a major break in the Eastern Mediterranean. *Mar. Biol.* 339-350.

Park, G.M., Chung, E.Y. & Hur, S.B. (2002). Genetic characters of the Korean shortnecked clam *Ruditapes philippinarum* based on ITS2 and mtCOI gene sequences. *Korean J. Genet.* 24, 51-58.

Parker, M.S., Jumars, P.A. & Leclair, L.L. (2003). Population genetics of two bivalve species (*Protothaca staminea* and *Macoma balthica*) in Puget Sound, Washington. *J. Shellfish Res.* 22, 681-688.

Passamonti, M., Mantovani, B. & Scali, V. (1997). Allozymic characterization and genetic relationships among four species of Tapetinae (Bivalvia, Veneridae). *Ital. J. Zool.* 64, 117-124.

Passamonti, M., Mantovani, B. & Scali, V. (1998). Characterization of a highly repeated DNA family in *Tapetinae species* (Mollusca bivalvia : Veneridae). *Zool. Sci.* 15, 599-605.

Pereira, J.C., Chaves, R., Leitao, A., Matias, D. & Guedes-Pinto, H. (2011). Genetic analysis of two Portuguese populations of *Ruditapes decussatus* by RAPD profiling. *Helgol. Mar. Res.* 65, 361-367.

Pereira, S., Arias, A., Méndez, J., Insua, A. & Freire, R. (2010). Isolation of twelve microsatellite markers in the pullet carpet shell *Venerupis pullastra* (Bivalvia: Veneridae). *Conserv. Genet. Resour.* 2, 201-203.

Pesch, G. (1972). Isozymes of lactate dehydrogenase in the hard clam, *Mercenaria mercenaria*. *Comp. Biochem. Physiol. B.* 43, 33-38.

Pesch, G. (1974). Protein polymorphisms in the hard clams *Mercenaria mercenaria* and *Mercenaria campechiensis*. *Biol. Bull.* 146, 393-403.

Piferrer, F., Beaumont, A., Falguiere, J., Flajshans, M., Haffray, P. & Colombo L. (2009). Polyploid fish and shellfish: production, biology and applications to aquaculture for performance improvement and genetic containment. *Aquaculture.* 293, 125-156.

Rasotto, M., Altieri, D. & Colombera, D. (1981). I chromosomi spermatocitari di 16 specie appartenenti alla classe Pelecypoda. *Atti dei congressi della società malacologica italiana*. pp. 113-127, Salice Terme, Italia.

Rogers, J.S. (1972). Measures of genetic similarity and genetic distance. In: *Studies in Genetics VII,* publication No. 7213. pp. 145-153, University of Texas Publication.

Saiki, R.K., Scharf, S.J., Faloona, F., Mullis, K.B., Horn, G.T., Erlich, H.A. & Arnheim, N. (1985). Enzymatic amplification of b-globin genomic sequences and restriction analysis for diagnosis of sickle cell anemia. *Science.* 230, 1350-1354.

Shpigel, M. & Spence, B. (1996). Performance of diploid and triploid Manila clams (*Tapes philippinarum*, Adams and Reeve) at various levels of tidal exposure in the UK and in water from fish ponds at Eilat, Israel. *Aquaculture*. 141, 159-171.

Skibinski, D.O., Gallagher, C. & Beynon, C.M. (1994). Mitochondrial DNA inheritance. *Nature*. 368, 817-818.

Slatkin, M. & Hudson, R.R. (1991). Pairwise comparisons of mitochondrial DNA sequences in stable and exponentially growing populations. *Genetics*. 129, 555-562.

Slattery, J.P., Vrijenhoek, R.C. & Lutz, R.A. (1991). Heterozygosity, growth, and survival of the hard clam, *Mercenaria mercenaria*, in seagrass vs sandflat habitats. *Mar. Biol.* 111, 335-342.

Southern, E.M. (1975). Detection of specific sequences among DNA fragments separated by gel electrophoresis. *J. Mol. Biol.* 98, 503-517.

Tanguy, A., Bierne, N., Saavedra, C., Pina, B., Bachère, E., Kube, M., Bazin, E., Bonhomme, F., Boudry, P., Boulo, V., Boutet, I., Cancela, L., Dossat, C., Favrel, P., Huvet, A., Jarque, S., Jollivet, D., Klages, S., Lapègue, S., Leite, R., Moal, J., Moraga, D., Reinhardt, R., Samain, J.F., Zouros, E. & Canario, A. (2008). Increasing genomic information in bivalves through new EST collections in four species: Development of new genetic markers for environmental studies and genome evolution. *Gene*. 408, 27-36.

Teixeira de Sousa, J., Matías, D., Joaquim, S., Ben-Hamadou, R. & Leitão, A. (2011). J. Exp. Mar. Biol. Ecol. 406, 46-53.

Thiriot-Quievréux, C. & Ayraud, N. (1982). Les caryotypes de quelques espèces de bivalves et de gastéropodes marins. *Mar. Biol.* 70, 165-175.

Thiriot-Quievréux, C. (2002). Review of the literature on bivalve cytogenetics in the last ten years. *Cah. Biol. Mar.* 43, 17-26.

Thorpe, J.P. (1983). Enzyme variation, genetic distance and evolutionary divergence in relation to levels of taxonomic separation. In: G.S. Oxford, D. Rollinson (Eds.), *Protein polymorphism: adaptive and taxonomic significance*. pp. 131-152, Academic Press, London.

Tsai, H.P., Lin, M.D. & Chao, N.H. (1996). Karyotypes of four species of aquaculture animals, loach *Misgurnus angulicaudatus*, zebra fish *Brachydanio rerio*, oyster *Crassostrea gigas* and hard clam *Meretrix lusoria*. *J. Taiwan Fish. Res.* 4, 63-76.

Unseld, M., Beyermann, B., Brandt, P. & Hiesel, R. (1995). Identification of the species origin of highly processed meat products by mitochondrial DNA sequences. *PCR Meth. Applic.* 4, 241-243.

Utting, S. & Child, A. (1994). Genetic manipulation of the Manila clam (*Tapes philippinarum*) using cytochalasin B to induce triploidy. *Aquaculture*. 120, 271-282.

Utting, S. & Doyou, J. (1992). The increased utilization of egg lipid reserves following induction of triploidy in the Manila clam (*Tapes philippinarum*). *Aquaculture*. 103, 17-28.

Utting, S.D., Millican, P.F. & Laing I. (1996). The breeding potential and biochemical composition of triploid Manila clams, *Tapes philippinarum* Adams and Reeve. *Aquaculture Res*. 27, 573-580.

Vargas, K., Asakura, Y., Ikeda, M., Taniguchi, N., Obata, Y., Hamasaki, K., Tsuchiya, K. & Kitada, S. (2008). Allozyme variation of littleneck clam *Ruditapes philippinarum* and genetic mixture analysis of foreign clams in Ariake Sea and Shiranui Sea off Kyushu Island, Japan. *Fish. Sci.* 74, 533-543.

Wan, Q-H., Wu, H., Fujihara, T. & Fang, S.G. (2004). Which genetic marker for which conservation genetics issue? *Electrophoresis*. 25, 2165-2176.

Wang, C., Wang, H., Li, Y. & Liu B. (2011). Identification of a fructose-1,6 bisphosphate aldolase gene and association of the single nucleotide polymorphisms with growth traits in the clam *Meretrix meretrix*. *Mol. Biol. Rep.* 39, 5017-5024.

Wang, C., You, Y., Wang, H. & Liu, B. (2012). Genetic diversity of the sulfotranferase-like gene and one nonsynonymous SNP associated with growth traits on clam, *Meretrix meretrix*. *Mol. Biol. Rep.* 39, 1323-1331.

Wang, Y. & Guo, X. (2001). Chromosomal mapping of the vertebrate telomeric sequence (TTAGGG)N in four biva.ve molluscs by fluorescence in situ hybridization. *J. Shellfish Res.* 20, 1187-1190.

Wang, Y. & Guo, X. (2004). Chromosomal rearrangement in Pectinidae revealed by rRNA loci and implications for bivalve evolution. *Biol. Bull.* 207, 247-256.

Wang, Y. & Guo, X. (2007). Chromosomal mapping of major ribosomal rRNA genes in the hard clam (*Mercenaria mercenaria*) using fluorescence *in situ* hybridization. *Mar. Biol.* 150, 1183-1189.

Wang, Y., Wang, A. & Guo, X. (2010). Development and characterization of polymorphic microsatellite markers for the Northern Quahog *Mercenaria mercenaria* (Linnaeus, 1758). *J. Shellfish Res.* 29, 77-82.

White, M.J.D. (1978). *Modes of speciation*. V.H. Freeman, San Francisco.

Wright, S. (1978). *Evolution and the genetics of populations*. Chicago, University of Chicago Press.

Yamakawa, A.Y., Yamaguchi, M. & Imai, H. (2008). Genetic relationships among species of Meretrix (Mollusca : Veneridae) in the western Pacific Ocean. *Pac. Sci.* 62, 385-394.

Yang, H. & Guo, X. (2006). Tetraploid induction by inhibiting mitosis I with heat shock, cold shock, and nocodazole in the hard clam *Mercenaria mercenaria* (Linnaeus, 1758). *Mar. Biotechnol.* 8, 501-510.

Yasuda, N., Nagai, S., Yamaguchi, S., Lian, C.L. & Hamaguchi, M. (2007). Development of microsatellite markers for the Manila clam *Ruditapes philippinarum*. *Mol. Ecol. Notes.* 7, 43-45.

Zhao, Y., Li, Q., Kong, L. & Mao, Y. (2009). Genetic and morphological variation in the Venus clam *Cyclina sinensis* along the coast of China. *Hydrobiologia.* 635, 227-235.

Zhao, Y., Li, Q., Kong, L., Bao, Z. & Wang, R. (2007). Genetic diversity and divergence among clam *Cyclina sinensis* populations assessed using amplified fragment length polymorphism. *Fish. Sci.* 73, 1338-1343.

Zouros, E. & Mallet A. (1989). Genetic explanations of the growth/heterozygosity correlation in marine mollusks. *Int. Symp. Ser.* 317-324.

Zouros, E., Ball, A.O., Saavedra, C. & Freeman, K.R. (1994). Mitochondrial DNA inheritance. *Nature.* 368, 818.

In: Clam Fisheries and Aquaculture
Editor: Fiz da Costa González

ISBN: 978-1-62257-518-3

Chapter 5

CLAM SYMBIONTS

C. López,[1*] S. Darriba[2] and J. I. Navas[3]
[1]Centro de Investigacións Mariñas (CIMA),
Vilanova de Arousa (Pontevedra), Spain
[2]Instituto Tecnolóxico para o Control do Medio
Mariño de Galicia (INTECMAR), Vilagarcía
de Arousa (Pontevedra), Spain
[3]IFAPA Centro Agua del Pino, Cartaya (Huelva), Spain

ABSTRACT

This chapter is a review of symbiont organisms affecting adults of different species of clams. The term "symbiont" includes mutualism, commensalism and parasitic relationships. We include within the term "clams" any bivalve mollusc other than oysters, mussels and scallops. The symbionts infecting clams reported in this chapter belong to viruses, fungi, prokaryotes, protozoans, and metazoan organisms.

5.1. INTRODUCTION

Pathology is the precise study and diagnosis of a disease, understanding disease as a negative deviation from the normal state (health) of a living organism. Negative deviation is damage quantifiable in terms of a reduction in ecological potential (e.g. survival, growth, reproduction, energy procurement); it can be provoked by a single cause or by multiple causes acting in concert. Negative deviation requires that the effect of the agent surpasses a certain threshold: a) increase in the number of given agents per host individual, b) increase of

[1*] Corresponding author: C. López. Centro de Investigacións Mariñas (CIMA). Consellería do Mar. Pedras de Corón, s/n. Apdo. 13, 36620, Vilanova de Arousa (Pontevedra), Spain.E-mail address: clopez@cimacoron.org.
[2] S. Darriba: Instituto Tecnolóxico para o Control do Medio Mariño de Galicia (INTECMAR). Consellería do Mar. Peirao de Vilaxoán s/n., 36611, Vilagarcía de Arousa (Pontevedra), Spain.
[3] J. I. Navas: IFAPA, Centro Agua del Pino. Consejería de Agricultura, Pesca y Medio Ambiente, Junta de Andalucía. Ctra. El Rompido-Punta Umbría, km 4, 21459, Cartaya (Huelva), Spain.

destructive potential (virulence), c) concomitant activities of several agents from different taxa, d) decrease in host resistance due to physiological state, environmental stress or to other additively effective disease-promoting or disease-causing conditions and e) changes in the quantitative relations between host and agent.

Causes of diseases can be classified in several groups: a) internal circumstances (innate or genetic diseases), b) nutritional disorders, c) abiotic environmental factors (salinity, temperature, oxygen) and natural and man-made pollutants, d) physical injuries, e) biotic agents (viruses, bacterial, fungi, protozoans and metazoans) and f) a combination of some or all the above-mentioned causes (Kinne, 1980).

Basic research on marine animal diseases needs to take into account that restricting itself to species of commercial interest is insufficient for an understanding of the essential ecological dynamics of life in oceans and coastal waters, for developing sound environmental protection measures, and for an understanding of the principles of biotic diseases. The species which happens to be of immediate commercial interest are embedded in a multitude of interrelations with numerous other coexisting organisms (Kinne, 1980).

The present chapter, entitled "clam symbionts", is limited to a review of current knowledge about symbiont organisms (biotic agents) - referring to "symbiont" in its original meaning, which includes mutualist, commensalist and parasitic relationships - affecting adults of different species of clams. It is important to note that the presence of potentially pathogenic symbiontic organisms does not lead directly to disease.

The term "clam" will be considered in this chapter as being any bivalve mollusc other than oysters, mussels and scallops. There are different reviews concerning bivalve pathology, including clams (Lauckner, 1983; Sindermann, 1990; McGladdery, 1999; Figueras, 2004; Gestal et al. 2008; Bower, 2010a; López et al 2011). Symbionts reported to infect clams belong to organisms within the classes of viruses, fungi, prokaryotes, protozoans, and metazoans. The present chapter is organized following the above classification, concluding with other less important symbionts.

5.2. Viruses

Viruses that infect bivalve molluscs have been researched by several authors (Farley, 1978; Lauckner, 1983; Johnson, 1984; Sindermann, 1990; Elston, 1997; McGladery, 1999; Renault and Novoa 2004; Bower, 2010b). The first works of research on viruses in bivalves were based on morphological descriptions at ultrastructural level, due to the absence of stable cell lines; later, some authors used fish cell lines to isolate and study the cytopathic effects of viruses (Bower, 2010b). Nearer to the present day, molecular techniques have been applied in order to improve bivalve virus identification (Renault and Novoa, 2004; Romalde et al. 2007; Davison, 2009; Renault et al. 2011).

All the viruses detected in bivalve molluscs can tentatively be assigned to families of viruses found in vertebrates (Farley, 1981), with the exception of the family of Malacoherpesviridae, which was created for herpesvirus that infect molluscs (Davison, 2009). The main viral agents affecting bivalve molluscs belong to the following families: Papillomaviridae, Polyomaviridae, Paramyxoviridae, Malacoherpesviridae, Togaviridae, Retroviridae, Reoviridae, Birnaviridae, Iridoviridae, Picornaviridae and Baculoviridae (Chang et al., 2002; Renault and Novoa, 2004; Garcia et al., 2006; Cheslett et al., 2009; Davison et al., 2009; Meyers et al., 2009). It is reported that viruses of some of the following families infect different species of adult clams: Papillomaviridae, Polyomaviridae, Birnaviridae, Picornaviridae and Retroviridae.

Papillomaviruses and Polyomaviruses (previously named Papovaviruses) are non-enveloped double-stranded circular DNA with icosahedral symmetry. Papillomaviruses and Polyomaviruses are about 60 nm and 40-50 nm in diameter, respectively. An intranuclear inclusion in cells resembling amebocytes in *Mya arenaria*, with viruses similar to Papovaviruses, were reported in Chesapeake Bay (US) (Harshbarger et al., 1979). A Papillomavirus-like organism was detected in populations of the soft-shell clam *M. arenaria* in the US, infecting gills and associated with inflammation and hyperplasia (Koepp, 1984). Polyoma-like virus infection was found in the connective tissue, haemocytes and gill epithelium of *M. arenaria* in the US (Sindermann 1990), and in the cytoplasm and nucleus of numerous cell types in *Venerupis philippinarum* (=*Tapes semidecussatus*) in Spain, as an opportunistic viral infection with *Perkinsus olseni* (=*Perkinsus atlanticus*), causing hypertrophy in the case of gill infections (Montes et al., 2001). Viruses similar to Papillomavirus and Polyomavirus were detected in the digestive glands of *Ensis magnus* (=*Ensis arcuatus*) in Spain (Figure 1 and 2) (Ruíz et al., 2011).

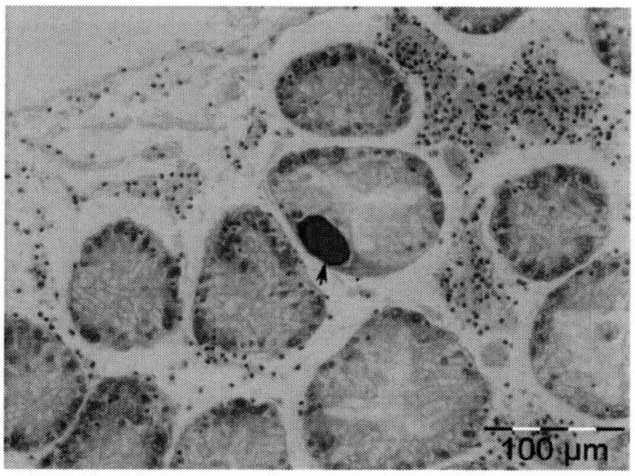

Figure 1. Histological section of *E. magnus* digestive gland with basophilic inclusion (arrow) of viral origin in the epithelium (Harris' Haematoxylin and Eosin, HHE).

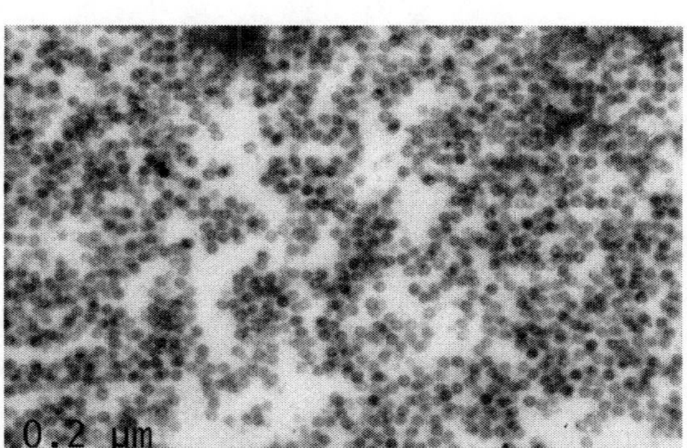

Figure 2. Ultrastructure of viral particles inside basophilic inclusions located in the epithelium of *E. magnus* digestive ducts.

Birnaviruses are non-enveloped bi-segmented double-stranded RNA with icosahedral symmetry, about 60 nm in diameter. Birnavirus-like organisms were detected and associated with high mortalities in *Meretrix lusoria* in Taiwan (reviewed by Bower, 2010) and, without association with mortality, in *Angulus tenuis* (=*Tellina tenuis*), *Mercenaria mercenaria, V. philippinarum* (=*Tapes japonica*) and *Sinonovacula constricta* (McGladdery, 1999; Suzuki and Nojima, 1999).

Picornaviruses are non-enveloped positive-sense singled-stranded RNA viruses with icosahedral symmetry, measuring around 25-45 nm. Picornavirus-like organisms were observed in the secretory cells of the digestive glands of *Paphies ventricosa* in New Zealand, in connective tissue of *Venerupis decussata* (=*Ruditapes decussatus*) and in *Ceratoderma edule* in Spain (Hine and Wesney, 1997; Novoa and Figueras, 2000; Carballal et al., 2003). In the case of *P. ventricosa* and *V. decussata*, the presence of viruses was associated with mortalities (Renault and Novoa, 2004). Virus-like particles similar to picornaviruses and parvoviruses have been associated with brown muscle disease in *V. philippinarum* (=*Ruditapes philippinarum*) populations with mortalities, in France (Dang et al., 2009).

Retroviruses are enveloped dimers of single-stranded RNA viruses around 100 nm in diameter. These RNA viruses have a reverse transcriptase enzyme, which allows the transcription of their RNA into DNA, after which the retroviral DNA is integrated into the chromosomal DNA of the host cell to express there. Retrovirus-like elements were observed in neoplastic cells of *M. arenaria* in the US (Oprandy et al., 1981; House et al., 1998) and in *C. edule* in Spain (Romalde et al., 2007).

Virus particles have been reported hyper-parasitizing pathogenic agents that infect bivalves, including clams. They commonly infect prokaryotic organisms (Figure 3), but also some protozoans and metazoans (Elston, 1997; Darriba et al., 2011). Farley (1981) suggested that the evolution of virus groups generally progresses alongside that of their host, and these bacterial viruses precede and evolve into the phylogenetically advanced viruses of eukaryotic organisms.

Mortality associated with viral infection (herpes-like and irido-like viruses) is very frequent in the larval stages of bivalves. The pathogenicity of viruses in adult clams is not clarified, but environmental and physiological stressors can increase host susceptibility and may result in outbreaks of mortality (reviewed by Renault and Novoa, 2004).

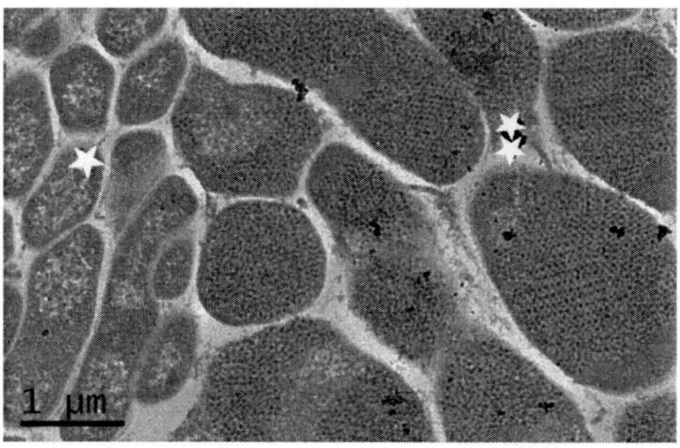

Figure 3. Ultrathin section of rickettsial organisms of *P. virgineus*, infected by virus (**) and uninfected (*).

Montes et al. (2001) suggested that an opportunistic viral infection could contribute to the mortality of clams infected by *P. olseni.*

Several authors have reported an association between viral infection and mortality in clam populations (Renault and Novoa, 2004; Bower et al., 2010b).

Some viruses are considered potentially oncogenic in vertebrates and invertebrates. The viral etiology of disseminated neoplasia was reviewed by Renault and Novoa (2004). The evidence of retroviral etiology for disseminated neoplasia in *C. edule* was reported using electron microscopy and quantification of reverse transcriptase activity (Medina et al., 1993; House et al., 1998; Romalde et al., 2007). However, Aboelkhair et al. (2009) suggested that reverse transcriptase activity is not related to retrovirus, but might be due to the activation of endogenous retro-elements.

5.3. PROKARYOTES

No bivalves are free of bacteria because they feed by filtration, and consequently exhibit a very high bacterial load, composed mainly of Gram-negative bacteria, and, to a lesser extent, Gram-positive bacteria. Bacterial diseases are frequently found in molluscs during their larval stages, but seem to be relatively insignificant in populations of adult animals (with a few notable exceptions). This fact may be explained partially as being due to the primary defence mechanisms of molluscs, phagocytosis and encapsulation, which are particularly well-equipped to fight against small-sized pathogens, and whose resistance may be age-related (Sindermann, 1990).

One of the most important bacterial diseases in clams is "brown ring disease", provoked by *Vibrio tapetis* in the *Venerupis* species (Paillard, 2004). *Vibrio splendidus* and *Vibrio alginolyticus* (Gómez-León et al., 2005), *Vibrio breoganii* and *Vibrio gallaecicus* (Beaz-Hidalgo et al., 2009a, b), *Aliivibrio finisterrensis* (Beaz-Hidalgo et al. 2010), *Vibrio atlanticus* and *Vibrio artabrorum* (Diéguez et al., 2011), and *Vibrio celticus* (Beaz-Hidalgo et al., 2011) have also been identified in clams in Spain. Bacterial infection in clams is dealt with in Chapter 7.

Bacteria belonging to the orders Rickettsiales (phylum Proteobacteria, Alphaprocteobacteria class), Chlamydiales (phylum Chlamydiae, Chlamydiae class) and Mycoplasmatales (phylum Tenericutes, Mollicutes class) were reported for the first time in marine bivalves by Harshbarger et al. (1977) and later reviewed by Fryer and Lannan (1994) and Bower (2004a). The orders Rickettsiales and Chlamydiales include small, Gram-negative prokaryotic organisms and, generally, obligate intracellular parasites. They possess many of the metabolic functions of bacteria but require exogenous co-factors from host cells. Most of them grow readily in the yolk sac of embryonated eggs and in cell cultures. Few species have been grown in artificial media. Rickettsiales organisms are pleomorphic and coccobacillary, and use binary fission for replication, unlike Chlamydiales, which are coccoid and have a complex developmental cycle. Mycoplasmatales are very small prokaryotic organisms, devoid of a cell wall, pleomorphic and Gram-negative; replication is by binary fission, and budding forms and chains of beads can also be observed (Holt et al., 1994). In the literature on bivalve pathology, rickettsia-like and chlamydia-like inclusions, were frequently identified on the basis of light microscopy observations to refer to prokaryotic-like intracellular inclusion bodies, generally basophilic and with a granular aspect. Inclusions located in the epithelium of the digestive gland (Figure 4) and gill (Figure 5) of clams were reported as being chlamydia-like and/or rickettsia-like in *Venerupis philippinarum, Venerupis decussata, Venerupis corrugata* (=*Venerupis pullastra*), *Polititapes virgineus* (=*Venerupis rhomboides*), *Ensis magnus, Ensis siliqua, Solen marginatus, Callista chione, Cerastoderma edule, Isognomon isognomum, Pinna bicolor, Pinna deltodes, Pitar rostratus* and *Leukoma staminea* (=*Prothotaca staminea*) in different countries (Navas et al. 1992; Villalba et al., 1999; Hine and Thorne, 2000; Bower, 2004a; Cremonte et al., 2005a; Delgado et al., 2007; López et al., 2011). These inclusion bodies were located in kidney, haemocytes, gonad and ova in *Mercenaria mercenaria* and *L. staminea* and (Fries and Grant, 1991; Marshall et al., 2003) and were intranuclear in gonad in *Siliqua patula* (Elston, 1986b).

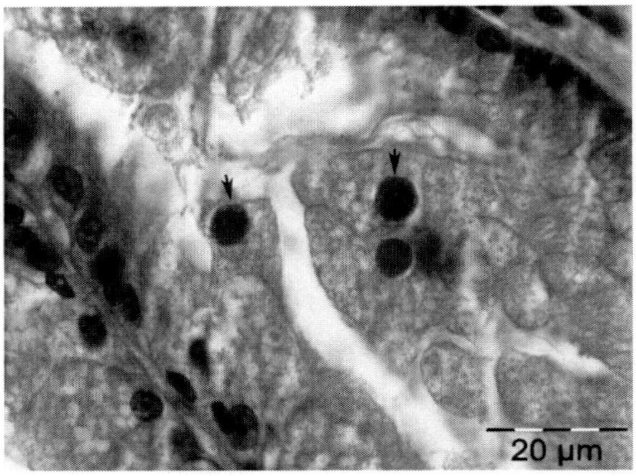

Figure 4. Histological section of V. decussata digestive gland, with intracellular inclusion bodies (arrow), containing prokaryotic-like organism in epithelial cells of digestive tubules (HHE).

Figure 5. Histological section of *V. decussata* gill, with intracellular inclusion bodies containing prokaryotic-like organisms (HHE).

Nevertheless, without ultrastructural data it is only possible to name these inclusions as "intracellular inclusion bodies" or "intracellular prokaryotic-like inclusions". Ultrastructural techniques allow classification on the orders Rickettsiales or Chlamydiales, as in *V. philippinarum*, *V. decussata*, *P. virgineus*, *M. mercenaria*, *Mya arenaria*, *Hippopus hippopus*, *Angulus tenuis*, *Loripes lucinalis*, *Tridacna crocea*, *Meretrix lusoria*, *Donax trunculus* and *S. patula* (Harshbarger et al., 1977; Buchanan, 1978; Comps and Raimbault, 1978; Meyers, 1979, 1981; Joly and Comps, 1980; Comps, 1983a; Elston and Peacock, 1984; Mialhe et al., 1987; Goggin and Lester, 1990; Fries and Grant, 1991; Fries and Grant, 1992; Norton et al., 1993a; Wen et al., 1994; Johnson and Le Pennec, 1995; Villalba et al., 1999).

Some authors provided a speculative approach towards these intracellular procaryotic organisms found in bivalves with rickettsial organisms belonging to genus *Wolbachia* spp., *Coxiella* and *Ehrlichia* (Comps and Raimbault, 1978; Comps, 1983a; Elston, 1986a; Fries and Grant, 1992), but without molecular data it is impossible to achieve this rapprochement. Serological studies have established a group antigenic relationship of chlamydial-like organisms in *M. mercenaria* with known chlamydia of mammalian and avian origin (Meyers, 1979). On the other hand, Elston and Peacock (1984) applied a specific immunofluorescent test for some *Rickettsia* spp. and *Chlamidia* spp. over impression smears of infected digestive gland of *S. patula*. All were negative. In vitro cultures, isolation of these organisms in invertebrates has not been achieved, or they are very limited in the number of passages (Buchanan, 1978). Rickettsial organisms have been identified, such as "*Candidatus* Xenohaliotis *californiensis*" in the gastropod mollusc *Haliotis cracherodii*,(Friedman et al., 2000) and in *Haliotis tuberculata* (Balseiro et al., 2006), using molecular techniques. Further similar studies are necessary in clams to enhance our knowledge of these intracellular prokaryotic inclusions.

The rickettsial and chlamydial infections detected in clams are frequently benign and with no obvious host response (Otto et al., 1979; Elston and Peacock, 1984; Bower, 2004a). Nevertheless, branchial rickettsia-like infections with host reaction have occasionally been associated with high rates of mortality in bivalve populations, including clams (Norton et al., 1993a; Villalba et al., 1999), particularly when the host is under stress. The study performed by Villalba et al. (1999) in *P. virgineus* suggested that mortality was associated to rickettsia-like infection, dismissing environmental factors as a cause.

Virus particles hyperparasitizing prokaryotic organisms may provide a tool for the study of chlamydial and rickettsial genetics as well as provide a mechanism for the control of pathogenic chlamydial organisms. (Otto et al., 1979)

Concerning mycoplasma infections in clams, the only one reported was an unusual branchial mycoplasma-like one in *C. edule* (Azevedo, 1993).

Cysts of prokaryotic-like organisms were detected in gills by light microscopy in several bivalve species including clams. They are basophilic and Gram-negative, and their intracellular location is not clearly defined by light microscopy; they are coated by an eosinophilic fibrous cover (Figure 6), which is interpreted by Gulka and Chang (1984b) as an encapsulation host response. These cysts were reported in *P. virgineus, C. edule, E. magnus* and *S. marginatus* in Spain (Villalba et al., 1999; Carballal et al., 2001; Darriba et al., 2010; López et al., 2011). Despite their intracellularity being unclear, these cysts were reported as rickettsial-like organisms in gills of *Mytilus edulis* and *T. crocea* (Gulka and Chang, 1984b; Goggin and Lester, 1990).

Intracellular bacteria, different from rickettsiales and chlamydiales organisms, were reported in epithelium cells of the siphonal mantle and gill of *Tridacna gigas* clams in Australia and associated with mortalities (Norton et al., 1993b). Montes et al. (2001) reported an opportunistic intracellular bacterial infection with *Perkinsus olseni* in gill epithelial cells of *V. philippinarum* (=*Tapes semidecussata*) in Spain, causing hypertrophy and dysplasia. Colonies of large intracellular rod-shaped bacteria, basophilic and predominantly Gram-positive, were observed in gill epithelial cells in *L. staminea* in Canada (Marshall et al., 2003). Intranuclear prokaryotic inclusions were reported in gill cells in *V. decussata* in Portugal (Azevedo, 1989a). Intranuclear pathogens similar to prokaryotic organisms associated with massive mortalities of *S. patula* were reported by Elston (1986b).

"Hinge ligament disease", caused by a *Cytophaga*-like bacteria, was detected in *M. mercenaria*, *V. philippinarum* and *S. patula* in the US.

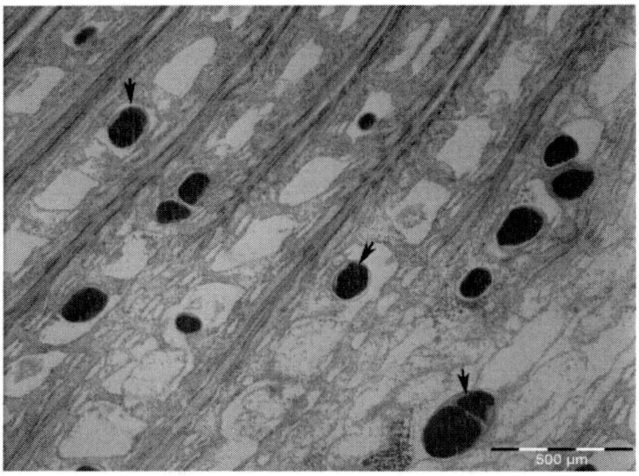

Figure 6. Histological section of *V. decussata* gill, with extracellular basophilic cysts (arrow) of prokaryotic-like organism located in water tubules (HHE).

The disease has little or no effect on healthy growing juveniles (Bower et al., 1994a). Basophilic, Feulgen-positive and Gram-negative inclusion bodies were observed in nearly every tissue of *L. staminea*, bound by haemocytes flattened against the infected cell, forming a thick eosinophilic membrane; the authors suggest that it is an infected, extremely hypertrophied haemocyte (Marshall et al., 2003).

5.4. Fungi

Few cases of fungi infections have been described in marine bivalve molluscs, and those that seem most important in natural populations attack their external parts. One of the severest fungal diseases in marine bivalves was "shell disease", caused by *Ostracoblade implexa* (phycomycete) (Alderman and Jones, 1971). *O. implexa* was observed in shells of dead *Solen* sp. and *Cerastoderma edule* (=*Cardium edule*) in Europe (Lauckner, 1983; McGladdery, 1999). The disease caused by an "Unknown Quahog Parasite" (QPX) was initially reported from wild and hatchery stocks of *Mercenaria mercenaria* in Canada, and also in the US. The taxonomic position of the pathogen agent to the phylum Labyrinthulomycota (=Labrinthomorpha), of the Thraustochytriidae family, was confirmed by molecular analysis (Ragan et al., 2000). QPX infects gills, palps, digestive gland, gonad and mantle, and was associated with 80-90% of the mortalities in *M. mercenaria* juveniles (up to 30 mm in shell length) in a nursery, and up to 100% in hatchery broodstock. QPX disrupts connective tissue throughout the body and is associated with gross lesions (swellings and nodules) in the mantle of infected quahogs and necrotic haemocytes. However, QPX has also been observed in apparently healthy wild adult populations from Atlantic Canada and at some locations in Virginia (Bower, 2010c).

5.5. Protozoa

5.5.1. Phylum Amoebozoa

An unidentified amoeba was reported in gills of *Cerastoderma edule* in Portugal by Azevedo (1997). This infection was associated with haemocytic infection and the presence of necrotic cells, suggesting a pathogenic effect on the cockle population (Azevedo, 1997).

5.5.2. Phylum Apicomplexa

Protozoa of phylum Apicomplexa, Classes Coccidia and Gregarinia, were described as parasites of marine bivalve molluscs in different countries (Lauckner, 1983). The life cycles of these parasites are not well known and there is ongoing examination of their systems. All organisms belonging to Phylum Apicomplexa have gamogonic and sporogonic stages in their life cycles, and some have a merogonic stage. There are species that complete their life cycle in just one host (monoxenous), while in other cases they experience host alternation (heteroxenous) (Lauckner, 1983; Vivier and Desportes, 1990; Desser and Bower, 1997; Desser et al., 1998).

Coccidia are usually detected in kidney tubules, although their presence is reported in gonads, intestine, digestive gland and gill (Lauckner, 1983; Desser and Bower, 1997). Coccidian species reported in clams belong to the *Pseudoklossia* or *Margolisiella* genus (Bower, 2007a; López et al., 2011). According to Desser and Bower (1997), species belonging to *Margolisiella* have monoxenous cycles and spend the gamogonic, sporogonic and merogonic phases in the same host; furthermore, their sporocyst has 2 to 4 sporozoites. Species pertaining to the *Pseudoklossia* genus have heteroxenous cycles (gamogonial and sporogonial in one host and merogonial in another), and the sporocyst has 2 sporozoites. Several stages of their life cycle were detected in different clams (merozoite, macrogamont, microgamont, oocyst, sporocyst and sporozoites) and the stages most commonly observed are the microgamont and macrogamont (Figure 7). *Peudoklossia glomerata* were reported in the kidneys of *Tapes floridus* and *Polititapes virgineus* (=*Tapes rhomboides*) in the Mediterranean Sea; *Pseudoklossia pelneneeri* in *Tellina* sp. and *Donax* sp. in France; *Pseudoklossia* sp. in *C. edule* and *Ensis magnus*; *Margolisiella tellinae* in *Angulus tenuis* in Scotland; *Margolisiella kabatai* in *Leukoma staminea* in Canada; and an unidentified coccidia in kidney was reported in *Scrobicularia plana* in Germany, in *Ensis siliqua* in Spain, in *L. staminea* in Canada, in *Amarilladesma mactroides* (=*Mesodesma mactroides*) in Argentina, and in *Pitar rostratus* in Uruguay (Cremonte and Figueras, 2004; Cremonte et al., 2005a;

Bower, 2007a; López et al., 2011; Montes, 2011). Heavy infection by coccidians may cause kidney damage, epithelial cells become hypertrophied, while kidney tubules could become filled with different stages of coccidians. No mortalities were reported associated with coccidian infections. Bower (2007b) reported an apicomplexa parasite similar to coccidians but located in connective tissue around the gut in *L. staminea* in Canada.

Gregarine species of the Porosporidae family have a life cycle with alternation between two hosts: a crustacean and a mollusc (Lauckner, 1983). *Nematopsis* spp. use bivalve molluscs as intermediate hosts. Our lack of knowledge of the complete life cycle of these species makes their identification difficult. Ovocysts and/or spores of different species of *Nematopsis* genus were reported in the following clam species from different parts of the world: *C. edule*, *Venerupis decussata*, *Cardium lamarcki*, *Hiatella* (=*Saxicava*) *rugosa*, *Tellina* spp., *Venerupis philippinarum*, *L. staminea*, *Nuttallia obscurata*, *Clinocardium nuttallii*, *Saxidomus gigantea*, *E. magnus*, *E. siliqua*, *Solen marginatus* (=*S. vagina*), *Chamelea gallina, Callista chione, P. virgineus, P. rostratus, Paphia undulata* and *Anomalocardia flexuosa* (=*Anomalocardia brasiliana*) (Soto et al., 1996; Villalba et al., 1999; Berrilli et al., 2000; Canestri-Trotti et al., 2000; Tuntiwaranuruk et al., 2004; Cremonte et al., 2005a; Bower, 2007b; Boehs et al., 2010; López et al., 2011; Montes, 2011). The stage most commonly observed is the oocyst, free or intracellular in all tissues (labial palps, gill, mantle, digestive gland, gonad, kidney and pericardial gland) (Figure 8). Oocysts (resistant phase) have one or two sporozoites (infective stage) inside, depending on the species.

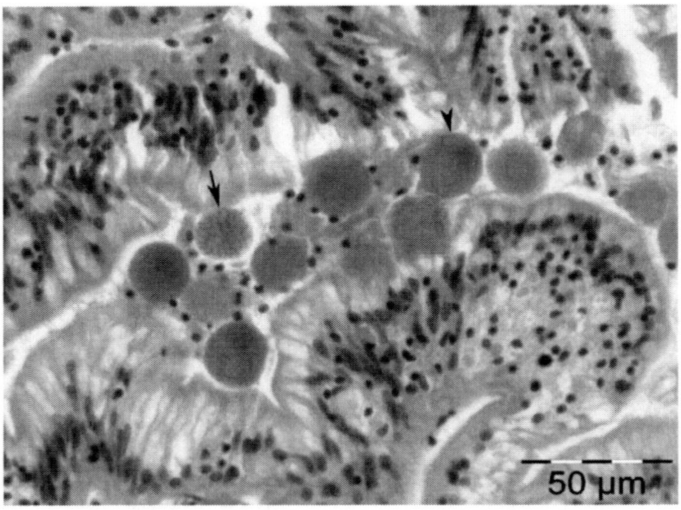

Figure 7. Histological section of *E. magnus* kidney with presence of coccidia. Macrogamonts (arrowhead) and microgamonts (arrow) in the lumen of kidney tubules (HHE).

Usually, *Nematopsis spp.* has no important pathological effects and was only associated to mortalities in the *C. edule* population in Portugal (Azevedo and Cachola, 1992). Oocysts of *Porospora* sp. were reported in *C. gallina* and in *C. chione* in Italy (Berrilli et al., 2000; Canestri-Trotti et al., 2000).

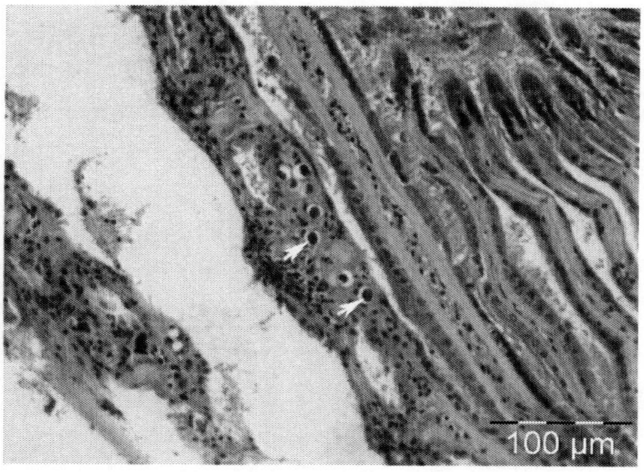

Figure 8. Histological section of *P. virgineus* gill, with oocysts of *Nematopsis* sp. (arrow) (HHE).

A non-identified gregarine was detected in intestinal epithelium and surrounding connective tissue (Figure 9) in *P. virgineus C. edule, E. magnus* and *E. siliqua* in Spain (Villalba et al., 1999; Carballal et al., 2001; López et al., 2011; Montes, 2011), in *V. philippinarum* in Canada (Bower et al., 1992) and in *A. mactroides* in Argentina (Cremonte and Figueras, 2004). A similar protozoan was identified by other authors as a coccidium; however, it is impossible to distinguish whether it is a gregarine or a coccidium, since only one phase of its life cycle has been observed (López et al., 2011).

5.5.3. Phylum Perkinsozoa

Genus *Perkinsus* is a major threat to shellfish farming. The parasitisation of bivalves of bivalves and gastropods by *Perkinsus* species associated with mass mortalities is well documented (Villalba et al., 2004).

Specifically in clams, *Perkinsus* spp. has been associated with mass mortalities of *V. decussata* from the south coast of Portugal, and with epizootic mortalities of *V. philippinarum* in Korea, China and Japan (Villalba et al., 2004; Choi and Park, 2005).

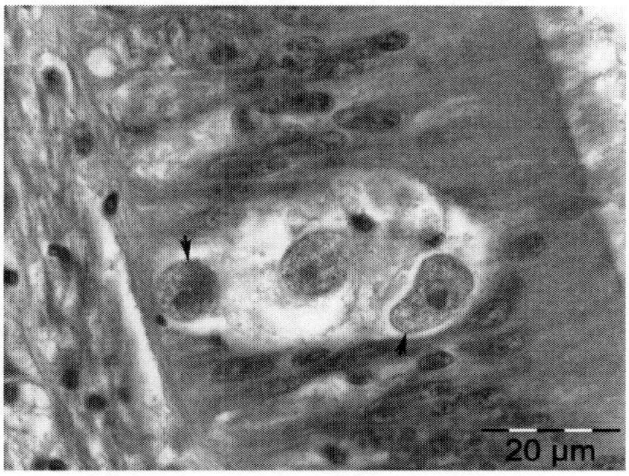

Figure 9. Histological section of *V. decussata* digestive gland. Unidentified gregarine (arrow) at the base of the epithelium of a digestive gland (HHE).

Since the first description of *Perkinsus marinus* as a fungus in 1950, *Dermocystidium marinum* (Mackin et al., 1950), the taxonomic position of *Perkinsus* spp. has been changing according to ultrastructural and genetic studies. Species of the genus *Perkinsus* have been correspondingly included in different taxa: Ascomycetes, Entomophthorales, Saprolegniales, Haplosporidia, Labyrinthulomycetes, Apicomplexa and Dinoflagellate (Villalba et al., 2004). Perkinsids are currently considered a group close to Dinoflagellates, both sharing a common node with Apicomplexa (Saldarriaga et al., 2003). *Perkinsus* spp. are included within the phylum Perkinsozoa, proposed by Norén et al. (1999), which today also includes genera *Rastrimonas* and *Parvilucifera*, both parasites of micro-eukaryotes (Robledo et al., 2011).

Ten species of *Perkinsus* have been described, but only seven have been accepted (Villalba et al., 2004; Moss et al., 2008). Three species are currently recognised as affecting clams and cockles: *Perkinsus olseni* was initially described in *Haliotis rubra rubra* (=*Haliotis ruber*) (Lester and Davis, 1981), *Perkinsus chesapeaki* in *Mya arenaria* (McLaughlin et al., 2000a) and *Perkinsus honshuensis* in *V. philippinarum* (Dungan and Reece, 2006). Two other species, *Perkinsus atlanticus*, described in *V. decussata* (Azevedo, 1989c), and *Perkinsus andrewsi* in *Macoma balthica* (Coss et al., 2001) are considered synonyms of *P. olseni* (Murrell et al., 2002) and *P. chesapeaki* (Burreson et al., 2005) respectively. However, the synonymy *P. chesapeaki / andrewsi* is still under discussion (Pecher et al., 2008).

P. olseni (=*P. atlanticus*) is known for its wide host range, including bivalves and gastropods, and its widespread distribution. Susceptible clam species are *Anadara trapezia, Austrovenus stutchburyi, V. decussata, V. philippinarum, Venerupis corrugata, Polititapes aureus* (=*Paphia aurea*) (reviewed by Villalba et al., 2004), *P. undulata* (Leethochavalit et al., 2003; Leethochavalit et al., 2004), *P. rostratus* (Cremonte et al., 2005), *Protothaca jedoensis* (Park et al., 2006), and *Tridacna crocea* (Sheppard and Phillips, 2008).

The distribution of *P. olseni* includes the European Atlantic coast, Mediterranean Sea (reviewed by Villalba et al., 2004), including the Adriatic Sea (Abollo et al., 2006). *P. olseni* is also present in Australia, Japan (reviewed by Villalba et al., 2004), New Zealand (Dungan et al., 2007b), Uruguay (Cremonte et al., 2005), Thailand (Leethochavalit et al., 2003, 2004), China (Zhang et al., 2005), Korea (Park et al., 2005, 2006) and Vietnam (Sheppard and Phillips, 2008).

P. chesapeaki (=*P. andrewsi*) has been described in *M. arenaria* (McLaughlin et al., 2000), and also affects other clam species: *M. balthica*, *Tagelus plebeius*, Macoma itchelli, Mercenaria mercenaria, *Mulinia lateralis*, *Rangia cuneata*, *Cyrtopleura costata* (Burreson et al., 2005; Reece et al., 2008). *P. chesapeaki* (=*P. andrewsi*) is present in Chesapeake Bay (McLaughlin and Faisal, 2000b) and Delaware Bay (Bushek et al., 2008).

P. honshuensis has been described, together with *P. olseni*, as infecting Japanese Manila clams *V. philippinarum* (Dungan and Reece, 2006). Based on the hypothesis that *P. olseni* could be transferred to Europe by infected Asian *V. philippinarum* introduced during the 1970's (Le Borgne, 1996), Dungan and Reece (2006) suggested that *P. honshuensis* may also have been imported to Europe.

This hypothesis involves systematic surveying of *P. honshuensis* in those potential European host populations where *Perkinsus* sp. has been described (Dungan and Reece, 2006).

Perkinsus spp. life cycle is direct, without an intermediate host (reviewed by Villalba et al., 2004). The proposed life cycles (Villalba et al., 2004; Robledo et al., 2011) are based on the three stages of the parasite observed: i) intra- or extra-cellular vegetative propagation within host tissues (trophozoite) (Figure 10), ii) the transformation into pre-zoosporangia (hypnospore) after its release to the environment via faeces (Bushek et al., 2002) or death host (Ragone-Calvo et al., 2003) (Figure 11), and iii) subsequent zoosporulation and release of zoospores (Figure 11).

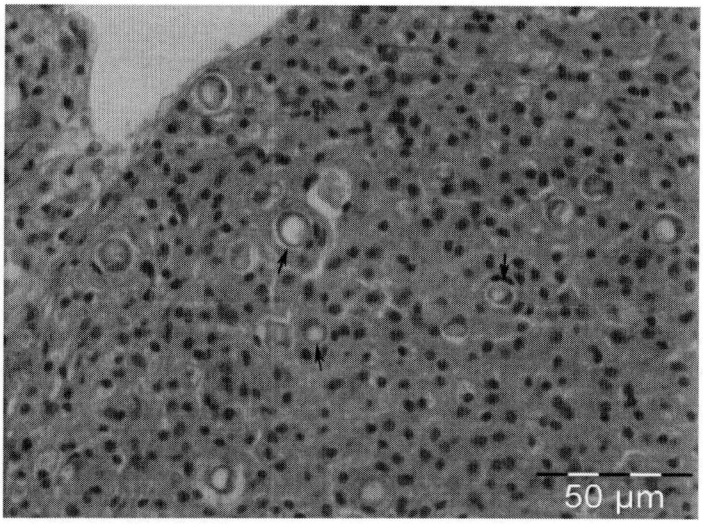

Figure 10. Histological section of *V. decussata* gill, with trophozoites (arrow) of *P. olseni* (HHE).

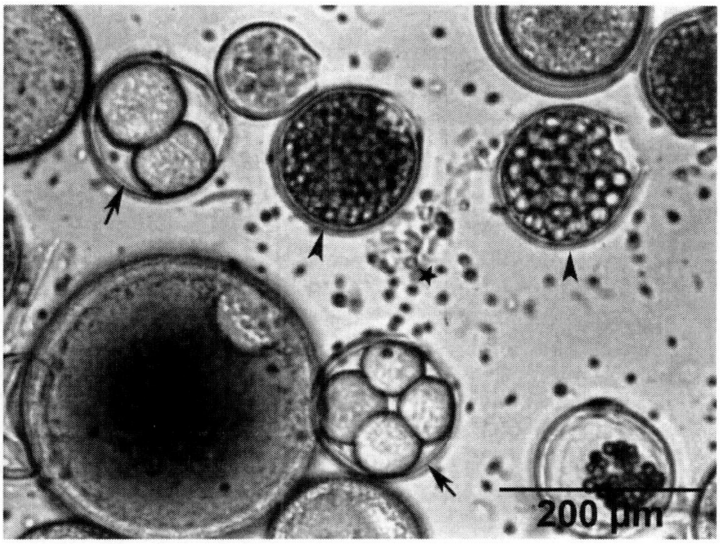

Figure 11. Micrographs with progression of zoosporulation of *P. olseni*. Zoosporangia starting sporulation (arrow), zoosporangia with spores (arrowhead), and free spores (*).

Trophozoites, hypnospores and zoospores are infective (Volety and Chu, 1994; Rodríguez et al., 1994; Ford et al., 2002). While the natural infecting phase is not well known (Chu, 1996), recognition mechanisms and phagocytosis by haemocytes play a crucial role in the initiation and development of the infection (Tasumi and Vasta, 2007). After phagocytosis, the *Perkinsus* trophozoite remains alive, multiplying and causing the destruction of the haemocytes, with the consequent spread of infection (Robledo et al., 2011).

A sexual phase was suggested by Perkins (1996), after observing the behaviour of hypnospores *in vitro*. Recent microsatellite analysis suggests sexual and asexual reproduction in *P. marinus* (Thompson et al., 2011) and *P. olseni* (Vilas et al., 2011).

Classical histological diagnosis is based on direct observation of free, phagocyted or encapsulated trophozoites in the connective tissue of the host. Parasite cells exhibit typical large vacuoles showing a "signet ring" appearance (Figure 10). Due to the clam inflammatory reaction, trophozoites can be embedded in dense, fibrous extracellular material (Chagot et al., 1987; Azevedo, 1989; Montes et al., 1995a). Montes et al. (1995b, 1996) demonstrated that the main component of the embedding material is a non-glycosylated polypeptide *P. olseni*/*P. atlanticus* infecting *V. decussata* and *V. philippinarum*. This polypeptide is secreted by the surrounding haemocytes and is likely to destroy the parasite. These cell aggregations are visible as whitish cysts on the gills, labial palps and mantle of heavily infected clams. These granulocytomes can produce severe injuries, the loss of tissular structure and organ dysfunction, and finally the death of the host. Sub-lethal effects caused by *Perkinsus* spp. infections include decreased growth, loss of condition and inhibition of gametogenesis. These negative effects can result in lower recruitment and scarcity of seed, with consequent implications on natural beds and the shellfish industry (Villalba et al., 2004).

Transmission electron microscopy (TEM) has been used to describe the cellular ultrastructure in different phases of *Perkinsus* spp. and host reaction (reviewed by Villalba et al., 2004). However, this technique cannot really be considered as a valid diagnostic method due to the very small size of the sample (Garcia et al., 2008).

The most widespread method of diagnosis is the incubation of host tissues in saline fluid thioglycollate medium (Ray's FTM) (Ray, 1966). Different tissues can be used (heart, rectum, gill, mantle, labial palp or haemolymph), with the gill as the one preferred for diagnosis in clams. After an incubation time of 7 days in Ray´s FTM, the enlarged trophozoites transformed into hypnospores can be stained with Lugol's iodine, and observed as blue-black spheres under light microscopy (Figure 12). This simple method allows a semi-quantitative assessment of the intensity of infection, and has been adopted as a standard diagnostic procedure for *Perkinsus* spp. (Villalba et al., 2004). However, Ray´s FTM method cannot be used to distinguish between *Perkinsus* species (Elandaloussi et al., 2009). In addition, more specific and sensitive molecular diagnostic methods have been developed for *Perkinsus* spp.

Polyclonal and monoclonal antibodies against *Perkinsus* spp. or their extracellular products have been developed for immunohistochemical diagnosis (reviewed by Villalba et al., 2004), environmental analysis (Ragone Calvo et al., 2003; Park et al., 2010) and host-parasite interactions (Earnhart et al., 2005). However, some antigens showed cross-reaction between *Perkinsus* spp. and with dinoflagellate species (Bushek et al., 2002b); therefore, specificity has to be evaluated by appropriate validation assays (Villalba et al., 2004).

The characterization of the ribosomal RNA locus of *Perkinsus* spp. has proved to be essential to design genus- or species-specific primers for PCR diagnosis and probes for *in situ* hybridisation (ISH). Several primers and probes have been used for specific diagnosis, and sequences have also been used to identify new species and establish phylograms (Villalba et al., 2004; Bower, 2011).

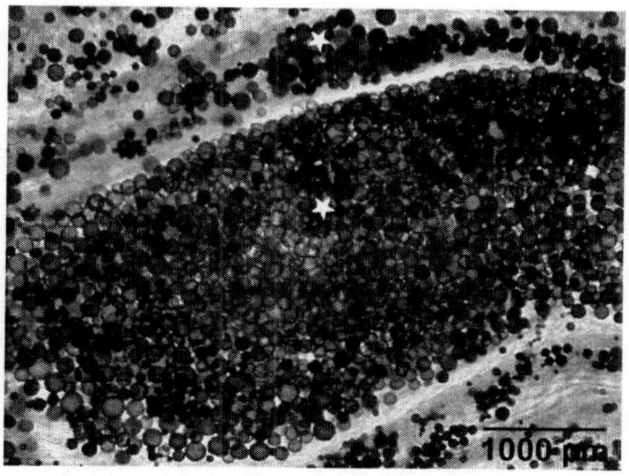

Figure 12. Micrographs of *V. decussata* gill after incubation in Ray´s FTM, showing hypnospores (*) (Lugol's iodine).

Several procedures have been described for *P. olseni* diagnosis by PCR (de la Herran et al., 2000; Robledo et al., 2000; Moss et al., 2006), and *in situ* hybridisation (ISH) (Moss et al., 2006).

Coss et al. (2001) and Pecher et al. (2008) have established PCR diagnoses for *P. chesapeaki* (=*P. andrewsi*). Generic *Perkinsus* sp. PCR diagnosis (Robledo et al., 2002) and ISH detection (Elston et al., 2004) have also been developed. The PCR technique has also been combined with immunodetection (ELISA) for diagnosis of *P. marinus* and *P. olseni* (=*P. atlanticus*) (Elandalloussi et al., 2004), or with restriction fragment length polymorphism (RFLP) for the differential diagnosis of *P. chesapeaki*, *P. marinus*, *P. olseni* and *P. mediterraneus* (Abollo et al., 2006).

There are no specific methods for controlling *Perkinsus* spp. infection in clams. Consequently, it is essential to carry out a control of broodstocks and seeds prior to introducing them into culturing areas. Different strategies have been proposed to control *P. marinus* infections in *Crassostrea virginica* and mitigate their consequences in shellfish industries (Andrews and Ray, 1988; Krantz and Jordan, 1996). Epidemiological data are essential to decide modifications in management procedures. Low salinity waters hinder the spread of, and warm temperatures have a large impact on, transmission and intensification of the infection and mollusc mortality (Villalba et al., 2004). Selective breeding to obtain resistant strains, triploid culturing, genetic engineering and the development of chemotherapeutants are the main research lines in the fight against *Perkinsus* (Villalba et al., 2004).

5.5.4. Phylum Haplosporidia

Important diseases in bivalve molluscs (bonamiosis, haplosporidiosis and mikrocitosis) are caused by parasites belonging to the Phylum Haplosporidia. The haplosporidian species reported in clams belong to *Minchinia* and *Haplosporidium* genus and some unidentified haplosporidia (Bower, 2007c; López et al., 2011). The haplosporidian stages described in bivalves are multinucleate plasmodia (Figure 13), sporocysts and spores (Figure 14). Specific characters of spores (ornamentation) observed with transmission and scanning electron microscopy can be used to distinguish the two genera. Species with spore ornamentation composed of epispore cytoplasm extensions must be placed in genus *Minchinia*, and species with spore ornamentation derived from the spore wall placed in *Haplosporidium* genus (Burreson et al., 2004). Recent molecular phylogenetic analysis (Reece et al., 2004) supports the importance of the ontogenic origin of spore ornamentation.

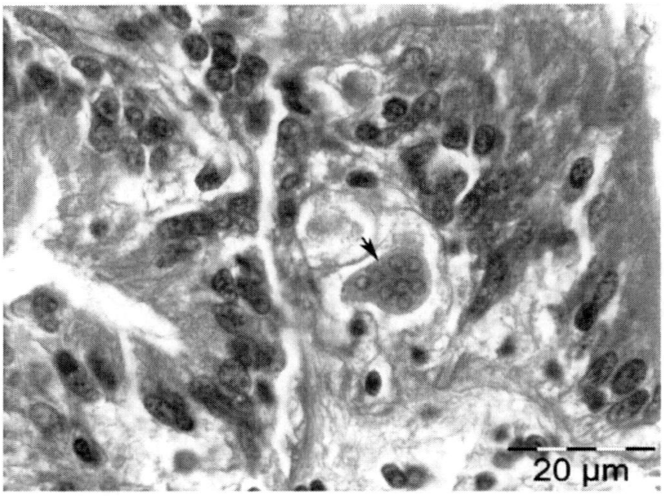

Figure 13. Histological section of *V. decussata* digestive gland. Plasmodium (arrow) of a non-identified haplosporidian in the base of the epithelium of a digestive duct (HHE).

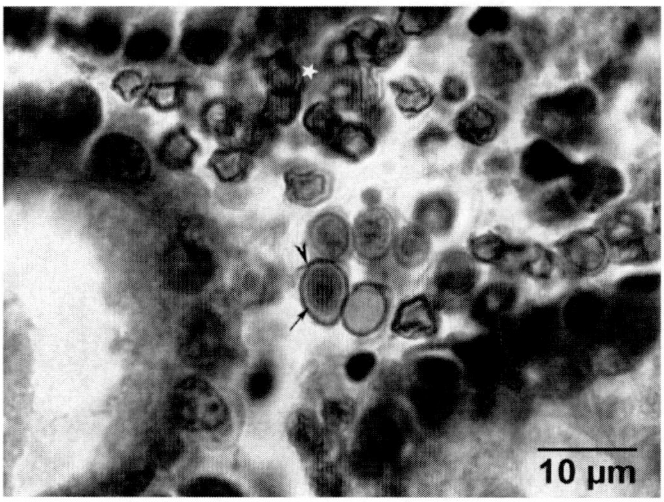

Figure 14. Histological section of *V. decussata* digestive gland with *Minchinia tapetis* spores (arrow) with operculum (arrowhead) and free spores (*).

Minchinia-like, *Haplosporidium*-like and unidentified haplosporidia were reported in different clam species: *V. decussata*, *P. aureus* (=*Venerupis aureus*), *V. philippinarum*, *C. edule*, *Tresus capax*, *E. magnus* and *S. marginatus* from Spain and Japan (Bower, 2007c; López et al., 2011). These authors have generally used only light microscopy, and most of them have observed only the stage of plasmodium. Spores found in different tissues (gill, mantle and foot tissues) of *V. decussata* from Portugal were identified as *Minchinia tapetis* (=*Haplosporidium tapetis*) (Azevedo, 2001), using transmission and scanning electron microscopy. Using the same techniques Azevedo et al. (2003) have identified *Haplosporidium edule* in *C. edule* from Spain.

Haplosporidium spp. (*H. costale* and *H. nelsoni*) have been associated with considerable mortalities in oyster populations in the US (Burreson and Ford, 2004). Azevedo (2001) detected *Minchinia tapetis* in a *V. decussata* population with high mortalities, but information about the factor causing the mortality was not provided.

Recently, *Mikrocytos* sp. infecting *Donax trunculus* were associated to mortality episodes in France (García et al., 2011).

5.5.5. Phylum Paramyxea

Paramyxean parasites are an important group of protozoa infecting marine invertebrates, particularly those included in genera *Marteilia* and *Marteilioides*. Marteliosis is a disease of the digestive gland of bivalve molluscs caused by *Marteilia refringens*. It is a disease listed as non-exotic, according to European legislation (Directive 2006/88/ CE), and of obligatory notification according to Directive 2006/88/CE and the aquatic code of the World Organisation for Animal Health (OIE, 2011). The first studies of *Marteilia* genus placed it in Phylum Ascetospora, together with haplosporidia (Paramyxea class and Stellatosporea class respectively) (reviewed by Desportes and Ginsburger-Vogel, 1981), whilst in later studies it was found necessary to create a new Phylum Paramyxea (reviewed by Desportes and Perkins, 1990; Berthe et al., 2000). One characteristic of the species that belong to this new Phylum is internal cleavage to produce cells within cells during sporulation. There are references to the presence of different *Marteilia* species in different bivalve molluscs worldwide (Lauckner, 1983; Berthe et al., 2004; López-Flores et al., 2008a; Bower, 2010a). With reference to clams, *Marteilia christenseni* was described in *Scrobicularia plana* (=*S. piperata*) in France (Comps, 1983b); *M. refringens* in *S. marginatus* and *C. gallina* in Spain (López-Flores et al., 2008a, b). Unidentified *Marteilia* were reported in *C. edule*, *P. virgineus*, *V. corrugata* (=*Tapes pullastra*), *Ensis minor* and *E. siliqua*, *V. philippinarum* and *Tridacna maxima* in Europe and Asia (Berthe et al., 2004; Bower and Itoh, 2011). Recently, a molecular study of *Marteila* affecting *C. edule* on the Spanish Mediterranean coast suggests the possibility of there being a new species (Carrasco et al, 2012).

Light microscopy allows observation of initial (Figure 15) and more advanced stages in clam tissues (Figure 16), whilst with electron microscopy it is possible to observe the internal cleavage process which produces cells within cells during sporulation (Figure 17).

This disease has caused very high mortalities in the European flat oyster *Ostrea edulis* populations. However no mortalities in clam populations were associated with *Marteilia* spp. The location of the *Marteilia*-like infection in the kidney of *T. maxima* is of interest since *Marteilia* spp. infections consistently occur in the digestive gland (Norton et al., 1993c). Norton et al. (1993c) suggested that a *Marteilia*-like infection in *T. maxima* is potentially pathogenic due to the tissue damage observed.

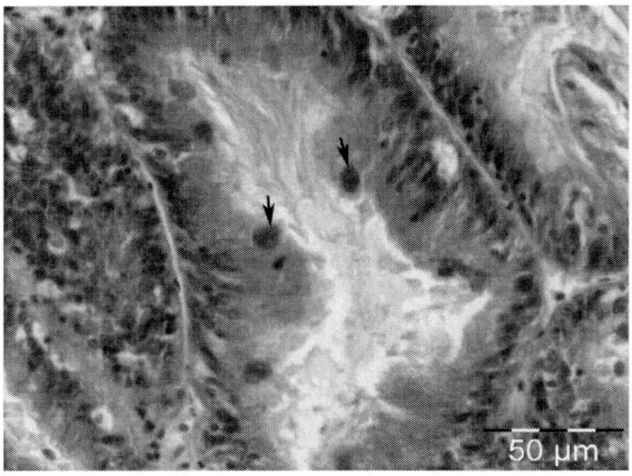

Figure 15. Histological section of *S. marginatus* digestive gland parasitized by initial stages of *Marteilia* sp. (arrow) (HHE).

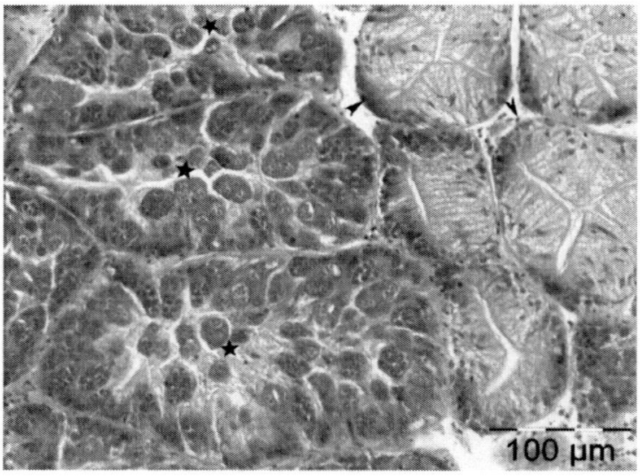

Figure 16. Histological section of *S. marginatus* digestive gland parasitized by advanced stages of *Marteilia* sp. Digestive tubules infected by *Marteilia* sp. (asterisk) and uninfected digestive tubules (arrowhead) (HHE).

Despite the *Marteilia* spp. life cycle being as yet unknown, recent studies have shown that the dynamics of *Marteilia refringens* and zooplankton appear to be linked (Audemard et al., 2002; Carrasco et al., 2007). They suggest complementary transmission experiments to prove the role of copepods as potential intermediate hosts, while other non-planktonic species may be taken into consideration as potential hosts, such as Cnidaria and Nematoda species, because these were also found to be infected.

Marteilioides sp., phylum Paramyxea, was also reported in *V. philippinarum* in Korea and Japan, infecting ovocytes, though prevalence was low and no host response was observed in either case (Bower and Itoh, 2005). Park et al. (2003) suggested that infection by *Marteiloides chungmuensis* in oyster ovocytes may have an adverse impact on metabolic recovery after spawning.

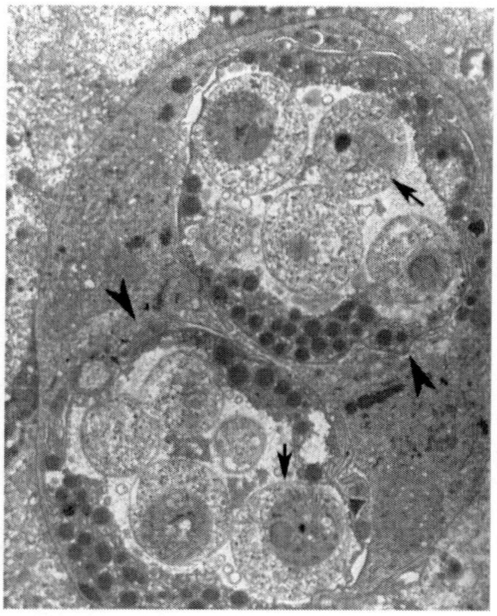

Figure 17. Ultrastructure of *Marteilia* sp. in *S. marginatus*. Sporangium (arrowhead) with spores (arrow) (3000x).

5.5.6. Phylum Microspora

Few members of the phylum Microspora are known to parasitize bivalve molluscs. In relation to clams, unidentified microsporidians were reported in epithelial cells of the digestive gland in *V. decussata* and *P. virgineus* in Spain (Navas et al., 1992; Villalba et al., 1999) and in *C. edule* in France (Bower et al., 1994b). A *Steinhausia*-like microsporidian (Figure 18) was observed in ovocytes of *Macoma baltica* in the US (Bower et al., 1994b) and of *C. edule* in France and Spain (Bower, 2007d). In general, there is no great prevalence or intensity; however, high intensity could affect the viability of infected ovocytes and hence fecundity. *Urosporidum* sp. were observed hyperparasitizing *Paravortex cardii* in *C. edule* in Spain (Carballal et al. 2005).

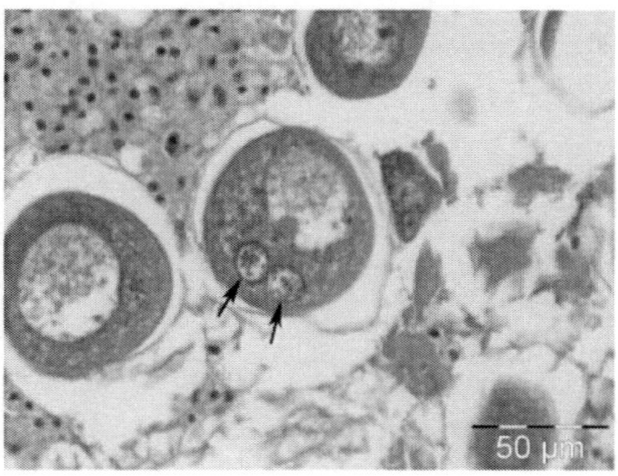

Figure 18. Histological section of *V. philippinarum* gonad with sporoblasts (arrow) of *Steinhausia*-like microsporidian inside an ovocyte (HHE).

5.5.7. Phylum Ciliophora

Ciliates in marine bivalves can be classified into three types adapted to live in the pallial cavity: 1) filter feeders utilizing the suspended food items made available by the ciliary currents of the hosts, 2) particle feeders collecting food and/or mucus from the gills or mantle epithelium, or 3) parasites feeding on the contents of the epithelial cells of the gills of hosts (Lauckner, 1993). The paleal cavity of clams is frequently infected by ciliates. Several species belonging to order Rhynchodida, such as *Ancistrocoma* spp., were observed in gills of *Mya* spp., in *Cryptomya californica* and in *Meretrix meretrix* (Bower et al., 1994c; Xu et al., 2011); *Sphenophrya* spp. and other unidentified rhynchodida-like species (Figure 19) were reported in gills of *Mya* spp., *C. edule*, *C. lamarcki*, *Dosinia exoleta*, *Timoclea ovata*, *Corbula gibba*, *Spisula solidissima*, *Sphaerium corneum*, *V. philippinarum* and *Pinna bicolor* (Bower et al., 1994d; Carballal et al., 2001; Hine and Thorne, 2000). *Trichodina* spp. and other unidentified trichodinid ciliates (order Mobilida) (Figure 20) were observed in *C. edule*, *C. lamarcki*, *Mya arenaria*, *M. balthica*, *C. gallina*, *A. mactroides*, *V. philippinarum*, *V. decussata*, *V. corrugata*, *P. aureus*, *Ensis* spp., *Solen* spp. and *T. plebeius* (Navas et al., 1992; Bower et al., 1994e; Berrilli et al., 2000; Cremonte and Figueras, 2004; Vázquez et al., 2006; Lopez et al., 2011). *Ancistrum* sp. and *Boveria* sp. (order Thigmotrichida) were detected in *C. gallina* (Berrilli et al., 2000). *Planeticovorticella paradoxa* (order Sessilida) was reported in gills of *M. meretrix* (Xu et al., 2011).

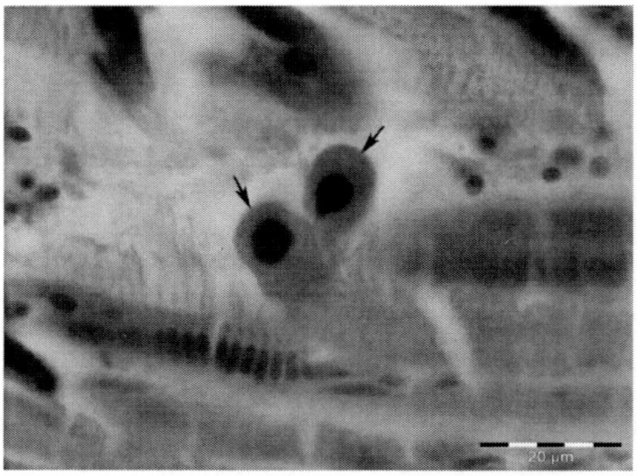

Figure 19. Histological section of *V. philippinarum* gill with *Rynchodida*-like ciliates (arrow) attached to branchial filaments (HHE).

Unidentified ciliates (Figure 21) were reported in *V. decussata*, *V. philippinarum*, *V. corrugata*, *P. aureus*, *P. virgineus*, *S. marginatus* and *T. plebeius* located in gills and pallial cavity (Navas et al., 1992; Villalba et al., 1999; Vázquez et al., 2006; Lopez et al., 2011) and located in the epithelium of the digestive gland in *P. virgineus* (Villalba et al., 1999). In general, no obvious host response was described associated to ciliates; however, some authors suggested that they may have an effect under adverse growing conditions in the case of *Ancistrocoma* spp. (Bower et al., 1994c); mortalities of *C. edule* associated with the presence of *Trichodina* ciliates were referred in both the German and Dutch Wadden Sea (Bower et al., 1994e).

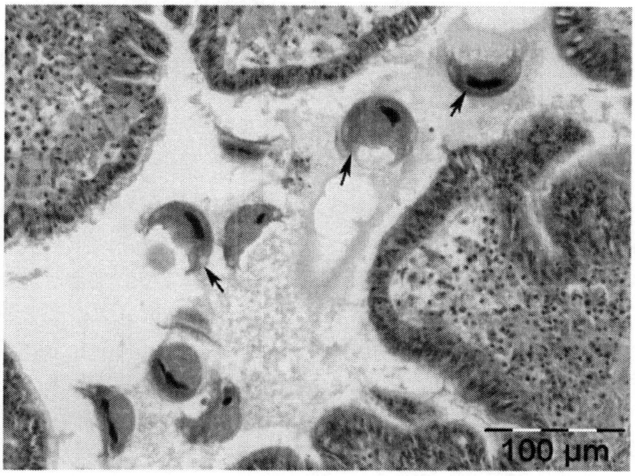

Figure 20. Histological section of *Trichodina* sp. ciliates (arrow) in *S. marginatus* pallial cavity (HHE).

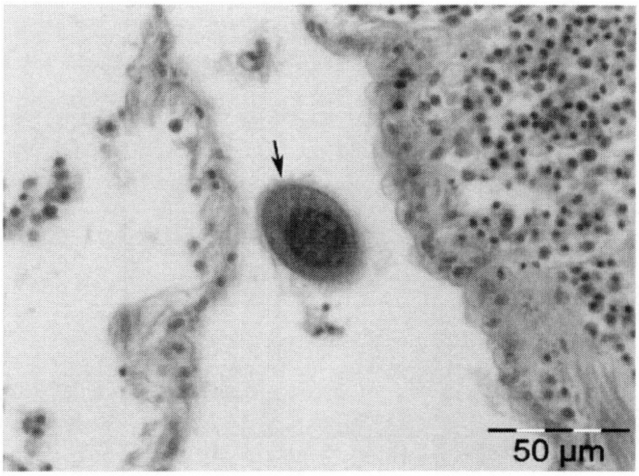

Figure 21. Histological section of an unidentified ciliate (arrow) in *S. marginatus* pallial cavity (HHE).

5.5.8. Protophyta

There are associations between marine bivalves and microalgae, ranging from mutualism to true parasistism. Lauckner (1983) reported some associations between burrowing algae and clams, as in the case of the mutualist association of cyanophytes with *C. edule* and *Mya arenaria.* Some algae can invade soft parts of clams, as in the case of zooxanthellae (dinoflagellate) invading *Tridacna* sp. (Lauckner, 1983). However, an alga similar to *Coccomyxa parasitica* (chlorophyceae) appears to be parasitic (facultative) and was reported in *C. nuttallii* (Lauckner, 1983) and in *Panopea abbreviata* (Vázquez et al., 2010). Macroscopically soft parts of clams infected by *Coccomyxa* present a green color, indicating algal colonization. Algae enter the bivalve by way of normal feeding, and they are resistant to the digestion process. Host phagocytosis contributes to the spread of algal cells.

5.6. METAZOA

5.6.1. Phylum Plathelmintha

5.6.1.1. Turbellaria

Turbellaria are predominantly free-living predators, different species being present in marine molluscs, including clams (Lauckner, 1983). Members of Rhabdocoela and Alloeocoela are the most commonly reported in clams, located in the lumen of the digestive gland and mantle cavity, respectively (reviewed by Lauckner 1983; Bower, 2004b). In many cases, these species are considered as commensals rather than authentic parasites (Lauckner, 1983). *Urastoma*-like species (Figure 22) were observed in the cavity of mantle and gill in *Ensis magnus* from Spain (Darriba et al., 2010) and *Urastoma cyprinae* in *Tridacna gigas* and *Tridacna maxima* from Australia (reviewed by Bower, 2004b). *Paravortex cardii* were reported in the gut lumen of *Cerastoderma edule* in Spain by Carballal et al. (2001). A *Paravortex*-like species (Figure 23) was detected in the lumen of the digestive gland of *Venerupis decussata, Venerupis corrugata, Venerupis philippinarum, Polititapes aureus, Polititapes virgineus* and *Solen marginatus*, from Spain (Navas et al., 1992; Villalba et al., 1999; López et al., 2011).

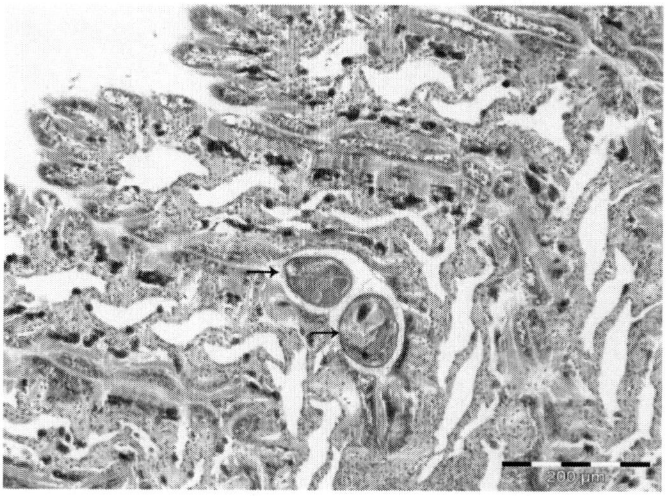

Figure 22. Histological section of *E. magnus* gill with *Urastoma*-like turbellaria (arrow) (HHE).

An unidentified Rhabdocoela turbellaria was observed in the gut lumen of *V. philippinarum* from Canada and in numerous clam species (Bower et al., 1992; Bower, 2004b), as well as in *Amarilladesma mactroides* from Argentina (Cremonte and Figueras, 2004). An unidentified turbellaria was reported in the gut lumen of *V. philippinarum*, and two turbellaria in kidney tubules of *Nuttallia obscurata* in Canada (Marshall et al., 2003). Turbellaria are considered to be mid-way between endocommensal and parasitic, with no known damaging effect on hosts (Lauckner, 1983).

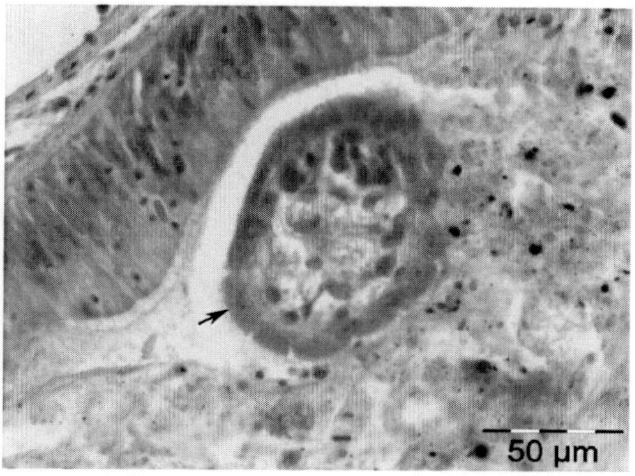

Figure 23. Histological section of *S. marginatus* digestive gland with *Paravortex*-like turbellaria (arrow) inside gut (HHE).

5.6.1.2. Trematoda

The life cycles of trematoda are very complex, with sexual and asexual reproduction phases. Larval stages parasitize a minimum of two hosts, (one of them frequently being a mollusc) which usually act as primary hosts - infected by rediae or sporocysts containing cercariae larval stages (Figure 24) - or as secondary hosts - infected by metacercariae larval stages (Figure 25). At times, they can also be the definitive hosts (Lauckner, 1983).

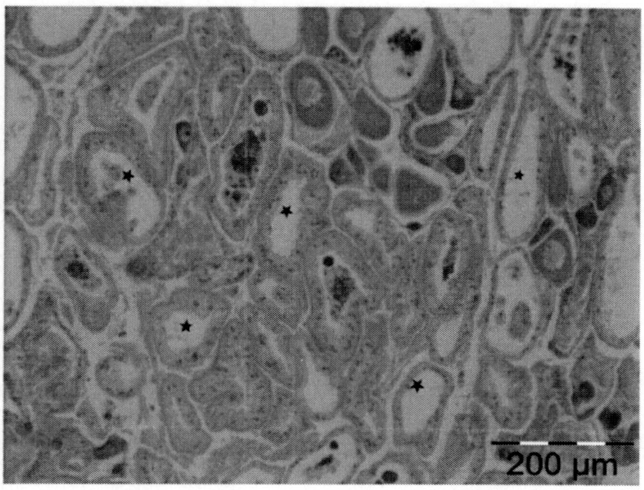

Figure 24. Histological section of *S. marginatus* gonad. Sporocysts (*) of trematode causing gonad castration (HHE).

Larval stages of digenean trematoda have been described in almost all marine bivalve species, so it is no exaggeration to consider this group the most important metazoan parasite affecting these molluscs (Lauckner, 1983).

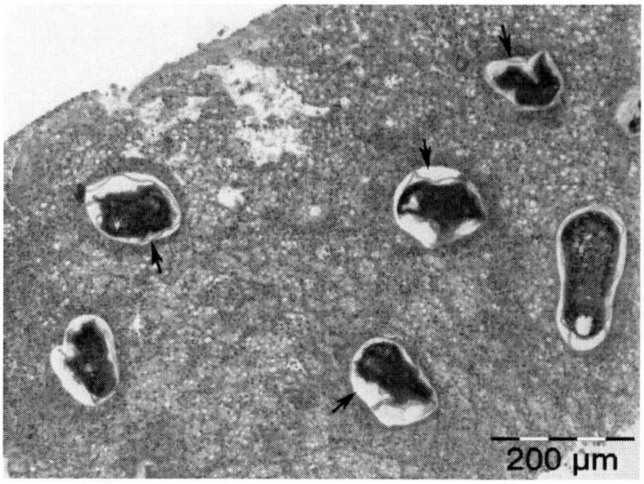

Figure 25. Histological section of *S. marginatus* pericardial gland. Encysted metacercaria (arrow) (HHE).

Larval stages of Digenea trematodes belonging to Bucephalidae, Sanguinicolidae, Monorchiidae, Fellodistomidae, Gymnophallidae, Lepocreadiidae, Zoogonidae, Echinostomatidae, Psilostomatidae and Renicolidae families have been described (Lauckner, 1983; Abdul-Salam and Sreelatha,1998; Bower, 2007e), affecting several clam species. In other papers, sporocysts and cercariae of Bucephalidae trematoda were referred in *Tridacna crocea* from Australia (Shelley et al., 1988), in *Pleuromeris armilla*, *Rangia cuneata*, *Anadara brasiliana* and *Periploma margaritaceum* from the US (Wardle, 1988, 1990), in *Anomalocardia flexuosa* from Brazil (Boehs et al., 2010) and in *E. magnus* from Spain (Darriba et al., 2010; Ruíz et al., 2012). Larval stages of species belonging to the family of Fellodistomidae were reported, such as *Bacciger bacciger* in *Donax trunculus* from Spain (Ramón et al., 1999) and *Proctoeces* sp. in *Malleus meridianus* from Australia (Hine and Thorne, 2000); unidentified fellodistomid and gymnophallid cercariae were detected in *Tagelus plebeius* from Argentina (Vázquez et al., 2006). Sporocysts of an unidentified trematode located in gonad, mantle, gills and kidney were reported in *Ameghinomya antigua* (=*Protothaca (=Venus) antiqua*) from Argentina (Cremonte et al., 2005b). Metacercariae of Gymnophallidae trematoda were detected in *A. antiqua* and *T. plebeius* from Argentina (Cremonte et al., 2005b; Vázquez et al., 2006); species of *Meiogymnophallus minutus*, *Meiogymnophallus strigatus* were reported in *C. edule*, *Meiogymnophallus fossarum* and *Meiogymnophallus rebecqui* in *C. glaucum* in western Europe and the Mediterranean (Bowers et al., 1996); and *Parvatrema* sp. in *T. plebeius* in Brazil (da Silva et al., 2009). Metacercariae of *Acanthoparyphium tyosenense* (Echinostomatidae) were reported in *Mactra quadrangularis* (=*M. veneriformis*) and *Solen grandis* in Korea (Chai et al., 2001). Metacercariae of *Curtuteria arguinae* (Echinostomatidae), infecting mantle and foot, were reported in *C. edule* in France (Desclaux et al., 2006) and metacercariae similar to *Curtutelia* were reported in *S. marginatus* in Spain, infecting the pericardial gland (Rodríguez et al., 2009). Montaudouin et al. (2009) have studied the distribution and identification of several digenean trematodes infecting *C. edule* along the north-eastern Atlantic shoreline, and they suggested that the distribution pattern could serve for future studies dealing with the effects of climate change on parasite-host systems.

Unidentified metacercariae were detected in different locations (mantle, siphons, digestive gland and pericardial space) of *V. philippinarum*, *V. decussata*, *V. corrugata*, *P. aureus*, *A. flexuosa* and *Iphigenis brasiliana* from Spain, Canada and Brazil (Navas et al., 1992; Marshall et al., 2003; Boehs et al., 2010), and in connective tissue of *Pitar rostratus* from Uruguay (Cremonte et al., 2005a). Sporocysts of a non-identified digenean trematode were observed in *V. philippinarum*, *V. decussata*, *V. corrugata*, *P. aureus*, *P. virgineus* and *S. marginatus* from Spain (Navas et al., 1992; Villalba et al., 1999; Rodríguez et al., 2009).

Sporocyst infections in the gonad provoke castration due to a mechanical effect or depletion of energy reserves (Lauckner, 1983), but they have been also associated with mortality events in cockles (Jonsson and André, 1992; Thieltges, 2006). Metacercariae stages do not normally provoke castration; however, they can cause weakness, deformities in the shell, hypertrophy or atrophy of tissues affected (Lauckner, 1983). Desclaux et al. (2004), after a study of a *C. edule* population infected by *Himasthla quissetensis*, concluded that both host growth rate and water temperature are significant factors in the initiation of parasite infection, and also that the intensity of infection and its effect on host mortality greatly depend on host growth and environmental factors.

5.6.1.3. Cestoda

Cestodes are characterized by complex life cycles with distinct larval and adult stages, and they use bivalves as intermediary hosts. The presence of cestoda, infecting all organs on bivalves, including clams, was reviewed by Lauckner (1983). A larva of *Tylocephalum* sp. was detected in the connective tissue of the digestive gland and mantle of *Malleus malleus*, *M. meridianus*, *Pinna bicolor* and *Pinna deltodes* from Australia (Hine and Thorne, 2000) and in the digestive gland of *Anomala brasiliana* and *I. brasiliana* from Brazil (Boehs et al., 2010). A plerocercoid larva of a tetraphyillidian was reported in the intestinal lumen of *A. antiqua* in Argentina (Cremonte et al., 2005b) and similar larvae were also detected in *E. magnus* and *S. marginatus* in Spain (Darriba et al., 2010; López et al., 2011) (Figure 26). Identification of larval species of marine cestodes on the basis of morphology is difficult; nevertheless, they can be linked to taxonomically known adult stages using mollecular techniques (Holland et al., 2009). These last authors identified *Acanthobothrium brevissime* infecting *Ensis minor*.

Most records of larval cestodes from marine bivalves are descriptive; there are few studies on the pathology of infestation in bivalves. However, there are references to heavy infestations causing physiological stress which may affect growth and reproduction (Lauckner, 1983).

5.6.2. Phylum Annelida

Infections by sedentary, burrowing polychaetes (segmented worms) such as *Polydora* spp. and *Boccardia* spp., are reported in *Chione fluctifraga* from Mexico and *Chione stutchburyi* in New Zealand (reviewed by Bower, 2002).

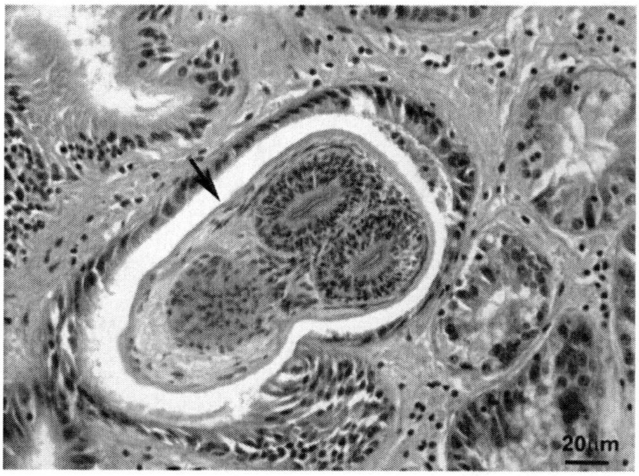

Figure 26. Histological section of *E. magnus* digestive gland. Unidentified cestode-like metazoan (arrow) in the lumen of a digestive duct (HHE).

Polydora ciliata were reported in *V. philippinarum* in Italy (Boscolo and Giovavardi, 2002). Most infections are innocuous, affecting only the shell (shell-boring). Polydora species live in a U-shaped tube inside a hole buried in the shell on the host, worms surviving in the shell after the clam is dead (Bower, 2002).

5.6.3. Phylum Mollusca

Ectoparasitic species of gastropod of family Pyramidellidae have been observed in clams. *Boonea* spp. and *Odostomia* spp. were reported in *Mya arenaria*, *C. edule*, *Cardium lamarcki*, *Hiatella rugosa* and *Tellina* spp. Pyramidellidae attach directly to the siphon tips of clams and penetrate the soft tissues using a stylet apparatus. They can produce tissue irritation in high-intensity cases, but mortalities have not been associated (reviewed by Lauckner, 1983; Bower et al., 1994f).

5.6.4. Phylum Arthropoda

Copepoda species are the commonest arthropod organisms reported as parasites of marine bivalve molluscs, including clams (Lauckner, 1983; Bower, 2009, 2010d), normally located within the lumen of the gut and pallial cavity. Copepoda similar to *Mytilicola orientalis*, originally described in oysters and mussels in Japan, have been identified in clam species around the world.

Their pathological effects are minimal in most cases; however this can vary, depending on the organ affected and level of intensity (reviewed by Bower, 2010d). *Mytilicola intestinalis*, originally described in mussels, were reported in clam species; there are no reports of pathogenicity to clams (reviewed by Bower, 2009).

Unidentified copepoda (Figure 27) were reported in *P. virgineus*, *E. magnus* and *S. marginatus* in Spain (Villalba et al., 1999; Darriba et al., 2010; López et al., 2011).

Lauckner (1983) reviewed other arthropod organisms in association with bivalves, and some belonging to Amphipoda, Isopoda, Cirripedia, Pantopoda and Decapoda were reported in clams. The presence of Decapoda in clams was also reviewed by Bower and MacGladdery (2004).

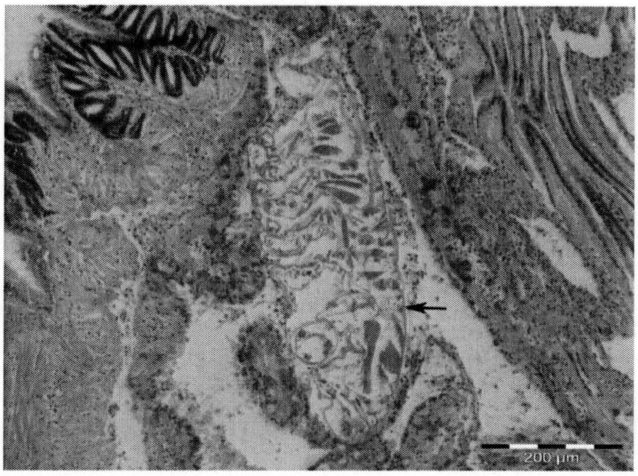

Figure 27. Histological section of *V. decussata*, showing a copepod (arrow) attached to the gill (HHE).

5.7. OTHER SYMBIONTS

Bivalves which live intertidally or in shallow subtidal beaches can be affected by large growths of several species of macroalgae, and in some cases macroalgae have induced bivalve mortality. Different kinds of brown, green and red algae were observed on the shells of *Donax vittatus* populations with mortalities in Scotland; nevertheless, the precise physical conditions which favour the development of algal colonies and which lead to mass mortality remain unknown (Ansell et al., 1988).

Several species of sponges were found attached to shell clams such as *Nucula* spp., *Tellina* sp. and *Macoma* sp.

Many hydroids (Cnidaria) have a mutualism association with marine bivalves. Most of the polyps are attached to the outer surfaces of valves, but several species are located inside the mantle cavity. Lauckner (1983) reviewed cnidarians associated with bivalves, including clams *Nucula* spp., *Tellina* sp., *Macoma* sp., *Mya arenaria*, *Hiatella arctica* (=*Saxicava arctica*), *Venerupis decussata* and *Cerastoderma edule*. Clams with hydroids had reduced shell lengths and their gonad development was retarded. Other cnidarians were found attached to the soft tissues of some clam species, as in the case of unidentified Campanularia on the siphons of *M. arenaria* and *H. arctica*; *Eugymnanthea* sp. were observed inside the mantle cavity of *V. decussata* and *C. edule*.

Bivalve molluscs are common hosts for larval stages of nematodes, and generally the pathology associated with their presence in molluscan tissues is moderate. Lauckner (1983) reported the presence of larval nematodes in *Spisula solidissima*, *Mercenaria* spp., *M. arenaria* and *Venerupis philippinarum*. *Sipirulina* larval nematodes were reported in *Tagelus plebeius* in Argentina, infecting muscles, labial palps and mantle; their presence can give rise to the formation of a thick capsule formed by haemocyte aggregation (Vázquez et al., 2006).

Nemerteans of the genus *Malacobdella* are reported in the mantle cavity of *Mercenaria* spp., *M. arenaria* and *C. edule*. Local irritation of infected individuals was observed, and impairment of the clams' filtration capacity must be expected (Lauckner, 1983).

Encrusting bryozoans are members of the fouling community living on molluscan shells. These organisms have only a minor negative effect on the shell, and were reported from the Cardiidae and Limidae families (Lauckner, 1983).

REFERENCES

Abollo, E., Casas, S. M., Ceschia, G., and Villalba, A. (2006). Differential diagnosis of *Perkinsus* species by polymerase chain reaction-restriction fragment length polymorphism assay. *Mol. Cell. Probes.* 20, 323-329.

Abdul-Salam, J. and Sreelatha, B. S. (1998). A list of larval digenetic trematodes patasitizing some marine invertebrates of Kuwait Bay. *Kuwait J. Sci. and Eng.* 25, 409-433.

Aboelkhair, M., Siah, A., Clark, K. F., Mckenna, P., Pariseau, J., Greenwood, S. J., Berthe, F. C. J., and Cepica, A. (2009). Reverse transcriptase activity associated with haemic neoplasia in the soft-shell clam *Mya arenaria*. *Dis. Aquat. Org.* 84, 57-63.

Andrews, J. D. and Ray, S. M. 1988 Epizootiology of the disease caused by the oyster pathogen *Perkinsus marinus* and its effects on the oyster industry. *Am. Fish. Soc. Spec. Publ.* 18, 47-63.

Ansell, A. D., Robb, L. and Powell, H. T. (1988). Algal-induced dislodgement as a cause of bivalve mortality on some scottish beaches. *J. Mar. Biol. Ass. U. K.* 68, 219-233.

Audermard, C., Le Roux, F., Barnaud, A., Collins, C., Sautour, B., Sauriau, P. G., de Montaudouin, X., Coustau, C., Combes, C., and Berthe, F. C. J. (2002). Needle in a haystack: involvement of the copepod *Paracartia grani* in the life-cycle of the oyster pathogen *Marteilia refringens*. *Parasitology.* 124, 315-323.

Azevedo, C. (1989a). Fine structure of endonucleobiotic bacteria in the gill epithelium of *Ruditapes decussatus*. *Mar. Biol.* 100, 339- 341.

Azevedo, C. (1989b). Ultrastructural observations of *Cryptosporidium* sp. parasite of *Ruditapes decussatus* (Mollusca, Bivalvia). *J. Invertebr. Pathol.* 54, 23-27.

Azevedo, C. (1989c). Fine structure of *Perkinsus atlanticus* n-sp. (Apicomplexa, Perkinsea) parasite of the clam *Ruditapes decussatus* from Portugal. *J. Parasitol.* 75(4), 627-635.

Azevedo, C. and Cachola, R. (1992). Fine-structure of the apicomplexa oocyst of *Nematopsis* sp. of two marine bivalve mollusks. *Dis. Aquat. Org.* 14, 69-73.

Azevedo, C. (1993). Occurrence of an unusual branchial *Mycoplasma*-like Infection in cockle *Cerastoderma edule* (Mollusca, Bivalvia). *Dis. Aquat. Org.* 16, 55-59.

Azevedo, C. (1997). Some observations on an amoeba, destructive agent of the gills of *Cerastoderma edule* (Mollusca, Bivalvia). *Dis. Aquat. Org.* 30, 223-225.

Azevedo, C. (2001). Ultrastructural description of the spore maturation stages of the clam parasite *Minchinia tapetis* (Vilela, 1951) (haplosporida: Haplosporidiidae). *Syst. Parasitol.* 49, 189-194.

Azevedo, C., Conchas, R. F. and Montes, J. (2003). Description of *Haplosporidium edule* n. sp. (Phylum Haplosporidia), a paradite of *Cerastoderma edule* (Mollusca, Bivalvia) with complex spore ornamentation. *J. Exp. Mar. Biol. Ecol.* 220, 47-65.

Beaz-Hidalgo, R., Cleenwerck, I., Balboa, S., Prado, S., de Vos, P., Romalde, J. L. (2009a). *Vibrio breoganii* sp. nov., a non-motile, alginolytic, marine bacterium within the *Vibrio halioticoli clade*. *Int. J. Syst. Evol. Microbiol.* 59, 1589-1594.

Beaz-Hidalgo, R., Doce, A., Pascual, J., Toranzo, A. E., Romalde, J. L. (2009b). *Vibrio gallaecicus* sp nov isolated from cultured clams in north-western Spain. *Syst. Appl. Microbiol.* 32, 111-117.

Beaz-Hidalgo, R., Dieguez, A. L., Cleenwerck, I., Balboa, S., Doce, A., de Vos, P., Romalde, J. L. (2011). *Vibrio celticus* sp. nov., a new *Vibrio* species belonging to the Splendidus clade with pathogenic potential for clams. *Syst. Appl. Microbiol.* 33, 311-315.

Berrilli, X. F., Ceshia, G., de Liberato, C., di Cave, D., and Orecchia, P. (2000). Parasitic infections of *Chamelea gallina* (Mollusca, Bivalvia) from commercially exploited banks of the Adriatic Sea. *Bull. Eur. Ass. Fish Pathol.* 20, 199-205.

Berthe, F. C. J., Le Roux, F., Peyretaillade, E., Peyret, P., Rodríguez, D., Gouy, M., and Vivarès, C. P. (2000). The existence of the FiloParamixea Desportes and Perkins, 1990 is validated by the phylogenetic analysis of the *Marteilia refringens* small subunit ribosomal RNA. *J. Eukaryotic Microbiol.* 47, 288-293.

Berthe, F. C. J., Le Roux, F., Adlard, R. D., and Figueras, A. (2004). Marteiliosis in molluscs: A review. *Aquat. Living Resour.* 17, 433-448.

Boehs, G., Villalba, A., Ceuta, L. O., and Luz, J. R. (2010). Parasites of three commercially exploited bivalve mollusc species of the estuarine region of the Cachoeira river (Ilheus, Bahia, Brazil). *J. Invertebr. Pathol.* 103, 43-47.

Boscolo, R. and Giovanardi, O. (2002). Polydora ciliata shell infestation in *Tapes philippinarum* Manila clam held out of the substrate in the Adriatic sea, Italy. *J. Invertebr. Pathol.* 79, 197-198.

Bower, S. M., Blackkbourn, J. and Meyer, G. R. (1992). Parasite and symbiont fauna of japanese littlenecks, *Tapes philippinarum* (Adams and Reeve, 1850), in British Columbia. *J. Shellfish Res.* 11, 13-19.

Bower, S. M., McGladdery, S. E. and Price, I. M. (1994a). Synopsis of Infectious Diseases and Parasites of Commercially Exploited Shellfish: Hinge ligament disease of juvenile clams. Available from: <http://www.pac.dfo-mpo.gc.ca/science/species-especes/shellfish-coquillages/diseases-maladies/hldcc-eng.htm.

Bower, S. M., McGladdery, S. E. and Price, I. M. (1994b). Synopsis of Infectious Diseases and Parasites of Commercially Exploited Shellfish: Microsporidiosis of clams. Available from: <http://www.pac.dfo-mpo.gc.ca/science/species-especes/shellfish-coquillages/diseases-maladies/msporcc-eng.htm.

Bower, S. M., McGladdery, S. E. and Price, I. M. (1994c). Synopsis of Infectious Diseases and Parasites of Commercially Exploited Shellfish: *Ancistrocoma pelseneeri* and *A. myae* ciliates of clams. Available from: <http://www.pac.dfo-mpo.gc.ca/science/species-especes/shellfish-coquillages/diseases-maladies/apamcc-eng.htm.

Bower, S. M., McGladdery, S. E. and Price, I. M. (1994d). Synopsis of Infectious Diseases and Parasites of Commercially Exploited Shellfish: *Sphenophrya-* like ciliated of clams and cockles. Available from: <http://www.pac.dfo-mpo.gc.ca/science/species-especes/shellfish-coquillages/diseases-maladies/spccc-eng.htm.

Bower, S. M., McGladdery, S. E. and Price, I. M. (1994e). Synopsis of Infectious Diseases and Parasites of Commercially Exploited Shellfish: Gill trichodina of clams and cockles. Available from: <http://www.pac.dfo-mpo.gc.ca/science/species-especes/shellfish-coquillages/diseases-maladies/gilltrcc-eng.htm.

Bower, S. M., McGladdery, S. E. and Price, I. M. (1994f). Synopsis of Infectious Diseases and Parasites of Commercially Exploited Shellfish: Siphon snails of clams and cockles. Available from: <http://www.pac.dfo-mpo.gc.ca/science/species-especes/shellfish-coquillages/diseases-maladies/csscc-eng.htm.

Bower, S. M. (2002). Synopsis of Infectious Diseases and Parasites of Commercially Exploited Shellfish: Shell-boring polychaetes of clams. Available from: <http://www.pac.dfo-mpo.gc.ca/science/species-especes/shellfish-coquillages/diseases-maladies/sbpcc-eng.htm.

Bower, S. M. (2004a). Synopsis of Infectious Diseases and Parasites of Commercially Exploited Shellfish: Rickettsia-like and chlamydia-like organisms of clams and cockles. Available from: <http://www.pac.dfo-mpo.gc.ca/science/species-especes/shellfish-coquillages/diseases-maladies/rlclocc-eng.htm.

Bower, S. M. (2004b). Synopsis of Infectious Diseases and Parasites of Commercially Exploited Shellfish: Turbellaria of clams. Available from: <http://www.pac.dfo-mpo.gc.ca/science/species-especes/shellfish-coquillages/diseases-maladies/turbelcc-eng.htm.

Bower, S. M. and McGladdery, S. E. (2004). Synopsis of Infectious Diseases and Parasites of Commercially Exploited Shellfish: Pea crabs in clams and cockles. Available from: <http://www.pac.dfo-mpo.gc.ca/science/species-especes/shellfish-coquillages/diseases-maladies/pcpcc-eng.htm.

Bower, S. M. and Itoh, N. (2005). Synopsis of Infectious Diseases and Parasites of Commercially Exploited Shellfish: *Marteilioides* sp. of clams. Available from: http://www.pac.dfo-mpo.gc.ca/science/species-especes/ shellfish-coquillages/diseases-maladies/pages/martioidcc-eng.htm.

Bower, S. M. (2007a). Synopsis of Infectious Diseases and Parasites of Commercially Exploited Shellfish: Kidney coccidia of clams. Available from: <http://www.pac.dfo-mpo.gc.ca/science/species-especes/shellfish-coquillages/diseases-maladies/kidcoccc-eng.htm.

Bower, S. M. (2007b). Synopsis of Infectious Diseases and Parasites of Commercially Exploited Shellfish: Gregarine parasitism of clams and cockles. Available from: <http://www.pac.dfo-mpo.gc.ca/science/species-especes/shellfish-coquillages/diseases-maladies/gregpcc-eng.htm.

Bower, S. M. (2007c). Synopsis of Infectious Diseases and Parasites of Commercially Exploited Shellfish: Haplosporidian infection of clams. Available from: <http://www.pac.dfo-mpo.gc.ca/science/species-especes/shellfish-coquillages/diseases-maladies/hapinfcc-eng.htm.

Bower, S. M. (2007d). Synopsis of Infectious Diseases and Parasites of Commercially Exploited Shellfish: *Steinhausia* sp. (cockle egg disease). Available from: <http://www.pac.dfo-mpo.gc.ca/science/species-especes/shellfish-coquillages/diseases-maladies/cedcc-eng.htm.

Bower, S. M. (2007e). Synopsis of Infectious Diseases and Parasites of Commercially Exploited Shellfish: Trematode metacercariae of clams and cockles. Available from: <http://www.pac.dfo-mpo.gc.ca/science/species-especes/shellfish-coquillages/diseases-maladies/tremetcc-eng.htm.

Bower, S. M. (2009). Synopsis of Infectious Diseases and Parasites of Commercially Exploited Shellfish: *Mytilicola intestinalis* (red worm disease) of clams and cockles. Available from: <http://www.pac.dfo-mpo.gc.ca/science/species-especes/shellfish-coquillages/diseases-maladies/mytintcc-eng.htm.

Bower, S. M. (2010a). Synopsis of Infectious Diseases and Parasites of Commercially Exploited Shellfish. Available from: <http://www.pac.dfo-mpo.gc.ca/science/species-especes/shellfish-coquillages/diseases-maladies/index-eng.htm.

Bower, S. M. (2010b). Synopsis of Infectious Diseases and Parasites of Commercially Exploited Shellfish: Virus infection of clams. Available from: <http://www.pac.dfo-mpo.gc.ca/science/species-especes/shellfish-coquillages/diseases-maladies/papovcc-eng.htm.

Bower, S. M. (2010c). Synopsis of Infectious Diseases and Parasites of Commercially Exploited Shellfish: QPX, a Thraustochytrid-like Disease of Clams. Available from: <http://www.pac.dfo-mpo.gc.ca/science/ species-especes/shellfish-coquillages/diseases-maladies/chydcc-eng.htm.

Bower, S. M. (2010d). Synopsis of Infectious Diseases and Parasites of Commercially Exploited Shellfish: *Mytilicola* orientalis (red worm) of clams and cockles. Available from: <http://www.pac.dfo-mpo.gc.ca/ science/species-especes/shellfish-coquillages/diseases-maladies/mytoricc-eng.htm.

Bower, S. M. 2011. Synopsis of Infectious Diseases and Parasites of Commercially Exploited Shellfish: *Perkinsus* of clams and cockles. Available from: <http://www.pac.dfo-mpo.gc.ca/science/species-especes/shellfish-coquillages/diseases-maladies/perkincc-eng.htm.

Bower, S. M. and Itoh, N. (2011). Synopsis of Infectious Diseases and Parasites of Commercially Exploited Shellfish: Marteliosis of clams and cockles. Available from: http://www.pac.dfo-mpo.gc.ca/science/species-especes/shellfish-coquillages/diseases-maladies/ pages/martcc-eng.htm.

Bowers, E. A., Bartoli, P., Russel-Pinto, F., and James, B. L. 1996. The metacercariae of sibling species of *Meiogymnophallus*, including *M. rebecqui* comb. nov. (Digenea: Gymnophallidae), and their effects on closely related *Cerastoderma* host species (Mollusca: Bivalvia). *Parasitol. Res.* 82, 505-510.

Buchanan, J. S. (1978). Cytological studies on a new species of rickettsia found in association with a phage in the digestive gland of the marine bivalve mollusk, *Tellina tenuis* (Da Costa). *J. Fish Dis.* 1, 27-43.

Burreson, E. M. and Ford, S. E. (2004). A review of recent information on the Haplosporidia, with special reference to *Haplosporidium nelsoni* (MSX disease). *Aquat. Living Resour.* 17, 499-517.

Burreson, E. M., Reece, K. S. and Dungan, C. F. (2005). Molecular, morphological, and experimental evidence support the synonymy of *Perkinsus chesapeaki* and *Perkinsus andrewsi*. *J. Eukaryot. Microbiol.* 52, 258-270.

Bushek, D., Ford, S. E. and Chintala, M. M. (2002a). Comparison of in vitro-cultured and wild-type *Perkinsus marinus*. III. Fecal elimination and its role in transmission. *Dis. Aquat. Org*. 51, 217-225.

Bushek, D., Dungan, C. F. and Lewitus, A. J. (2002b). Serological affinities of the oyster pathogen *Perkinsus marinus* (Apicomplexa) with some dinoflagellates (Dynophyceae). *J. Eukaryotic Microbiol.* 49, 11-16.

Bushek, D., Landau, B. and Scarpa, E. (2008). *Perkinsus chesapeaki* in stout razor clams *Tagelus plebeius* from Delaware Bay. *Dis. Aquat. Org*. 78, 243-247.

Canestri-Trotti, X. G., Baccarani, E. M., Paesanti, F., and Turolla, E. (2000). Monitoring of infections by protozoa of the genera *Nematopsis*, *Perkinsus* and *Porospora* in the smooth venus clam *Callista chione* from the North-Western Adriatic Sea (Italy). *Dis. Aquat. Org*. 42, 157-161.

Carballal, M. J., Iglesias, D., Santamarina, J., Ferro-Soto, B., and Villalba, A. (2001). Parasites and pathological conditions of the cockle *Cerastoderma edule* populations of the coast of Galicia (NW Spain). *J. Invertebr. Pathol.* 78, 87-97.

Carballal, M. J., Villalba, A., Iglesias, D., and Hine, P. M. (2003). Virus-like particles associated with large foci of heavy hemocytic infiltration in cockles *Cerastoderma edule* from Galicia (NW Spain). *J. Invertebr. Pathol.* 84, 234-237.

Carballal, M. J., Díaz, S. and Villalba, A. (2005). *Urosporidium* sp. hyperparasite of the turbellarian *Paravortex cardii* in the cockle *Cerastoderma edule*. *J. Invertebr. Pathol.* 90, 104-107.

Carnegie, R. B., Cochennec-Laureau, N. (2004). Microcell parasites of oysters: Recent insights and future trends. *Aquat. Living Resour.* 17, 519-528.

Carrasco, N., Lopez-Flores, I., Alcaraz, M., Furones, M. D., Berthe, F. C. J., and Arzul, I. (2007). Dynamics of the parasite *Marteilia refringens* (Paramyxea) in *Mytilus galloprovincialis* and zooplankton populations in Alfacs Bay (Catalonia, Spain). *Parasitology*. 134, 1541-1550.

Carrasco, N., Roque, A., Andree, K. B., Rodgers, Ch., Lacuesta, B., Furones, M. D. (2012). Molecular characterization of the *Marteilia* parasite infecting the common edible cockle *Cerastoderma edule* in the Spanish Mediterranean coast. A new *Marteilia* species affecting bivalves in Europe? *Aquaculture*. 324-325, 20-26.

Cremonte, F., Figueras, A. and Burreson, E. M. (2005). A histopathological survey of some commercially exploited bivalve molluscs in Northern Patagonia, Argentina. *Aquaculture*. 249, 23-33.

Chagot, D., Comps, M., Boulo, V., Ruano, F., and Grizel, H. (1987). Histological study of a cellular reaction in *Ruditapes decussatus* infected by a protozoan. *Aquaculture*. 67, 260-261.

Chai, J. Y., Han, E. T., Park, Y. K., Guk, S. M., and Lee, S. H. (2001). *Acanthoparyphium tyosenense*: the discovery of human infection and identification of its Source. *J. Parasitol.* 87, 794-800.

Chang, Y. J., Huh, M.-D., Oh, M.-J., and Sugawara, Y. (2002). Baculovirus-like particles in epithelial cell of digestive divertivula of the scallop, *Patinopecten yessoensis*. *J. Shellfish Res.* 21, 109-112.

Cheslett, D., Mckiernan, F., Hickey, C., and Collins, E. (2009). Viral gametocytic hypertrophy of the pacific oyster *Crassostrea gigas* in Ireland. *Dis. Aquat. Org.* 83, 181-185.

Choi, K.-S. and Park, K.-I. (2005). Current status of *Perkinsus* infection in Korean waters. In: Walker, P. J., R. G., Lester, M. G. Bondad-Reantaso (Eds.), *Diseases in Asian Aquaculture V. Proceedings of the 5th Symposium on Diseases in Asian Aquaculture. Fish Health Section, Asian Fisheries Society*, pp. 263-274, Manila, Philippines.

Comps, M. (1983a). Etude morphologique d´une infection rickettsienne de la palourde *Ruditapes philippinarum* Adam et Reeves. *Rev. Trav. Inst. Pêches Marit.* 46, 141-145.

Comps, M. (1983b). Etude morphologique de *Marteilia christenseni* sp. n. parasite du lavignon *Scrobicularia piperata* P. (mollusque pelecypode). *Rev. Trav. Inst. Pêches Marit.* 47, 99-104.

Comps, M. and Raimbault, R. (1978). Infection rickettsienne de la glande digestive de *Donax trunculus* L. *Sciencs et Pêche, Bull. Inst. Pêches Marit.* 281, 442-443.

Coss, C. A., Robledo, J. A. F., Ruiz, G. M., and Vasta, G. R. 2001. Description of *Perkinsus andrewsi* n. sp. isolated from the baltic clam (*Macoma balthica*) by characterization of the ribosomal RNA locus, and development of a species-specific PCR-based diagnostic assay. *J. Eukaryotic Microbiol.* 48, 52-61.

Cremonte, F. and Figueras, A. (2004). Parasites as possible cause of mass mortalities of the critically endangered clam *Mesodesma mactroides* on the Atlantic coast of Argentina. *Bull. Eur. Ass. Fish Pathol.* 24, 166-171.

Cremonte, F., Balseiro, P. and Figueras, A. (2005a). Occurrence of *Perkinsus olseni* (Protozoa: Apicomplexa) and other parasites in the venerid commercial Clam *Pitar rostrata* From Uruguay, Southwestern Atlantic Coast. *Dis. Aquat. Org.* 64, 85-90.

Cremonte, F., Figueras, A. and Burreson, E. M. (2005b). A histopathological survey of some commercially exploited bivalve molluscs in northern Patagonia, Argentina. *Aquaculture*. 249, 23-33.

Dang, C., Gonzalez, P., Mesmer-Dudons, N., Bonami, J. R., Caill-Milly, N., and de Montaudouin, X. (2009). Virus-like particles associated with brown muscle disease in manila clam, *Ruditapes philippinarum*, in Arcachon Bay (France). *J. Fish Dis.* 32 (7), 577-584.

Darriba, S., Iglesias, D., Ruiz, M., Rodriguez, R., and López, C. (2010). Histopathological survey of symbionts and other conditions in razor clam *Ensis arcuatus* (Jeffreys, 1865) (Pharidae) of the coast of Galicia (NW Spain). *J. Invertebr. Pathol.* 104, 23-30.

Darriba, S., Ruiz, M. and López C. (2011). Phage particles infecting intracellular prokaryotic organisms in *Venerupis rhomboides* (Bivalvia) from Galicia (NW Spain). *Proceedings of the XV Conference of EAFP*, pp. 282, Split, Croacia.

Davison, J. A., Eberle, R., Ehlers, B., Hayward, G. S., McGeoch, D. J., Minson, A. C., Pellett, P. E., Roizman, B., Studdert, M. J., and Thiry, E. (2009). The Order Herpesvirales. *Arch. Virol.* 154, 171-177.

De la Herrán, R., Garrido-Ramos, M. A., Navas, J. I., Ruiz Rejón, C., and Ruiz Rejón, M. (2000). Molecular characterization of the ribosomal RNA gene region of *Perkinsus atlanticus*: its use in phylogenetic analysis and as a target for a molecular diagnosis. *Parasitology*. 120, 345-353.

Delgado, M., Carrasco, N., Elandaloussi, L., Furones, D., and Roque A. (2007). A mortality event of the venerid bivalve *Callista chione* (Linnaeus, 1758) in a hatchery system - a case study. *Bull. Eur. Ass. Fish Pathol.* 27, 214-222.

Desclaux, C., De Montaudouin, X. and Bachelet, G. (2004). Cockle *Cerastoderma edule* population mortality: role of the digenean parasite *Himasthla quissetensis*. *Mar. Ecol. Prog. Ser.* 279, 141-150.

Desclaux, C., Russell-Pinto, F., de Montaudouin, X. and Bachelet, G. 2006. First record and description of metacercariae of *Curtuteria arguinae* n. sp. (Digenea: Echinostomatidae), parasite of cockles *Cerastoderma edule* (Mollusca: Bivalvia) in Arcachon Bay, France. *J. Parasitol.* 92, 578-587.

Desportes, I. and Ginsburger-Vogel, T. (1981). Étude ultraestructurale de la sporulation de *Paramyxa paradoxa* Chatton (Paramyxida) parasite de l'annélide polychète *Poecilochaetus serpens*. *Protistologica.* 17, 365-386.

Desportes, I. and Perkins, F. O. (1990). Phylum Paramyxea. In: L. Margulis, J. O. Corliss, M. Melkonian, D. J. Chapman (Eds.), *Handbook of Protoctista*. pp. 30, Jones and Bartlett Publising, Boston.

Desser, S. S. and Bower, S. M. (1997). *Margolisiella Kabatai* gen. et sp. n. (Apicomplexa: Eimeriidae), a parasite of native littleneck clams, *Protothaca staminea*, from British Columbia, Canada, with a taxonomic revision of the coccidian parasites of bivalves (Mollusca: Bivalvia). *Folia Parasitol.* 44, 241-247.

Desser, S. S., Bower, S. M. and Hong, H. (1998). *Pseudoklossia Semiluna* n. sp. (Apicomplexa: Aggregatidae): a coccidian parasite of the kidney of blue mussels, species of *Mytilus*, from British Columbia, Canada. *J. Soc. Francaise Parasitol.* 5, 17-22.

Diéguez, A. L., Beaz-Hidalgo, R., Cleenwerck, I., Balboa, S., de Vos, P., Romalde, J. L. (2011). *Vibrio atlanticus* sp. nov. and *Vibrio artabrorum* sp. nov. isolated from the clams *Ruditapes philippinarum* and *Ruditapes decussatus*. *Int. J. Syst. Evol. Microbiol.* 61, 2406-2411.

Dungan, C. F. and Reece, K. S. (2006). In vitro propagation of two *Perkinsus* spp. parasites from japanese manila clams *Venerupis philippinarum* and description of *Perkinsus honshuensis* n. sp. *J. Eukaryotic Microbiol.* 53, 316-326.

Dungan, C. F., Reece, K. S., Moss, J. A., Hamilton, R. M., and Diggles, B. K. (2007b). *Perkinsus olseni* in vitro isolates from the New Zealand Clam *Austrovenus stutchburyi*. *J. Eukaryotic Microbiol.* 54, 263-270.

Earnhart, C. G., Gauthier, D. T, Vogelbein, W. K., and Kaattari, S. L. (2005). Monoclonal antibody analysis of *Perkinsus marinus* extracellular products. *Int. J. Parasitol.* 35, 171-84.

Elandalloussi, L. M., Leite, R. M., Afonso, R., Nunes, P. A., Robledo, J. A. F., Vasta, G. R., and Cancela, M. L. (2004). Development of a PCR-ELISA assay for diagnosis of *Perkinsus marinus* and *Perkinsus atlanticus* infections in bivalve molluscs. *Mol. Cell. Probes.* 18, 89-96.

Elandaloussi, L. M., Carrasco, N., Roque, A., Andree, K., and Furones, M. D. (2009). First record of *Perkinsus olseni*, a protozoan parasite infecting the commercial clam *Ruditapes decussatus* in Spanish Mediterranean waters. *J. Invertebr. Pathol.* 100, 50-53.

Elston, R. A. and Peacock, M. G. (1984). A rickettsiales-like infection in the pacific razor clam, *Siliqua patula*. *J. Invertebr. Pathol.* 44, 84-96.

Elston, R. (1986a). Occurrence of branchial rickettsial-like infections in two bivalve mollusks, *Tapes japonica* and *Patinopecten yessoensis*, with comments on their significance. *J. Fish Dis.* 9, 69-71.

Elston, R. A. (1986b). An intranuclear pathogen (nuclear inclusion X (NIX) associated with massive mortalities of the pacific razor clam, *Siliqua patula*. *J. Invertebr. Pathol.* 47, 93-104.

Elston, R. A. (1997). Special topic review: Bivalve mollusc virases. *World J. Microbiol. Biotechnol.* 13, 393-403.

Elston, R. A., Dungan, C. F., Meyers, T. R., and Reece, K. S. (2004). *Perkinsus* sp. infection risk for Manila clams, *Venererupis philippinarum* (Adams and Reeve, 1850) on the Pacific coast of north and central America. *J. Shellfish Res.* 23, 101-105.

Farley, C. A. (1978). Viruses and viruslike lesions in marine molluscs. *Mar. Fish. Rev.* 40, 18-20.

Farley, C. A. (1981). Phylogenetic relationships between viruses, marine invertebrates and neoplasia. In: C. J. Dawe, J. C. Harshbarger, S. Kondo, T. Sugimura, S. Takayama (Eds.), *Phyletic Approaches to Cancer*. pp. 75-87.

Ford, S. E., Chintala, M. M. and Bushek, D. (2002). Comparison of in vitro-cultured and wild-type *Perkinsus marinus*. I. Pathogen virulence. *Dis. Aquat. Org.* 51, 187-201.

Freire-Santos, F., Oteiza-López, A. M., Vergara-Castiblanco, C. A., Ares-Mazás, E., Álvarez-Suárez, E., and García-Martín, O. (2000). Detection of *Cryptosporidium* oocysts in bivalve molluscs destined for human consumption. *J. Parasitol.* 86, 853-854.

Friedman, C. S., Andree, K. B. and Hedrick, R. P. (2000). *Candidatus Xenohaliotis californiensis*, a newly described pathogen of abalone, *Haliotis* spp., along the West Coast of North America. *Int. J. Syst. Evol. Microbiol.* 50, 847-855.

Figueras, A. (2004). Bivalve Diseases. *Aquat. Living Resour.* 17, 395-530.

Fries, C. R. and Grant, D. M. (1991). Rickettsiae in gill epithelial cells of the hard clam, *Mercenaria mercenaria*. *J. Invertebr. Pathol.* 57, 166-171.

Fries, C. R. and Grant, D. M. (1992). *Ehrlichia*-like microorganisms in hemocytes in the gills of the marine bivalve, *Mercenaria mercenaria*. *J. Invertebr. Pathol.* 59, 210-211.

Fryer, J. L. and Lannan, C. N. (1994). Rickettsial and chlamydial infections of freshwater and marine fishes, bivalves, and crustaceans. *Zool. Stud.* 33, 95-107.

Garcia, C., Robert, M., Arzul, I., Chollet, B., Joly, J. P., Miossec, L., Comtet, T., and Berthe, F. (2006). Viral gametocytic hypertrophy of *Crassostrea gigas* in France: From Occasional Records to Disease Emergence? *Dis. Aquat. Org.* 70, 193-199.

Garcia, C., Leite, R. and Arzul, I. (2008). Parasites of the genus *Perkinsus*. In: A. Villalba (Ed.), *Workshop for the analysis of the impact of Perkinsosis to the European Shellfish Industry*. pp. 19-31, Centro de Investigacións Mariñas, Consellería de Pesca e Asuntos Marítimos da Xunta de Galicia, Vilanova de Arousa, Spain. Centro Tecnológico del Mar - Fundación CETMAR, Vigo, Spain.

Garcia, C., Arzu, I., Joly, J. P., Guichard, B., Choller, B., Omnes, E., Haond, C., Robert, M., Lupo, C., and François, C. (2011). Detection of *Microcytos* like protozoans during mortality events of the shellfish *Donax trunculus*, in France. *Proceedings of the XV Conference of EAFP*, pp. 39, Split, Croatia.

Gestal, C., Roch, P., Renault, T., Pallavicini, A., Paillard, C., Novoa, B., Oubella, R., Venier, P., and Figueras, A. (2008). Study of diseases and the immune system of bivalves using molecular biology and genomics. *Rev. Fish. Sci.* 16, 131-154.

Goggin, C. L. and Lester, R. J. G. (1990). Rickettsiales-like infection in the gills of *Tridacna crocea* from the Great Barrier Reef. *J. Invertebr. Pathol.* 56, 135-138.

Gómez-León, J., Villamil, L., Lemos, M. L., Novoa, B., Figueras, A. (2005). Isolation of *Vibrio alginolyticus* and *Vibrio splendidus* from aquacultured carpet shell clam (*Ruditapes decussatus*) larvae associated with mass mortalities. *Appl. Environ. Microbiol.* 71, 98-104.

Gulka, G. and Chang, P. W. (1984b). Pathogenicity and infectivity of a rickettsia- like organism in the sea scallop, *Placopecten magellanicus. J. Fish Dis.* 8, 309-318.

Harshbarger, J. C., Chang, S. C. and Otto, S. V. (1977). Chlamydiae (with phages), Mycoplasmas, and Rickettsiae in Chesapeake Bay bivalves. *Science.* 196, 666-668.

Harshbarger, J. C., Otto, S. V. and Chang S.-CH. (1979). Proliferative disorders in *Crassostrea virginica* and *Mya arenaria* from the Chesapeake Bay and intranuclear virus-like inclusions in *Mya arenaria* with germinomas from a maine oil spill site. *Haliotis.* 8, 243-248.

Hine, P. M. and Wesney, B. (1997). Virus-like particles associated with cytopathology in the digestive gland epithelium of scallops *Pecten Novaezelandiae* and toheroa *Paphies ventricosum*. *Dis. Aquat. Org*. 29, 197-204.

Hine, P. M. and Thorne, T. (2000). A Survey of some parasites and diseases of several species of bivalve mollusc in northern western Australia. *Dis. Aquat. Org*. 40, 67-78.

Hine, P. M. and Thorne, T. (2002). *Haplosporidium* sp. (Alveolata: Haplosporidia) associated with mortalities among rock oysters *Saccostrea cuccullata* in north Western Australia. *Dis. Aquat. Org*. 51, 123-133.

Holland, N. D. and Wilson, N. G. (2009). Molecular identification of larvae of a Tetraphyllidean tapeworm (Platyhelminthes: Eucestoda) in a razor clam as an alternative intermediate host in the life cycle of *Acanthobothrium brevissime. J. Parasitol.* 95, 1215-1217.

Holt, J. G., Krieg, N. R., Sneath, P. H. A., Staley, J. T., and Williams, S. T. (1994). Grup 30. The mycoplasmas (or Mollicutes): cell wall-less bacteria. In: W. R. Hensyl (Ed.), *Bergey`s manual of determinative bacteriology*. pp. 705-718, Baltimore, Maryland.

House, M. L., Kim, C. H. and Reno, P. W. (1998). Soft shell clams *Mya arenaria* with disseminated neoplasia demonstrate reverse transcriptase activity. *Dis. Aquat. Org.* 34, 187-192.

Joly, J. P. and Comps, M. (1980). Etude d´un micro-organisme de type chlamydien chez la palourde *Ruditapes decussatus* L. *Rev. Trav. Inst. Pêches Marit.* 44, 285-287.

Johnson, P. T. (1984). Viral diseases of marine invertebrates. *Helgolander Meeresunters.* 37, 65-98.

Johnson, M. A. and Le Pennec, M. (1995). Association between the mollusk bivalve *Loripes lucinalis* and a chlamydia-like organism, with comments on its pathogenic impact, life-cycle and possible mode of transmission. *Mar. Biol.* 123, 523-530.

Jonsson, R. and André, C. (1992). Mass mortality of the bivalve *Cerastoderma edule* on the swedish west coast caused by infestation with the digenean trematode *Cercaria cerastodermae*. *Ophelia*, 36, 151-157.

Kinne, O. (1980). Chapter 2. Diseases of Marine Animals: General Aspects. In: O. Kinne (Ed.), *Vol. I. Diseases of Marine Animals*. pp. 13-64, Wiley, Salisbury.

Koepp, S. J. (1984). Detection of DNA virus within an Upper York bay soft-shell clam population. *Coast. Ocean Pollut. Assess. News.* 3, 26-28.

Krantz, G. E. and Jordan, S. J. 1996, Management alternatives for protecting *Crassostrea virginica* fisheries in *Perkinsus marinus* enzootic and epizootic areas. *J. Shellfish Res.* 15, 167-176.

Lauckner, G. (1983). Chapter 13. Diseases of Mollusca: Bivalvia. In: O. Kinne (Ed.), *Vol. II. Diseases of Marine Animals*. pp. 477-879, Biologische Anstalt Helgoland, Hamburg.

Le Borgne, Y. (1996). Echanges internationaux des mollusques bivalves et situation actuelle en France et en Europe. *Rev. Sci. Tech. Off. Int. Epiz.* 15, 491-498.

Leethochavalit, S., Upatham, E. S., Choi, K. S., Sawangwong, P., Chalermwat, K., and Kruatrachue, M. (2003). Ribosomal RNA characterization of non-transcribed spacer and two internal transcribed spacers with 5.8S ribosomal RNA of *Perkinsus* sp. found in undulated surf clams (*Paphia undulata*) from Thailand. *J. Shellfish Res.* 22, 431-434.

Leethochavalit, S., Chalermwat, K., Upatham, E. S., Choi, K. S., Sawangwong, P., and Kruatrachue, M. (2004). Occurrence of *Perkinsus* sp. in undulated surf clams *Paphia undulata* from the Gulf of Thailand. *Dis. Aquat. Org*. 60, 165-171.

López, C., Darriba, S., Iglesias, D., Ruiz, M., and Rodríguez, R. (2011). Chapert 7. Pathology of sword razor shell (*Ensis arcuatus*) and grooved razor shell (*Solen marginatus*). In: A. Guerra, C. Lodeiros, M. Gaspar, F. da Costa (Eds.), *Razor Clams: biology, aquaculture and fisheries*. pp. 161-168, Xunta de Galicia, Santiago de Compostela, Spain.

López-Flores, I., Robles, F., Valencia, J. M., Grau, A., Villalba, A., de la Herran, R., Garrido-Ramos, M. A., Ruiz-Rejon, C., Ruiz-Rejon, M., and Navas, J. I. (2008a). Detection of *Marteilia refringens* using nested PCR and *in situ* hybridisation in *Chamelea gallina* from the Balearic Islands (Spain). *Dis. Aquat. Org*. 82, 79-87.

López-Flores, I., Garrido-Ramos, M. A., De La Herran, R., Ruiz-Rejon, C., Ruiz-Rejon, M., and Navas, J. I. (2008b). Identification of *Marteilia refringens* infecting the razor clam *Solen Marginatus* by PCR and *in situ* hybridization. *Mol. Cell. Probes*. 22, 151-155.

Mackin, J. G., Owen, H. M. and Collier, A. 1950. Preliminary note on the occurrence of a new protistan parasite, *Dermocystidium marinum* n. sp. in *Crassostrea virginica* (Gmelin). *Science*. 111, 328-329.

Marshall, W. L., Bower, S. M. and Meyer, G. R. (2003). A comparison of the parasite and symbiont fauna of cohabiting native (*Protothaca staminea*) and introduced (*Venerupis philippinarum* and *Nuttalia obscurata*) clams in British Columbia. *J. Shellfish Dis*. 22, 185-192.

McGladdery, S. E. (1999). Chapter 20. Shellfish Diseases (Viral, Bacterial and Fungal). In: P. T. K. Woo and D. W. Bruno (Eds.), *Vol. 3. Fish Diseases and Disorders*. pp. 723-860, New Brunswick, Canada.

McLaughlin, S. M., Tall, B. D., Shaheen, A., Elsayed, E. E., and Faisal, M. (2000a). Zoosporulation of a new *Perkinsus* species isolated from the gills of the softshell clam *Mya arenaria*. *J. Soc. Francaise Parasitol.* 7, 115-122.

McLaughlin, S. M. and Faisal, M. (2000b). Prevalence of *Perkinsus* spp. in Chesapeake Bay soft-shell clams, *Mya arenaria* Linnaeus, 1758 during 1990-1998. *J. Shellfish Res.* 19, 349-352.

Medina, D. J., Paquette, G. E., Sadasiv, E. C., and Chang, P. W. (1993). Isolation of infectious particles having reverse transcriptase activity and producing hematopoietic neoplasia in *Mya arenaria*. *J. Shellfish Res.* 12, 112-113.

Meyers, T. R. (1979). Preliminary studies on a chlamydial agent in the digestive diverticular epithelium of hard clams *Mercenaria mercenaria* (L) from Great South Bay, New York. *J. Fish Dis.* 2, 179-189.

Meyers, T. R. (1981). Endemic diseases of cultured shellfish of Long Island, New York: Adult and juvenile american oysters (*Crassostrea virginica*) and hard clams (*Mercenaria mercenaria*). *Aquaculture.* 22, 305-330.

Meyers, T. R., Burton, T., Evans, W., and Starkey, N. (2009). Detection of viruses and virus-like particles in four species of wild and farmed bivalve molluscs in Alaska, US, from 1987 to 2009. *Dis. Aquat. Org.* 88, 1-12.

Mialhe, E., Chagot, D., Boulo, V., Comps, M., Ruano, F., and Grizel, H. (1987). An infection of *Ruditapes decussatus* (Bivalvia) by *Rickettsia*. *Aquaculture.* 67, 258-259.

Montes, J. F., Durfort, M. and García-Valero, J. (1995a). Cellular defence mechanism of the clam *Tapes semidecussatus* against infection by the protozoan *Perkinsus* sp. *Cell Tissue Res.* 279, 529-538.

Montes, J. F., Durfort, M. and García-Valero, J. (1995b). Characterization and localization of an Mr 225 kDa polypeptide specifically involved in the defence mechanims of the clam *Tapes semidecussatus*. *Cell Tissue Res.* 280, 27-37.

Montes, J. F., Durfort, M. and García-Valero, J. (1996). When the venerid clam *Tapes decussatus* is parasitized by the protozoan *Perkinsus* sp. it synthetizes a defensive polypeptide that is closely related to p225. *Dis. Aquat. Org.* 26, 149-157.

Montes, J. F., Durfort, M. and Garcia-Valero, J. (2001). Parasitism by the protozoan *Perkinsus atlanticus* favours the development of opportunistic infections. *Dis. Aquat. Org.* 46, 57-66.

Montes, J. (2011). Chapert 6. Pathologies in pod razor clam (*Ensis siliqua*). In: A. Guerra, C. Lodeiros, M. Gaspar, F. da Costa (Eds). *Razor Clams: biology, aquaculture and fisheries*. pp. 161-168, Xunta de Galicia, Santiago de Compostela.

Moss, J. A., Burreson, E. M. and Reece, K. S. (2006). Advanced *Perkinsus marinus* infections in *Crassostrea ariakensis* maintained under laboratory conditions. *J. Shellfish Res.* 25, 65-72.

Moss, J. A., Xiao, J., Dungan, C. F. and Reece, K. S. (2008). Description of *Perkinsus beihaiensis* n. sp., a new *Perkinsus* sp. parasite in oysters of Southern China. *J. Eukaryotic Microbiol.* 55, 117-130.

Murrell, A., Kleeman, S. N., Barker, S. C., and Lester, R. J. G. (2002). Synonymy of *Perkinsus olseni* Lester and Davis, 1981 and *Perkinsus atlanticus* Azevedo, 1989 and an update on the phylogenetic position of the genus *Perkinsus*. *Bull. Eur. Ass. Fish Pathol.* 22, 258-265.

Navas, J. I., Castillo, M. C.,Vera, P., and Ruiz- Rico, M. (1992). Principal parasites observed in clams, *Ruditapes decussatus* (L), *Ruditapes philippinarum* (Adams-et-Reeve), *Venerupis pullastra* (Montagu) and *Venerupis aureus* (Gmelin), from the Huelva coast (SW Spain). *Aquaculture.* 107, 193-199.

Norén, F., Moestrup, Ø. and Rehnstam-Holm, A. S. 1999. *Parvilucifera infectans* Norén et Moestrup gen. et sp. nov. (Perkinsozoa phylum nov.): a parasitic flagellate capable of killing toxic microalgae. *Eur. J. Protistol.* 35, 233-254.

Norton, J. H., Shepherd, M. A., Abdon-Naguit, M. R., and Lindsay, S. (1993a). Mortalities in the giant clam *Hippopus hippopus* associated with Rickettsiales-like organisms. *J. Invertebr. Pathol.* 62, 207-209.

Norton, J. H., Shepherd, M. A. and Prior, H. C. (1993b). Intracellular bacteria associated with winter mortality in juvenile giant clams, *Tridacna gigas*. *J. Invertebr. Pathol.* 62, 204-206.

Norton, J. H., Perkins, F. P. and Ledua, E. (1993c). *Marteilia-* like infection in a giant clam, *Tridacna maxima*, in Fiji. *J. Invertebr. Pathol.* 61, 328-330.

Novoa, B. and Figueras, A. (2000). Virus-like particles associated with mortalities of the carpet-shell clam *Ruditapes decussatus*. *Dis. Aquat. Org.* 39, 147-149.

O. I. E. (2011). Aquatic Animal Health Code. Office Internationale des Epizooties, Paris. Available from: http://www.oie.int/eng/normes/ fcode/en_sommaire.htm.

Oprandy, J. J., Chang, P. W., Pronovost, A. D., Cooper, K. R., Brown, R. S., and Yates, V. J. (1981). Isolation of a viral agent causing hematopoietic neoplasia in the soft-shell clam, *Mya arenaria*. *J. Invertebr. Pathol.* 38, 45-51.

Otto, S. V., Harshbarger, J. C. and Chang, S.-C. (1979). Status of selected unicellular eucaryote pathogens, and prevalence and histopathology of inclusions containing obligate procaryote parasites, in commercial bivalve mollusks from Maryland estuaries. *Haliotis*. 8, 285-295.

Paillard, Ch. (2004). A short-review of brown ring disease, a vibriosis affecting clams, *Ruditapes philippinarum* and *Ruditapes decussatus*. *Aquat. Living. Resour.* 17, 467-475.

Park, M. S., Kang, C. K., Choi, D. L., and Jee, B. Y. (2003). Appearance and Pathogenicity of ovarian parasite *Marteilioides chungmuensis* in the farmed pacific oysters, *Crassostrea gigas*, in Korea. *J. Shellfish Res.* 22, 475-479.

Park, K. I., Ngo, T. T. T., Choi, S. D., Cho, M., and Choi, K. S. (2006). Occurrence of *Perkinsus olseni* in the venus clam *Protothaca jedoensis* in Korean waters. *J. Invertebr. Pathol.* 93, 81-87.

Park, K.-I., Yang, H.-S., Kang, H.-S., Cho, M., Park, K.-J., and Choi, K.-S. (2010). Isolation and identification of *Perkinsus olseni* from feces and marine sediment using immunological and molecular techniques. *J. Invertebr. Pathol.* 105, 261-269.

Pecher, W. T., Alavi, M. R., Schott, E. J., Fernandez-Robledo, J. A., Roth, L., Berg, S. T., and Vasta, G. R. (2008). Assessment of the northern distribution range of selected Perkinsus species in eastern oysters (*Crassostrea virginica*) and hard clams (*Mercenaria mercenaria*) with the use of PCR-based detection assays. *J. Parasitol.* 95, 410-422.

Perkins, F. O. (1996). The structure of *Perkinsus marinus* (Mackin, Owen and Collier, 1950) Levine, 1978 with comments on taxonomy and phylogeny of *Perkinsus* spp. *J. Shellfish Res.* 15, 67-87.

Ragan, M. A., MacCallum, G. S., Murphy, C. A., Cannone, J. J., Gutell, R. R., and McGladdery, S. E. (2000). Protistan parasite QPX of hard-shell clam *Mercenaria mercenaria* is a member of the Labyrinthulomycota. *Dis. Aquat. Org.* 42, 185-190.

Ragone Calvo, L. M., Dungan, C. F., Roberson, B. S., and Burreson, E. M. (2003). Systematic evaluation of factors controlling *Perkinsus marinus* transmission dynamics in lower Chesapeake Bay. *Dis. Aquat. Org.* 56, 75-86.

Ramón, M., Gracenea, M. and González-Moreno, O. (1999). *Bacciger bacciger* (Trematoda, Fellodistomidae) infection in commercial clams *Donax trunculus* (Bivalvia, Donacidae) from the sandy beaches of the western Mediterranean. *Dis. Aquat. Org.* 35, 37-46.

Ray, S. M. (1966). A review of the culture method for detecting *Dermocystidium marinum*, with suggested modifications and precautions. *Proc. Nat. Shellfish. Assoc.* 54, 55-69.

Reece, K. S., Siddall, M. E., Stokes, N. A., and Burreson, E. M. (2004). Molecular phylogeny of the Haplosporidia based on two independent gene sequences. *J. Parasitol.* 90, 1111-1122.

Reece, K. S., Dungan, C. F. and Burreson, E. M. (2008). Molecular epizootiology of *Perkinsus marinus* and *P. chesapeaki* infections among wild oysters and clams in Chesapeake Bay, US. *Dis. Aquat. Org.* 82, 237-248.

Renault, T. and Novoa, B. (2004). Viruses infecting bivalve molluscs. *Aquat. Living Resour.* 17, 397-409.

Renault, X. T., Faury, N., Barbosa-Solomieu, V., and Moreau, K. (2011). Suppression substractive hybridisation (SSH) and real time PCR reveal differential gene expression in the Pacific cupped oyster, *Crassostrea gigas*, challenged with *Ostreid herpesvirus* 1. *Dev. Comp. Immunol.* 35, 725-735.

Robledo, J. A. F., Coss, C. A. and Vasta, G. R. (2000). Characterization of the ribosomal RNA locus of *Perkinsus atlanticus* and development of a polymerase chain reaction-based diagnostic assay. *J. Parasitol.* 86, 972-978.

Robledo, J. A. F., Nunes, P. A., Cancela, M. L., and Vasta, G. R. (2002). Development of an *in vitro* clonal culture and characterization of the rRNA gene cluster of *Perkinsus atlanticus*, a protistan parasite of the clam *Tapes decussatus*. *J. Eukaryotic Microbiol.* 49, 414-422.

Robledo, J. A. F., Caler, E., Matsuzaki, M., Keeling, P. J., Shanmugam, D., Roos, D. S., and Vasta, G. R. (2011). The search for the missing link: a relic plastid in Perkinsus? *Int. J. Parasitol.* 41, 1217-29.

Rodríguez, R., Iglesias, D., Ruiz, M., and López, C. (2009). Digenean trematodes affecting razor clam Solen marginatus populations from Galicia (NW Spain) natural beds. *Proceedings of the XIV Conference of EAFP*. pp. 68, Prague, Czech Republic.

Rodríguez, F., Godoy, T. and Navas, J. I. 1994. Cross-infection with *Perkinsus atlanticus* in *Ruditapes decussatus*, *Ruditapes philippinarum* and *Venerupis pullastra*. *Bull. Eur. Ass. Fish Pathol.* 14, 24-27.

Romalde, J. L., Vilariño, M. L., Beaz, R., Rodríguez, J. M., Díaz, S., Villalba, A., and Carballal, M. J. (2007). Evidence of retroviral etiology for disseminated neoplasia in cockles (Cerastoderma edule). *J. Invertebr. Pathol.* 94, 95-101.

Ruiz, M., Darriba, S., Rodriguez, R., Iglesias, D., Lee, R., and Lopez, C. (2011). Viral basophilic inclusions in the digestive gland of razor clams *Ensis arcuatus* (Pharidae) in Galicia (NW Spain). *Dis. Aquat. Org.* 94, 239-241.

Ruiz, M., Iglesias, D., Darriba, S., Rodríguez, R., and López. C. (2012). Epidemiological survey of digenean trematodes affecting razor clams, *Ensis arcuatus* (Jeffreys, 1865), from Galicia (NW Spain). *Bull. Eur. Ass. Fish Pathol.* 32, 1-11

Saldarriaga, J. F., McEwan, M. L., Fast, N. M., Taylor, F. J., and Keeling, P. J. (2003). Multiple protein phylogenies show that *Oxyrrhis marina* and *Perkinsus marinus* are early branches of the dinoflagellate lineage. *Int. J. Syst. Evol. Microbiol.* 53, 355-365.

Sheppard, B. J. and Phillips, A. C. 2008. *Perkinsus olseni* detected in Vietnamese aquacultured reef clams *Tridacna crocea* imported to the US, following a mortality event. *Dis. Aquat. Org.* 79, 229-35.

Shelley, C. C., Glazebrook, J. S., Turak, E., Winsor, L., and Denton, G. R. W. (1988). Trematode (Digenea: Bucephalidae) infection in the burrowing clam *Tridacna crocea* from the Great barrier Reef. *Dis. Aquat. Org*. 4, 143-147.

Da Silva, P. M., Cremonte, F., Sabry, R. C., Rosa, R. D., Cantelli, L., and Barranco, M. A. (2009). Presence and histopathological effects of the *Parvatrema* sp. (Digenea, Gymnophallidae) in the stout razor clam *Tagelus plebeius* (Bivalvia, Psammobiidae). *J. Invertebr. Pathol.* 102, 14-20.

Sindermann, C. J. (1990). Part II. Diseases of Shellfish caused by Microbial Pathogens and Animal Parasites. In: Sindermann (Ed.), *Principal Diseases of Marine Fish and Shellfish*, pp. 9-198. Academic Press, San Diego.

Soto, M., Pascual, S., Rodríguez, H., Gestal, C., Abollo, E., Arias, C., and Estévez, J. (1996). *Nematopsis* spp. Schneider, 1892 (Apicomplexa: Gregarinida) in bivalve molluscs off ría de Vigo (Galicia, NW Spain). *Bull. Eur. Ass. Fish Pathol.* 16, 157-160.

Suzuki, S. and Nojima, M. (1999). Detection of marine birnavirus in wild molluscan shellfish species from Japan. *Fish Pathol.* 34, 121-125.

Tasumi, S. and Vasta, G. R. (2007). A galectin of unique domain organization from hemocytes of the Eastern oyster (*Crassostrea virginica*) is a receptor for the protistan parasite *Perkinsus marinus. J. Immunol.* 179, 3086-3098.

Thieltges, D. W. (2006). Parasite induced summer mortality in the cockle *Cerastoderma edule* by the trematode *Gymnophallus choledochus*. *Hydrobiologia.* 559, 455-461.

Thompson, P. C., Rosenthal, B. M. and Hare, M. P. (2011). An evolutionary legacy of sex and clonal reproduction in the protistan oyster parasite *Perkinsus marinus*. *Infect. Genet. Evol.* 11, 598-609.

Tuntiwaranuruk, C., Chalermwat, K., Upatham, E. S., Kruatrachue, M., and Azevedo, C. (2004). Investigation of *Nematopsis* spp. oocysts in 7 species of bivalves from Chonburi Province, Gulf of Thailand. *Dis. Aquat. Org*. 58, 47-53.

Vázquez, N. N., Ituarte, C., Navone, G. T., and Cremonte, F. (2006). Parasites of the stout razor clam *Tagelus plebeius* (Psammobiidae) from the southwestern Atlantic ocean. *J. Shellfish Res.* 25, 877-886.

Vázquez, N., Rodríguez, F., Ituarte, C., Klaich, J., and Cremonte, F. (2010). Host-parasite relationship of the geoduck *Panopea abbreviata* and the green alga *Coccomyxa parasitica* in the Argentinean Patagonian coast. *J. Invertebr. Pathol.* 105, 254-260.

Vilas, R., Cao, A., Pardo, B. G., Fernández, S., Villalba, A., and Martínez, P. (2011). Very low microsatellite polymorphism and large heterozygote deficits suggest founder effects and cryptic structure in the parasite *Perkinsus olseni*. *Infect. Genet. Evol.* 11, 904-11.

Villalba, A., Carballal, M. J., López, C., Cabada, A., Corral, L., and Azevedo, C. (1999). Branchial rickettsia-like infection associated with clam *Venerupis rhomboides* mortality. *Dis. Aquat. Org.* 6, 53-60.

Villalba, A., Reece, K. S., Ordás, M. C., Casas, S. M., and Figueras, A. (2004). Perkinsosis in molluscs: A review. *Aquat. Living. Resour.* 17, 411-432.

Vivier, E. and Desportes, I. (1990). Filo Apicomplexa. In: L. Margulis, J. O. Corliss, M. Melkoniam and D. J. Chapman (Eds.), *Handbook of protoctista*. pp. 549-573, Jones and Bartlett Learning, London.

Volety, A. K. and Chu, F. L. E. (1994). Comparison of infectivity and pathogenicity of meront (trophozoite) and prezoosporangiae stages of the oyster pathogen *Perkinsus marinus* in eastern oysters, *Crassostrea virginica* (Gmelin, 1791). *J. Shellfish Res.* 13, 521-527.

Wardle, W. J. (1988). A bucephalid larva, *Cercaria pleuromerae* n. sp. (Trematoda: Digenea), parasitizing a deepwater bivalve from the Gulf of Mexico. *J. Parasitol.* 74, 692-694.

Wardle, W. J. (1990). Larval bucephalids (Trematoda: Digenea) parasitizing bivalve molluscs in the Galveston Bay Area, Texas. *J. Helminthol. Soc. Wash.* 57, 5- 11.

Wen, C.-M., Kou, G.-H., Chen, S.-N. (1994). Rickettsiaceae-like microorganisms in the gill and digestive gland of the hard clam, *Meretrix lusoria* Röding. *J. Invertebr. Pathol.* 64, 138-142.

Zhang, X. C., Liang, Y. B., Fan, J. F., Zhang, W., Pu, H. Y., Liang, B., Chen, H. X., and Song, L. C. (2005). Identification of *Perkinsus*-like parasite in Manila clam, *Ruditapes philippinarum* using DNA molecular marker at ITS region. *Acta Oceanol. Sin.* 24, 139-144.

Xu, K. D., Song, W. B. and Warren, A. (1999). Thrichodinic ectoparasites (Ciliophora: Peritrichia) from the gills of mariculture molluscs in China, with the descriptions of four new species of *Trichodina* Ehrenberg, 1838. *Syst. Parasitol.* 42, 229-237.

In: Clam Fisheries and Aquaculture
Editor: Fiz da Costa González

ISBN: 978-1-62257-518-3

Chapter 6

NEOPLASMS IN CLAMS

M. Ruiz and C. López[*]
Centro de Investigacións Mariñas (CIMA),
Vilanova de Arousa (Pontevedra), Spain

ABSTRACT

The purpose of this chapter is to review the neoplasms (or tumors) affecting different species of clams. We include as "clams" any bivalve mollusc other than oysters, mussels and scallops. Two types of neoplasm have been described in clams: disseminated neoplasms and germinoma (one histotype of gonadal neoplasms). Disseminated neoplasm, the type prevalent in bivalve molluscs, including clams, involves the extensive proliferation of circulating abnormal cells (neoplastic cells) of unknown origin, through the tissues of the organisms. Germinomas comprise the abnormal proliferation of altered immature germ cells (neoplastic cells). Some clam species and populations present epidemic levels of neoplasms (*Mya arenaria*, *Mercenaria* spp., *Cerastoderma edule* and *Venerupis aurea*), while others are less affected (*Venerupis decussata* (=*Ruditapes decussatus*), *Ensis magnus* (=*E. arcuatus*) and *Ensis siliqua*). The etiology of neoplasms is unknown, some studies suggesting the implication of an infectious agent and other factors in the case of disseminated neoplasms. By contrast, only stress factors seem to be involved in gonadal neoplasms. In recent years, many studies have been carried out focusing on the molecular basis of this disease.

6.1. INTRODUCTION

There are references in the literature to the presence of different types of neoplasms in bivalve molluscs, the predominant one being the disseminated neoplasm, followed by the germinoma (one type of gonadal neoplasm). Elston et al. (1992) describe the confusion that has arisen in the terminology used to describe these conditions: terms cited include "disseminated neoplasia", "hemocytic neoplasia", "disseminated neoplasm", "disseminated

[*] Corresponding author: C. López. Centro de Investigacións Mariñas (CIMA). Consellería do Mar. Pedras de Corón, s/n. Apdo. 13, 36620, Vilanova de Arousa (Pontevedra), Spain. E-mail address: clopez@cimacoron.org.

sarcoma", "gonadal neoplasia" and "gonadal neoplasm", among others. The lack of knowledge regarding the tissue of origin contributes to this confusion, the terms "disseminated" and "hemocytic" being used to refer to the suspected hematopoietic origin, due to their occurrence in the circulatory system. The meanings of the terms "neoplasia", "neoplasm" and "sarcoma" are different in current medical usage. A neoplasm (or tumor) is an abnormal mass of tissue caused by an uncontrolled and progressive cell proliferation which continues after the initiating stimulus has disappeared. Sarcoma and carcinoma are neoplasms with non-epithelial and epithelial origin, respectively, while neoplasia is the process of an abnormal proliferation of cells in a tissue or organ leading to a neoplasm. A neoplasm can be benign (well-circumscribed, slow-growing and well-differentiated cells) or malignant (invasive fast-growing, anaplastic or undifferentiated cells). Since the late nineteen-sixties, tumors have been reported with increasing frequency in populations of several marine bivalve molluscs around the world, including clams. Barber (2004) and Bower (2006a,b) have updated previous reviews of bivalve tumors (Pauley, 1969; Peters, 1988; Sindermann, 1990; Elston et al., 1992; Peters et al., 1994).

6.2. Disseminated Neoplasms

Disseminated neoplasms are characterized by the presence of circulating hypertrophic cells (usually two to four times the diameter of normal cells) with a hyperchromatic and, frequently, pleomorphic nucleus, and with one or more patent nucleoli. Mitotic figures are frequently present (Figure 1). Neoplastic cells are detected invading the connective tissue, blood vessels, sinuses of the visceral mass, muscle, mantle tissue and gills of bivalves (Peters, 1988; Elston et al., 1992). Ultrastructural studies of cells have supported the diagnosis of these abnormal cells as neoplastic cells. In their revision of disseminated neoplasms, Elston et al. (1992) reported the existence of different neoplastic cell types (I and II) in *Mya arenaria* clams. Type I closely coincides with the previous description; type II neoplastic cells present intermediate characteristics, that is, between type I cells and normal cells, including round, medium-sized nuclei, rare mitotic figures and less prominent nucleoli. Two types of neoplastic cells were also reported in the cockle *Cerastoderma edule* (Carballal et al., 2001).

The question of whether cellular type II is an intermediate stage between normal and neoplastic cells (type I and II from same cellular line) or whether the two cellular types belong to two different cellular lines, has yet to be resolved. The first report of disseminated neoplasms in bivalves was in *Crassostrea gigas* and *C. virginica* (Farley, 1969a). In clams, disseminated neoplasms were reported for the first time in the US in *Mya arenaria* (Yevich and Barszcz, 1976) and in Europe in *Cerastoderma edule* from Ireland (Twomey and Mulcahy, 1984).

Other clam populations affected by this neoplasm have been reported in North America, *Macoma calcarea* and *Mya truncata* in Canada, *Macoma nasuta, Macoma irus* and *Macoma inquinata* in the US, *Mya arenaria* in different states of the US and Canada (Peters, 1988; Elston et al., 1992; Bower, 2006a). In Europe, *Cerastoderma edule* in France and Spain, *Venerupis decussata* (=*Ruditapes decussatus*), *Venerupis aurea*, *Solen marginatus* and *Ensis siliqua* in Spain and *Mya arenaria* in Poland (Poder and Auffret, 1986; Villalba et al., 1995, 2001; Wolowicz et al., 2005; Iglesias et al., 2007; López et al., 2011).

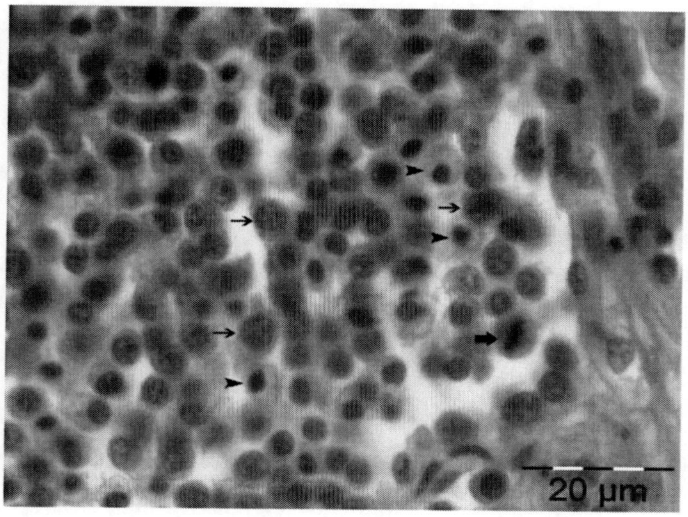

Figure 1. Disseminated neoplasm in gill of *Solen marginatus*. Neoplastic cells (thin arrows), mitotic figure (thick arrow) and normal cells (arrowheads).

In *Mya arenaria* in North America (Farley et al., 1986, 1991; McGladdery et al., 2001) and in *Cerastoderma edule* in Europe (Twomey and Mulcahy, 1988; Villalba et al., 2001), this proliferative disease was seen to reach epidemic levels, and was associated to mortalities. More recently, the presence of disseminated neoplasms has also been reported in *V. aurea* with high prevalence, and is associated with mortalities (Iglesias et al., 2007). Disseminated neoplasms can appear in *M. arenaria* all year round, but two different seasonal patterns were reported with the highest prevalence being during autumn (Leavitt et al., 1990) and summer (Weinberg et al., 1997). In *C. edule*, high prevalence occurs throughout the year, although a seasonal pattern with two peaks of high prevalence (late spring-early summer and late autumn-early winter) were reported by Twomey and Mulcahy (1988). Epidemiological studies have showed that prevalence can fluctuate seasonally due to mortality of the individuals with high intensities of disseminated neoplasms. In some instances, disease prevalence has been correlated with seasonal reproductive activity, the sex of individuals, or environmental conditions. Additionally, the disseminated neoplasm is capable of affecting the population structure by selectively killing particular age groups (Barber et al., 2004).

The classic diagnostic method for disseminated neoplasm diseases is histology. This technique allows us to observe all the organs and thus determine disease intensity on the basis of the degree of neoplastic cell propagation. In addition, histology provides a large amount of information, allowing us to detect other disease agents or physiological problems. The obvious limitation of histology is that, in the absence of biopsy methods, it is necessary to sacrifice the animal. Other non-lethal techniques based on hemolymph extraction have been developed, such as hemocytological smears, flow cytometry (based on the measurement of hemolymph DNA content) and immunology (using monoclonal antibodies specific for neoplastic cells) (Elston et al., 1992). Some of these techniques have been reviewed by Barber (2004) and Smolarz et al. (2005a); they reported the advantages and disadvantages with regard to the sensitivity and accuracy of the diagnosis made with some of these techniques. Barber (2004) also describes different quantitative and qualitative scales to study disease progression.

Before the use of flow cytometry techniques, Lowe and Moore (1978) made a systematic study of DNA levels using quantitative micro-densitometry to measure the DNA of Feulgen-stained cells in histological preparations. Flow cytometry used to measure ploidy status in bivalves showed that neoplastic cells contain more DNA than normal cells (Barber, 2004); hyperdiploid (>2n) to heptadecaploide (17n) neoplastic cell populations were detected in different marine bivalves (Bower, 1989; Elston et al., 1990; Smolarz et al., 2005a,b; Da Silva et al., 2005; Díaz et al., 2005; Delaporte et al., 2008a). Ploidy variation is related to cellular morphological changes and disease progression (Elston et al., 1990). Delaporte et al. (2008b) developed a double staining protocol for flow cytometry, using propidium iodide for hemocyte cycle analysis and a specific monoclonal antibody of neoplastic cells, finding a positive correlation between tetraploid cells and antibody-stained cells. Galimany and Sunila (2008), using flow cytometry, observed that neoplastic cells have less fagocytic ability (this is in concordance with the observation that they lose the ability to emit pseudopods) and more apoptotic activity than normal cells.

As regards the ontogenesis of disseminated neoplasms, there are several hypotheses, though the majority of authors postulate a hematopoietic origin, due mainly to the morphological similarities existing between neoplastic cells and granular hemocytes and hyalinocytes, and to the fact that they are initially detected in the circulatory system (Barber, 2004). There is an experimental study that support this theory (Smolowitz et al., 1989), though it is hard to confirm, due to the fact that we do not know the origin of hematopoiesis in bivalves.

About the etiology of disseminated neoplasms, the ultrastructural examination of neoplastic cells in different species failed to reveal the presence of particles of viral origin in the nucleus of neoplastic cells. However, there are studies which do support the hypothesis of a viral etiology, as it was observed that diseases were transmitted between individuals, and reverse transcriptase activity was also noticed in neoplastic cells (Collins and Mulcahy, 2003; Barber, 2004; Romalde et al., 2007).

AboElkhair et al. (2009a,b) also detected a greater reverse transcriptase activity in *Mya arenaria*, with disseminated neoplasia, but viral particles were not observed, so it is thought that this increase in reverse transcriptase activity may be due to an endogenous source and not to the presence of an exogenous retrovirus.

It is important to draw attention to the study realized by Walker et al. (2009), in which satisfactory *in vitro* trials of maintenance of neoplastic cells in *Mya arenaria* were first carried out. The development of a molluscan cell line may provide information regarding the hematopoiesis and transformation of neoplastic cells, besides making it easier to carry out experiments aimed to determine the possible viral etiology of disseminated neoplasms.

Oprandy et al. (1981) managed to isolate a viral agent, but doubt has been cast on their work, because their results were not reproduced in later trials (Elston et al., 1992).

Barber (2004) concludes that the likeliest hypothesis is that the etiological agent of disseminated neoplasms has an infectious origin, though the disease may be modulated by different stress factors, such as environmental contaminants, biotoxins, carcinogenic transformations or other factors.

With regard to the pathogenesis of the disease, the neoplastic cells usually proliferate to occupy all hemolymph spaces, causing the displacement of normal tissues and necrosis in advanced cases (Elston et al., 1988). In some clam species (*M. arenaria*, *C. edule* and *V. aurea*) and also other marine bivalves, the disease is progressive and fatal, but some individuals are able to develop a host response, resulting in remission, at least temporarily. This remission phenomenon is characterized by the entrapment of neoplastic cells in a fibrous extracellular matrix material, probably secreted by normal hemocytes (Elston et al., 1988).

The observations of remission and attendant host response thus showed that the immune systems of certain individual bivalves are capable of recognising neoplastic cells and can retard or reverse the advance of the disease (Elston et al., 1992).

a)

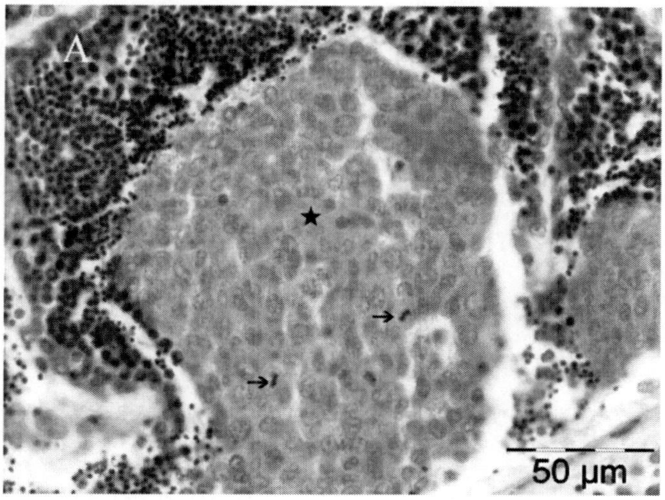

b)

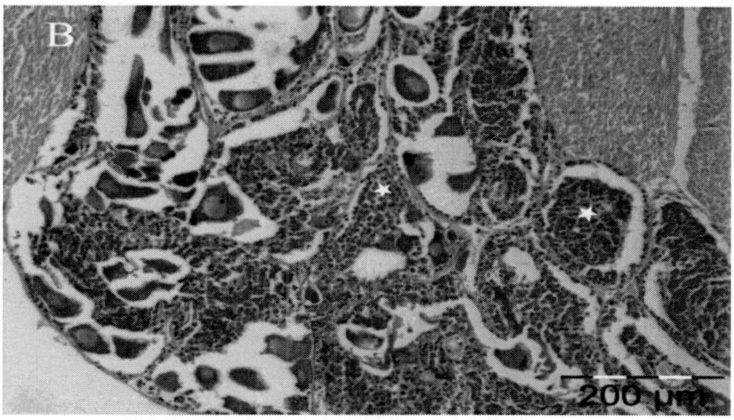

Figure 2. Germinoma of *Ensis magnus*. A) Germinoma in male (star) with mitotic figures (arrows). B) Germinoma in female (stars).

6.3. Gonadal Neoplasms

There are three gonadal neoplasm histotypes: 1) germinoma, the most frequent, with germ cell origin; 2) gonadal stromal neoplasm, of stromal origin, and 3) gonadoblastoma, of germinal and stromal origin (Peters et al., 1994). The morphological characteristics of germinoma vary, depending on the clam species (Barber, 2004).

In *Mercenaria* spp., *Ensis magnus* (=*E. arcuatus*) and *Ensis siliqua*, the germinoma appeared as a neoplastic cellular mass characterized by the replacement of normal germinal epithelium with large, polyhedral cells with a clear eosinophilic cytoplasm (Figure 2) (Barber, 2004; López et al., 2011).

Nuclei are large, with clumped, granular, occasionally marginated chromatin and with prominent nucleoli in the case of *Mercenaria* spp. germinoma (Peters et al., 1994; Barber, 2004); however, in *E. magnus* and *E. siliqua* nucleoli were seldom apparent (Darriba et al. 2006; Ruiz et al., submitted). By contrast, in *Mya arenaria* the germinoma is a neoplastic cell mass characterized by small, basophilic, undifferentiated germ cells, eccentric nuclei, with the nucleolus frequently being difficult to distinguish (Barber, 2004). Mitotically active cells were observed in the neoplastic cell mass of germinoma, suggesting rapid growth, although not as commonly as in some cases of disseminated neoplasms (Barber, 2004; López et al., 2011).

The first report of germinoma in clam species was in *Mercenaria mercenaria* in the US (Yevich and Barry, 1969). Several studies have reported other clam populations affected by this tumor: *Arctica islandica*, *Mercenaria campechiensis* and *Mya arenaria* in US, *Macoma calcarea* in Canada, *Cerastoderma edule* in Ireland (Peters et al., 1994). Most recently, it was reported in *Ensis magnus* (Darriba et al., 2006) and in *E. siliqua* in Spain (Ruiz et al. submitted).

The little fieldwork carried out indicates that germinoma displays slow progression, with a low mortality rate, and that the most significant damage produced is a reduction in the capacity to reproduce. A significantly higher percentages of females than males of *Mya arenaria* and in *Mercenaria* spp., have been reported to be affected, may be due to a greater vulnerability to this disease among females, or to the fact that this disease affects sexual differentiation (Barber, 2004).

Nevertheless, in *E. magnus* and *E. siliqua* the prevalence of germinoma was higher in males (Darriba et al., 2006; Ruiz et al., submitted). Hesselman et al. (1988) found higher prevalences of germinoma in summer in *Mercenaria* spp.; however Barber et al. (1996) did not find a seasonal pattern in *Mya arenaria.* The data reported by Darriba et al. (2006) and Ruiz et al. (submitted) were insufficient to establish a statistically valid temporal pattern.

Histological examination is the most frequently-used technique for gonadal neoplasm detection; the fact that neoplastic germ cells are not prevalent in the circulatory system excludes hemocytological techniques as valid diagnostic tests. The fact that non-lethal techniques cannot be applied makes difficult to determine the evolution of the disease (progression and/or remission) in a single animal, or estimate epidemiological parameters as incidence rates of gonadal neoplasms in the populations.

As regards the etiology of germinoma, the fact that greater prevalence has been found in hybrid species of *Mercenaria* spp., as against pure species, suggests a possible genetic etiology (Bert et al., 1993). The fact that transmission of germinoma from affected individuals to healthy ones has not been demonstrated, either in fieldwork or in laboratory experiments, would seem to support the hypothesis that no infectious agent is involved in the disease's development (Barber et al., 2002). Several authors have investigated the possible association between gonadal neoplasia and the presence of high levels of different contaminants, such as fuel oil and herbicides (Yevich and Barszcz, 1977; Gardner et al., 1991).

However, since it has not been possible to induce the disease, and this does not appear in all the populations exposed to contaminants, and alse cases of germinoma have been detected in areas free of contamination, it is considered that environmental contaminants are not the factor that triggers the disease, although they may be indirectly involved. Some authors suggest other stress factors, such as tenvironmental and genetic factors as etiological agents of gonadal neoplasia (Barber et al., 2002).

Later, regarding the possible role of environmental contaminants in inducing neoplasia in molluscs, Butler et al. (2004) carried out an assay in *M. arenaria*, whereby clams were exposed to the 2,3,7,8-tetrachlorodibenzo-p-dioxin (TCDD) in order to determine whether this product induced germinoma. Germinoma was not developed after six months of exposure to TCDD; however, alterations in gametogenesis were detected, characterized by a larger percentage of sexually non-differentiated individuals, though it is not known whether these alterations represent a pre-tumoral state. Later, Pariseau et al. (2009) carried out a trial in order to determine whether there was a correlation between the exposure of *M. arenaria* individuals to two fungicides and the development of disseminated neoplasia, it being observed that although the disease did not develop with chronic or acute exposure, in the case of acute exposure, high concentrations of toxic compounds in mantle tissue were reached after a expositon during 24 or 48 hours (depending on the type of fungicide used), the concentration descending progressively until it reached undetectable levels at the end of the trial (total duration of the trial: 72 hours). The authors conclude that, even though the accumulation of fungicides in the mantle tissue is transitory, the fact that the active ingredients of the fungicides tested provoke oxidative stress may induce some kind of cell dysfunction, making necessary to study the effects of these fungicides in proteins such as p53.

Concerning the pathogenesis of germinoma, the neoplastic cells can appear initially as small foci in one or more gonadal follicles, multiplying to completely fill most or all follicles. The intensities reported were higher in *Mercenaria* spp. and *M. arenaria* than in *Ensis* spp. (Barber, 2004; Darriba et al., 2006; Ruiz et al., submitted). Metastasis invading interfollicular connective tissue, body wall, epibranchial chamber, and genital ducts were reported in *Mercenaria* spp. and *M. arenaria*; based on all these characteristics, germinoma in these species are considered malignant (Barber, 2004). In *Ensis spp.* it was not observed that metastasis invaded other tissues (Darriba et al. 2006; Ruiz et al. submitted), though brown cells associated with the germinoma as a host defence process were suggested (Figure 3) (Ruiz et al., submitted), as was suggested in other bivalve species (Peters et al., 1994).

6.4. Other Types of Neoplasms

Gill carcinoma (a tumor with epithelial origin) reported in *Macoma balthica* (Christensen et al., 1974; Farley, 1976) is often confused with disseminated neoplasms. Smolarz et al. (2006), carried out an ultrastructural study of neoplastic cells in *M. balthica*, describing two types of neoplastic cells, the first characterized by rounded cells, and the second characterized by fusiform cells similar to myofibroblasts, with some intracytoplasmic filaments, and with nuclei radially-oriented in "palisade" formation.

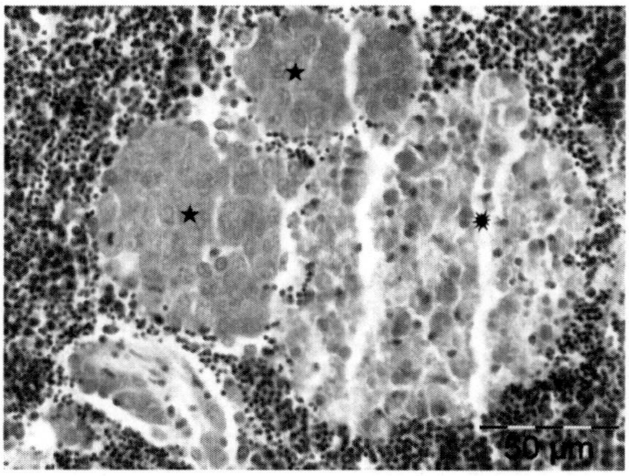

Figure 3. Germinoma (stars) in male of *Ensis siliqua*, with brown cells (asterisk).

The first type of neoplastic cells was detected in all the individuals affected, while the second was only observed in five of the eleven cases diagnosed as neoplastic. This last study is the first description of the coexistence of two different types of neoplastic cells, suggesting two possible kinds of cancer, the disseminated neoplasm and probably a fibrosarcoma.

6.5. Genetic Alterations Associated with Neoplasms in Bivalves

Proto-oncogenes and tumor suppressor genes are involved in the regulation of cell division. The mutation of proto-oncogenes, called oncogenes, encodes proteins (oncoproteins) which increase cell division, decrease cell differentiation and inhibit cell death. The tumor suppressor genes (anti-oncogenes) encode proteins which inhibit the tumorigenic process. The mutation of these genes induces normal cells to become tumoral cells.

Proto-oncogenes and tumor suppressor genes are well-known both in vertebrates and invertebrates, including bivalves (Elston et al., 1992). Barker et al. (1997) detected a mutation in protein p53 within neoplastic cells in *Mya arenaria*, which proceeded from an area with a high degree of xenobiotic contamination. It is thought that the expression of the mutated form of p53 could be used in the future as a biomarker of xenobiotic effects in a marine environment. Stephens et al. (2001) studied the differences, at protein level, between neoplastic cells and normal hemocytes in *Mya arenaria*, observing that neoplastic cells, detected by specific monoclonal antibodies, expressed a very hydrophobic 252 kDa glycoprotein, probably a transmembrane protein. However, normal hemocytes expressed a 185 kDa glycoprotein. Another difference found between the two cell types was the absence of tubulin in normal hemocytes. Furthermore, these last authors analyzed the expression of p53 gene family members in both cell types, finding significant differences in expression of p97 and p73 gene families. In normal hemocytes, it was predominantly the p97 gene that was expressed, while the p73 gene was only expressed in neoplastic cells. Shifts in the ratio of p97/p73/p53 may precede cell transformation, and thus these proteins may serve as useful biomarkers of genetic activation.

In recent years, work has been in progress on the study of molecular bases in the development of neoplastic cells. Ras and p53 genes have been sequenced in several bivalve species, such as *Mytilus edulis* (Ciocan and Rotchell, 2005; Muttray et al., 2005), *Mytilus trossulus* (Muttray et al., 2005; Ciocan et al., 2006), *Spisula solidissima* (Cox et al., 2003), *Crassostrea gigas* (Genbank Accession Number AM236465) and *Crassostrea rhizophorae* (Genbank Accession Number AY442309). The ras gene presents a conserved sequence, so it is believed to be able to develop functions similar to those observed in vertebrates (Ciocan and Rotchell, 2005). Ciocan et al. (2006) studied the ras gene sequence in *Mytilus trossulus*, observing an over-expression of this gene in neoplastic individuals compared with healthy ones; furthermore, polymorphic variations were observed in the ras gene in some neoplastic individuals, which would indicate a high degree of genome instability.

St-Jean et al. (2005), in a study carried out on *Mytilus edulis* in Nova Scotia (Canada), compared the expression of the p53-family protein in neoplastic cells with that in normal hemocytes, by means of monoclonal and polyclonal antibodies developed for these proteins, significant differences being observed in the expression patterns of p63/p73 and p97/p120 proteins, which appeared as over-expressed in neoplastic cells.

Later, Walker et al. (2006) observed that in neoplastic cells in *M. arenaria*, the p53 and mortalin proteins were found to form a complex in the cytoplasm, rather than in the nucleus, where p53 develops its function as a regulator of the cell cycle, which has been described previously in several types of human cancers. These writers detected that when individuals with disseminated neoplasia were treated with rhodacyanine (MKT-077), a rupture was produced in the mortalin-p53 complex, part of the p53 protein, which was found to be sequestered in the cytoplasm, being displaced to the interior of the nucleus, where it activated apoptosis mechanisms.

Siah et al. (2008), quantified the expression of the p53, p73 and mortalin genes of hemocytes of *M. arenaria,* in relation with the ploidy status. They observed that clams with moderate percentage of tetraploid cells present significant high levels of p53 and p73 expression. Furthermore, they found that the expression of p53 and mortalin displayed a strong correlation, which supports the theory that a p53 cytoplasmic sequestration mechanism is produced in neoplastic cells, by means of the formation of a complex with mortalin.

Walker and Böttger (2008), continued the research initiated by Walker et al. (2006) concerning p53 sequestration in the cytoplasm in neoplastic cells, showing that treatment with a nuclear pore blocker wheat germ agglutinin (WGA), followed by treatment with rhodocyanin dye (MKT-077), produces the displacement of p53 from the cytoplasm to the interior of the mitochondria, likewise triggering tumor cell apoptosis. This mechanism is similar to that previously described in mammals.

Muttray et al. (2008) researched the expression of different isoforms of the p53 family (p53, TAp63/73 y ΔNp63/73) in *Mytilus trossulus* with disseminated neoplasia, and observed that all the isoforms were expressed in both healthy and neoplastic groups, though levels of mRNA of p53 and ΔNp63/73 were significantly higher in those individuals with disseminated neoplasia. It is believed that the over-regulation of ΔNp63/73 could impose an oncogenic activity that interferes with the suppressor function of p53 and ΔNp63/73, which would be consistent with what has been observed in several human cancers, where the activity of p53 sometimes appears to be modulated by the presence of other isoforms of the p53 family.

References

AboElkhair, M., Siah, A., Clark, K. F., McKenna, P., Pariseau, J., Greenwood, S. J., Berthe, F. C. J., and Cepica, A. (2009a). Reverse transcriptase activity associated with haemic neoplasia in the soft-shell clam *Mya arenaria*. *Dis. Aquat. Org*. 84, 57-63.

AboElkhair, M., Synard, S., Siah, A., Pariseau, J., Davidson, J., Johnson, G., Greenwood, S. J., Casey, J. W., Berthe, F. C. J., and Cepica, A. (2009b). Reverse transcriptase activity in tissues of the soft shell clam *Mya arenaria* affected with haemic neoplasia. *J. Invertebr. Pathol*. 102, 133-140.

Barber, B. J. (1996). Effects of gonadal neoplasms on oogenesis in soft- shell clams, *Mya arenaria*. *J. Invertebr. Pathol.* 67, 161-168.

Barber, B. J., MacCallum, G. S., Robinson, S. M. C., and McGladdery, S. E. (2002). Occurrence and lack of transmissibility of gonadal neo- plasia in softshell clams, *Mya arenaria*, in Maine (US) and Atlantic Canada. *Aquat. Living Resour.* 15, 319-326.

Barber, B. J. (2004). Neoplastic diseases of commercially important marine bivalves. *Aquat. Living Resour*. 17, 449-466.

Barker, C. M., Calvert, R. J., Walker, C. W., and Reinisch, C. L. (1997). Detection of mutant p53 in clam leukemia cells. *Exp. Cell Res*. 232, 240-245.

Bert, T. M., Hesselman D. M., Arnold W. S., Moore W. S., Cruz-Lopez, H., and Marelli D. C. (1993). High frequency of gonadal neoplasia in a hard clam (*Mercenaria* spp.) hybrid zone. *Mar. Biol*. 117, 97-104.

Bower, S. M. (1989). The summer mortality syndrome and haemocytic neoplasia in the blue mussels (*Mytilus edulis*) from British Columbia. *Can. Espc. Publ. Fish. Aquat. Sci.* 1703.

Bower, S. M. (2006a). Synopsis of Infectious Diseases and Parasites of Commercially Exploited Shellfish: Haemocytic Neoplasia of Clams. Available from <http://www.pac.dfo-mpo.gc.ca/science/species-especes/shellfish-coquillages/diseases-maladies/pages/hncc-eng.htm.

Bower, S. M. (2006b). Synopsis of Infectious Diseases and Parasites of Commercially Exploited Shellfish: Gonadal Neoplasia of Clams. Available from <http://www.pac.dfo-mpo.gc.ca/science/species-especes/shellfish-coquillages/diseases-maladies/pages/gonncc-eng.htm.

Butler, R. A., Kelley, M. L., Olberding, K. E.,Gardner, G. R., and Van Beneden, R. J. (2004). Aryl hydrocarbon receptor (AhR)-independent effects of 2,3,7,8-tatrechlorodibenzo-p-dioxin (TCDD) on softshell clam (*Mya arenaria*) reproductive tissue. *Com. Biochem. Physiol*. 138, 375-381.

Carballal, M. J., Iglesias, D., Santamarina, J., Ferro-Soto, B., and Villalba, A. (2001). Parasites and pathologic conditions of the cockle *Cerastoderma edul*e populations of the coast of Galicia (NW Spain). *J. Invertebr. Pathol*. 78, 87-97.

Ciocan, C. M. and Rotchell, J. M. (2005). Conservation of cancer genes in the marine invertebrate *Mytilus edulis*. *Environ. Sci. Technol*. 39, 3029-3033.

Ciocan, C. M., Moore, J. D. and Rotchell, J. M. (2006). The role of ras gene in the development of haemic neoplasia in *Mytilus trossulus*. *Mar. Environ. Res*. 62, 147-150.

Collins, C. M. and Mulcahy, M. F. (2003). Cell-free transmission of a haemic neoplasm in the cockle *Cerastoderma edule*. *Dis. Aquat. Org*. 54, 61-67.

Cox, R. L., Stephens, R. E. and Reinisch, C. L. (2003). p63/73 homologues in surf clam: novel signling motifs and implications for control of expression. *Gene*. 320, 49-58.

Christensen, D. J., Farley, C. A. and Kern, F. G. (1974). Epizootic neoplasms in the clam *Macoma balthica* (L.) from Chesapeake Bay. *J. Natl. Cancer Inst*. 52, 1739-1749.

Da Silva, P. M., Soudant, P., Carballal, M. J., Lambert, L. C., and Villalba, A. (2005). Flow cytometric DNA content analysis of neoplastic cells in haemolymph of the cockle *Cerastoderma edule*. *Dis. Aquat. Org*. 67, 133-139.

Darriba, S., Iglesias, D., Harshbarger, J. C., and López, C. (2006). Germinoma in razor clam *Ensis arcuatus* (Jeffreys, 1865) in Galicia (NW Spain). *J. Invertebr. Pathol*. 93, 140-142.

Delaporte, M., Synard, S., Pariseau, J., McKenna, P., Tremblay, R., Davidson, J., and Berthe, F. (2008a). Assessment of haemic neoplasia in different soft shell clam *Mya arenaria* populations from eastern Canada by flow cytometry. *J. Invertebr. Pathol*. 98, 190-197.

Delaporte, M., McKenna, P., Siah, A., and Berthe, F. C. J. (2008b). Immunophenotyping of *Mya arenaria* neoplastic hemocytes using propidium iodide and a specific monoclonal antibody by flow cytometry. *J. Invertebr. Pathol*. 99, 120-122.

Díaz, S., Soudant, P., da Silva, P. M., Carballal, M. J., Lambert, C., and Villalba, A. (2005). Flow cytometric DNA content analyses of normal haemocytes and neoplastic cells in the haemolymph of the cockle *Cerastoderma edule*. *Proceedings of the XII Conference of EAFP*, pp. 207, Copenhagen, Denmark.

Díaz, S., Cao, A., Villalba, A., and Carballal, M. J. (2010). Expression of the mutant protein p53, Hsp70 and Hsp90 chaperons in cockles *Cerastoderma edule* affected by neoplasia. *Dis. Aquat. Org*. 90, 219-226.

Elston, R. A., Kent, M. L. and Drum, A. S. (1988). Progression, lethality and remission of hemic neoplasia in the bay mussel *Mytilus edulis*. *Dis. Aquat. Org*. 4, 135-142.

Elston, R. A., Drum, A. S. and Allen, S. K. Jr. (1990). Progressive development of circulating polyploid cells in *Mytilus* with hemic neoplasia. *Dis. Aquat. Org*. 8, 51-59.

Elston, R. A., Moore, J. D. and Brooks, K. M. (1992). Disseminated neoplasia of bivalve molluscs. *Rev. Aquat. Sci*. 6, 405-466.

Farley, C. A. (1969). Probable neoplastic disease of the hematopoietic system in oysters, *Crassostrea virginica* and *Crassostrea gigas*. *National Cancer Institute Monographs*. 31, 541-555.

Farley, C. A. (1976). Proliferative disorders in bivalve mollusks. *Mar. Fish. Rev*. 38, 30-33.

Farley, C. A., Otto, S. V. and Reinisch, C. L. (1986). New occurrence of epizootic sarcoma in Chesapeake Bay soft shell clams, *Mya arenaria*. *Fish. Bull*. 84, 851-857.

Farley, C. A., Plutschak, D. L. and Scout, R. F. (1991). Epizootiology and distribution of transmissible sarcoma Maryland soft shell clams, *Mya arenaria*, 1984-1988. *Environ. Health. Perspect*. 90, 35-41.

Galimany, E. and Sunila, I. (2008). Several cases of disseminated neoplasia in mussels *Mytilus edulis* (L.) in Western Long Island Sound. *J. Shellfish Res*. 27, 1201-1207.

Gardner, G. R., Yevich, P. P., Hurst, J., Thayer, P., Benyi, S., Harshbarger, J. C., and Pruell, R. J. (1991). Germinomas and teratoid siphon anomalies in soft shell clams, *Mya arenaria*, environmentally exposed to herbicides. *Environ. Health Perspect*. 90, 43-51.

Hesselman, D. M., Blake, N. J. and Peters, E. C. (1988). Gonadal neoplasms in hard shell clams *Mercenaria* spp., from the Indian River, Florida: Occurrence, prevalence, and histopathology. *J. Invertebr. Pathol*. 52, 436-446.

Iglesias, D., Carballal, M. J. and Villalba, A. (2007). An epizootic proliferative disorder detected in a natural population of the clam, Venerupis aurea in Galicia (NW Spain). *Proceedings of the XIII Conference of EAFP*, pp. 231, Grado, Italy.

Leavitt, D. F., Capuzzo, J. M., Smolowitz, R. M., Miosky, D. L., and Lancaster, B. A. (1990). Hematopoietic neoplasia in *Mya arenaria*: prevalence and indices of physiological condition. *Mar. Biol.* 105, 313-321.

López, C., Darriba, S., Iglesias, D., Ruiz, M., and Rodríguez, R. (2011). Chapert 7. Pathology of sword razor shell (*Ensis arcuatus*) and grooved razor shell (*Solen marginatus*). In: A. Guerra, C. Lodeiros, M. Gaspar, F. da Costa (Eds.). *Razor clams: biology, aquaculture and fisheries*. pp. 161-168, Xunta de Galicia, Santiago de Compostela.

Lowe, D. M. and Moore, M. N. (1978). Cytology and Quantitative Cytochemistry of a Proliferative Atypical Hemocytic Condition in *Mytilus edulis* (Bivalvia, Mollusca). *J. Natl. Cancer Inst.* 60, 1455-1459.

McGladdery, S. E., Reinisch, C. L., MacCallum, G. S., Stephens, R. E., Walker, C. L., and Davidson, J. T. (2001). Haemic neoplasia in soft-shell clams (*Mya arenaria*): recent outbreaks in Atlantic Canada and discovery of a p53 gene homologue associated with the condition. *Bull. Aquac. Assoc. Can.* 101, 21-22.

Muttray, A. F., Cox, R. L., St-Jean, S., van Poppelen, P., Reinisch, C. L., and Baldwin, S. A. (2005). Identification and phylogenetic comparison of p53 in two distinct mussel species (*Mytilus*). *Comp. Biochem. Physiol. C. Toxicol. Pharmacol.* 140, 237-250.

Muttray, A. F., Schulte, P. M. and Baldwin, S. A. (2008). Invertebrate p53-like mRNA isoforms are differentially expressed in mussel haemic neoplasia. *Mar. Environ. Res.* 66, 412-421.

Oprandy, J. J., Chang, P. W., Pronovost, A. D., Cooper, K. R., Brown, C. W., and Yates, V. J. (1981). Isolation of a viral agent causing hematopoietic neoplasia in the soft-shell clam, *Mya arenaria*. *J. Invertebr. Pathol.* 38, 45-51.

Pariseau, J., Saint-Louis, R., Delaporte, M., AboElkhair, M., McKenna, P., Tremblay, R., Davidson, T. J, Pelletier, E., and Berthe, F. C. J. (2009). Potential link between exposure to fungicides chlorothalonil and mancozeb and haemic neoplasia development in the soft-shell clam *Mya arenaria*: A laboratory experiment. *Mar. Pollut. Bull.* 58, 503-514.

Pauley, G. B. (1969). A critical review of neoplasias and tumor-like lesions in mollusks. *Nat. Cancer Inst. Monogr.* 31, 509-529.

Peters, E. C. (1988). Recent Investigations on the disseminated sarcomas of marine bivalve molluscs. *Am. Fish. Soc. Special Publ.* 18, 74-92.

Peters, E. C., Yevich, P. P., Harshbarger, J. C., and Zaroogian, G. E. (1994). Comparative histopathology of gonadal neoplasms in marine bivalve molluscs. *Dis. Aquat. Org.* 20, 59-76.

Poder, M. and Auffret, M. (1986). Sarcomatus lesion in the cockle *Cerastoderma edule* L. I. Morphology and population survey in Brittany, France. *Aquaculture*. 58, 1-8.

Romalde, J. L., Vilariño, M. L, Beaz, R., Rodríguez, J. M., Díaz, S., Villalba, A., and Carballal, M. J. (2007). Evidence of retroviral etiology for disseminated neoplasia in cockles (*Cerastoderma edule*). *J. Invertebr. Pathol.* 94, 95-1001.

Ruiz, M., Darriba, S., Iglesias, D., Rodríguez, R., and López, C. (submitted). Histological survey of symbionts and other conditions of pod razor clam *Ensis siliqua* (Linnaeus, 1758) in Galicia (NW Spain). *J. Invertebr. Pathol.*

Siah, A., Delaporte, M., Pariseau, J., McKenna, P., and Berthe, F. C. J. (2008). Patterns of p53, p73 and mortalin gene expression associated with haemocyte polyploidy in the soft-shell clam, *Mya arenaria*. *J. Invertebr. Pathol.* 98, 148-152.

Sindermann, C. J. (1990). Part III. Noninfectious Diseases of Shellfish. In: C. J. Sindermann (Ed.), *Principal Diseases of Marine Fish and Shellfish*. pp. 209-242, Academic Press, San Diego.

Smolarz, K., Renault, T., Soletchnik, P., and Wolowicz, M. (2005a). Neoplasia detection in *Macoma balthica* from the Gulf of Gdansk: comparison of flow cytometry, histology and chromosome analysis. *Dis. Aquat. Org.* 65, 187-195.

Smolarz, K., Reanult, T., Soletchnik, P., and Wolowicz, M. (2005b). Survey for neoplasia in *Macoma balthica* from the Gulf of Gdansk by flow cytometry. *Dis. Aquat. Org.* 66, 41-46.

Smolarz, K., Renault, T. and Wolowicz, M. (2006). Ultrastructural study of neoplastic cells in *Macoma balthica* (Bivalvia) from the Gulf of Gdansk (Poland). *J. Invertebr. Pathol*, 92: 79-84.

Smolowitz, R. M., Miosky, D. and Reinisch, C. L. (1989). Ontogeny of leukemic cells of the soft shell clam. *J. Invertebr.Pathol.* 53, 41-51.

Stephens, R. E., Walker, C. W. and Reinisch, C. L. (2001). Multiple protein differences distinguish clam leukemia cells from normal haemocytes: evidence for the involvement of p53 homologues. *Comp. Biochem. Physiol. C.* 129, 329-338.

St-Jean, S. D., Stephens, R. E., Courtenay, S. C., and Reinisch, C. L. (2005). Detecting p53 family proteins in haemocytic leukemia cells of Mytilus edulis from Pictou Harbour, Nova Scotia, Canada. *Can. J. Fish. Aquat. Sci.* 62, 2055-2066.

Twomey, E. and Mulcahy, M. F. (1984). A proliferative disorder of possible hemic origin in the common cockle, *Cerastoderma edule*. *J. Invertebr. Pathol.* 44, 109-111.

Twomey, E. and Mulcahy, M. F. (1988). Epizootiological aspects of a sarcoma in the cockle *Cerastoderma edule*. *Dis. Aquat. Org.* 5, 225-238.

Villalba, A., Peters, E. C., López, M. C., and Carballal, M. J. (1995). Disseminated sarcoma in the clam Ruditapes decussatus in Galicia (NW Spain). *J. Invertebr. Pathol.* 65, 76-78.

Villalba, A., Carballal, M. J. and López, M. C. (2001). Disseminated neoplasia and large foci indicating heavy haemocytic infiltration in cockles *Cerastoderma edule* from Galicia (NW Spain). *Dis. Aquat. Org.* 46, 213-216.

Walker, C., Böttger, S. and Low, B. (2006). Mortalin-based cytoplasmic sequestration of p53 in a nonmammalian cancer model. *Am. J. Pathol.* 168, 1526-1530.

Walker, C. W. and Böttger, S. A. (2008). A naturally occurring cancer with molecular connectivity to human diseases. *Cell Cycle*. 7, 2286-2289.

Walker, C., Böttger, A., Mulkern, J., Jerszyk, E., Litvaitis, M., and Lesser, M. (2009). Mass culture and characterization of tumor cells from a naturally occurring invertebrate cancer model: applications for human and animal disease and environmental health. *Biol. Bull.* 216, 23-39.

Weinberg, J. R., Leavitt, D. F., Lancaster, B. A., and McDowell Capuzzo, J. (1997). Experimental Field Studies with *Mya arenaria* (Bivalvia) on the Induction and Effect of Hematopoietic Neoplasia. *J. Invertebr. Pathol.* 69, 183-194.

Wolowicz, M., Smolarz, K. and Sokolowski, A. (2005). Neoplasia in estuarine bivalves: Effect of feeding behaviour and pollution in the golf of Gdansk (Baltic Sea, Poland).

The comparative Roles of Suspension-Feeders in Ecosystems. NATO Science Series: IV: Earth and Environmental Sciences. 47, 165-182.

Yevich P. P. and Barry, M. M. (1969). Ovarian tumours in the quahog *Mercenaria mercenaria. J. Invertebr. Pathol.* 14, 266-267.

Yevich, P. P. and Barszcz, C. A. (1976). Gonadal and hematopoietic neoplasms in *Mya arenaria. Mar. Fish. Rev.* 38, 42-43.

Yevich P. P., Barszcz C. A. (1977). Neoplasia in soft-shell clams (*Mya arenaria*) collected from oil-impacted sites. *Ann. N.Y. Acad. Sci.* 298, 409-426.

In: Clam Fisheries and Aquaculture
Editor: Fiz da Costa González

ISBN: 978-1-62257-518-3

Chapter 7

ADVANCES IN THE KNOWLEDGE OF THE MICROBIOTA ASSOCIATED WITH CLAMS FROM NATURAL BEDS

J. L. Romalde,[1*] A. L. Diéguez,[1] A. Doce,[1] A. Lasa,[1] S. Balboa,[1] C. López[2] and R. Beaz-Hidalgo[3]

[1]Departamento de Microbiología y Parasitología, CIBUS-Facultad de Biología, Universidad de Santiago de Compostela, Santiago de Compostela (A Coruña), Spain

[2]Centro de Investigacións Mariñas, Vilanova de Arousa (Pontevedra), Spain

[3]Departament de Ciències Mèdiques Bàsiques, Facultat de Medicina I Ciènces de la Salut. IISPV, Universitat Rovira I Virgili. Reus (Tarragona), Spain

ABSTRACT

The culture of clams represents an important marine resource and is of great economic importance for many coastal areas worldwide, including the coast of Galicia (NW Spain). In some of these areas, natural beds of authoctonous clams have been harvested beyond their maximum sustainable yield, which has led to the introduction of the foreing species to cover the consumer demands, as was the case in Galicia with the authoctonous carpet-shell clam (*Venerupis* [=*Ruditapes*] *decussata*) and the foreing Manila clam (*Venerupis* [=*Ruditapes*] *philippinarum*). Due to their filter-feeding habit, bivalves normally accumulate a rich bacterial microbiota, composed of various species belonging to different genera like *Vibrio, Pseudomonas, Acinetobacter, Photobacterium, Moraxella, Aeromonas*, *Micrococcus* and *Bacillus*.

One of the main problems in the culture of clams, and other bivalve molluscs, is the repeated episodes of mortality, many of them atributed to bacterial infections, that reduce

[*] Corresponding author: J. L. Romalde. Departamento de Microbiología y Parasitología, CIBUS-Facultad de Biología, Universidad de Santiago de Compostela. 15782 Santiago de Compostela, Spain. E-mail address: jesus.romalde@usc.es.

the production and cause high economical losses. Although some of the pathogens could be introduced associated to the imports of foreing shellfish, it is plausible to think that most of them could be part of the normal microbiota. Despite this fact, very little is known about the normal microbiota associated with clams from natural environments and its role in the appearance of mortality.

In this chapter, a deep literature review together with results obtained by our research group in the last eight years will be presented in order to draw the current state of knowledge on this subject.

7.1. INTRODUCTION

The study of the microbiota associated to cultured populations of different clam species is important to know their sanitary status, as well as to determine the pathobiological bases of the periodic disease outbreaks affecting these populations.

In the literature, studies analyzing the diversity, distribution and density of marine bacteria associated with bivalve molluscs are scarce. These studies date back to the 1960´s and in general, results agree in the dominance of Gram negative over Gram positive bacteria in molluscs, as well as in the high abundance of bacteria belonging to the genus *Vibrio* (Colwell and Liston, 1960; Beenson and Johnson, 1967; Kueh and Chan, 1985; Prieur et al., 1990; Pujalte et al., 1999).

As a matter of fact, most studies have been focused on the characterization of bacteria with pathogenic potential for mollusc, or specifically for clams. Thus, the genus *Vibrio* has been extensively studied, since it is known to be one of the most important bacterial genus affecting the culture of bivalves.

Brown Ring Disease (BRD) caused by *V. tapetis* has received special attention since it is the only disease with bacterial etiology described in adult clams. In addition, it is considered one of the main limiting factors for the culture of Manila clams (*Venerupis philippinarum*) in Europe (Borrego et al., 1996; Paillard et al., 2004), being also recently detected in Korea (Park et al., 2006).

The impact of the infectious diseases in the mollusc production has implied a growing interest in the study of intracellular prokaryotes (Lauckner, 1983). Morphologically, these prokaryotes have been related with the *Chlamydiales* (Moulder, 1984) or the *Rickettsiales* (Weiss and Moulder, 1984). In most cases of important mortality, the detected prokaryotes have been associated with *Rickettsia*-like organisms (RLO).

On the other hand, marine bacteria with beneficial effects (i.e. inhibitory activity against clam pathogens) have been described associated with the microbiota associated to hatcheries of different molluscs (Prado et al., 2009). Their presence in juvenile and adult populations would play a role in a kind of protection against pathogenic bacteria and be indicative of a good condition of the bivalve population.

In this chapter, and in order to present the current state of knowledge on this subject, a deep literature review is presented together with results obtained by our research group in the last eight years in different aspects related with the study of the microbiota associated with clams cultured in natural beds, including description of species and the determination of their pathogenic or probiotic capabilities.

7.2. State of the Art

Two kinds of microbiota associated with marine organisms have been described. The "resident" microbiota which is stable in the host without influence of the environment, and the "transition" microbiota, which vary according to the environmental conditions. Jorquera et al. (2001) suggested that the bivalve molluscs only present "transition" microbiota. A variety of functions have been assigned to these bacterial communities, including food, source of vitamins and growth factors, a role in the defence mechanisms to prevent the colonization by bacterial pathogens, or the elimination of toxic substances (McHenery and Birkbeck, 1985; Prieur et al., 1990; Seguineau et al., 1996).

Due to their filter-feeding habit, bivalves normally accumulate a rich bacterial microbiota, composed of various species belonging to different genera like *Vibrio, Pseudomonas, Acinetobacter, Photobacterium, Moraxella, Aeromonas*, *Micrococcus* and *Bacillus* (Murchelano and Brown, 1970; Kueh and Chan, 1985; Prieur et al., 1990), which are part of the natural bacterial populations in the aqutic environment. However, few studies have compared the microbiota of the mollusc with the bacterial communities present in the surrounding marine environment.

The first study analyzing the normal microbiota of a bivalve species was performed in 1960 by Colwell and Liston with the Pacific oyster (*Crassostrea gigas*), founding a high proportion of Gram negative bacteria (>80%), with a predominance of the genera *Pseudomonas* and *Vibrio*, but also of *Flavobacterium* and *Achromobacter*. In general, the detected bacteria could be considered as typical psychrophilic marine bacteria, physiologically adapted to survive within the bivalve. Similar results were obtained in further studies of the same group (Colwell and Sparks, 1967; Lovelace et al., 1968).

Sugita et al. (1981) studied the bacteria associated with six bivalve species including *V. philippinarum*, *Mactra quadrangularis* (=*M. veneriformis*), *Mytilus coruscus*, *C. gigas*, *Dosinia nipponica* (=*Phacosoma japonicum*) and *Anadara* (=*Scapharca*) *broughtonii*, as well as with the water and sediments in the Tokyo bay. In general, the predominant groups belonged to the facultative anaerobic bacteria, mainly the genera *Vibrio*, *Aeromonas*, *Pseudomonas*, *Moraxella* and *Micrococcus*. These authors found the same composition in the molluscs than in the sediments and water, although with different abundance, concluding that the microbiota of the mollusc was greatly influenced by the environmental conditions.

Unfortunately, and probably due to the tedious work involved in the analysis of numerical taxonomy for the study of the bacterial communities, further studies on the microbiota of clam species in natural beds are scarce. It is interesting to point out that all the studies discussed here were based on cultured-dependent methods and therefore, biased towards the culturable bacteria. The inclusion of metagenomic approaches to the study of the clam microbiota would help to fill this gap in the future.

One of the main problems in the culture of bivalve molluscs is the repeated episodes of mortality due to bacterial infections that reduce the production and cause high economical losses. Some members of the genus *Vibrio* as well as some oxidative bacteria have been described as the main causal agents of diseases affecting all life stages of bivalve molluscs: larval, juveniles, and adults (Liu et al., 2000, 2001; Allam et al., 2002; Waechter et al., 2002; Lee et al., 2003; Anguiano-Beltrán et al., 2004; Estes et al., 2004; Gay et al., 2004a,b; Paillard et al., 2004; Gómez-León et al., 2005, 2008; Prado et al., 2005; Labreuche et al., 2006 a,b; Garnier et al., 2007, 2008).

Tubiash et al. (1965) were the first to perform research studies on pathogenic vibrios in juvenile and adult bivalves. Their studies associated moribund adult bivalves (*Mercenaria mercenaria, Crassostrea virginica, Mytilus edulis* and *Mya arenaria*) with the species *V. tubiashii* and *V. alginolyticus*. Elston and Leibovitz (1980) described further cases of vibriosis in oysters, detecting anomalies in the shell and alterations in the function of the ligament and digestive processes.

The diversity of *Vibrio* species associated with bivalves in different geographical areas has also been the subject of various studies (Montilla et al., 1994; Hariharan et al., 1995; Arias et al., 1999; Pujalte et al., 1999; Maugeri et al., 2000; Castro et al., 2002; Cavallo and Stabili, 2002; Guisande et al., 2004; Beaz-Hidalgo et al., 2008; Lafisca et al., 2008).

Studies on the diversity of *Vibrio* species found that the predominating species associated with bivalves from different geographical locations (Spain, Canada, Italy or Brazil) were either *V. splendidus, V. alginolyticus, V. harveyi* or any combination of these species. Regarding this, Pujalte et al. (1999) and Arias et al. (1999) found *V. splendidus* and *V. harveyi* as the prevailing species in wild bivalves. Montilla et al. (1995) agreed in the dominance of *V. splendidus*, but also recovered a high proportion of *V. tubiashii* strains, as did Castro et al. (2002) who also recognized *V. harveyi* as an abundant species. However, other studies have found *V. alginolyticus* as the dominant species (Hariharan et al., 1995; Cavallo and Stabili, 2002; Lafisca et al., 2008). Other species such as *V. fluvialis, V. vulnificus* and *V. mimicus* have also been associated with mussels (*Mytilus galloprovincialis*) (Maugeri et al., 2000; Cavallo and Stabili, 2002).

It is well known that environmental parameters, such as variations in the water temperature and salinity, can influence the diversity of *Vibrio* spp. in the environment as well as the physiological state of the bivalve and its susceptibility to bacterial infections (Arias et al., 1999; Pujalte et al., 1999; Maugeri et al., 2000; Paillard et al., 2004; Garnier et al., 2007). Paillard et al. (2004) recognized that the emergence of vibrios as etiological agents in cultured bivalves is likely to increase over the coming years due to ocean warming.

In those mentioned studies, the prevalence of *Vibrio* species was established using phenotypic identification methods, which are known to underestimate the real diversity present in bivalves. For instance, the high variability among the phenotypic characteristics within the *V. splendidus*-like and the *V. harveyi*-like groups makes virtually impossible to discriminate the many species masked under these groups (Thompson et al., 2005; Le Roux and Austin, 2006; Pascual et al., 2010). The introduction of molecular techniques such as the amplified fragment length polymorphism (AFLP) and multilocus sequence analysis (MLSA) have allowed a more precise identification of *Vibrio* species which were previously masked under other taxa (Thompson et al., 2001, 2005; Beaz-Hidalgo et al., 2008; Pascual et al., 2010). In this sense, phenotypically identified *V. harveyi* strains were re-classified as *V. campbellii* by AFLP, DNA-DNA hybridization and MLSA (Gómez-Gil et al., 2004; Thompson et al., 2007). Furthermore, molecular studies have demonstrated the genetic diversity and the polyphyletic nature of *V. splendidus* (Thompson et al., 2001, 2005; Le Roux et al., 2002) and have enabled many new species to be described, such as *V. kanaloae, V. pomeroyi* and *V. chagasii* (Thompson et al., 2003a,b,c,d).

Until now, the only disease of bacterial etiology affecting adult clams (*V. phillipinarum* and *V. decussata*) is the Brown Ring Disease (BRD). The disease, caused by *Vibrio tapetis,* has been widely studied and the pathogenic etiology in adult clams has been reviewed previously (Paillard et al., 1994; Paillard, 2004). Bivalve susceptibility to *V. tapetis* infections is species specific, causing greater physiological disturbances and mortality in *V. philippinarum* than in other species of clam (*V. decussata* and *M. mercenaria*) or in the oyster *C. virginica* (Allam et al., 2001, 2006). This microorganism alters the clam's haemocytes decreasing their viability and phagocytic activity in experimental studies (Allam and Ford, 2006; Jeffroy and Paillard, 2011). The disease is characterized by the alteration of the calcification process of the inner surface of the valves and the appearance of a characteristic brown deposit consisting of conchiolin between the edge of the shell and the pallial line (Paillard and Maes, 1990; Borrego *et al.,* 1996). This brown deposit resembles that which appears in the Juvenile Oyster Disease (or *Roseovarius* Oyster Disease) produced by the bacterial species *Roseovarius crassostreae*, although in this case the conchiolin is deposited over the entire mantle and the attachment of the adductor muscle (Paillard et al., 2004; Boardman et al., 2008).

BRD is considered one of the main limiting factors when culturing Manila clams (*V. philippinarum*) on many European coasts: France, Italy, Spain, Portugal, England, Ireland and has been recently introduced in Norway (Paillard and Maes, 1990; Figueras et al., 1996; Allam et al., 2000; Drummond et al., 2007; Paillard et al., 2008). The pathogen has also been recently detected in Manila clams cultured in Korea (Park et al., 2006). Environmental factors (i.e. temperature and salinity) play a role in the development of BRD, being more frequent in the spring and winter as the optimum growth temperature for *V. tapetis* is 15°C (Paillard et al., 1994; 2004). Reid et al. (2003) demonstrated when challenging BRD infections on Manila clams, that the disease was more severe when performed at 20 ppt salinity than at 40 ppt. The diagnostic of BRD is currently based on the examination of the characteristic brown ring on the inner edge of the shell. For an early diagnosis in the absence of the brown ring, specific PCR detection protocols targeting the 16S rRNA gene of *V. tapetis* have been designed (Paillard et al., 2006; Park et al, 2006; Romalde et al., 2007).

Regarding the intracellular prokaryotes, specifically the taxon *Rickettsiales*, they have been described in a variety of groups of organisms, although only those related with human, animal or plant pathologies and transmited by haematophag or phytophag insects or aracnids have been deeply investigated (Weiss and Dasch, 1981; Weiss, 1982). However, *Rickettsia*-like organisms (RLO) have been described in a great variety of aquatic organisms including fish (Fryer et al., 1990, 1992; Carcés et al., 1991), molluscs (Comps et al., 1977; Harshbarger et al., 1977; Buchanan, 1978; Chang et al., 1980; Gulka et al., 1983; LeGall et al., 1988; Fries and Grant, 1991; Wen et al., 1994; Gardner et al., 1995; Wu and Pan, 1997, 1999a,b, 2000; Moore et al., 2000; Wu, 2003a), and crustaceans (Bonami and Papalardo, 1980; Johnson, 1983). Since their first description in marine molluscs in 1977 (Harshbarger et al., 1977), RLO infections have been described in approximately 25 mollusc species (Wu, 2003b).

Histologically, RLO are characterized by the formation of microcolonies in epithelial cells of the mantle, digestive gland, gill and hepatopancreas, as well as in cells from conective tissue in mantle and hepatopancreas (Gulka et al., 1983; Renault and Cochennec, 1994; Wu and Pan, 1999a,b). In addition, in cases of severe infections, they have also been observed in vesicular conective tissue and in endothelial cells from small capilar in the hepatopancreas (Wu and Pan, 1999c). On the other hand, cytoplasmatic inclusions containing RLO are usually bigger in gill than in digestive gland or mantle, and those observed in conective tissue are bigger than those in epithelial cells from infected tissues. These inclusions or phagosomes can contain not only RLO, but also celular debriss and structures similar to mielin (Cvitanich et al., 1991; Wu and Pan, 1999a). It has been suggested that the bacteria can multiply within phagosomes and that the celular debriss dissapear in madure stages of these structures (Sun and Wu, 2004). It has also been observed the presence of RLO cells in the cytoplasm outside these vacuoles and, therefore, the existence of stages of the multiplication cycle outside the phagosomes cannot be discarded (Wu and Pan, 1999a).

Some RLO imply an important risk of disease and mortality in molluscs such as clam *Polititapes virgineus* (=*Venerupis rhomboides*)(Villalba et al., 1999), scallops (*Placopecten magellanicus* and *Pecten maximus*) (Gulka et al., 1988; LeGall et al., 1988), giant clam (*Hippopus hippopus*)(Norton et al., 1993a,b), and oyster (Wu and Pan 1999a,c, 2000). For other authors, RLO only cause mild infections (Elston, 1986; Fryer and Lannan, 1994), although the infected animals could act as reservoirs in the marine environment.

7.3. Comparative Study of the Microbiota Associated to Venerupis Decussata and V. Philippinarum

As mentioned above, despite the importance of the study of the microbiota associated to cultured populations of different clam species, such studies are scarce. In Galicia (NW Spain), an area where the culture of bivalve molluscs is of great economic importance, no systematic studies have been performed to determine the bacterial communities associated to clams cultured in natural beds, or the role of these bacteria in the appearance of the disease outbreaks periodically affecting those clam populations.

To cover this gap, our research group have started eight years ago an ambitious project to analyze the microbiota associated to different clam populations in Galicia, in order to know the diversity of bacterial species, possible differences between clam species and/or locations, seasonal distribution, etc. The presence in the clam microbiota of bacteria with pathogenic or probiotic capacities has been also determined with the aim to get a better knowledge of the dynamics of health/disease within the populations.

7.3.1. Bacterial Counts

Sampling was performed bimonthly on four different clam culture natural beds in Galicia for more than 5 years. Populations of carpet shell clam (*V. decussata*) and Manila clam (*V. philippinarum*) were analysed in all locations. Colonies on Marine Agar (MA) and Tiosulphate-Citrate-Bilis-Sucrose (TCBS) media were counted, and the bacteria present in clams were isolated. Results of bacterial density data over time indicated a seasonal relationship increasing during warmer temperatures (Figure 1).

Values were similar for all locations analysed ranging 10^5-10^7 cfu/g of clam tissue in MA and 10^4-10^6 cfu/g in TCBS. These values agree with those of previous reports in the Ebro delta and in the Mediterranean Sea (Ortigosa et al., 1989; Montilla et al., 1994; Arias et al., 1999), in which the highest incidence of *Vibrio* and total bacteria in molluscs was registered in summer and beginning of autumn.

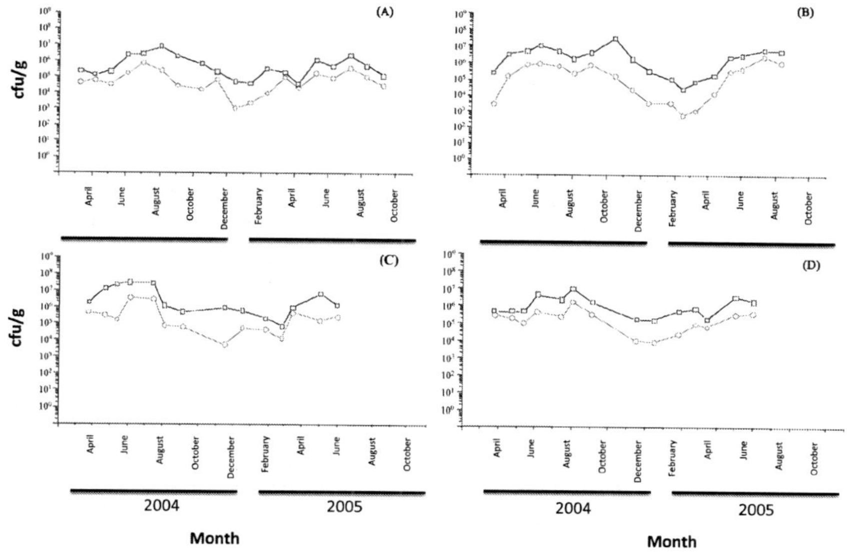

Figure 1. Total bacteria (black line) and vibrios (gray line) counts during 2004 and 2005 obtained for four different locations (A to D) in Galicia.

Generally, the bacterial density in the organs (hepatopancreas, mantle, gonad and gill) analyzed, as well as the extrapalial fluid was similar and temperature dependent. However, higher levels of total bacteria and vibrios were observed in the gonad in clams of certain localities. The presence of pathogenic species in this organ could be important due to the possibility of vertical transmission to the progeny. Therefore, this organ could be considered as a primary target for sampling in the studies of bacterial diversity as well as screening of potential pathogens.

Bacterial diversity was found similar among the four geographic locations. These results are similar those obtained by Arias et al. (1999), which although observing a greater incidence of certain species during summer months, and other species being dominant throughout the year in bivalves in the Mediterranean Sea, they were not able to relate specific bacterial diversity with any geographic location.

7.3.2. Characterization of *Vibrios*

Proportion of vibrios in the different locations and clam species were similar during the sampling, accounting for approximately 60% of the total isolates. More tan 1400 *Vibrio* strains were isolated and their phenotypic profiles obtained during these years (Beaz-Hidalgo et al., 2008, 2010a). Numerical taxonomy was performed and clustering on the basis of the Unweighted Pair Group Method with Arithmetic mean (UPGMA) grouped the isolates in 32 phena (Figure 2).

In the dendrogram, the phenum F1 was the biggest and identified as *V. splendidus* biovar I. The phenum was separated into 2 groups, as a consequence of variation in the growth at different temperatures and salinities. Species belonging to this group are very similar phenotypically and practically impossible to differentiate by classical characterization. Therefore genetic techniques would be required for a proper identification to species level.

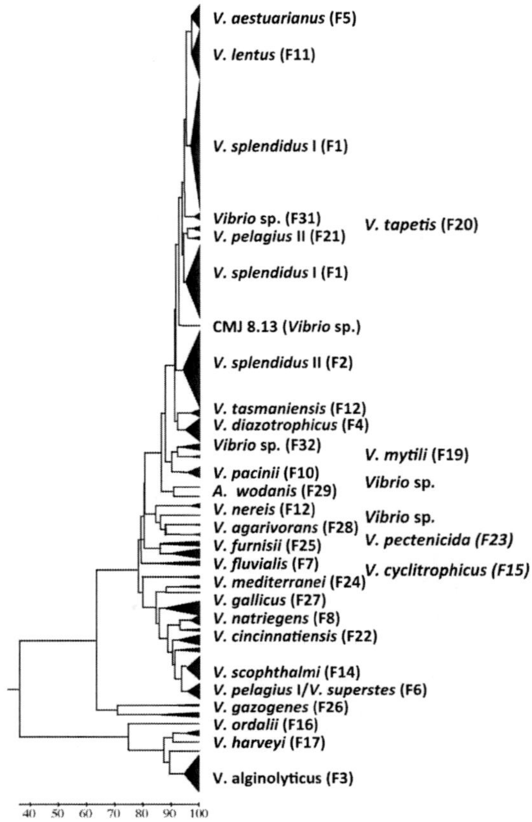

Figure 2. Clustering of vibrios isolated from clams on the basis of their phenotypic characteristics.

Phenum F2 was identified as *V. splendidus* biovar II, differentiated from phenum F1 in the arginine dihidrolase test. Phenum F3 contained *V. alginolyticus* strains, the position in the dendrogram appeared distant in comparison with the rest of the vibrios due to several unusual characteristics in the genus, such as resistance to the Vibriostatic agent or tolerance to extreme temperatures and salinities. Phenum F4 correspondeding to the species *V. diazotrophicus*, close to *V. tasmaniensis* but could be differentiated in the growth at 6% of NaCl and the arginine dihydrolase character.

Phenum 5 presented the *V. aestuarianus* strains and phenum F11 was identified as *V. lentus*. Both species are phenotypically similar to *V. splendidus* biovar I but could be differentiated in the amylase test and resistance to the Vibriostatic agent, respectively. Phenum F6 grouped a total of 20 strains identified as *V. pelagius* I / *V. superstes,* the limited phenotypic features could not discriminate between these two species and were treated as one phenum. Phenua F7 and F25 were identified as *V. fluvialis* and *V. furnisii*, respectively. Both species are very similar in their phenotypic features and grow in a wide temperature and salinity range, however they could be differentiated in the production of gas, which is positive for *V. furnisii*.

Phena F8, F10, F12, F13 and F14 were identified as *V. natriegens, V. pacinii, V. tasmaniensis, Aliivibrio fischeri* and *V. scophthalmi,* respectively. *V. natriegens* and *A. fischeri* could be differentiated in the decarboxylation of lysine; and *V. pacinii* and *V. mytili* in the production of gelatinase and amylase, as well as in the growth at 4ºC temperature.

Other phena with less number of isolates were F15, *V. cyclitrophicus,* F16, *V. ordalii,* F17, *V. harveyi*, F18, *V. nereis*, F20, *V. tapetis* and F21, *V. pelagius* II. The phenotypic profiles of *V. tapetis* and *V. pelagius* II were very similar and were distinguished by the production of gelatinase and growth at 4ºC temperature. The minoritary phena corresponded to *V. cincinnatiensis, V. pectenicida, V. mediterranei, V. gazogenes, V. gallicus, V. agarivorans* and *Alliivibrio wodanis*. The species *V. mediterranei* and *V. gallicus* were phenotypically similar but could be differentiated in the motility and the decarboxylation of lysine. The species *V. agarivorans* and *V. pectenicida,* both in a close position in the dendrogram were differentiated in the production of amylase and growth on TCBS media, at 4ºC temperature and at 0.5% and 8% NaCl concentration.

In the phenotypic analysis a total of 43 strains could not be assigned to any known species, indicating the possibility of the presence of new *Vibrio* species phenotypically similar to the close clusters in the dendrogram.

In summary, a great diversity was observed with dominance of certain species, mainly *V. splendidus* (53.92%). Other predominant species were *V. lentus* (6.4%), *V. alginolyticus* (6.0%), *V. diazotrophicus* (4.7%), *V. aestuarianus* (4.2%) and *V. pelagius* I / *V. superstes* (2.6%) (Figure 3). *Vibrio* species composition obtained in Galician clams was similar to the results obtained in other studies (Ortigosa et al., 1989; Montilla et al., 1994; Pujalte et al., 1999; Castro et al., 2002). Recently, Noguerola and Blanch (2007) have proposed a set of dichotomous keys for rapid identification of *Vibrio* isolates. With minor exceptions, our results were in agreement with the identification schemes proposed in their work. Discrepancies were observed for some traits in few species, such as growth at differeent temperatures and salinities or Voges-Proskauer test in *V. diazotrophicus*, *V. scophthalmi* and *V. natriengens*. These discrepancies were probably due to the fact that those authors studied only the type strains of the different species instead of a large number of environmental isolates as included in our work. Selected representatives of the different phenotypical groups were subjected to genetic analysis by amplified fragment length polymorphisms (AFLP) analysis including also type and reference strains of practically all known species of the genus *Vibrio*.

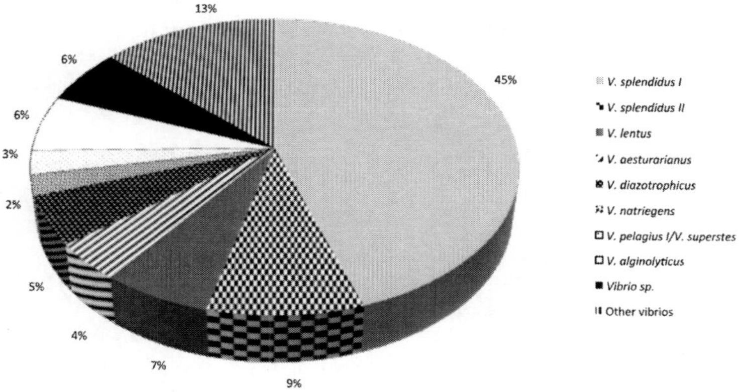

Figure 3. Distribution of clam isolates in the different *Vibrio* species as determined by their phenotypic characterization.

AFLP fingerprinting has been suggested as a valuable tool for classification and identification of *Vibrionaceae* strains (Thompson et al., 2001), even as an alternative to DNA-DNA hybridization (Janssen et al., 1996). Data of AFLP have been used in the descriptions of novel *Vibrio* species such as *V. coralliilyticus*, *V. neptunius*, *V. fortis*, *V. hispanicus*, *V. xuii*, *V. hepatarius*, *V. rotiferianus*, and *V. tasmaniensis*, among other (Borrego et al., 1996; Thompson et al., 2001, 2003a,b,c,d, 2005; Gómez-Gil et al., 2003, 2004). A total of 94 clusters and 81 unclustered strains were obtained (Figure 4).

Only a part of the Galician clam isolates were identified to species level and distributed in 13 clusters. Species identified were *V. cyclitrophicus*, *V. splendidus*, *V. alginolyticus*, *V. diabolicus*, *V. crassostreae*, *V. chagasii*, *V. mediterranei*, *V. ichthyoenteri*, *V. parahaemolyticus*, *V. pectenicida* and *V. lentus*. Most clam strains remained unidentified by AFLP, since they did not cluster with any type strain. Fifty nine of them were distributed over 16 clusters, while 29 were unclustered. A wide geographical distribution was observed for most phenotypic and AFLP groups established, being not possible any association among groups and a specific site.

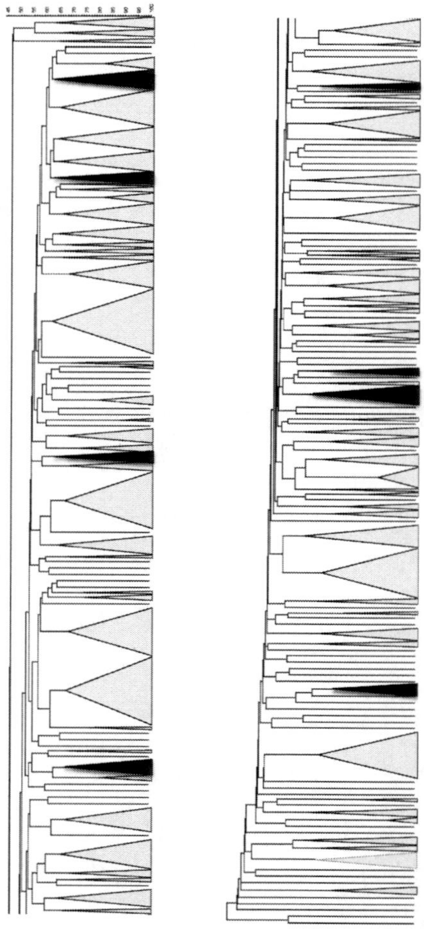

Figure 4. AFLP dendrogram obtained with the vibrios isolated from clams. In solid black are indicated the unidentifed groups, which were further described as new *Vibrio* species.

A comparison of the phenotypic and the AFLP results showed that, for isolates that could be identified at species level by both procedures, identification was coincident only in 29.82% of the cases. Major discrepancies were observed in the heterogeneous species, such as *V. splendidus* (Gómez-León et al., 2005). Thus, some isolates considered as *V. splendidus* based on phenotypic results, were identified as *V. cyclitrophicus*, *V. crassostreae*, *V. chagasii* or were unidentified by AFLP.Further studies using the 16S rRNA gen e and a MLSA with 4 genes (*rpo*D, *rec*A, *pyr*H and *atp*A) enabled us to recognize eight new species within the family *Vibrionaceae*: *Aliivibrio finisterrensis*, *Vibrio breoganii*, *V. gallaecicus*, *V. celticus*, *V. atlanticus*, *V. artabrorum*, *V. toranzoniae* and *V. ponsveterensis* (Beaz-Hidalgo et al., 2009a,b, 2010b,c; Diéguez et al., 2011; Doce et al., 2011; Lasa et al., 2011) (Figure 5). The AFLP analyses performed in this study has revealed a high genetic diversity among Galician clam strains and indicates, together with the sequencing results, the existence of novel taxa within the genus *Vibrio*. This study has proved that phenotypic data gives insufficient information for species delineation and that genetic approaches can help to uncover the prokaryotic diversity. Clearly the molecular diversity and microevolution of vibrios deserve an in-depth investigation in order to better understand the ecological role of these bacteria in the aquatic environment and their possible pathogenic effect on clams and other economically important molluscs.

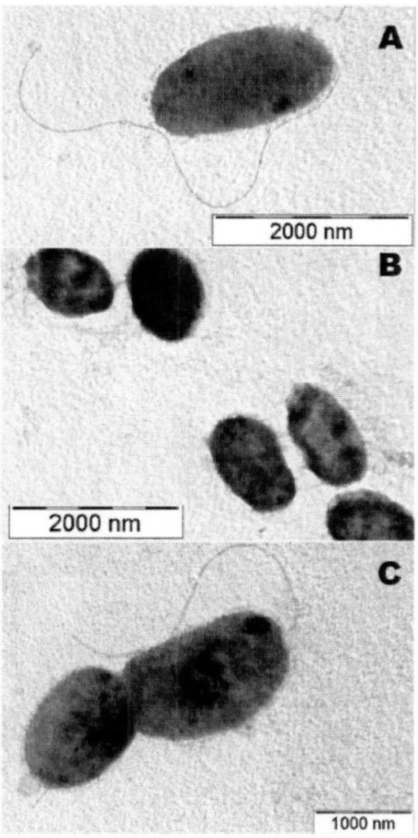

Figure 5. Transmission electronmicrographs of *Vibrio gallaecicus* sp. nov. (A), *Vibrio breoganii* sp. nov. (B), and *Aliivibrio finisterrensis* sp. nov. (C), members of the clam microbiota.

7.3.3. Characterization of Oxidative Bacteria

It was already mention in this chapter that some studies on the natural bacterial biota of bivalve molluscs have been carried out, mainly focusing different oyster and mussel species (Colwell and Liston, 1960; Murchelano and Brown, 1968; Kueh and Chan, 1985; Pujalte et al., 1999; Beleneva et al., 2003; Cavallo et al., 2009).

To our knowledge, besides the works of Kueh and Chan (1985) on arkshell clam (*Anadara* [=*Scapharca*] *cornea*) and Beenson and Johnson (1967) on bean clam (*Donax gouldii*), such studies have not been done for clam species including Manila clam (*V. philippinarum*) or carpet shell clam (*V. decussata*).

Regarding the characterization of the oxidative bacteria, with aerobic metabolism, more than 800 isolates were obtained during the sampling periods. A great diversity was observed after the phenotypic characterization, being the isolates grouped in more than 25 different clusters. However, the identification at species level of these aerobic bacteria by phenotypic characterization is difficult due to the lack in the literature of appropriated identificative keys, which can be related with the high diversity of these bacteria, even within the same species and/or genus.

Therefore, the chosen approach to analyze the taxonomical diversity obtained was the sequencing of the 16S rRNA gene of isolates from the different groups. This technique showed that the predominant genera were *Pseudoalteromonas* (52.8%) and *Shewanella* (16%) (Figure 6). Only 57.8% of the pseudoalteromonads could be assigned to a defined species, including *P. carrageenovora*, *P. marina*, *P. atlantica*, *P. nigrifaciens*, *P. nigrifaciens*, *P. prydzensis*, *P. agarivorans*, *P. tetraodonis* or *P. piscicida*, among others up to 17 different species. The rest of isolates remained identified as *Pseudoalteromonas* sp., constituting probably new species within the genus. In the case of *Shewanella*, some of the species detected were *S. colwelliana*, *S. japonica*, *S. donghaensis*, *S. hafniensis* or *S. algidipiscicola*. As in the case of *Pseudoalteromonas*, 25.8% of the *Shewanella* isolates could not be assigned to any known species of the genus.

Other genera detected in lower proportion were *Tenacibaculum* (*T. mesophilum*, *T. gallaicum* and *T. discolor*), *Psychrobacter* (*P. maritimus*, *P. Nivimaris* and *P. piscatorii*), *Alteromonas* (*A. stellipolaris*, *A. fuliginea* and *A. genovensis*), *Polaribacter*, *Luteimonas*, *Marinomonas*, *Lacitrinux* or *Cobetia*, among others (Figure 6). Finally, some isolates belonged to Gram positive genera such as *Nesterenkonia* or *Arthrobacter*. Again, no correlation was found among the phenotypical groups and the sequencing results, and it seems that an important number of new species could be described within these bacterial genera. The great diversity observed for the aerobic bacteria from clams is in agreement with previous results for this bacterial group in other bivalve species such as oysters and mussels (Pujalte et al., 1999; Beleneva et al., 2003; Cavallo et al., 2009). In addition, and as in these works, the genus *Pseudoalteromonas* constitutes one of the major groups in the clam microbiota, although some seasonal variations could be observed.

7.3.4. *Rickettsia*-Like Organisms

Rickettsia-like organisms (RLO) were described in marine molluscs for the first time in 1977 (Harshbarger et al., 1977). Since then, they have been detected in approximately 25 mollusc species (Wu, 2003b), including clams (Villalba et al., 1999), scallops (Gulka et al., 1988; LeGall et al., 1988), giant clams (Norton et al., 1993a,b), and oysters (Wu and Pan 1999a,c, 2000).

For some authors, RLO imply an important risk of disease and mortality in molluscs, whereas for other authors, they only cause mild infections (Elston, 1986; Fryer and Lannan, 1994).

In the project developed with Galician clams, a great prevalence of intracellular prokaryotes presumptively identified as *Rickettsia*-like organisms was observed. Diverse types of bacterial inclusions were observed in different clam tissues, including gills, digestive gland, and labial palp (Figure 7).

Intracellular bacterial inclusions were detected in both *V. decussata* and *V. phillipinarum* in all the sampling sites, showing higher prevalences in *V. decussata*. No pathological effects were observed in the host, with the exception of possible mechanical lessions due to the size of the bacterial inclusion. Only in the case of heavy infestations, a reduction of the filtering capacity would be possible. In addition, no inflammatory response by the host was detected. These findings support the hypothesis of a low pathogenicity of these bacteria, which would cause only mild infections in the clams (Elston, 1986; Fryer and Lannan, 1994; Paillard et al., 2004). Attemps were made to genetically characterize the intracellular prokaryotes detected, using PCR procedures with universal primers for the Domain *Bacteria* and/or Order *Rickettsiales*.

Although amplification was obtained with the bacterial universal primers, no signal was detected when specific primers for rickettsias were utilized, indicating that the clam RLO would probably constitute new taxa within this bacterial group. Further studies are needed to clarify their correct taxonomic position.

7.3.5. Pathogenicity of Bacterial Isolates

Under intensive culture conditions, marine species are exposed to various stressors, including bacterial pathogens. Most of these, including vibrios, are probably oportunistic pathogenic bacteria, able to cause infection under disfavourable conditions.

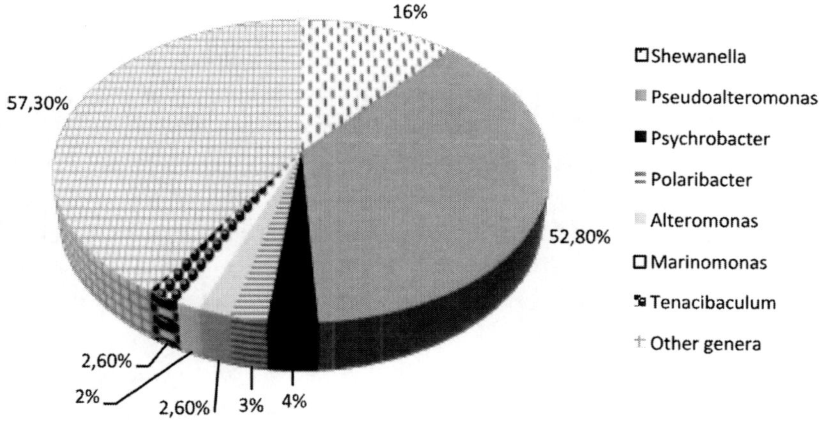

Figure 6. Distribution of aerobic clam isolates among the different genera detected.

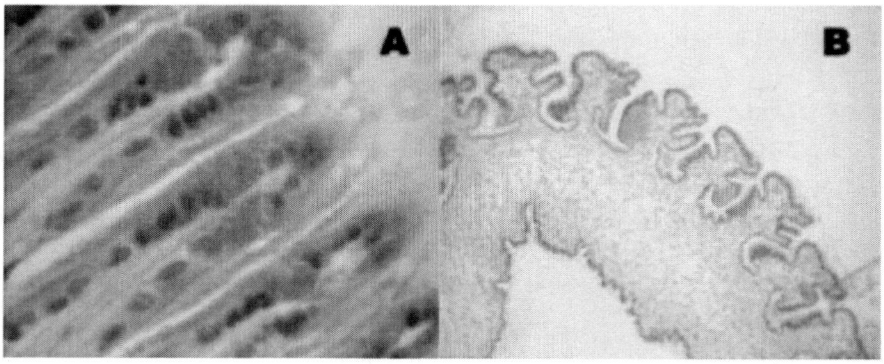

Figure 7. Inclusions of RLO observed in gills (A) and labial palp (B) of *V. decussata*.

Therefore, it seems necessary the performance of infection assays *in vivo* as well as the study of virulence factors of new species associated with the culture of clams to determine their pathogenic potential. The main characteristics that determine bacterial pathogenicity are the capability of bacteria to fix and colonize specific host sites, as well as the synthesis of harmful substances such as the extracelullar products (ECP). ECP may include extracelular enzymes (gelatinase, elastase, lecitinase, lipase, etc.), citolysines, citotoxines, hemolysines, hemagglutinines and thermostable toxins (Baffone et al., 2000).

Although most of the studies on ECP of *Vibrio* spp. have been related mainly to their virulence on fish (Romalde et al., 1990; Santos et al., 1991; Toranzo and Barja, 1993; Biosca and Amaro, 1996), there are also some works in which the virulence of pathogenic *Vibrio* spp. in bivalve molluscs could be related to their ability to produce ECP (Elston and Leibovitz, 1980; Nottage and Birbeck, 1986). According to Maeda and Yamamoto (1996), these ECP, mostly consisting of proteases, could facilitate the propagation of the bacteria by causing extensive host tissue damage. Furthermore, ECP could also counteract the host defense system by degrading immunoglobulins and other components.

Gómez-León et al. (2005) performed an *in vivo* experiment with *V. splendidus* and *V. alginolyticus* strains on clam larvae (*V. decussata*). The mortalities were explained as a consequence of bacterial infection. They demonstrated that the ECPs of both species have cytotoxic activity and reduce the clam hemocyte survival after 24 h of incubation. They suggested that the ECP of both species interfere in the bacterial pathogenesis and that *V. alginolyticus* strains were more virulent as a consequence of siderophore activity.

Representatives of the different bacterial groups detected in the regular samplings, both facultative anaerobic and aerobic bacteria, were selected to study their extracelular and cytotoxic properties. In general, API ZYM enzymatic profiles were highly variable and could not be related to any specific cluster. A positive relationship was found between specific enzymatic activities performed on agar plates, proteolytic activity and cytotoxicity on the cell line SAF-1 (Sea bream cell line) (Béjar et al., 2005). Positive enzymatic responses and high values of proteolytic activity corresponded to cytotoxic activity causing total or partial destruction of the cell line.

Cytotoxic effects were found in strains identified as *V. cyclitrophicus, V. diabolicus, V. mediterranei, V. crassostreae, V. chagasii* and *V. alginolyticus*. Although some these species have not been described as pathogenic in bivalve species yet, for the two latest their pathogenicicity for bivalves has already been demonstrated (Riquelme et al., 1996; Waechter et al., 2002). With respect to the new *Vibrio* species, ECPs of *V. breoganii* and *V. gallaecicus* strains had practically no enzymatic and low proteolytic activities, although produced partial destruction on the cell line SAF-1. Variable results were obtained for *A. finisterrensis* strains in enzymatic, proteolytic and cytotoxic activities. Strains from *V. celticus* and *V. artabrorum* also displayed specific enzymatic activities on agar plates and caused total and partial destruction of the cell line respectively. Regarding the aerobic isolates, the majority of isolates showing cytotoxic activity belonged to the genera *Pseudoalteromonas* and *Shewanella*, although cytotoxic isolates could also be assigned to *Alteromonas*, *Cobetia*, *Marinomonas*, *Lacinutrix*, *Psychrobacter*, *Kangiella* or *Idiomarina*.

Virulence was evaluated by inoculation of larvae and/or adult Manila and carpet shell clams (*V. philippinarum* and *V. decussata*). Results obtained demonstrated that the majority of strains inoculated at two different densities (10^6 and 10^4 cel/ml) were not virulent. However, infection assays with the type strain of *Vibrio celticus* sp. nov. produced 100% mortality after 4 days for Manila clams, and after 3 days for carpet shell clams. The strain was reisolated from dead organisms as pure culture, fulfilling the Koch's postulates for etiological agents. Six aerobic isolates were able to cause mortality in the experimental infections, being identified by sequencing of the 16s rRNA gene as *Alteromonas genovensis*, *Tenacibaculum mesophilum*, *Pseudoalteromonas tetraodonis*, *P. citrea*, *P. carrageenovora* and *P. piscicida*. Some of these bacterial species have been described as pathogens for different marine organisms, including *P. tetraodonis* for sea cucumber (*Apostichopus japonicus*) (Liu et al., 2010) and *P. pisicida* for fish (Bein, 1954). The other species never have been reported as pathogenic for clams or other organisms. These results indicate the possibility of the emergence of new diseases in the near future, as well as support the utility of the studies on mollusc microbiota in the knowledge of the pathogenic potential of marine bacteria.

The relationship between the cytotoxic activity and virulence of a bacterial strain has been a matter of controversy. Thus, some authors have found it for certain fish pathogens (Fouz, 1993; Balebona et al., 1998), but in a number of cases it could not be established (Toranzo et al., 1983; Santos et al., 1991). In agreement with these previous studies, the correlation between the presence of virulence factors, including cytotoxicity and exotoxin production, and potential pathogenicity for clam it was only possible for the new species *V. celticus*, as well as for some possible new species within the oxidative group, belonging to the genera *Pseudoalteromonas*, *Alteromonas* and *Tenacibaculum*.

It is important to consider that bacteria may not produce the same virulence factors *in vitro* as *in vivo* (Toranzo and Barja, 1993). Also, proteins expressed during growth in the laboratory culture media may not be the same than those expressed inside the host species (Smith, 1984). In addition, it must be taken into account that a disease is the result of an interaction pathogen-host-environment, and the physiological and immunological state of the host will play a determinant role in the development of the bacterial disease. Natural habitat conditions are impossible to obtain in laboratory experiments, and may affect the virulence capacity of the bacterial strains examined.

7.3.6. Study of the Inhibitory Effects of the Isolates against Pathogenic Bacteria

Several works have described inhibitory effects against pathogens from members of the fish and shellfish microbiota (Dopazo et al., 1988; Westerdahl et al., 1991; Bower et al., 1994; Riquelme et al., 1996; Ruiz et al., 1996; Sugita et al., 1996). The inhibitory effect of bacteria is due to the action, separately or in combination, of the following factors: production of antibiotics, bacteriocins and siderophores, synthesis of enzymes like lysozymes and proteases, alteration of pH values, etc. (Sugita et al., 1997; Gibson, 1999; Castro et al., 2002).

We have tested more than 352 isolates, mainly with oxidative metabolism, against four fish and shellfish pathogenic vibrios, including *V. anguillarum* ATCC 43308^T, *V. neptunius* 145.98, *V. crassostreae* LMG 22240^T and *V. pectenicida* DSM 1985^T, as well as against the Gram positive fish pathogen *Streptococcus parauberis* RM 207.1.

Thirty-six (10%) of those isolates showed inhibitory effects against at least one of the pathogenic bacteria (Figure 8). The percentage of active isolates is in the range previously reported in other works focused in the search of probiotic bacteria in shellfish hatchery environment (Prado et al., 2009). Active isolates were identified as belonging to the genera *Pseudoalteromonas*, *Tenacibaculum*, *Shewanella*, *Psychrobacter*, *Flammeovirga*, *Salinibacterium*, *Aquimarina*, *Pseudomonas* or *Lacinutrix*. The description of *Pseudoalteromonas* species (*P. citrea*) as probiotics or antibiotic producers is well documented (Gauthier, 1977). The potential use of these isolates in aquaculture as probiotics, as well as their role in the environment of the clam natural beds must be clarified in further studies.

7.4. Concluding Remarks

In nature, there are no mollusk (including clams) free of bacteria, but the distinction between non pathogenic species or strains and those really pathogenic is not easy, as well as the analysis of the factors determining the virulence capacity of a species or strain. In some cases, there are evidences that a quantitative factor is important, since only high densities of a strain can cause the death of the organisms, while low densities are tolerated without causing mortality or disease symptoms (Elston and Leibovitz, 1980; Paillard et al., 2004). Stress is other important factor in the clam susceptibility to bacterial infections (Lipp et al., 1976; Paillard et al., 2004).

It has been suggested that the most needed studies are polyphasic approaches which correlate phenotype and genotype of potential pathogens, as well as the evaluation of virulence factors, and the develolment of diagnostic procedures (Paillard et al., 2004). In recent years, sequencing of the ribosomal and other housekeeping genes allowed to clarify the filogenetic relationship among microorganisms and to restructure the bacterial systematic within a polyphasic frame. Consequently, a high number of reclassifications were proposed, supported by the lack of coherence of the results obtained with those of the phenotypic taxonomy. This new methodology also allows to describe new taxa in a more reliable way. Thus, there are numerous descriptions of new species within the Splendidus clade in the genus *Vibrio*, which until few years ago was considered as one single species. A good example are the new species described within the clam microbiota such as *V. gallaecicus*, *V. atlanticus*, *V. artabrorum* or *V. celticus* (Beaz-Hidalgo et al., 2009b, 2010c; Dieguez et al., 2011), the latest with pathogenic potential for molluscs.

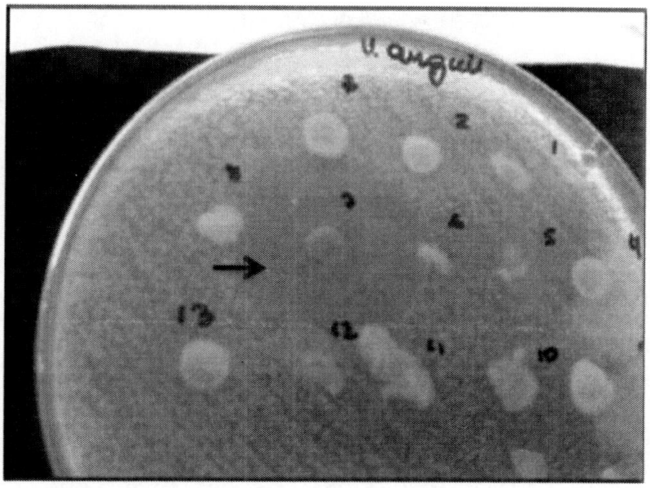

Figure 8. Inhibitory effects of clam isolates against *V. anguillarum* ATCC 43308^T. Arrow notes the halo of no-growth.

On the other hand, a number of marine bacteria with inhibitory effects against pathogens have been described, often associated with the oyster and clam cultures (Prado et al., 2009). It seems that, in the case of marine bivalves, the balance between pathogenic and saprophytic microorganisms determines the success of the culture as well as the susceptibility of the mollusc to bacterial infections (Jeanthon et al., 1988; Castro et al., 2002). In fact, future work is needed in order to determine the physiological status of the active isolates in the environment of the natural clam beds and their putative role in the maintenance of the healthy condition of the cultured clam populations.

Finally, some authors have suggested the possibility of the use of probiotic bacteria to control diseases in aquaculture on the basis of the experimental results at laboratory scale (Kesarkodi-Watson et al., 2008; Prado et al., 2009). The application of such probiotic organisms at different stages of clam culture constitutes one of the most promising emerging research areas.

ACKNOWLEDGMENTS

The studies of the University of Santiago reviewed in this chapter were supported in part by grants AGL2003-09307-C02-01, AGL2006-13208-C02-01, and AGL2010-18438 from the Ministerio de Ciencia y Tecnología, and grant PGIDIT04PXIC20001PN from the Xunta de Galicia (Spain). A.D., A.L. and S.B. acknowledge the Xunta de Galicia and the Ministerio de Ciencia y Tecnología (Spain) for research fellowships.

REFERENCES

Allam, B., Paillard, C., Howard, A., and Le Pennec, M. (2000). Isolation of the pathogen *Vibrio tapetis* and defence parameters in brown ring diseased Manila clams *Ruditapes philippinarum* cultivated in England. *Dis. Aquat. Org.* 41, 105-113.

Allam, B., Ashton-Alcox, A. and Ford, S. E. (2001). Hemocyte activities associated with resistance to brown ring disease in *Ruditapes* spp. clams. *Dev. Comp. Immunol.* 25, 365-375.

Allam, B., Paillard, C. and Ford, S. E. (2002). Pathogenicity of *Vibrio tapetis*, the etiological agent of brown ring disease in clams. *Dis. Aquat. Org.* 48, 221-231.

Allam, B. and Ford, S. E. (2006). Effects of the pathogenic *Vibrio tapetis* on defence factors of susceptible and non-susceptible bivalve species: I. Haemocyte changes following in vitro challenge. *Fish Shellfish Immunol.* 20, 374-383.

Allam, B., Paillard, C., Auffre, M., and Ford, S. E. (2006). Effects of the pathogenic *Vibrio tapetis* on defence factors of susceptible and non-susceptible bivalves species: II. Cellular and biochemical changes following in vivo challenges. *Fish Shellfish Immunol.* 20, 384-397.

Anguiano-Beltrán, C., Lizárraga-Partida, M. L. and Searcy-Bernal, R. (2004). Effect of *Vibrio alginolyticus* on larval survival of the blue mussel *Mytilus galloprovincialis*. *Dis. Aquat. Org.* 59, 119-123.

Arias, C. R., Macián, M. C., Aznar, R., Garay, E., and Pujalte, M. J. (1999). Low incidence of *Vibrio vulnificus* among *Vibrio* isolates from sea water and shellfish of the western Mediterranean coast. *J. Appl. Microbiol.* 86, 125-134.

Baffone, W., Pianetti, A., Bruscolini, F., Barbieri, E., and Cittero, B. (2000). Ocurrence and expression of virulence-related properties of *Vibrio* species isolated from widely consumed seafood products. *Int. J. Food Microbiol.* 54, 9-18.

Balebona, M. C., Andreu, M., Bordas, M. A., Zorrilla, I., Moriñigo, M. A., and Borrego, J. J. (1998). Pathogenicity of *Vibrio alginolyticus* for cultured gilt-head sea bream (*Sparus aurata* L.). *Appl. Environ. Microbiol.* 64, 71-78.

Beaz-Hidalgo, R., Cleenwerck, I., Balboa, S., De Watcher, M., Thompson, F. L., Swings, J., De Vos, P., and Romalde, J. L. (2008). Diversity of vibrios with reared clams in Galicia (NW Spain). *Syst. Appl. Microbiol.* 31, 215-222.

Beaz-Hidalgo, R., Cleenwerck, I., Balboa, S., Prado, S., de Vos, P., and Romalde, J. L. (2009a). *Vibrio breoganii* sp. nov., a non motile alginolytic marine bacterium within the *Vibrio halioticoli* clade. *Int. J. Syst. Evol. Microbiol.* 59, 1589-1594.

Beaz-Hidalgo, R., Doce, A., Pascual, J., Toranzo, A. E., and Romalde, J. L. (2009b). *Vibrio gallaecicus* sp. nov., isolated from cultured clams in north-western Spain. *Syst. Appl. Microbiol.* 32, 111-117.

Beaz-Hidalgo, R., Balboa, S., Romalde, J., and Figueras, M. J. (2010a). Diversity and pathogenecity of *Vibrio* species in cultured bivalve molluscs. *Environ. Microbiol. Reports.* 2, 34-43.

Beaz-Hidalgo, R., Doce, A., Balboa, S., Barja, J. L., and Romalde, J. L. (2010b). *Aliivibrio finisterrensis* sp. nov. isolated from Manila clam, *Ruditapes philippinarum* and emended description of the genus *Aliivibrio*. *Int. J. Syst. Appl. Microbiol.* 60, 223-228.

Beaz-Hidalgo, R., Diéguez, A. L., Cleenwerck, I., Balboa, S., Doce, A., De Vos, P., and Romalde, J. L. (2010c). *Vibrio celticus* sp. nov., a new species within the Splendidus clade with pathogenic potential for clam. *Syst. Appl. Microbiol.* 33, 311-315.

Beenson, R. J. and Johnson, P. T. (1967). Natural bacterial flora of the bean clam, *Donax gouldi*. *J. Invertebr. Pathol.* 9, 104-110.

Bein, S. J. (1954). A study of certain chromogenic bacteria isolated from "red tide" water with a description of a new species. *Bull. Mar. Sci. Gulf Caribb.* 4, 110-119.

Béjar, J., Porta, J., Borrego, J. J., and Álvarez, M. C. (2005). The piscine SAF-1 cell line: genetic stability and labeling. *Mar. Biotechnol.* 7, 389-395.

Beleneva, I. A., Zhukova, N. V. and Maslennikova, E. F. (2003). Comparative study of microbial communities from cultured and natural populations of the mussel *Mytilus trossulus* in Peter the Great Bay. *Microbiology.* 72, 472-477.

Biosca, E. G. and Amaro, C. (1996). Toxic and enzymatic activities of *Vibrio vulnificus* Biotype 2 with respect to host specificity. *Appl. Envion. Microbiol.* 62, 2331-2337.

Boardman, C. L., Maloy, A. and Boettcher, K. (2008). Localization of the bacterial agent juvenile oyster disease (*Roseovarius crassostreae*) within affected eastern oysters (*Crassostrea virginica*). *J. Invertebr. Pathol.* 97, 150-158.

Bonami, J. R. and Pappalardo, R. (1980). Rickettsial infection in marine crustacea. *Experientia.* 36, 180-181.

Borrego, J. J., Castro, D., Luque, A., Paillard, C., Maes, P., García, M., and Ventosa, A. (1996). *Vibrio tapetis* sp. nov., the causative agent of the brown ring disease affecting cultured clams. *Int. J. Syst. Bacteriol.* 46, 480-484.

Bower, S. M., McGladdery, S. E. and Price, I. M. (1994). Synopsis of infectious disease and parasites of commercially exploited shellfish. *Ann. Rev. Fish Dis.* 4, 1-200.

Buchanan, J. S. (1978). Cytological studies on a new species of *Rickettsia* found in association with a ohage in the digestive gland of the marine bivalve mollusc, *Tellina tenuis* (da Costa). *J. Fish Dis.* 1, 27-43.

Carcés, L. H., Larenas, J. J., Smith, P. A., Sandino, S., Lannan, C. N., and Fryer, J. L. (1991). Infectivity of a rickettsia isolated from coho salmon *Oncorhynchus kisutch*. *Dis. Aquat. Org.* 11, 93-97.

Castro, D., Pujalte, M. J., Lopez-Cortes, L., Garay, E., and Borrego, J. J. (2002). Vibrios isolated from the cultured manila clam (*Ruditapes philippinarum*): numerical taxonomy and antibacterial activities. *J. Appl. Microbiol.* 93, 438-447.

Cavallo, R. A. and Stabili, L. (2002). Presence of vibrios in seawater and *Mytillus galloprovincialis* (Lam.) from the Mar Piccolo of Taranto (Ionian Sea). *Water Res.* 36, 3719-3729.

Cavallo, R. A., Acquaviva, M. I. and Stabili, L. (2009). Culturable heterotrophic bacteria in seawater and *Mytilus galloprovincialis* from a Mediterranean area (Northern Ionian Sea-Italy). *Environ. Monit. Assess.* 149, 465-475.

Chang, S. C., Harshbarger, J. C. and Otto, S. V. (1980). Status of cytoplasmic prokaryote infections and neoplasms in bivalve mollusks. *6th Food and Drug Science Symposium on Aquaculture: public health, regulatory and management aspects.* pp. 12-14. New Orleans, LA, US.

Colwell, R. R. and Liston, J. (1960). Microbiology of shellfish. Bacteriological study of the natural flora of Pacific oysters (*Crassostraea gigas*). *Appl. Microbiol.* 8, 104-109.

Colwell, R. R. and Sparks, A. K. (1967). Properties of *Pseudomonas enalia*, a marine bacterium pathogenic for the invertebrate *Crassostrea gigas* (Thunberg). *Appl. Microbiol.* 15, 980-986.

Comps, M., Bonami, J. R. and Vago, C. (1977). Rickettsial infection in oysters. *C.R. Acad. Sci. Paris.* 285, 427-429.

Cvitanich, J. D., Gárate, N. O. and Smith, C. E. (1991). Isolation of a rickettsial-like organism causing diseases and mortality in Chilean salmonids and its confirmation by Koch's postulates. *J. Fish Dis.* 14, 121-145.

Diéguez, A. L., Beaz-Hidalgo, R., Cleenwerck, I., Balboa, S., De Vos, P., and Romalde, J. L. (2011). *Vibrio atlanticus* sp. nov. and *Vibrio artabrorum* sp. nov. isolated from clam (*Ruditapes decussatus* and *R. philippinarum*). *Int. J. Syst. Evol. Microbiol.* 61, 2406-2411.

Doce, A., Balboa, S. and Romalde, J. L. (2011). *Vibrio ponsveterensis* sp. nov., isolated from carpet-shell clams (*Ruditapes decussatus*) in Ría de Pontevedra (NW Spain). *Proceedings of the Fourth Conference on the Biology of Vibrios.* pp. 142. Santiago de Compostela, Spain.

Dopazo, C. P., Lemos, M. L., Lodeiros, C., Bolinches, J., Barja, J. L., and Toranzo, A. E. (1988). Inhibitory activity of antibiotic-producing marine bacteria against fish pathogens. *J. Appl. Bacteriol.* 65, 97-101.

Drummond, L. C., Balboa, S., Beaz, R., Mulcahy, M. F, Barja, J. L., Culloty, S. C., and Romalde J. L. (2007). The susceptibility of Irish-grown and Galician-grown Manila clams, *Ruditapes philippinarum*, to *Vibrio tapetis* and brown ring disease. *J. Invertebr. Pathol.* 95, 1-8.

Elston, R. A. (1986). Occurrence of a branchial rickettsiales-like infections in two bivalve molluscs, *Tapes japonica* and *Patinopecten yessoensis*, with comments on their significance. *J. Fish Dis.* 9, 69-71.

Elston, R. and Leibovitz, L. (1980). Pathogenesis of experimental vibriosis in larval American oysters, *Crassostrea virginica*. *Can. J. Fish Aquat. Sci.* 37, 964-978.

Estes, R. M., Friedman, C. S., Elston, R. A., and Herwig, R. P. (2004). Pathogenicity testing of shellfish hatchery bacterial isolates on Pacific oyster *Crassostrea gigas* larvae. *Dis. Aquat. Org.* 58, 223-230.

Figueras, A., Robledo, J. A. F. and Novoa, B. (1996). Brown ring disease and parasites in clams (*Ruditapes decussatus* and *Ruditapes philippinarum*) from Spain and Portugal. *J. Shellfish Res.* 15, 363-368.

Fouz, B. R. (1993). Caracterización de *Vibrio damselae*: Nuevo agente patógeno en acuicultura marina. *Ph.D Thesis. Universidad de Santiago de Compostela.*

Fries, C. R. and Grant, D. F. M. (1991). Rickettsiae in gill epithelial cells of the hard clam, *Mercenaria mercenaria*. *J. Invertebr. Pathol.* 57, 166-171.

Fryer, J. L. and Lannan, C. N. (1994). Rickettsial and chlamydial infections of freshwater and marine fishes, bivalves and crustaceans. *Zool. Stud.* 33, 95-107.

Fryer, J. L., Lannan, C. N., Carcés, L. H., Larenas, J. J., and Smith, P. A. (1990). Isolation of a rickettsiales-like organism from diseased coho salmon (*Oncorhynchus kisutch*) in Chile. *Fish Pathol.* 25, 107-114.

Fryer, J. L., Lannan, C. N., Giovannoni, S. J., and Wood, N. D. (1992). *Piscirickettsia salmonis* gen. nov., the causative agent of an epizootic disease in salmonid fishes. *Int. J. Syst. Bacteriol.* 42, 120-126.

Gardner, G. R., Harshbarger, J. C., Lake, J. L., Sawyer, T. K., Price, K. L., Stephenson, M. D., Haaker, P. L., and Togstad, H. A. (1995). Association of prokaryotes with symptomatic appearance of withering syndrome in black abalone *Haliotis cracherodii*. *J. Invertebr. Pathol.* 66, 111-120.

Garnier, M., Labreuche, Y., García, C., Robert, M., and Nicolas, J. L. (2007). Evidence for the involvement of pathogenic bacteria in summer mortalities of the Pacific oyster *Crassostrea gigas*. *Microbiol. Ecol.* 53, 187-196.

Garnier, M., Laubreche, Y. and Nicolas, J. L. (2008). Molecular and phenotypic characterization of *Vibrio aestuarianus* subsp. *francensis* subsp. nov., a pathogen of the oyster *Crassostrea gigas*. *Syst. Appl. Microbiol.* 31, 358-365.

Gauthier, M. J. (1977). *Alteromonas citrea*, a new Gram-negative, yellow-pigmented species from seawater. *Int. J. Syst. Bacteriol.* 27, 349-354.

Gay, M., Berthe, F. C. J. and Le Roux, F. (2004a). Screening of *Vibrio* isolates to develop an experimental infection model in the Pacific oyster *Crassostrea gigas*. *Dis. Aquat. Org.* 59, 49-56.

Gay, M., Renault, T., Pons, A. M., and Le Roux, F. (2004b). Two *Vibrio splendidus* related strains collaborate to kill *Crassostrea gigas*: taxonomy and host alterations. *Dis. Aquat. Org.* 62, 65-74.

Gibson, L. F. (1999). Bacteriocin activity and probiotic activity of Aeromonas media. *J. Appl. Microbiol. Symposium Supplement.* 85, 243S-248S.

Gómez-Gil, B., Thompson, F. L., Thompson, C. C., and Swings, J. (2003). *Vibrio rotiferienus* sp. nov. isolated from cultures of the rotifer *Brachionus plicatilis*. *Int. J. Syst. Evol. Microbiol.* 53, 239-243.

Gómez-Gil, B., Soto-Rodríguez, S., García-Gasca, A., Roque, A., Vázquez-Juárez, A., and Swings, J. (2004). Molecular identification of *V. harveyi*-related isolates associated with diseased aquatic organisms. *Microbiology.* 150, 1769-1777.

Gómez-León, J., Villamil, L., Lemos, M. L., and Novoa, B. (2005). Isolation of *Vibrio alginolyticus* and *Vibrio splendidus* from aquacultured carpet shell clam (*Ruditapes decussatus*) larvae associated with mass mortalities. *Appl. Environ. Microbiol.* 71, 98-104.

Gómez- León, J., Villamil, L., Salger, S. A., Sallum, R. H., Remacha-Triviño, A., Leavitt, D. F., and Gómez-Chiarri, M. (2008). Survival of eastern oysters *Crassostrea virginica* from three lines following experimental challenge with bacterial pathogens. *Dis. Aquat. Org.* 79, 95-105.

Guisande, J. A., Montes, M., Farto, R., Armada, S. P., Pérez, M. J., and Nieto T. P. (2004). A set of tests for the phenotypic identification of culturable bacteria associated with Galician bivalve mollusc production. *J. Shellfish Res.* 23, 599-610.

Gulka, G., Chang, P. W. and Marti, K. A. (1983). Prokaryiotic infection associated with a mass mortality of the sea scallop, Placopecten magellanicus. *J. Fish Dis.* 6, 355-364.

Hariharan, H., Giles, J. S., Heaney, S. B., Arsenault, G., Mcnair, N., and Rainnie, D. J. (1995). Bacteriological studies on mussels and oysters from six river systems in Prince Edward Island, Canada. *J. Shellfish Res.* 14, 527-532.

Harshbarger, J. C., Chang, S. C. and Otto, S. V. (1977). Chlamydiae (with phages), mycoplasmas and richettsiae in Chesapeake Bay bivalves. *Science.* 196, 666-668.

Janssen, P., Coopman, R., Huys, G., Swings, J., Bleeker, M., Vos, P., Zabeau, M., and Kersters, K. (1996). Evaluation of the DNA fingerprinting method AFLP as a new tool for bacterial taxonomy. *Microbiology*. 142, 1881-1893.

Jeanthon, C., Prieur, D. and Cochard, J. C. (1988). Bacteriological survey of antibiotic treated seawaters in a *Pecten maximus* hatchery. *Aquaculture*. 29, 201-226.

Jeffroy, F. and Paillard, C. (2011). Involvement of nitric oxide in the *in vitro* interaction between Manila clam, *Ruditapes philippinarum*, hemocytes and the bacterium *Vibrio tapetis*. *Fish Shellfish Immunol.* 31, 1137-1141.

Johnson, P. T. (1983). Diseases caused by viruses, rickettsiae, bacteria, and fungi. *Biol. Crustacea*. 6, 1-78.

Jorquera, M. A., Silva, F. R. and Riquelme, C. E. (2001). Bacteria in the culture of the scallop *Argopecten purpuratus* (Lamarck, 1819). *Aquacult. Int.* 9, 285-303.

Kesarkodi-Watson, A., Kaspas, H., Lategan, M. J., and Gibson, L. (2008). Probiotics in aquaculture: the need, principles and mechanisms of action and screening processes. *Aquaculture*. 274, 1-14.

Kueh, C. S. W. and Chan, K. Y. (1985). Bacteria in bivalve shellfish with special reference to the oyster. *J. Appl. Bacteriol.* 59, 41-47.

Labreuche, Y., Lambert, C., Soudant, P., Boulo, V., Huvet, A., and Nicolas, J. L. (2006a). Cellular and molecular hemocyte responses of the Pacific oyster, *Crassostrea gigas*, following bacterial infection with *Vibrio aestuarianus* strain 01/32. *Microbes Infect.* 8, 2715-2724.

Labreuche, Y., Soudant, P., Goncalves, M, Lambert, C., and Nicolas, J. L. (2006b). Effects of extracellular products from the pathogenic *Vibrio aestuarianus* strain 01/32 on lethality and cellular immune responses of the oyster *Crassostrea gigas*. *Dev. Comp. Immunol.* 30, 367-379.

Lafisca, A., Pereira C. S., Giaccone, V., and Rodriguez, D. D. P. (2008). Enzymatic characterization of *Vibrio alginolyticus* strains isolated from bivalves harvested at Venice lagoon (Italy) and Guanabara Bay (Brazil). *Rev. Inst. Med. Trop. S. Paulo*. 50, 199-202.

Lauckner, G. (1983). Diseases of Mollusca: Bivalvia. In: O. Kinne (Ed.), *Diseases of marine animals, Vol. II*. pp. 477-961, Biologische Anstalt Helgoland, Hamburg.

Lasa, A., Diéguez, A. L. and Romalde, J. L. (2011) Description of *Vibrio toranzoniae* sp. nov. *Proceedings of the Fourth Conference on the Biology of Vibrios*. pp. 147, Santiago de Compostela. Spain.

Lee, K. K., Ping-Chung, L. and Huang, C. Y. (2003). *Vibrio parahaemolyticus* infections for both humans and edible mollusc abalone. *Microbes Infect.* 5, 481-485.

LeGall, G., Chagot, D., Mialhe, E., and Grizel, H. (1988). Branchial rickettsiales-like infection associated with a masss mortality of sea scallop, *Pecten maximus*. *Dis. Aquat. Org.* 4, 229-232

Le Roux, F. and Austin, B. (2006). *Vibrio splendidus*. In: F. L. Thompson, B. Austin, J. Swings (Eds), *The Biology of Vibrios*. pp. 285-296, ASM Press edition, Washington.

Le Roux, F., Gay, M., Lambert, C., Waechter, M., Poubalanne, S., Chollet, B., Nicolas, J. L., and Berthe, F. C. (2002). Comparative analysis of *Vibrio splendidus* related strains isolated during *Crassostrea gigas* mortality events. *Aquat. Living Resour.* 15, 251-258.

Lipp, P. R., Brown, B., Liston, J., and Chew, K. (1976). Recent findings on the summer diseases of Pacific oysters. *Proc. Nat. Shellfish Ass.* 65, 9-10.

Liu, H., Zheng, F., Sun, X., Hong, X., Dong, S., Wang, B., Tang, X., and Wang, Y. (2010). Identification of the pathogens associated with skin ulceration and peristome tumescence in cultured sea cucumbers *Apostichopus japonicus* (Selenka). *J. Invertebr. Pathol.* 105, 236-242.

Liu, P. C., Chen, C. Y., Huang, C. Y., and Lee, K. K. (2000). Virulence of *Vibrio parahaemolyticus* isolated from cultured small abalone, *Haliotis diversicolor supertexta*, with withering syndrome. *Lett. Appl. Microbiol.* 31, 433-437.

Liu, P. C., Chen, Y. C. and Lee, K. K. (2001). Pathogenicity of *Vibrio alginolyticus* isolated from diseased small abalone *Haliotis diversicolor supertexta. Microbios.* 104, 71-77.

Lovelace, T. E., Tubiash, H. and Colwell, R. R. (1968). Quantitative and qualitative commensal bacterial flora of *Crassostrea virginica* in Chesapeake Bay. *Proc. Nat. Shellfish Ass.* 58, 82-87.

Maeda, H. and Yamamoto, T. (1996). Pathogenic mechanisms induced by microbial proteases in microbial infections. *Biol. Chem. Hoppe Seyler.* 377, 217-226.

Maugeri, T. L., Caccamo, D. and Gugliandolo, C. (2000). Potentially pathogenic vibrios in brackish waters and mussels. *J. Appl. Microbiol.* 89, 261-266.

Mchenery, J. G. and Birbeck, T. H. (1986). Inhibition of filtration in *Mytilus edulis* L. by marine vibrios. *J. Fish Dis.* 9, 257-261.

Montilla, R., Palomar, J., Santmarti, M., Fuste, C., and Viñas, M. (1994). Isolation and characterization of halophilic *Vibrio* from bivalves bred in nurseries at the Ebre Delta. *J. Invertebr. Pathol.* 63, 178-181.

Montilla, R., Palomar, J., Fuste, M. C., and Viñas, M. (1995). Taxonomy and protein fingerprinting of halophilic *Vibrio* isolates from bivalves of the Ebre delta. *Can. J. Microbiol.* 41, 64-69.

Moore, J. D., Robbins, T. T. and Friedman, C. S. (2000). The role of a Rickettsia-like prokaryote in withering syndrome in California red abalone, *Haliotis rufescens. J. Shellfish Res.* 19, 525-526.

Moulder, J. W. (1984). The rickettsias and chlamidias. Order II. *Chlamydiales*. In: N. R. Krieg, J. G. Holt (Eds.), *Bergey's manual of systematic bacteriology, Vol. 1.* pp. 729-739, Williams and Wilkins, Baltimore.

Murchelano, R. A. and Brown, C. (1968). Bacteriological study of the natural flora of the eastern oyster, *Crassostrea virginica. J. Invertebr. Pathol.* 11, 520-521.

Murchelano, R. A. and Brown, C. (1970). Heterotrophic bacteria in Long Island Sound. *Mar. Biol.* 7, 1-6.

Noguerola, I. and Blanch, A. R. (2008). Identification of *Vibrio* spp. with a set of dichotomous keys. *J. Appl. Micriobiol.* 105, 175-185.

Norton, J. H., Shepherd, M. A., Long, H. M., and Prior, H. C. (1993a). Parasites of the giant clams (*Tridacnidae*). Biology and mariculture of giant clams. pp. 18-23, Australian Centre for International Agricultural Research. Canberra, a.c.t. (Australia).

Norton, J. H., Shepherd, M. A., Abdon-Nagutt, M. R., and Lindsay, S. (1993b). Mortalities in the giant clam *Hippopus hippopus* associated with rickettsiales-like organisms. *J. Invertebr. Pathol.* 62, 207-209.

Nottage, A. S. and Birkbeck T. H. (1987). The role of toxins in *Vibrio* infections of bivalve molluscs. *Aquaculture.* 67, 244-246.

Ortigosa, M., Esteve, C. and Pujalte, M. J. (1989). *Vibrio* species in seawater and mussels: abundance and numerical taxonomy. *Syst. Appl. Microbiol.* 12, 316-325.

Paillard, C. (2004). A short-review of brown ring disease, a vibriosis affecting clams, *Ruditapes philipinarum* and *Ruditapes decussatus*. *Aquat. Living Resour.* 17, 467-475.

Paillard, C. and Maes, P. (1990). Étiologie de la maladie de l'anneau brun chez *Tapes philippinarum*: Pathogenicité d'un *Vibrio* sp. *C.R. Acad. Sci. Paris* 310, 15-20.

Paillard, C., Maes, P. and Oubella, R. (1994). Brown ring disease in clams. *Ann. Rev. Fish Dis.* 4, 1-22.

Paillard, C., Le Roux, F. and Borrego, J. J. (2004). Bacterial disease in marine bivalves, a review of recent studies: Trends and evolution. *Aquat. Living Resour.* 17, 477-498.

Paillard, C., Gausson, S., Nicolas, J. L., Le Pennec, J. P., and Haras, D. (2006). Molecular identification of *Vibrio tapetis*, the causative agent of the BRD of *Ruditapes philippinarum*. *Aquaculture*. 253, 25-38.

Paillard, C., Korsnes, K., Le Chevalier, P., Le Boulay, C., Harkestad, L., Eriksen, A. G., Willanssen, E., Bergh, Ø., Bovo, C., Skår, C., and Mortensen, S. (2008). *Vibrio tapetis*-like strain isolated from introduced Manila clams *Ruditapes philippinarum* showing symptoms of brown ring disease in Norway. *Dis. Aquat. Org.* 81, 153-161.

Park, K. I., Paillard, C., Le Chevalier, P., and Choi, K. S. (2006). Report on the occurrence of brown ring disease (BRD) in Manila clam, *Ruditapes philippinarum*, on the west coast of Korea. *Aquaculture.* 200, 610-613.

Pascual, J., Macián, M. C., Arahal, D. R., Garay, E., and Pujalte, M. J. (2010). Multilocus Sequence Analysis of the central clade of the genus *Vibrio* using the 16S rRNA, *recA, pyrH, rpoD, gyrB, rctB* and *toxR* genes. *Int. J. Syst. Evol. Microbiol.* 60, 154-165.

Prado, S., Romalde, J. L., Montes, J., and Barja J. L. (2005). Pathogenic bacteria isolated from disease outbreaks in shellfish hatcheries. First description of *Vibrio neptunius* as an oyster pathogen. *Dis. Aquat. Org.* 67, 209-215.

Prado, S., Montes, J., Romalde, J. L., and Barja, J. L. (2009). Inhibitory activity of *Phaeobacter* strains against aquaculture pathogenic bacteria. *Int. Microbiol.* 12, 107-114.

Prieur, D., Mével, G., Nicolas, J. L., Plusquellec, A., and Vigneulle, M. (1990). Aerobic and facultative anaerobic heterotrophic bacteria associated to Mediterranean oysters and seawater. *Int. Microbiol.* 2, 259-266.

Pujalte, M. J., Ortigosa, M., Macián, M. C., and Garay, E. (1999). Aerobic and facultative anaerobic heterotrophic bacteria associated to Mediterranean oysters and seawater. *Int. Microbiol.* 2, 259-266.

Reid, H. I., Soudant, P., Lambert, C., Paillard, C., and Birkbeck, T. H. (2003). Salinity effects on immune parameters of *Ruditapes philippinarum* challenged with *Vibrio tapetis*. *Dis. Aquat. Org.* 56, 249-258.

Renault, T. and Cochennec, N. (1994). RLOs in the cytoplam of gill epithelial cells of the Pacific oyster *Crassostrea gigas*. *J. Invertebr. Pathol.* 64, 160-162.

Riquelme, C., Hayashida, G., Araya, R., Uchida, A., Santoni, M., and Ishida, Y. (1996). Isolation of a native bacterial strain from the scallop, *Argopecten purpuratus*, with inhibitory effects against pathogenic vibrios *J. Shellfish Res.* 15, 369-374.

Romalde, J. L., Barja, J. L. and Toranzo, A. E. (1990). Vibrios associated with red tides caused by *Mesodinium rubrum*. *Appl. Environ. Microbiol.* 56, 3615-3619.

Romalde, J. L., Rodríguez, J. M. and Borrego, J. J. (2007). Protocolo de PCR (reacción en cadena de la polimerasa) para detección e identificación de *Vibrio tapetis* a partir de cultivos puros del microorganismo y de tejidos de moluscos. Patent ES 2 265 707. Oficina Española de Patentes y Marcas.

Ruiz, C. M., Roman, G. and Sanchez, J. L (1996). A marine bacterial strain effective in producing antagonisms of other bacteria. *Aquacult. Int.* 4, 289-291.

Santos, Y., Lallier, R., Bandín, I., Lamas, J., and Toranzo, A. E. (1991). Susceptibility of turbot (*Scophthalmus maximus*), coho salmon (*Oncorhynchus kisutch*) and rainbow trout (*Oncorhynchus mykiss*) to the strains of *Vibrio anguillarum* and their exotoxins. *J. Appl. Ichthyol.* 7, 160-167

Seguineau, C., Laschi- Loquerie, A., Moal, J., and Samai, J. F. (1996). Vitamin requirements in great scallop larvae. *Aquacult. Int.* 4, 315-324.

Smith, H. (1984). The biochemical challenge of microbial pathogenicity. *J. Appl. Bacteriol.* 47, 395-404.

Sugita, H., Tanaami, H., Kobashi, T., and Deguchi, Y. (1981). Bacterial flora of coastal bivalves. *Bull. Jpn. Soc. Scient. Fish.* 47, 655-661.

Sugita, H., Matsuo, N., Shibuya, K., and Deguchi, Y. (1996). Antibacterial substance-producing ability of the intestinal bacteria isolated from coastal animals. *J. Mar. Biotech.* 4, 220-223.

Sugita, H., Matsuo, Hirose, Y., Iwato, M., and Deguchi, Y. (1997). *Vibrio* sp. strain NM10, isolated from the intestine of a Japanese coastal fish, has an inhibitory effect against *Pasteurella piscicida*. *Appl. Environ. Microbiol.* 63, 4986-4989.

Sun, J. F. and Wu, X. Z. (2004). Histology, ultrastructure, and morphogenesis of a rickettsia-like organism causing disease in the oyster, *Crassostrea ariakensis* Gould. *J. Invertebr. Pathol.* 86, 77-86.

Thompson, F. L., Hoste, B., Vandeneulebroecke, K., and Swings, J. (2001). Genomic diversity amongst *Vibrio* isolates from different source determined by fluorescent amplified fragment length polymorphism. *Syst. Appl. Microbiol.* 24, 520-538.

Thompson, F. L., Thompson, C. C. and Swings, J. (2003a). *Vibrio tasmaniensis* sp. nov. isolated from Atlantic salmon (*Salmo salar* L.). *Syst. Appl. Microbiol.* 26, 65-69.

Thompson, F. L., Li, Y., Gómez-Gil, B., Thompson, C. C., Hoste, B., Vandermeulebroecke, K., Rupp, G. S., Pereira, A., De Bem, M. M., Sorgeloos, P., and Swings, J. (2003b). *Vibrio neptunius* sp. nov., *Vibrio brasiliensis* sp. nov. and *Vibrio xuii* sp. nov., isolated from the marine aquaculture environment (bivalves, fish, rotifers and shrimps). *Int. J. Sys. Evol. Microbiol.* 53, 245-252.

Thompson, F. L., Thompson, C. C., Hoste, B., Vandemeulebroecke, K., Gullian, M., and Swings, J. (2003c). *Vibrio fortis* sp. nov. and *Vibrio hepatarius* sp. nov., isolated from aquatic animals and the marine environment. *Int. J. Sys. Evol. Microbiol.* 53, 1495-1501.

Thompson, F. L., Thompson, C. C., Li, Y., Gómez-Gil, B., Vanderberghe, J., Hoste, B., and Swings, J. (2003d). *Vibrio kanaloae* sp. nov., *Vibrio pomeroyi* sp. nov. and *Vibrio chagasii* sp. nov., from sea water and marine animals. *Int. J. Syst. Evol. Microbiol.* 53, 753-759.

Thompson, F. L., Gevers, D., Thompson, C. C., Dawyndt P., Naser, S., Hoste, B., Munn, C. B., and Swing, J. (2005). Phylogeny and molecular identification of vibrios on the basis of multilocus sequence analysis. *Appl. Environ. Microbiol.* 71, 5107-5115.

Thompson, F. L., Gómez-Gil, B., Ribeiro Vasconcelos, A. T., and Sawabe, T. (2007). Multilocus sequence analysis reveals that *Vibrio harveyi* and *V. campbelli* are distinct species. *Appl. Environ. Microbiol.* 73, 4279-4285.

Toranzo, A. E. and Barja, J. L. (1993). Virulence factors of bacteria pathogenic for coldwater fish. *Annu. Rev. Fish Dis.* 3, 5-36.

Toranzo, A. E., Barja, J. L., Colwell, R. R., Hetrick, F. M., and Crosa, J. H. (1983). Hemaglutinating, hemolityc and cytotoxic activities of *Vibrio anguillarum* and related vibrios isolated from striped bass on the Atlantic coast. *FEMS Microbiol. Lett.* 18, 257-262.

Tubiash, H. S., Chanley, P. E. and Leifson, E. (1965). Bacillary necrosis, a disease of larval and juvenile bivalve molluscs. I. Etiology and epizootiology. *J. Bacteriol.* 90, 1036-1044.

Villalba, A., Carballal, M. J., López, C., Cabada, A., Corral, L., and Azevedo, C. (1999). Branchial rickettsia-like infection associated with clam *Venerupis rhomboides* mortality. *Dis. Aquat. Org.* 36, 53-60.

Waechter, M., Le Roux, F., Nicolas, J. L., Marissal, E., and Berthe, F. (2002). Characterization of pathogenic bacteria of the cupped oyster *Crassostrea gigas. C. R. Biol.* 325, 231-238.

Wen, C. M., Sen, T., Kou, G. H., and Chen, S. N. 1994. Rickettsiaceae-like microorganisms in the cultured hard clam (*Mercenaria lusoria* Roding) in Taiwan. *J. Fish. Soc. Taiwan* 20, 347-356.

Weiss, E. (1982). The biology of *Rickettsiae. Ann. Rev. Microbiol.* 36, 345-370.

Weiss, E. and Dasch, G. A. (1981). The family *Rickettsieaseae*: pathogens of domestic animals and invertebrates; non-pthogenic arthropod symbiotes. In: M. P. Starr, H. Stolp., H. G. Trüper, A. Ballows, H. G. Schlegel (Eds.), *The procaryotes: a handbook on habitats, isolation and identification of bacteria. Vol. 2.* pp. 2161-2171, Springer-Verlag, Berlin.

Weiss, E. and Moulder, J. W. (1984). The rickettsias and chlamidias. Order I. Rickettsiales. In: N. R. Krieg, J. G. Holt (Eds.). *Bergey's man ual of systematic bacteriology, Vol. 1.* pp. 687-729, Williams and Wilkins, Baltimore.

Westherdahl, A., Olsson, J. C., Kjelleberg, S., and Conway, P. l. (1991). Isolation and characterization of turbot (*Scophthalmus maximus*)-associated bacteria with inhibitory effects against *Vibrio anguillarum. Appl. Environ. Microbiol.* 57, 2223-2228.

Wu, X. Z. (2003a). *Rickettsiae* infection in farmed variously colored abalone, *Haliotis diversicolor* Reeve (Mollusca: Gastropoda) from the Daya Bay, Guandong in China. In: P. Bruce, M. Bern, Y. Zhou (Eds.), *Proceedings of the 3rd World Fisheries Congress.* American Fisheries Society, Bethesda, MD, US.

Wu, X. Z. (2003b). Advances in the research of marine-cultured animal diseases in China. In: C. S. Lee (Ed.), *Status of Aquaculture in China.* pp. 29-54, The Oceanic Institute, Hawaii, US.

Wu, X. Z. and Pan, J. P. (1997). Studies on Rickettsia-like organism (RLO) diseases of tropical pearl oyster III. Morphology of RLO parasitized in *Pinctada fucata. Trop. Mar. Res.* 5, 110-117.

Wu, X. Z. and Pan, J. P. (1999a). Studies on Rickettsia-like organism diseases of tropical pearl oyster I: the fine structure and morphogenesis of *Pinctada maxima* pathogen rickettsia-like organism. *J. Invertebr. Pathol.* 73, 162-172.

Wu, X. Z. and Pan, J. P. (1999b). Studies on Rickettsia-like organism diseases of tropical pearl oyster- the morphology, morphogenesis and ultrastructure of RLO inclusions, an agent for *Pinctada maxima. Oceanol. Limnol. Sin.* 30, 73-80.

Wu, X. Z. and Pan, J. P. (1999c). Studies on Rickettsia-like organism diseases of tropical pearl oyster, Pinctada maxima and *P. fusata*: IV. On histo-cytopathology of RLO disease. *Acta Oceanol. Sin.* 21, 93-98.

Wu, X. Z. and Pan, J. P. (2000). An intracellular prokaryotic microorganism associated with lessions in the oyster, *Crassostrea ariakensis* Gould. *J. Fish Dis.* 23, 409-414.

In: Clam Fisheries and Aquaculture
Editor: Fiz da Costa González
ISBN: 978-1-62257-518-3

Chapter 8

STUDIES ON THE MICROBIOTA ASSOCIATED WITH CLAMS IN HATCHERIES

***S. Prado,*[*] *J. Dubert and J. L. Barja*[#]**
Department of Microbiology and Parasitology.
CIBUS-Faculty of Biology. University of Santiago de Compostela, Santiago de Compostela (A Coruña), Spain

ABSTRACT

Clams comprise a group of bivalve species of high commercial value. As with other bivalves such as oysters and scallops, overexploitation has led to the decline of wild populations. Hatchery culture is currently the only reliable means of providing spat to replenish natural beds or to support aquaculture activities.

Hatchery culture of clams such as *Mercenaria mercenaria* has been carried out since the last century. However, the number of threatened species in natural environments has increased in the last decade and therefore, increased the requirement for juveniles. As consequence protocols for hatchery culture of different species, including members of the Superfamilies Veneracea (*Venerupis decussata*, *V. philippinarum* and *V. corrugata*), Solenacea (*Ensis siliqua* and *Solen marginatus*) and Tellinacea (*Donax trunculus*), are currently being developed.

Among the variables that affect the success of spat production in hatcheries, the associated microbiota has been demonstrated to be critical for successful bivalve culture. Problems caused by the bacterial populations in larval cultures can lead to the loss of complete batches during production. Treatment is not possible once the first signs of infection have been detected, because of the rapid development of disease. The aetiological agents recognised to date are opportunistic pathogens which are common in the marine environment and which proliferate under favourable conditions. Most of those identified are Gram negative bacteria belonging to genus *Vibrio*.

---

[*] Corresponding author: S. Prado. E-mail address: susana.prado@usc.es.
[#] Department of Microbiology and Parasitology. CIBUS-Faculty of Biology. University of Santiago de Compostela, 15782, Santiago de Compostela (A Coruña), Spain.

Knowledge of the microbiota associated with clams in hatcheries is essential to reduce the risk of infection and mortality. In hatcheries, the seawater and phytoplankton used as food are used for all the species in culture and can be source of bacteria. However, the bacterial populations associated with the broodstock may differ among species, and this should be taken into account in optimizing the culture protocols for each species.

Each broodstock, which is usually obtained from natural environments, has its own associated microbiota. The dynamics of changes in the bacterial load during the conditioning process in the hatchery, as well as the composition of the bacterial populations associated with the broodstock may be key factors in controlling the entry of opportunistic pathogens into the system and the risk of vertical transmission to larvae.

In relation to the larval cultures, two different "compartments", and the dynamics of bacterial populations within and between these, should be considered. The first compartment comprises the larvae with their own microbiota, initially influenced by the broodstock. The second compartment comprises the seawater in the culture tanks, especially important because of the narrow interaction between animal and aquaculture environment.

In this chapter, knowledge about microbiota associated with different species of clams in hatcheries is presented, as regards both the broodstock and the larval cultures. Special emphasis is placed on members of the genus *Vibrio*, which includes the main larval pathogens for bivalves. This constitutes the first report of the bacterial populations present during culture of the genera *Ensis*, *Solen* and *Donax*.

We have demonstrated that, as with other bivalve species cultured in hatcheries, *Vibrio* spp. are an important part of the cultivable microbiota of clam broodstock and larvae. We have established that the broodstock constitutes a reservoir of bacteria. When the broodstock was first brought to the hatchery, it carried its own microbiota, with relatively high total numbers of heterotrophic bacteria (average values ranging from 10^5 to 10^7 cfu/g) and *Vibrio* spp. ($\approx 10^5$ cfu/g) associated with the gonad. The bacterial load changed during conditioning, with final mean numbers of total bacteria and *Vibrio* spp. in the range of 10^5-10^6 cfu/g and 10^3-10^5 cfu/g, respectively. However, there was no repetitive pattern of variation. We found large differences between broodstocks and even among between individuals in the same batch.

In larval cultures, *Vibrio* spp. were mainly found associated with larvae and less frequently with the seawater in the tanks. Populations of these species were associated with outbreaks of disease in clam cultures, as reported for other bivalves cultured in hatcheries. The use of antibiotic did not eliminate vibrios from larvae, and only controlled their proliferation in the seawater. This effect was not sufficient to prevent mortalities amongst the clams.

Respect to the identification of the cultivable microbiota associated with clams in hatchery, the Splendidus-clade predominated in the larval samples, regardless of the clam species, the use of antibiotic or clam survival. Interestingly, the presence of isolates belonging to Harveyi-clade, other than *V. alginolyticus*, was also reported. This is an important finding considering that both clades include known pathogens for hatchery cultures.

Finally the link between adult and larvae microbiota was demonstrated, with vertical transmission of vibrios from broodstock to larval cultures becoming evident. The results strongly indicate the need to continue studies on the microbiota associated with clams cultured in hatcheries, in order to elucidate the role of all these bacterial populations in the development and success of these cultures.

8.1. INTRODUCTION

Bivalve culture is an important part of aquaculture production worldwide. A primary requisite for successful culture of any species is an abundant, regular and inexpensive supply of juvenile specimens. However, natural stocks have been overexploited and the lack of reliable levels of natural recruitment has led to the consideration of hatchery culture as a means of meeting the seed requirements for bivalve aquaculture, including clam species (Helm and Bourne, 2004). Moreover, bivalve culture is now highly diversified and includes introduced as well as native species. Therefore, hatcheries must rear all these species to supply the spat needed, in order to counteract fishing pressures and the decline of natural beds, and also to expand commercial exploitation.

Shellfish hatcheries are hindered by a very high incidence of fatal outbreaks of disease among larval stocks. As a result of these diseases, large numbers of larvae die suddenly and entire culture batches can be lost.

Bacteria have been identified as the main cause of these outbreaks of disease in larval cultures in hatcheries. Tubiash et al. (1965) first described a disease, termed "bacillary necrosis", as being caused by bacterial isolates from moribund and dead bivalve larvae, later identified as members of the genus *Vibrio*. The authors found that each strain was pathogenic for larvae of homologous and heterologous species of all the lamellibranchs challenged (*Mercenaria mercenaria*, *Ostrea edulis*, *Aequipecten irradians* and *Teredo navalis*). On the basis of the results, these authors suggested that the etiological agents of bacillary necrosis normally exist as widely distributed saprophytes or commensals of marine organisms.

It has clearly been established that the causes of outbreaks of disease in bivalve hatcheries are mainly opportunistic pathogens, i.e. free-living microorganisms that require favourable conditions for proliferation in the husbandry system or in association with the host in order to cause disease. Such microorganisms may be common in the marine environment at certain locations and times. Moreover, they cause problems with production on a worldwide scale, affecting almost all species cultured in hatcheries (Elston, 1984).

Different species of the genus *Vibrio* have been described as opportunistic pathogens associated with episodes of mortalities during bivalve larval development. These species include *V. alginolyticus*, *V. anguillarum*, *V. pectenicida*, *V. neptunius*, *V. splendidus* and *V. tubiashii* (Tubiash et al., 1965; Brown, 1981; Jeffries, 1982; Riquelme et al., 1995a, 1996; Nicolas et al., 1996; Lambert et al., 1998; Sáinz et al., 1998; Sugumar et al., 1998; Luna-González et al., 2002; Anguiano-Beltrán et al., 2004; Gómez-León et al., 2005; Prado et al., 2005; Elston et al., 2008; Kesarcodi-Watson et al., 2009). The cited studies have described the disease of a wide range of bivalve species, including oysters (*Ostrea* and *Crassostrea*), clams (*Mercenaria* and *Venerupis*), scallops (*Pecten*, *Argopecten* and *Nodipecten*) and mussels (*Mytilus* and *Perna*).

Pathogenic agents can enter intensive husbandry systems via the following three main routes: seawater, broodstock and microalgae used as food. Seawater and the microalgae are common to all species cultured in hatcheries. However, the broodstock is a source of bacteria specific to each batch of larvae. Since the reproductive characteristics of clams imply the differentiation and release of gametes into the environment, the microbiota associated to the gonadal tissue of adults could be directly transferred to eggs-larvae. The knowledge about this specific subject, scarce until now, would be very useful to optimize the management of broodstock in hatcheries with the final aim of enhancing production.

The Centro de Cultivos Mariños (CCM, Ribadeo, Xunta de Galicia) is developing and improving the hatchery culture of different species of bivalves, with the main aim of increasing the sustainability of aquaculture on the Galician coast (NW Spain).

The grooved carpet shell, *Venerupis decussata*, and the pullet carpet shell, *V. corrugata*, are commercially very important in Galicia. However, the high economic value and market demand have led to overexploitation of these species. As a result, natural stocks have been greatly depleted and clam culture clearly limited by the availability of seed (Ojea et al., 2004). As consequence, the Manila clam, *V. philippinarum*, was introduced because of its high growth rate and its relatively high resistance to changes in environmental conditions.

Furthermore, the commercial value of razor clams, *Ensis* spp. and *Solen* spp., has increased in recent years, with the consequent increase in fishing pressure. Although natural clam beds have started to show signs of overexploitation, demand has continued to increase. Hence, these species are of interest to aquaculture for food production and to improve natural stocks (da Costa and Martínez-Patiño, 2009). A similar situation has been reported for the wedge-shell clam *Donax trunculus* (Ruiz-Azcona et al., 1996).

In this chapter we review the existing information about the cultivable microbiota associated with hatchery cultured clams, and present some results of our current works on this subject. These studies are focused on the bacterial populations associated with larval cultures and with the broodstock conditioned in hatchery installations. Special emphasis is given to bacteria belonging to the genus *Vibrio*, which includes most of the known pathogens of bivalve larvae in hatcheries.

8.2. Broodstock

Hatchery broodstock conditioning is an established technique. Adult specimens are collected from wild beds, placed in tanks and induced to spawn by artificial means. The physical environment and the nutrition of broodstock are modified to promote gonad development and gametogenesis, by e.g. manipulating seawater temperature and supplying adequate phytoplankton (Utting and Millican, 1997). Hatchery managers thus try to extend the production season or to accelerate gamete development in animals undergoing gametogenesis (Sastry, 1979; Helm and Bourne, 2004). The method is also a useful tool for optimizing the hatchery culture of new bivalve species.

Although some studies have investigated the influence of factors such as temperature and food availability on conditioning, there is a lack of information about the bacterial populations associated with broodstock while in the hatchery. One of the main routes of entry of bacteria to larval cultures of many different species in hatcheries is via the broodstock (Lodeiros et al., 1987; Riquelme et al., 1994, 1995b; Sáinz-Hernández and Maeda-Martínez, 2005). The bacterial load may include opportunistic pathogens that do not affect adult specimens but could cause severe infections in larvae (Tubiash et al., 1965; Tubiash and Otto, 1986). The bioaccumulative properties of filter-feeder organisms such as bivalves favour the entry of bacteria. The larval pathogens entering the system from seawater may accumulate, along with normal microbiota in broodstock. If passed on to larvae, such strains may proliferate in stressed individuals and infect healthy individuals, thus causing mass mortalities.

In our studies, the gonadal tissue was chosen for sampling because gamete differentiation and release to the environment are characteristic features of bivalves. In clams, the gonad occupies a portion of the visceral mass. The gametes arise by proliferation of germinal cells that line the follicle wall. The gonad undergoes continuous development until it becomes fully mature. The gametes are discharged into the open environment where fertilization occurs (Helm and Bourne, 2004). In previous studies, the presence of bacteria in the gonads of bivalve broodstock was confirmed for *Ostrea edulis* (Lodeiros et al., 1987), *Argopecten purpuratus* (Riquelme et al., 1994, 1995b), *A. ventricosus* (Sáinz-Hernández and Maeda-Martínez, 2005) and *Crassostrea gigas* (Sugumar et al., 1998). Moreover, bacteria belonging to the genus *Vibrio* were identified in the bacterial populations in all of these studies.

Broodstock of the different clams cultured in the CCM were processed individually for bacteriological studies. The studies included members of the Superfamilies Veneracea (genus *Venerupis*), Solenacea (genera *Ensis* and *Solen*) and Tellinacea (genus *Donax*). The clams were washed in sterile seawater (SSW), and pieces of gonad were excised aseptically, weighed and homogenized in SSW. Serial dilutions of these suspensions were made and spread on appropriate culture plates.

The media used were Marine Agar (MA, Pronadisa, Spain), for marine heterotrophic bacteria, and Thiosulphate Citrate Bile Sucrose (TCBS, Oxoid, UK), which is selective for vibrios. The plates were incubated at 23±2°C. Predominant colonies were selected and isolated after 24-48 hours (TCBS) and 7 days (MA), and further re-streaked on MA. The strains were preserved in Marine Broth with glycerol (15% v/v) by freezing at -80°C. The isolates were examined for their phenotypic characteristics. Molecular studies of selected bacterial strains were performed by sequencing of the 16S rDNA gene. All of the methods used are described elsewhere (Prado et al., 2005).

Bacterial counts were expressed as colony forming units per gram of gonad (cfu/g), corresponding to Total Viable Counts (TVC, in MA medium) and Presumptive Vibrio Counts (PVC, in TCBS medium).

8.2.1. Superfamily Veneracea. Family Veneridae

8.2.1.1. Venerupis corrugata (Pullet Carpet Shell)

The broodstock of the pullet carpet shell clam, *Venerupis corrugata* displayed a high bacterial load associated with the gonad, with mean initial counts of viable bacteria (TVC) of 7.9 x 10^5 cfu/g (Figure 1). The PVC in these samples were also high, with mean values of 1.6 x 10^5 cfu/g, indicating that vibrios constituted an important fraction of bacterial population (20.5%). After conditioning, the average values of TVC and PVC remained similar, 1.5 x 10^6 and 1.4 x 10^5 respectively, with only a slight reduction in the PVC/TVC ratio (9.2%) due to the overall increase in TVC. Analysis of variability among three batches revealed differences of 1 order of magnitude in TVC (10^5-10^6 cfu/g) and 2 orders of magnitude in PVC (10^3-10^5 cfu/g). The effect of conditioning varied in each stock. In the first batch the PVC/TVC ratio (expressed as a percentage) increased slightly, but the load of both total bacterial and vibrio populations decreased. The reduction in the PVC/TVC ratio in the second batch was due to the combined increase in TVC and decrease in PVC.

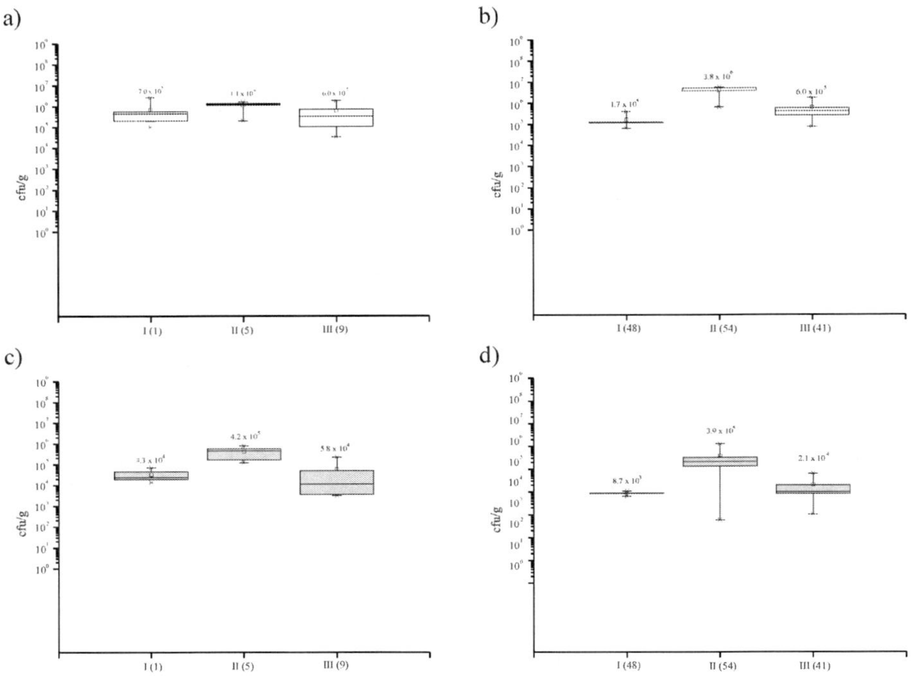

Figure 1. Heterotrophic bacteria (TVC) and vibrios (PVC) in the gonad of *Venerupis corrugata* broodstock before and after conditioning in hatchery. The results are expressed in colony forming units per gram of gonad (cfu/g). The upper graphs (a and b) represent the results for MA medium, the lower graphs (c and d) represent the results for TCBS medium. The graphs on the left (a and c) show the counts before conditioning, and those on the right (b and d) show the counts after conditioning. The numbers in parentheses indicate the days maintained in hatchery at the time of sampling.

In the third batch the TVC remained the same and the PVC decreased, with the subsequent reduction in the PVC/TVC ratio.

The variability among individuals within batches was low in the initial samples, between 0 and 2 orders of magnitude in both media. Nevertheless, the counts after conditioning revealed large differences among individuals, reaching up to 4 orders of magnitude in the PVC. These results indicated that conditioning exerted highly variable effects on the bacterial populations of pullet shell clam. Moreover, there was no large reduction in bacterial populations associated with gonad.

8.2.1.2. Venerupis decussata (Grooved Carpet Shell) and V. philippinarum (Manila Clam)

The bacterial loads in gonads of grooved carpet shell clam (*V. decussata*) and Manila clam (*V. philippinarum*) showed the lowest mean ratio of vibrios (0.1 and 3.5%, respectively) when they arrived in the hatchery, as a result of the large numbers of total bacteria (10^6-10^7 cfu/g), and in the case of *V. decussata* the low counts of PVC ($\approx 10^4$ cfu/g).

In *V. decussata*, the average bacterial counts after conditioning decreased by 1 order of magnitude in both media. However, the individual values were highly variable as a result of the different degree of reduction in bacterial populations, 2 orders of magnitude in TVC (10^4-10^6 cfu/g) and 3 orders of magnitude in PVC (below the limit of detection-10^3 cfu/g).

The TVC for *V. philippinarum* stabilized after conditioning in 10^5 cfu/g. The PVC counts showed, as in *V. decussata*, high variability among individuals, with values ranging from no detection to 10^4 cfu/g.

Again, the changes in bacterial populations after conditioning varied considerably among individuals, with the stabilization of TVC but different degrees of reduction in vibrios. In this first approach, the bacterial load of *V. decussata and V. philippinarum* broodstocks appeared to be lower than *V. corrugata* batches, mainly as regards the population of *Vibrio* spp. Further studies are required to establish the post conditioning variability among stocks of grooved carpet shell and Manila clams.

8.2.2. Superfamily Solenacea. Family Pharidae

8.2.2.1. Ensis siliqua (Pod Razor Clam)

The mean TVC for the broodstock of the pod razor clam *Ensis siliqua* was 6.7 x 10^6 cfu/g in the initial controls, and the mean PVCs were one order of magnitude lower (6.5 x 10^5 cfu/g) (Figure 2). The variability among the three batches was high, as in the other species studied. The initial PVC/TVC rate were 49.7, 30.2 and 5.2 % for each batch. As observed in Figure 2, the lowest counts in MA and TCBS were recorded for the first broodstock, both were of the same order of magnitude (10^4 cfu/g).

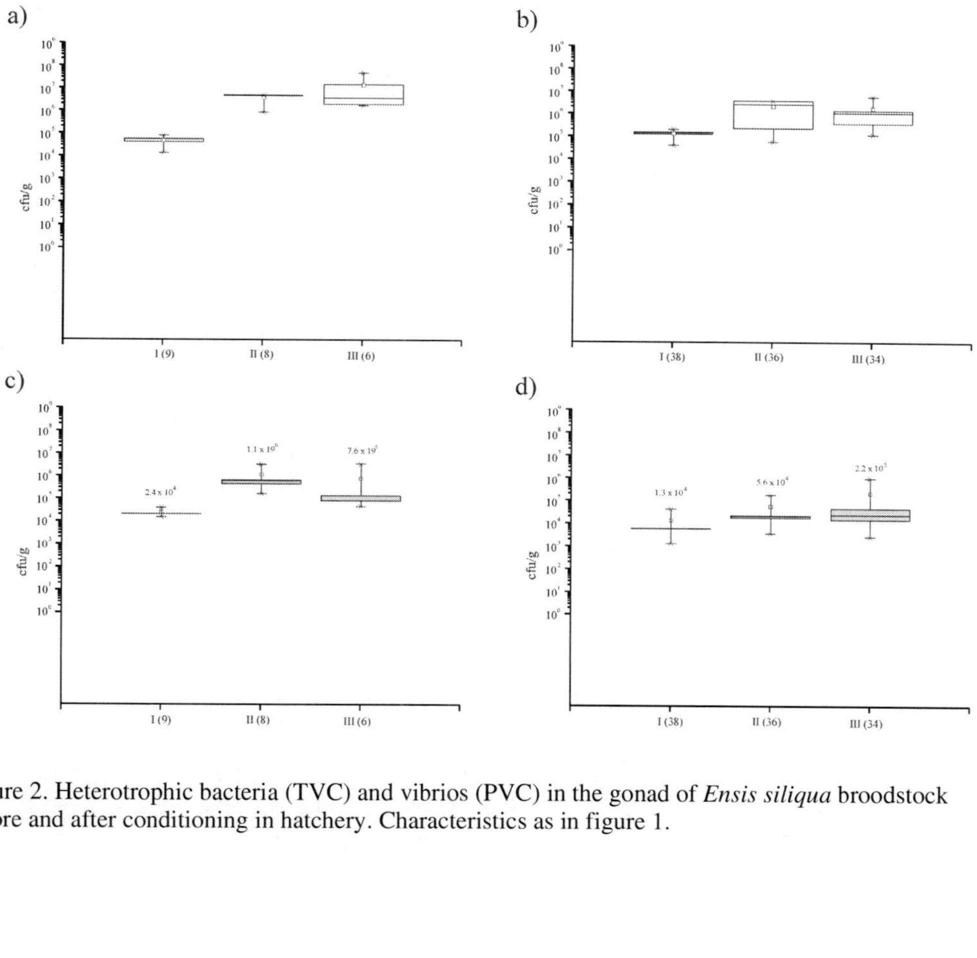

Figure 2. Heterotrophic bacteria (TVC) and vibrios (PVC) in the gonad of *Ensis siliqua* broodstock before and after conditioning in hatchery. Characteristics as in figure 1.

The main effect of conditioning was the stabilization of the PVC at slightly lower values than in the initial controls. However, the TVC increased in one case, remained the same in another and decreased in the third, tending to be more homogeneous in the final controls. The differences among individuals did not follow any pattern, reflecting the high variability in the changes in the bacterial populations associated with gonad, even in clams held under the same conditions.

8.2.3. Superfamily Solenacea. Family Solenidae

8.2.3.1. Solen marginatus (Grooved Razor Clam)

The highest PVC in the initial controls were those for the grooved razor clam *Solen marginatus* with also a high mean proportion of vibrios in the gonad.

Nevertheless, there was variability in PVC between batches, with the highest values in two of them ($\approx 10^6$ cfu/g) and the lowest in the other ($\approx 10^4$ cfu/g). This finding, combined with the relatively homogeneous TVC resulted in PVC/TVC ratios ranging from 98.2 to 1.3%. After conditioning the values were more homogeneous, due to stabilization of the vibrio load at around 10^4 cfu/g in all cases, together with the slight decrease in the contents of heterotrophic bacteria (10^5-10^6 cfu/g). Again, comparison between individuals of the same batch did not follow a clear pattern.

8.2.4. Superfamily Tellinacea. Family Donacidae

8.2.4.1. Donax trunculus (Wedge-Shell Clam)

The initial TVC was highest in the wedge-shell clam *Donax trunculus*, as reflected by the low PVC/TVC, despite the relatively high load of vibrios (Figure 3). The variability between batches was high. In the initial counts, the TVC varied greatly (10^5-10^8 cfu/g), but values became more similar after conditioning. On the other hand, the variability in the PVC increased, with higher or lower numbers of vibrios depending on the batch.

As result of these combinations, in one batch the proportion of *Vibrio* spp. increased, remained the same in another and decreased in another. The variability among individuals was higher in the initial values, mainly of PVC, which reached differences of 5 logs in one batch. As reported for the other species, the effect of conditioning varied among batches and did not follow a clear pattern.

8.3. Larval Cultures

The literature on the microbiota of clams in hatchery culture is scarce, being the hard shell clam or Northern quahog *Mercenaria mercenaria* the best studied species in this respect.

Guillard (1959) demonstrated that two strains of bacteria (*Vibrio* and *Pseudomonas*) isolated from infected hard clam larvae destroyed healthy larvae, and that the presence of living bacteria was required to cause disease. There were no differences in mortalities in studies carried out at different temperatures within the range 20-30ºC.

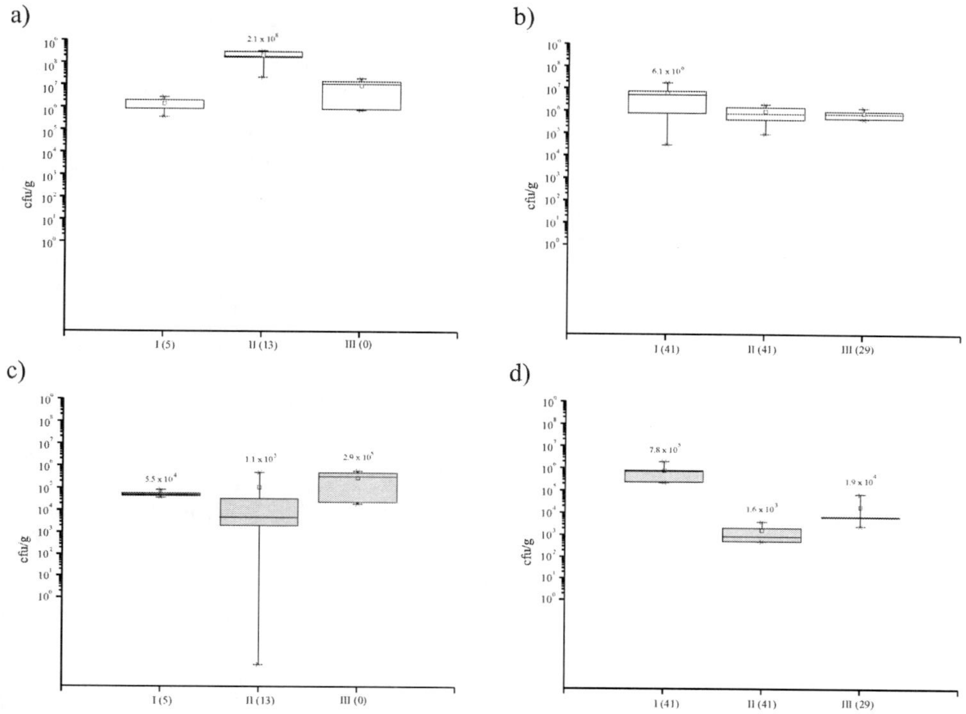

Figure 3. Heterotrophic bacteria (TVC) and vibrios. (PVC) in the gonad of *Donax trunculus* broodstock before and after conditioning in hatchery. Characteristics as in figure 1.

The hard clam was one of the species studied by Tubiash et al. (1965) in their description of "bacillary necrosis" caused by *V. tubiashii*, *V. alginolyticus* and *Vibrio* sp. Tubiash and Otto (1986) found bacterial isolates from infected cultures of *M. mercenaria* and the American oyster *Crassostrea virginica*, which were able to kill larval cultures from a wide range of hosts, including hard clams. The strains were identified as *V. anguillarum* and *V. alginolyticus*.

Brown (1974) found that a red-pigmented *Pseudomonas* caused the decrease in embryonic development or mortality of hard clam embryos. In 1988, Brown and Tettelbach demonstrated that a *V. anguillarum*-like isolated from moribund larvae of *M. mercenaria* was pathogenic to developing hard shell clams.

Apart from *M. mercenaria*, only the microbiota of hatchery cultured *Venerupis* (= *Ruditapes*) has been investigated in specific studies. Gómez-León et al. (2005) isolated strains of *V. splendidus* and *V. alginolyticus* as the causative agents of episodes of mass mortality of larvae and spat of *Venerupis decussata* (=*R. decussatus*) in a commercial hatchery.

Nicolas et al. (1992) reported the pathogenicity of a *Vibrio* strain involved in a recurrent outbreak of disease in *Venerupis philippinarum* (=*R. philippinarum*) larvae in a commercial hatchery. Oysters and scallop larvae were not affected by this strain, which the authors considered as the first specific pathogen. Moreover, the strain showed unusual traits because it did not grow on TCBS and did not survive very long in seawater, with survival times closer to those of freshwater bacteria than of marine bacteria.

The different clams cultured in the CCM include species traditionally cultured in Galicia (Spain), such as the grooved carpet shell and pullet carpet shell, and also species new to hatchery culture, such as wedge-shell clam or razor clams (native species) and the Manila clam (introduced species). Episodes of mortalities affecting larval cultures have been observed in the CCM. In the present study, the bacterial populations associated with different larval cultures (healthy or suffering disease) were studied in order to determine their influence on the larval survival.

In addition, antibiotic was used experimentally to elucidate the involvement of bacteria in these outbreaks. Chloramphenicol was chosen because it was previously the most commonly used in bivalve hatcheries as it has a broad spectrum of activity. The effects of antibiotic on bacterial populations of larval cultures have been reported in many studies, thus enabling comparison with the results obtained for other species of bivalves.

The larval cultures were established in the CCM, following the procedures developed in the institution.

Some batches were divided in two, and the antibiotic was added to one batch (Ab^+: 2.6 mg/l chloramphenicol) during the periodic changes of seawater, to evaluate its effect on the larval survival. The other batch was maintained as a control (Ab^-).

Microbiological samples were processed in the hatchery. Larvae or post-larvae were taken using a sterile inoculation loop (1 μl) and spread directly on MA and TCBS plates. Seawater from culture tanks was sampled at the same time. Appropriate dilutions were made in SSW and spread on plates. The plates were incubated and predominant colonies isolated and preserved (see above). The isolates were characterized phenotypically. The 16S rDNA genes of selected strains were sequenced for better identification. The methods used are described elsewhere (Prado et al., 2005).

Four larval cultures of *V. philippinarum* obtained throughout the season from the same broodstock were compared. Each batch was divided in two; antibiotic was added to one and the other batch was maintained as a control without treatment. The size and survival at settlement are detailed in table 1.

Batch 1 (June). At the start of the culture there was a high bacterial load (mainly fermentative bacteria) associated with the larvae.

Throughout culture, the antibiotic did not decrease the bacterial concentration in seawater or the vibrios in larvae. Survival at settlement was below 2% in both cultures.

Batch 2 (July). The initial bacterial load associated with larvae was low, but with presence of fermentative bacteria. There were no differences in the total bacteria in seawater between the two cultures. No *Vibrio* spp. were isolated from this source.

Table 1. Size and survival at settlement of 4 larval cultures of *Venerupis philippinarum* obtained from one broodstock

Batch	Settlement size (μm)		Survival (%)	
	Ab^-	Ab^+	Ab^-	Ab^+
1	238	236	1.6	1.9
2	–*	239	–*	13
3	203	205	28	25
4	219	221	33	51

*mortality (day 22).

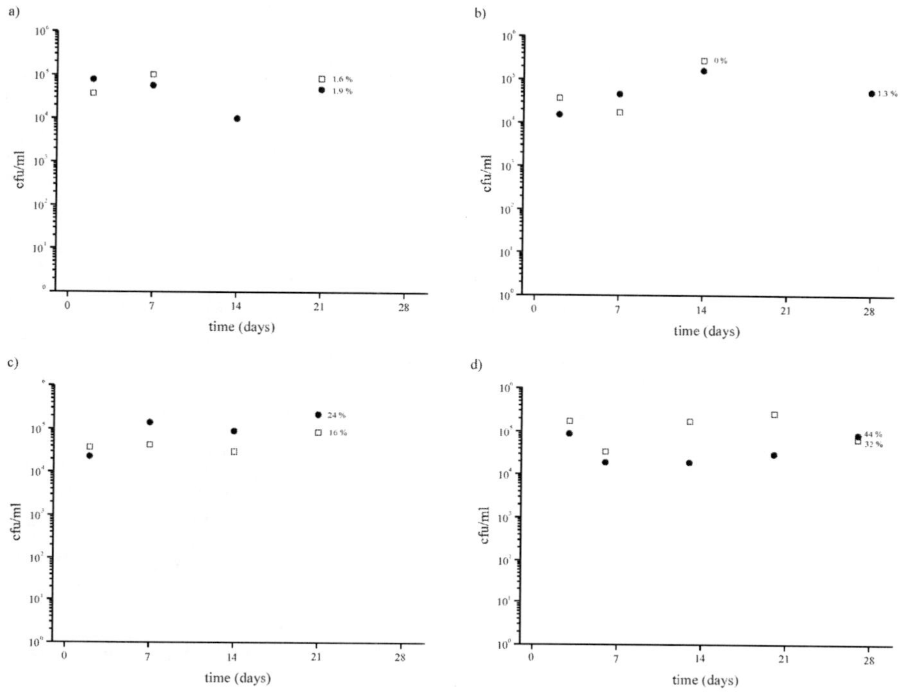

Figure 4. Counts of heterotrophic marine bacteria in the seawater of *Venerupis philippinarum* larval cultures, maintained without (white squares, Ab^-) and with antibiotic (black circles, Ab^+). The numbers indicate the survival.

Differences in survival were recorded on day 6 (24% Ab^-, 42% Ab^+), with a parallel increase vibrios associated with larvae Ab^-, absent in larvae Ab^+. Finally, the culture without antibiotic was discarded on day 22. The larval samples showed high loads of *Vibrio* spp.

Batch 3 (August). The bacterial load in seawater was similar, with slightly higher values in culture Ab^+. No vibrios were detected in these samples. Only fermentative bacteria were isolated from larvae, mainly Ab^-, in very low numbers. Survival was 28 and 25 %, in Ab^- and Ab^+ respectively.

Batch 4 (September). The initial bacterial load in larvae decreased substantially in both cultures after 6 days. Thereafter, vibrios were isolated only from Ab^- and were mainly associated with larvae. The antibiotic maintained the total bacteria counts in seawater 1 log lower during the larval culture. The survival of culture Ab^+ was higher (51%) than that of culture Ab^- (33%).

The total viable counts of heterotrophic bacteria in MA associated with the seawater in the culture tanks were similar. The effect of the antibiotic on total bacteria did not follow any clear pattern and no relationship with the larval survival was established. These findings ruled out this parameter as indicating problems.

High levels of vibrios in seawater samples are considered as indicative of bacterial disease and the absence of these species is desirable, although this does not guarantee the health of the culture, as observed in samples from batches 1 and 2. The fermentative isolates from batch 2 were characterized (RpL) in order to identify the members of genus *Vibrio* present and to compare the species associated with parallel cultures with and without antibiotic, the latter of which suffered heavy mortalities.

The members of Splendidus-clade were the most abundant in both cultures (Figure 5). Some types were isolated from cultures with and without antibiotic, suggesting that the antibiotic did not prevent the presence of these bacteria. This was the case for the *V. splendidus*-like strains (RpL-Ab$^+$ 3, RpL-Ab$^+$7, RpL-Ab$^-$ 9). Interestingly, one group of strains was only isolated from larvae cultured without antibiotic (RpL-Ab$^-$ 1, RpL-Ab$^-$ 4, RpL-Ab$^-$ 12). Their role in the mortalities should be determined further by pathogenicity assays, because they may be true pathogens or only secondary invaders of the weakened larvae. As regards their taxonomic position, these bacteria were located in a separate branch in the phylogenetic tree, suggesting that this may be a new species in this clade, not described until now. The presence of members of the Splendidus-clade was expected, as 13 members of this clade have been originally isolated from marine environments and eight of these from bivalves. *Vibrio crassostreae* (Faury et al., 2004) and *V. gigantis* (Le Roux et al., 2005) were isolated from haemolymph of diseased adults of *Crassostrea gigas*, and *V. lentus* from *Ostrea edulis* (Macián et al., 2001). *Vibrio kanaloae* was isolated from diseased *Ostrea edulis* larvae and *V. pomeroyi* from healthy larvae of the scallop *Nodipecten nodosus* (Thompson et al., 2003). Recently, four new species have been described, all isolated from clams (*V. philippinarum*, *V. decussata* and *V. corrugata*) cultured in Galicia, *V. gallaecicus*, *V. celticus*, *V. atlanticus* and *V. artabrorum* (Beaz-Hidalgo et al., 2009, 2010; Diéguez et al., 2011).

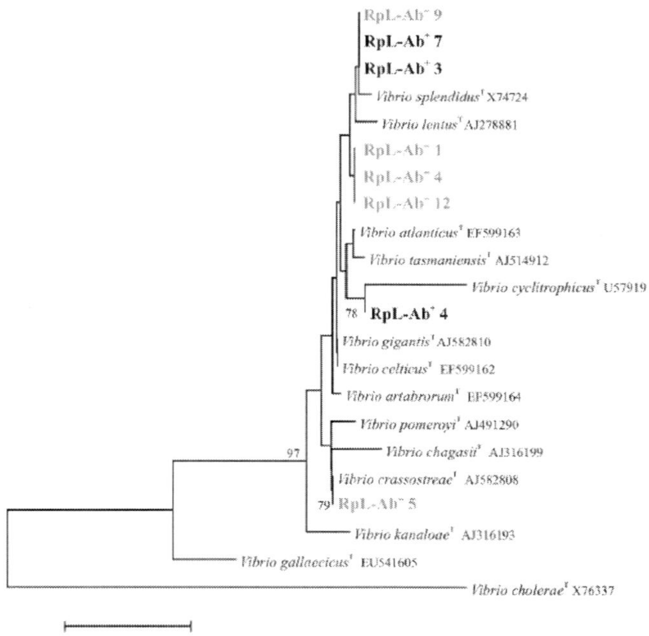

Figure 5. Phylogenetic tree based on partial 16S rDNA sequences of isolates from *V. philippinarum* (RpL, Batch 2) belonging to Splendidus-clade, constructed by the neighbour-joining method. *Vibrio cholerae* was used as an outgroup. GenBank sequence accession numbers are given with every type species. Bootstrap values (1000 replicates) higher than 50 are close to the corresponding branch. The bar represents substitutions per nucleotide position. The strains from the culture with antibiotic (Ab$^+$) are indicated in black bold type, and those one from the culture without antibiotic (Ab$^-$) in grey.

However, the high phenotypic diversity within this clade makes difficult to differentiate species on the basis of biochemical tests. Moreover, the sequences of the highly conserved 16S rRNA gene enable assignment of the isolates to the clade, but not to one defined species, and further molecular studies are necessary.

The dominance of this clade in samples makes knowledge of the role of their members in hatchery environment essential. This is reinforced by the reports of some *V. splendidus*-like isolates as pathogens of clams (Gómez-León et al., 2005), oysters, scallops and mussels in hatcheries (Jeffries, 1982, Sugumar et al., 1998; Kesarcodi-Watson et al., 2009).

In addition to the Splendidus-clade, members of the Harveyi-clade of *Vibrio* were also present (Figure 6). The strains in this clade were isolated from cultures with and without antibiotic.

Three groups were observed in the phylogenetic tree. One group included strains similar to *V. rotiferianus* - *V.campbellii* - *V. harveyi*, from larvae cultures without antibiotic (RpL-Ab$^-$ 2) and with it (RpL-Ab$^+$ 1, RpL-Ab$^+$ 6). Another group comprised isolates also from both sources, RpL-Ab$^+$ 2, RpL-Ab$^-$ 10, which could not be assigned to a species, but were closely related to *V. campbellii* - *V. rotiferianus*, *V. owensii* - *V. communis* and *V. azureus*.

As with the Splendidus-clade, one group was clearly located in a separate branch from any described *Vibrio* sp., again suggesting the possible presence of a new species. The strains included in this group were mainly isolated from cultures without antibiotic, and further characterization is required.

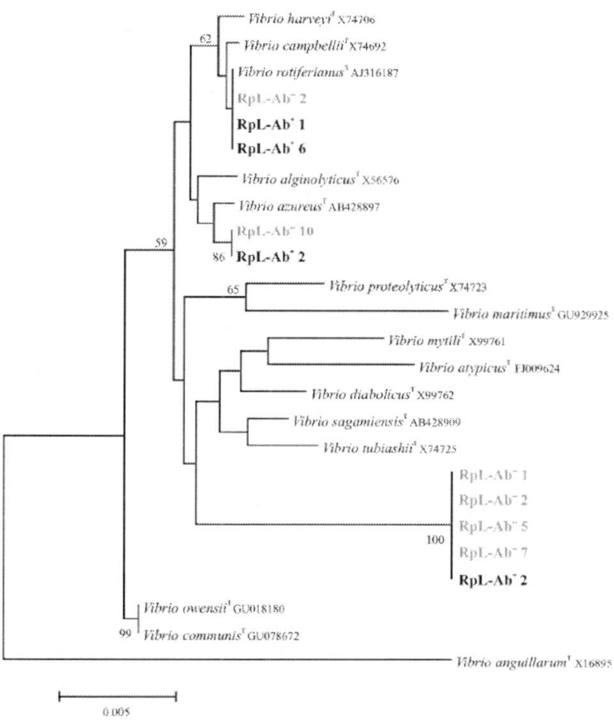

Figure 6. Phylogenetic tree based on partial 16S rDNA sequences of isolates from *V. philippinarum* larvae (RpL, Batch 2) belonging to Harveyi-clade and closely related strains. *Vibrio anguillarum* was used as outgroup. Characteristics as in figure 5.

Members of Harveyi-clade are commonly found in marine and estuarine surface waters and sediments, as commensals on the surface or within the intestinal microbiota of marine animals, as opportunistic pathogens, or as primary pathogens of many commercially farmed marine invertebrate and vertebrate species (O'Brien and Sizemore, 1979; Thompson et al., 2004).

This clade contains known pathogens of fish, corals, molluscs and humans, and has therefore been of concern as regards public health and in the aquaculture industry for the last 10 years (Hoffman et al., 2012). The species *V. harveyi* has been identified as a pathogen of fish, shrimps, lobsters, abalones, sea cucumber and corals (Nishimori et al., 1998; Nicolas et al., 2002; Austin and Zhang, 2006; Sawabe et al., 2007; Cano-Gómez et al., 2009), and in one case was associated with mortalities in the adult pearl oyster *Pinctada maxima* (Pass et al., 1987).

Vibrio campbellii has been described as a pathogen of Artemia and shrimps (Phuoc et al., 2009; Haldar et al., 2011) and the misidentification of many isolates originally assigned to the species *V. harveyi* is now under consideration (Lin et al., 2010), probably being *V. campbelli*. *Vibrio parahaemolyticus* has been found to cause death of shrimps (Zhang et al., 2009) and abalones (Liu et al., 2000; Lee et al., 2001; Cai et al., 2006, 2007).

In one case it was associated with problems in bivalves, in the adult clam *Meretrix meretrix* (Yue et al., 2010). A new species *V. owensii* (Cano-Gómez et al., 2010) was isolated from diseased crustaceans *Panulirus ornatus* and *Penaeus monodon* in culture, and demonstrated as highly virulent to prawn and lobster larvae (Pizzuto and Hirst, 1995). However, the species most closely associated with bivalve cultures in hatcheries is *V. alginolyticus*, a known pathogen of oysters, scallops (Tubiash et al., 1965; Jeffries, 1982; Tubiash and Otto, 1986; Riquelme et al., 1996; Luna-González et al., 2002) and clams such as *M. mercenaria* (Tubiash et al., 1965; Tubiash and Otto, 1986) and *V. decussata* (Gómez-León et al., 2005).

Considering the results for larval cultures of wedge-shell clam *D. trunculus*, the comparison with Manila clam isolates revealed similarities in the vibrios associated with both clam species.

The cultivable microbiota of wedge-shell clam in two different larval cultures indicated that Splendidus-clade was again the most abundant in this environment. The fermentative isolates obtained from one batch which suffered heavy mortalities (DtL-m) were compared with those from a healthy culture (DtL-h). In both cases the antibiotic was supplied at each change of seawater, to establish if this would control the outbreaks of disease previously recorded in the hatchery.

Batch A suffered mortalities after settlement. The samples of seawater and larvae cultured in MA medium showed a predominance of oxidative bacteria, with only one fermentative strain isolated from larvae in the final sample (suffering mortalities).

A total of 8 isolates from TCBS samples were fermentative bacteria, identified as *Vibrio* spp. These were all associated with larvae and were not detected in seawater.

Batch B did not suffer heavy mortalities. From this source, five out of seven isolates from TCBS and one from MA were fermentative bacteria, all from larvae mainly in the first samples. Most of them were identified as *Vibrio*, distributed in different species.

The sequencing of 16S genes from representative isolates of fermentative bacteria again showed the predominance of Splendidus-clade in both cultures.

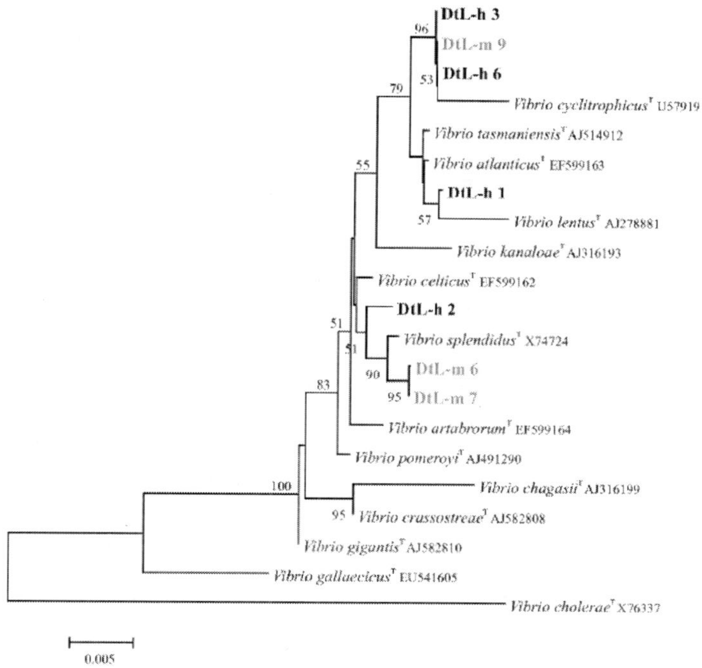

Figure 7. Phylogenetic tree based on partial 16S rDNA sequences of representative isolates from *D. trunculus* larvae (DtL) belonging to Splendidus-clade. The strains associated with mortalities (DtL-m) are highlighted in grey to distinguish them from the isolates of healthy cultures (DtL-h). Characteristics as in figure 5.

The main difference in the cultivable microbiota of these two batches was the presence of some strains closely related to the species *V. splendidus* associated with the larvae suffering mortalities (DtL-m 6, DtL-m 7). This type was isolated in all the larval samples, although only two representative strains were included in the phylogenetic tree. These strains all shared an identical phenotypic profile. This group appears to be a strong candidate for further studies, including pathogenicity tests to elucidate its role in outbreaks of disease that affect this species.

One group of strains was found in both cultures, similar to the species *V. atlanticus* - *V. tasmaniensis* (DtL-h 3, DtL-h 6, DtL-m 9). Interestingly these were very similar to the isolate RpL-Ab$^+$ 4 from Manila clam larvae (Figure 4). Their presence may have been enhanced by the antibiotic, possibly due to the elimination of other competitors.

However, this group did not appear to be involved in the mortalities. Confirming the hypothesis about the common environment in all the larval cultures maintained in one installation, a member of Harveyi-clade was found in the final samples of wedge-shell clams suffering mortalities, DtL-m 8, similar to isolates from the Manila clam, RpL-Ab$^+$ 2 and RpL-Ab$^-$ 10 (data not shown).

In the hatchery, the seawater and the microalgae were routinely controlled and ruled out as the source of vibrios. Therefore, the main candidate as the origin of these bacteria was the broodstock. In a study of the cultivable microbiota associated with hatchery cultured *Solen marginatus*, including larval cultures and the broodstock, vertical transmission of bacteria from conditioned adults to the larvae was demonstrated.

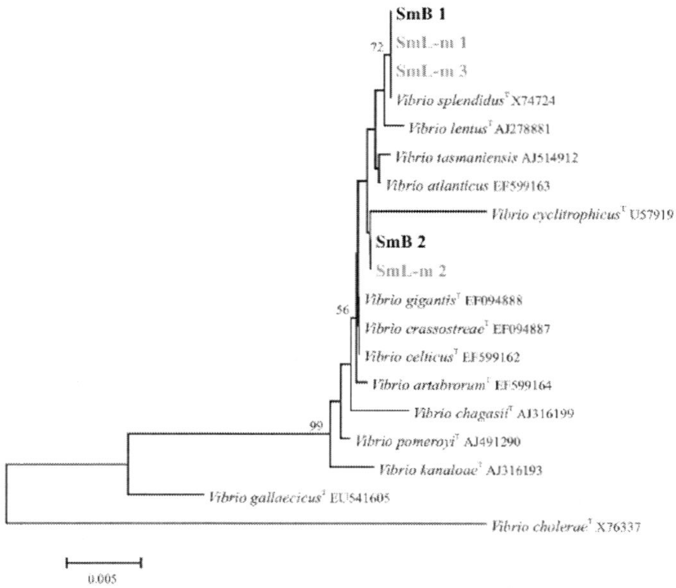

Figure 8. Phylogenetic tree based on partial 16S rDNA sequences of isolates, from *S. marginatus* broodstock (Sm-B) and the larval culture obtained from them, which suffered heavy mortalities (SmL-m, grey), belonging to Splendidus-clade. Characteristics as in figure 5.

As illustrated in the phylogenetic tree, isolates from razor clam larvae suffering mortalities in hatchery were similar to some strains from the broodstock. Two different types were identified. One corresponded to the group closely related to *V. splendidus*, also found in other bivalve larvae. The other one was similar to the isolates related to *V. atlanticus* - *V. tasmaniensis* - *V. cyclitrophicus* also described in Manila clam and wedge-shell clam.

The highly common microbial environment shared by all the bivalve species cultured in one hatchery appears to be demonstrated on the basis of the results obtained. The common taxonomic position of isolates from larval cultures of different species of clams is illustrated in figure 9.

8.4. General Discussion

Episodes of heavy mortalities have been observed in hatchery culture of different species of clams, as reported in other bivalve species. In addition to physiological and physico-chemical factors, the role of bacteria in clam culture should be elucidated, mainly those belonging to genus *Vibrio*, identified as aetiological agents in outbreaks of disease in other bivalves (see Introduction).

The first step in hatchery culture is to obtain and condition adult specimens. The risk of the presence of bacteria in the gonad and their transfer to larval cultures, i.e. the vertical transmission of bacteria from adults to larvae, was suggested by Lodeiros et al. (1987) in a study of bacillary necrosis in hatcheries of flat oyster caused by *Vibrio* spp. Riquelme et al. (1994, 1995b) demonstrated parental transfer of bacteria in *Argopecten purpuratus*, including *V. anguillarum*-related strains.

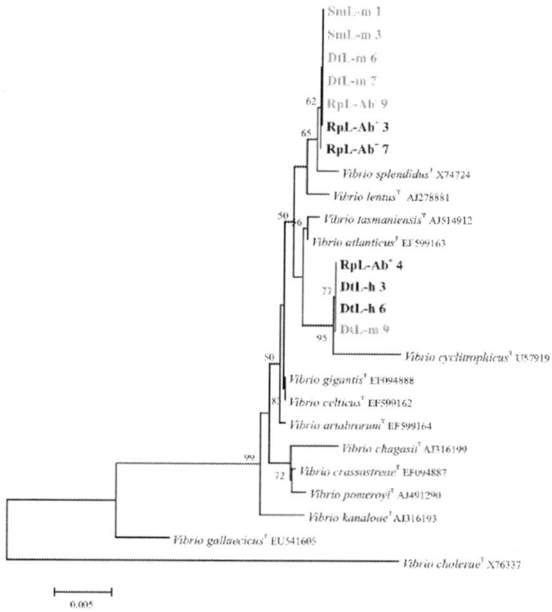

Figure 9. Phylogenetic tree including the 16S rDNA sequences of similar isolates, belonging to Splendidus-clade, from larval cultures of the different clam species (*V. philippinarum*, *D. trunculus* and *S. marginatus*). The strains associated with mortalities are highlighted in grey. Characteristics as in figure 5.

Sáinz-Hernández and Maeda-Martínez (2005) demonstrated that expulsion of gametes in *A. ventricosus* was accompanied by the release of *Vibrio* that proliferated at a high concentration within 24 h.

The results reported in this chapter demonstrate the existence of *Vibrio* spp. associated with the gonad of all species of clams studied, i.e. *V. corrugata*, *V. decussata*, *V. philippinarum*, *E. siliqua*, *S. marginatus* and *D. trunculus*. The lack of quantitative data in the literature makes comparison with other bivalve species difficult. Nonetheless, we have clearly established that the broodstock is a reservoir of bacteria, so that it is almost impossible to prevent the transfer of vibrios to gametes during spawning.

Despite the diversity of the natural environments where each clam species was obtained, all broodstock batches arrived at the hatchery with relatively high total numbers of heterotrophic bacteria and vibrios associated with the gonad, suggesting that it may be difficult to obtain broodstock with low numbers of bacteria, especially during the reproductive season. Therefore, the most practical approach appears to be optimization of the conditioning process.

The behaviour of the bacterial populations after conditioning was irregular. There was a high diversity between different broodstock batches, and also between individuals of the same batch maintained under the same conditions.

Appropriate conditioning of broodstock may be an effective procedure for reducing the risk of vertical transmission of pathogens, as observed by Riquelme et al. (1995b) with the decrease in bacterial concentration in treated organisms. Widman et al. (2001) recommended that *Vibrio* infestation in hatcheries can be reduced by purging molluscs for 24 h at 17-20ºC in filtered seawater, with several changes of water, before the animals are used for spawning.

However, Sáinz-Hernández and Maeda-Martínez (2005) demonstrated that this treatment is not sufficient to prevent the entry of vibrios into the system because these bacteria were detected in gonad tissues, far from the digestive tract.

The study of bacterial populations in broodstock maintained in hatchery, the effect of the shared seawater and food on them, and the factors involved in the variation in the composition of this microbiota will provide a useful tool for improving the microbiological quality of the broodstock and reducing the risk of entry of bacterial pathogens in the facility.

Vibrios are naturally present in marine environments, including clam hatcheries. Despite efforts to eliminate them from seawater or phytoplankton by different treatments, these bacteria are frequently detected in larval cultures. As reported above, the broodstock is the main source of contamination, although pathogens can also be introduced at any point of the system or persist in tanks forming biofilms (Elston, 1984; Prado et al., 2005). The main problem in relation to bivalve larvae is not strictly the presence of *Vibrio* spp., but rather the numbers in which they are present.

We have demonstrated the association of *Vibrio* with the larval cultures of different species of clams and their probable relationship with outbreaks of disease in the hatchery.

As observed in the four batches of *V. philippinarum* larvae, bacteria were present in the seawater in the culture tanks, even though antibiotic were used. Moreover, its use did not guarantee the survival of the batch treated. The results strongly suggest that the antibiotic was able to control the presence of *Vibrio* spp. in seawater, modifying the composition of the bacterial population, but that it was ineffective in eliminating the vibrios associated with larvae. We have observed that if the hatchery maintains a good microbiological quality, i.e., if the seawater and the phytoplankton are free of vibrios, that is, below 10^1 cfu/ml, high counts of *Vibrio* spp. will not usually be found in the water containing the larval cultures (data not shown). However, some authors consider high levels of vibrios in the seawater as a good indicator of vibriosis, though the numbers have not been established, with proposed thresholds of $>10^4$ cfu/ml (Widman et al., 2001) or $>10^3$ cfu/ml (Lodeiros et al., 1987). As every hatchery is located at a particular site, has its own design, operation systems and procedures, and focuses on particular species, it is difficult to establish standard values. In any case, although the presence of high numbers of *Vibrio* spp. in the seawater in culture tanks indicates that a problem is arising, absence of vibrios does not necessarily indicate a healthy culture. Therefore, the analysis of seawater appears insufficient for determining the microbiological status of a larval culture.

On the other hand, *Vibrio* spp. are commonly associated with larvae, probably because of the almost unavoidable transfer through the gametes. These observations were confirmed in the batches of Manila clam: vibrios were isolated from larval samples, not from seawater, mainly in cultures with high mortalities. The same was found with wedge-shell clam and razor clam, in cultures treated with antibiotic in an attempt to prevent mortalities.

Chloramphenicol has been reported as effective for controlling mortalities in larval clam cultures in laboratory (Tubiash et al., 1965; Brown 1974) as well as in a commercial hatchery (Gomez-León et al., 2005). However, the appearance of *Vibrio* sp. that is resistant to this antibiotic and cause recurrent outbreaks of disease in Manila clam in a commercial hatchery has also been reported (Nicolas et al., 1992). Chloramphenicol was previously one of the most commonly used antibiotics in aquaculture (Schwarz et al., 2004; Saptoka et al., 2008) and the main one in bivalve hatcheries for several years (see reviews D'Agostino, 1975; Le Pennec and Prieur, 1977; Prado, 2006). It is a potent, inexpensive broad-spectrum antibiotic, and does not require specific storage conditions. Nevertheless, its continued use has led to the above-mentioned chloramphenicol-resistance in bacteria from aquaculture environments, including *Vibrio* spp. (Ho et al., 2000; Roque et al., 2001; Tendencia and de la Peña, 2001; Hameed et al., 2003; Vaseeharan et al., 2005; Dang et al., 2006), with the subsequent risks for the facilities. Use of chloramphenicol in animals intended for food production has been banned in many countries because it has not been possible to establish a safe level of human exposure (Hernández-Serrano, 2005). However, its application in the treatment of larval and early juvenile bivalves in experimentally controlled systems probably should not represent a risk to consumers, as this involves the production of small sized seed, which will be later grown to commercial size during 1-2 years at sea (Uriarte et al., 2001). And these experiments could help to elucidate the role of bacteria in the success of the cultures.

The results demonstrated that the use of antibiotic is insufficient to guarantee the survival of clam cultures and favours the growth of some *Vibrio* spp. (probably resistant) over others. Alternatives to antibiotics should be found, on the basis of knowledge and control of opportunistic pathogens combined with the enhancement of beneficial bacteria (Prado et al., 2010). As regards the identification of the cultivable microbiota of clams in hatchery, the Splendidus-clade clearly predominated in all larval samples, regardless of the clam species, the use of antibiotic or survival. This indicates the diversity of the ecological roles of the members within this clade. Their role in the episodes of mortalities of the different strains should be elucidated with pathogenicity assays. Several taxonomic studies of the *V. splendidus* clade have been carried out in recent years and have led to the description of new species. However, the interest for hatcheries is mainly the ecological role of these bacteria and their interactions with e larvae. Further studies should focus on these aspects.

Something similar occurs with the Harveyi-clade. In this case, their presence in clam cultures in hatcheries had not been reported. The present results suggest that these bacteria may form part of the usual microbiota in clam larvae. Finally, the dynamics of the infection in hatchery environment should be studied in detail. Again, little is known about this subject, and although some hypotheses have been proposed, nothing has been demonstrated. The induction of the disease process remains poorly understood (Elston et al., 2008).

8.5. Final Remarks

From the results above presented, we can conclude:

1 The cultivable microbiota associated with hatchery culture of clams mainly comprised *Vibrio* spp., which were found in broodstock as well as in larvae.
2 The changes in the bacterial load of the gonad of broodstock during conditioning in hatchery did not follow a clear pattern. Further studies are required to establish the relationship between the modified environmental conditions and the dynamics of bacterial populations. This knowledge should help in the optimization of protocols for hatchery culture of the different species of clams.
3 In larval cultures, *Vibrio* spp. were found associated with larvae rather than in seawater. Members of Splendidus-clade have been the most abundant, regardless of the species of clam. Members of Harveyi-clade, other than *V. alginolyticus*, were

isolated, although scarcely reported until now in hatchery-cultured bivalves. The present results demonstrated the existence of microbiota shared by all the cultures maintained in one installation.

4 The presence of similar bacterial strains in larvae suffering mortalities suggests their involvement in the disease. Further studies, including pathogenicity tests, should be carried-out to elucidate their role in these episodes.

5 The vertical transmission of bacteria from broodstock to their progeny was clearly demonstrated in *Solen marginatus* for a *V. splendidus*-like bacterium. Interestingly, similar strains were isolated from cultures suffering mortalities of the different species of clams included in this study, namely *S. marginatus*, *D. trunculus* and *V. philippinarum*.

6 Difficulties in the assignment of many isolates to any species of genus *Vibrio* indicated the little that is known about the specific microbiota of clams in hatchery, but also the difficulties identifying "significant" putative pathogenic species. Taxonomic and ecological studies are required to establish the role of the bacterial populations on the larval cultures. In particular, the dynamics of infection and the virulence determinants of pathogenic species must be elucidated. Such information would be valuable in designing management protocols for hatcheries.

7 Use of the antibiotic chloramphenicol did not prevent the proliferation of vibrios associated to with larval clams. Furthermore, it did not guarantee the survival of larval cultures. These results, together with the risks of development of antibiotic-resistance and the legal restrictions for their use, make the search for alternatives means of controlling outbreaks of disease in hatcheries urgent. Such means may include the use of probiotics and procedures to establish a microbiota favourable to larval development.

All of those topics should be addressed in future research.

ACKNOWLEDGMENTS

This work was financially supported by projects from the Xunta de Galicia (Acción Especial 2010/CI775, Consellería do Mar), JACUMAR (ALMEJAS 2008-2010, Ministerio de Medio Ambiente y Medio Rural y Marino, Gobierno de España), Plan Nacional de Investigación Científica, Desarrollo e Innovación Tecnológica (AGL2011-29765, Ministerio de Ciencia e Innovación) and in part by the E.U. (REPROSEED, FP7-245119). The authors gratefully acknowledge the cooperation of the CCM-Ribadeo.

REFERENCES

Anguiano-Beltrán, C., Lizárraga-Partida, M. L. and Searcy-Bernal, R. (2004). Effect of *Vibrio alginolyticus* on larval survival of the blue mussel *Mytilus galloprovincialis*. *Dis. Aquat. Org.* 59, 119-123.

Austin, B. and Zhang, X.-H. (2006). *Vibrio harveyi*: a significant pathogen of marine vertebrates and invertebrates. *Lett. Appl. Microbiol.* 43, 119-124.

Beaz-Hidalgo, R., Doce, A., Pascual, J., Toranzo, A. E., and Romalde, J. L. (2009). *Vibrio gallaecicus* sp. nov. isolated from cultured clams in north-western Spain. *Sys. Appl. Microbiol.* 32, 111-117.

Beaz-Hidalgo, R., Diéguez, A. L., Cleenwerck, I., Balboa, S., Doce, A., de Vos, P., and Romalde, J. L. (2010). *Vibrio celticus* sp. nov., a new *Vibrio* species belonging to the Splendidus clade with pathogenic potential for clams. *Sys. Appl. Microbiol.* 33, 311-315.

Brown, C. (1974). A pigment-producing Pseudomonad which discolors culture containers of embryos of a bivalve mollusk. *Chesapeake Sci.* 15, 17-21.

Brown, C. (1981). A study of two shellfish-pathogenic *Vibrio* strains isolated from a Long Island hatchery during a recent outbreak of disease. *J. Shellfish Res.* 1, 83-87.

Brown, C. and Tettelbach, L. P. (1988). Characterization of a nonmotile *Vibrio* sp. pathogenic to larvae of *Mercenaria mercenaria* and *Crassostrea virginica*. *Aquaculture.* 74, 195-204.

Cai, J., Han, Y. and Wang, Z. (2006). Isolation of *Vibrio parahaemolyticus* from abalone (*Haliotis diversicolor supertexta* L.) postlarvae associated with mass mortalities. *Aquaculture.* 257, 161-166.

Cai, J., Li, J., Thompson, K. D., Li, C. and Han, H. (2007). Isolation and characterization of pathogenic *Vibrio parahaemolyticus* from diseased post-larvae of abalone *Haliotis diversicolor supertexta*. *J. Basic Microbiol.* 47, 84-86.

Cano-Gómez, A., Bourne, D. G., Hall, M. R., Owens, L., and Hoj, L. (2009). Molecular identification, typing and tracking of *Vibrio harveyi* in aquaculture systems: Current methods and future prospects. *Aquaculture.* 287, 1-10.

Cano-Gómez, A., Goulden, E. F., Owens, L., and Høj, L. (2010). *Vibrio owensii* sp. nov., isolated from cultured crustaceans in Australia. *FEMS Microbiol. Lett.* 302, 175-181.

D'Agostino, A. (1975). Antibiotics in cultures of invertebrates. In: W. L. Smith and M. H. Chanley (Eds.), *Culture of marine invertebrate animals.* pp. 109-133, Plenum Press, New York.

Da Costa, F. and Martínez-Patiño, D. (2009). Culture potential of the razor clam *Solen marginatus* (Pennánt, 1777). *Aquaculture.* 288, 57-64.

Dang, H., Song, L., Chen, M., and Chang, Y. (2006). Concurrence of *cat* and *tet* genes in multiple antibiotic-resistant bacteria isolated from a sea cucumber and sea urchin mariculture farm in China. *Microb. Ecol.* 52, 634-643.

Diéguez, A. L., Beaz-Hidalgo, R., Cleenwerck, I., Balboa, S., de Vos, P., and Romalde, J. L. (2011). *Vibrio atlanticus* sp. nov. and *Vibrio artabrorum* sp. nov. isolated from clams *Ruditapes philippinarum* and *R. decussatus*. *Int. J. Sys. Evol. Microbiol.* 61, 2406-2411.

Elston, R. (1984). Prevention and management of infectious diseases in intensive mollusc husbandry. *J. World Maricul. Soc.* 15, 284-300.

Elston, R. A., Hasegawa, H., Humphrey, K. L., Polyak, I. K., and Häse, C. C. (2008). Re-emergence of *Vibrio tubiashii* in bivalve shellfish aquaculture: severity, environmental drivers, geographic extent and management. *Dis. Aquat. Org.* 82, 119-134.

Faury, N., Saulnier, D., Thompson, F. L., Gay, M., Swings, J., and Le Roux, F. (2004). *Vibrio crassostreae* sp.nov., isolated from the hemolymph of oysters (*Crassostrea gigas*). *Int. J. Sys. Evol. Microbiol.* 54, 2137-2140.

Gómez-León, J., Villamil, L., Lemos, M. L., Novoa, B., and Figueras, A. (2005). Isolation of *Vibrio alginolyticus* and *Vibrio splendidus* from aquacultured carpet shell clam (*Ruditapes decussatus*) larvae associated with mass mortalities. *Appl. Environ. Microbiol.* 71, 98-104.

Guillard, R. R. L. (1959). Further evidence of the destruction of bivalve larvae by bacteria. *Biol. Bull.* 117, 258-266.

Haldar, S., Chatterjee, S., Sugimoto, N., Das, S., Chowdhury, N., Hinenoya, A., Asakura, M., and Yamasaki, S. (2011). Identification of *Vibrio campbellii* isolated from diseased farm-shrimps from south India and establishment of its pathogenic potential in an Artemia model. *Microbiology.* 157, 179-198.

Hameed, A. S. S., Rahaman, K. H., Alagan, A., and Yoganandhan, K. (2003). Antibiotic resistance in bacteria isolated from hatchery-reared larvae and post-larvae of *Macrobrachium rosenbergii. Aquaculture.* 217, 39-48.

Helm, M. M. and Bourne, N. (2004). Hatchery operation: Broodstock conditioning, spawning and fertilization. In: A. Lovatelli (Ed.), *Hatchery culture of bivalves. A practical manual.* pp. 59-83, FAO Fish. Tech. Paper, No. 471, Rome.

Hernández-Serrano, P. (2005). *Responsible use of antibiotics in aquaculture.* FAO Fish. Tech. Paper. No. 469. Rome.

Ho, S.-P., Hsu, T.-Y., Chen, M.-H., and Wang, W.-S. (2000). Antibacterial effect of chloramphenicol, thiamphenicol and florfenicol against aquatic animal bacteria. *J. Vet. Med. Sci.* 62, 479-485.

Hoffman, M., Monday, S. R., Fischer, M., and Brown, E. W. (2012). Genetic and phylogenetic evidence for misidentification of *Vibrio* species within the Harveyi clade. *Lett. Appl. Microbiol.* 54, 160-165.

Jeffries, V. E. (1982). Three *Vibrio* strains pathogenic to larvae of *Crassostrea gigas* and *Ostrea edulis. Aquaculture.* 29, 201-226.

Kesarcodi-Watson, A., Kaspar, H., Lategan, M. J., and Gibson, L. F. (2009). Challenge of New Zealand Greenshell™ mussel *Perna canaliculus* larvae using two *Vibrio* pathogens: a hatchery study. *Dis. Aquat. Org.* 86, 15-20.

Lambert, C., Nicolas, J. L., Cilia, V., and Corre, S. (1998). *Vibrio pectenicida* sp. nov., a pathogen of scallop (*Pecten maximus*) larvae. *Int. J. Syst. Bacteriol.* 48, 481-487.

Le Pennec, M. and Prieur, D. (1977). Les antibiotiques dans les élevages de larves de bivalves marins. *Aquaculture.* 12, 15-30.

Le Roux, F. Goubet, A., Thompson, F. L., Faury, N., Gay, M., Swings, J., and Saulnier, D. (2005). *Vibrio gigantis* sp. nov., isolated from the haemolymph of cultured oysters (*Crassostrea gigas*). *Int. J. Sys. Evol. Microbiol.* 55, 2251-2255.

Lee, K.-K., Liu, P.-C., Chen, Y.-C., and Huang, C.-Y. (2001). The implication of ambient temperature with the outbreak of vibriosis in cultured small abalone *Haliotis diversicolor supertexta* Lischke. *J. Thermal Biol.* 26, 585-587.

Lin, B., Wang, Z., Malanoski, A. P., O'Grady, E. A., Wimpee, C. F., Vuddhakul, V., Alves, N., jr., Thompson, F. L., Gómez-Gil, B., and Vora, G. J. (2010). Comparative genomic analyses identify the *Vibrio harveyi* genome sequenced strains BAA-1116 and HY01 as *Vibrio campbellii. Environ. Microbiol. Reports.* 2, 81-89.

Liu, P.-C., Chen, Y.-C., Huang, C.-Y., and Lee, K.-K. (2000). Virulence of *Vibrio parahaemolyticus* isolated from cultured small abalone, *Haliotis diversicolor supertexta*, with withering syndrome. *Lett. Appl. Microbiol.* 31, 433-437.

Lodeiros, C., Bolinches, J., Dopazo, C. P., and Toranzo, A. E. (1987). Bacillary necrosis in hatcheries of *Ostrea edulis* in Spain. *Aquaculture*. 65, 15-29.

Luna-González, A., Maeda-Martínez, A. N., Sáinz, J. C., and Ascencio-Valle, F. (2002). Comparative susceptibility of veliger larvae of four bivalve mollusks to a *Vibrio alginolyticus* strains. *Dis. Aquat. Org*. 49, 221-226.

Macián, M. C., Ludwig, W., Aznar, R., Grimont, P. A. D., Schleifer, K. H., Garay, E., and Pujalte, M. J. (2001). *Vibrio lentus* sp. nov., isolated from Mediterranean oysters. *Int. J. Syst. Evol. Microbiol.* 51,1449-1456.

Nicolas, J. L., Ansquer, D. and Cochard, J. C. (1992). Isolation and characterization of a pathogenic bacterium specific to Manila clam *Tapes philippinarum* larvae. *Dis. Aquat. Org*. 14, 153-159.

Nicolas, J. L., Corre, S., Gauthier, G., Robert, R., and Ansquer, D. (1996). Bacterial problems associated with scallop *Pecten maximus* larval culture. *Dis. Aquat. Org*. 27, 67-76.

Nicolas, J. L., Basuyaux, O., Mazurié, J., and Thébault, A. (2002). *Vibrio carchariae*, a pathogen of the abalone *Haliotis tuberculata. Dis. Aquat. Org*. 50, 35-43.

Nishimori, E., Hasegawa, O., Numata, T., and Wakabayashi, H. (1998). *Vibrio carchariae* causes mass mortalities in Japanese abalone, *Sulculus diversicolor supratexta. Fish Pathol.* 33, 495–502.

O'Brien, C. H. and Sizemore, R. K. (1979). Distribution of the luminous bacterium *Beneckea harveyi* in a semitropical estuarine environment. *Appl. Environ. Microbiol.* 38, 928-933.

Ojea, J., Pazos, A. J., Martínez, D., Nóvoa, S., Sánchez, J. L., and Abad, M. (2004). Seasonal variation in weight and biochemical composition of the tissues of *Ruditapes decussatus* in relation to the gametogenic cycle. *Aquaculture*. 238, 451-468.

Pass, D. A., Dybdahl, R. and Mannion, M. M. (1987). Investigations into the causes of mortality of the pearl oyster, *Pinctada maxima* (Jamson), in Western Australia. *Aquaculture*. 65, 149-169.

Phuoc, L. H., Defoirdt, T., Sorgeloos, P., and Bossier, P. (2009). Virulence of luminescent and non-luminescent isogenic vibrios towards gnotobiotic *Artemia franciscana* larvae and specific pathogen-free *Litopenaeus vannamei* shrimp. *J. Appl. Microbiol.* 106, 1388-1396.

Pizzuto, M. and Hirst, R. G. (1995). Classification of isolates of *Vibrio harveyi* virulent to *Penaeus monodon* larvae by protein profile analysis andM13 DNA fingerprinting. *Dis. Aquat. Org*. 21, 61-68.

Prado, S. (2006) Microbiota asociada a criaderos de moluscos: patogénesis y probiosis. *Ph.D Thesis. University of Santiago de Compostela, Spain.* (in Spanish). 310 pp.

Prado, S., Romalde, J. L., Montes, J., and Barja, J. L. (2005). Pathogenic bacteria isolated from disease outbreaks in shellfish hatcheries. First description of *Vibrio neptunius* as an oyster pathogen. *Dis. Aquat. Org*. 67, 209-215.

Prado, S., Romalde, J. L. and Barja, J. L. (2010). Review of probiotics for use in bivalve hatcheries. *Vet. Microbiol.* 145, 187-197.

Riquelme, C. E., Chávez, P., Morales, Y., and Hayashida, G. (1994). Evidence of parental bacterial transfer to larvae in *Argopecten purpuratus. Biol. Res.* 27, 129-134.

Riquelme, C., Hayashida, G., Toranzo, A. E., Vilches, J., and Chávez, P. (1995a). Pathogenicity studies on a *Vibrio anguillarum*-related (VAR) strain causing an epizootic in *Argopecten purpuratus* larvae cultured in Chile. *Dis. Aquat. Org*. 22, 135-141.

Riquelme, C., Hayashida, G., Vergara, N., Vasquez, A., Morales, Y., and Chavez, P. (1995b). Bacteriology of the scallop *Argopecten purpuratus* (Lamarck, 1819) cultured in Chile. *Aquaculture*. 138, 49-60.

Riquelme, C., Toranzo, A. E., Barja, J. L., Vergara, N., and Araya, R. (1996). Association of *Aeromonas hydrophila* and *Vibrio alginolyticus* with larval mortalities of scallop (*Argopecten purpuratus*). *J. Invertebr. Pathol.* 67, 213-218.

Roque, A., Molina-Aja, A., Bolán-Mejía, C., and Gómez-Gil, B. (2001). In vitro susceptibility to 15 antibiotics of vibrios isolated from penaeis shrimps in Northwestern Mexico. *Int. J. Antimicrob. Agents* 17, 383-387.

Ruiz-Azcona, P., Rodríguez-Sierra, R. and Martín, J. B. (1996). Culture of coquina clam, *Donax trunculus*, larvae. *Aquaculture* 139, 151-155.

Sáinz, J. C., Maeda-Martínez, A. N. and Ascencio, F. (1998). Experimental vibriosis induction with *Vibrio alginolyticus* of larvae of the Catarina scallop (*Argopecten ventricosus=circularis*) (Sowerby II, 1842). *Microb. Ecol.* 35, 188-192.

Sáinz-Hernández, J. C. and Maeda-Martínez, A. N. (2005). Sources of *Vibrio* bacteria in mollusc hatcheries and control methods: a case study. *Aquacult. Res.* 36, 1611-1618.

Sapkota, A., Sapkota, A. R., Kucharski, M., Burke, J., McKenzie, S., Walker, P., and Lawrence, R. (2008). Aquaculture practices and potential human health risks: Current knowledge and future priorities. *Environ. Int.* 34, 1215-1226.

Sastry, A. N. (1979). Pelecypoda (Excluding ostreidae). In: A. C. Giese and J. S. Pearse. (Eds.), *Reproduction of Marine Invertebrates. Volume V. Molluscs: Pelecypods and Lesser Classes*, pp. 113-292. Academic Press, New York.

Sawabe, T., Inoue, S., Fukui, Y., Yoshie, K., Nishihara, Y., and Miura, H. (2007). Mass mortality of Japanese abalone *Haliotis discus hannai* caused by *Vibrio harveyi* infection. *Microbes Environ.* 22, 300-308.

Schwarz, S., Kehrenberg, C., Doublet, B., and Cloeckaert, A. (2004). Molecular basis of bacterial resistance to chloramphenicol and florfenicol. *FEMS Microbiol. Rev.* 28, 519-542.

Sugumar, G., Nakai, T., Hirata, Y., Matsubara, D., and Muroga, K. (1998). *Vibrio splendidus* biovar II as the causative agent of bacillary necrosis of Japanese oyster *Crassostrea gigas* larvae. *Dis. Aquat. Org.* 33, 111-118.

Tendencia, E. A. and de la Peña, L. D. (2001). Antibiotic resistance of bacteria from shrimp ponds. *Aquaculture.* 195, 193-204.

Thompson, F. L., Thompson, C. C., Li, Y., Gómez-Gil, B., Vandenberghe, J., Hoste, B., and Swings, J. (2003). *Vibrio kanaloae* sp. nov., *Vibrio pomeroyi* sp. nov. and *Vibrio chagasii* sp. nov., from sea water and marine animals. *Int. J. Sys. Evol. Microbiol.* 53, 753-759.

Thompson, F. L., Iida, T. and Swings, J. (2004). Biodiversity of vibrios. *Microbiol. Mol. Biol. Rev.* 68, 403-431.

Tubiash, H. S. and Otto, S. V. (1986). Bacterial problems in oysters. A review. In: C. P. Vivarès, J.-R. Bonami and E. Jaspers (Eds.). *Pathology in Marine Aquaculture.* European Aquaculture Society, Special Publ. No. 9, pp. 233-242. Bredene.

Tubiash, H. S., Chanley, P. E. and Leifson, E. (1965). Bacillary necrosis, a disease of larval and juvenile bivalve mollusks. I. Etiology and epizootiology. *J. Bacteriol.* 90, 1036-1044.

Uriarte, I., Farías, A. and Castilla, J. C. (2001). Effect of antibiotic treatment during larval development of the Chilean scallop *Argopecten purpuratus*. *Aquacult. Eng*. 25, 139-147.

Utting, S. D. and Millican, P. F. (1997). Techniques for the hatchery conditioning of bivalve broodstocks and the subsequent effect on egg quality and larval viability. *Aquaculture*. 155, 45-54.

Vaseeharan, B., Ramasamy, P., Murugan, T., and Chen, J. C. (2005). In vitro susceptibility of antibiotics against *Vibrio* spp. and *Aeromonas* spp. isolated from *Penaeus monodon* hatcheries and ponds. *Int. J. Antimicrobial Agents*. 26, 285-291.

Widman, J. C., Choromanski, J., Robohm, R. A., Stiles, S., Wikfors, G. H., and Calabrese, A. (2001). Manual for hatchery culture of the bay scallop, *Argopecten irradians irradians*. Connecticut Sea Grant College Program, Milford, CT. 49 pp.

Yue, X., Liu, B., Xiang, J., and Jia, J. (2010). Identification and characterization of the pathogenic effect of a *Vibrio parahaemolyticus*-related bacterium isolated from clam *Meretrix meretrix* with mass mortality. *J. Invertebr. Pathol.* 103, 109-115.

Zhang, X. J., Chen, C. Z., Yan, B. L., Fang, H., Qin, G. M., and Xu, J. (2009). Phenotypic and molecular characterization of pathogenic *Vibrio parahaemolyticus* isolated from *Penaeus vannamei*. *Oceanol. Limnol. Sin*. 40, 654-662.

In: Clam Fisheries and Aquaculture
Editor: Fiz da Costa González

ISBN: 978-1-62257-518-3

Chapter 9

Clam Hatchery and Nursery Culture

F. da Costa,[1,2*] ***J. A. Aranda-Burgos,***[1]
A. Cerviño-Otero,[1] ***A. Fernández-Pardo,***[1] ***A. Louzán,***[1]
S. Nóvoa,[1]***J. Ojea***[1] ***and D. Martínez-Patiño***[1#]
[1]Centro de Cultivos Marinos de Ribadeo-CIMA, Ribadeo (Lugo), Spain
[2]Ifremer, Laboratoire de Physiologie des Invertébrés Marins, Station Expérimentale d'Argenton, Landunvez, France

Abstract

In the present chapter the hatchery and nursery culture of seven species of clams of commercial interest in Spain are reported and discussed, and data are published with reference to other species of clams in the world. These species belong to the families Veneridae (*Venerupis philippinarum*, *Venerupis decussata* and *Venerupis corrugata* (=*V. pullastra*)), Donacidae (*Donax trunculus*), Pharidae (*Ensis magnus* (=*E. arcuatus*) and *Ensis siliqua*) and Solenidae (*Solen marginatus*). The "state of the art" of seed production in land-based facilities is analyzed and the different factors affecting all stages of hatchery and nursery production are discussed (temperature, salinity, stocking density, food quantity and quality, etc.). In those species with short maturity periods it is possible to carry out broodstock conditioning, a procedure by which hatcheries are able to extend their production season by means of the control of factors such as temperature, food supply and photoperiod. An increase in the seawater temperature during broodstock conditioning helps the maturation process in most species, except for *E. magnus*, which is conditioned to ripeness at low temperatures. Ripe adult clams can be successfully induced to spawn using thermal shock in some species, while in *E. magnus* the only effective method is that of changing water levels by simulating tides. Once the gametes are released and collected, they are incubated in larval rearing tanks for 1 or 2 days

* Corresponding author: F. da Costa. Centro de Cultivos Marinos de Ribadeo-CIMA, Muelle de Porcillán, s/n, 27700, Ribadeo (Lugo), Spain. Present address: Ifremer, Laboratoire de Physiologie des Invertébrés Marins, Station Expérimentale d'Argenton, Presqu'île du Vivier, 29840, Landunvez, France. E-mail address: fiz.da.costa.gonzalez@ifremer.fr; mptea@cimacoron.org.

J. A. Aranda-Burgos, A. Cerviño-Otero, A. Fernández-Pardo, A. Louzán, S. Nóvoa, J. Ojea, D. Martínez-Patiño: Centro de Cultivos Marinos de Ribadeo-CIMA, Muelle de Porcillán, s/n, 27700, Ribadeo (Lugo), Spain.

without a supply of feed. The duration of larval culture varies between species, being very short in *S. marginatus* (8 days) and longer in *V. decussata* and *D. trunculus* (30 days). Larval survival also depends on the species reared, being high in *V. corrugata* and *S. marginatus*, and showing poor results in *V. decussata*, *E. magnus* and *D. trunculus*. Larval rearing temperature affects both larval growth and survival, optimal temperature being species-dependent. Nutrition is the dominant factor influencing larval and postlarval bivalve growth and survival. *V. philippinarum* nutrition has been widely studied due to the worldwide importance of this species, while in other species either information is very scarce or no studies have yet been carried out. In this chapter we review the different nutritional studies using both live and inert diets. The latter are of utmost importance in order to reduce production costs and remove the reliance on phytoplankton production in hatcheries and nursery facilities. Postlarval and seed survival varies in the different species studied, being high in *V. corrugata* and *V. philippinarum* and low, with very variable results, in *E. magnus*, *E. siliqua*, *S. marginatus*, *V. decussata* and *D. trunculus*.

9.1. Introduction

Worldwide clam aquaculture production in 2009 accounted for 4,437,786 metric tonnes and 4,335,914,000 US dollars, as appears in FAO statistics from the Fishstat Plus aquaculture production database. This data shows the importance both of captures of aquaculture-produced clams and the economic value of this activity. The data provided in these statistics not only includes clam seed produced in hatcheries but also production of gathered and stocked wild seed, as is the practice most commonly carried out in China, which accounts for the bulk of world clam production (>95% for the Manila clam) (Guo et al., 1999). The use of hatcheries is still limited in China (Zhang and Yan, 2006). Currently, other commercially important species are not yet produced on an industrial scale, so there is a limited availability of seed, production proceeding exclusively from natural recruitment and/or experimental production. Consequently, in this latter case, clam hatchery production must be developed in order to reinforce natural recruitment, restock unproductive areas and develop industrial production of these species in the future.

In this chapter we report on the hatchery and nursery culture of seven species of clams of commercial interest in Spain and we discuss it with regard to data published about other species of clams in the world. These clam species belong to the families Veneridae (*Venerupis philippinarum*, *V. decussata* and *Venerupis corrugata*), Donacidae (*Donax trunculus*), Pharidae (*Ensis magnus* and *E. siliqua*) and Solenidae (*Solen marginatus*). The present state of seed production in land-based facilities is analyzed, and the different factors affecting all stages of hatchery and nursery production are discussed (temperature, salinity, stocking density, food quantity and quality, etc.). We have borne in mind that it is unfeasible to review the whole corpus of scientific literature on clam hatchery and nursery culture currently available in the world. Moreover, despite the great importance of the culture of *Mercenaria mercenaria*, only specific studies on this species are included in this chapter due to the detailed review of *M. mercenaria* culture already carried out by Castagna (2001). In addition, this chapter does not aim to describe the procedures commonly used in clam hatcheries, as there are comprehensive manuals already published, such as the one compiled by Helm et al. (2004). Consequently, the aim of this chapter is to review the most important species within Veneridae and other species that have the word "clam" in their common name in English, and are similar to those species reared in our hatchery facilities.

9.2. Conditioning

Seed production in wild populations relies on that relatively short period of the year during which sexually mature adults are able to release gametes. Thus, the advantages of manipulating gonadal cycle and spawning period in hatchery are that adults can be spawned either earlier or later than occurs naturally (Ojea et al., 2008), maximizing the fecundity of parent animals, while maintaining egg quality and larval viability (Utting and Millican, 1997). This allows for a more efficient use of larval and nursery systems by extending the period when larvae are available. Moreover, obtaining larvae early in the season may allow juveniles with a suitable size to be transferred for on-growing in wild populations, when conditions which maximize both growth and survival are present. Factors that may have a key role in bivalve broodstock conditioning are: temperature, food quality and quantity, and photoperiod (which will be reviewed in the following sections). The maintenance system is also a factor that must be considered, in order to reduce or even wholly avoid stressful conditions that may hinder gonad maturation. Clam maintenance and conditioning is performed in our facilities in rectangular fiber glass tanks in open circuit, with a continuous supply of food (Figure 1A). In some species, such as razor clams, the requirement of a substrate for broodstock burrowing is another factor affecting the success of conditioning (da Costa et al., 2011a) (Figure 1B). Razor clams, which are usually buried in the substrate in nature, feed more efficiently if they are kept in a suitable substrate, and in this way shell gaping is also prevented.

9.2.1. The Effects of Temperature

Temperature is one of the main factors influencing gametogenic cycles in bivalves (Mann, 1979; Delgado and Pérez-Camacho, 2003). We must bear in mind that abrupt changes in temperature may lead to stress and distort the cycles of storage and utilization of energy reserves of the adults; consequently, this could mean gametes of inferior quality being obtained (Ojea et al., 2008).

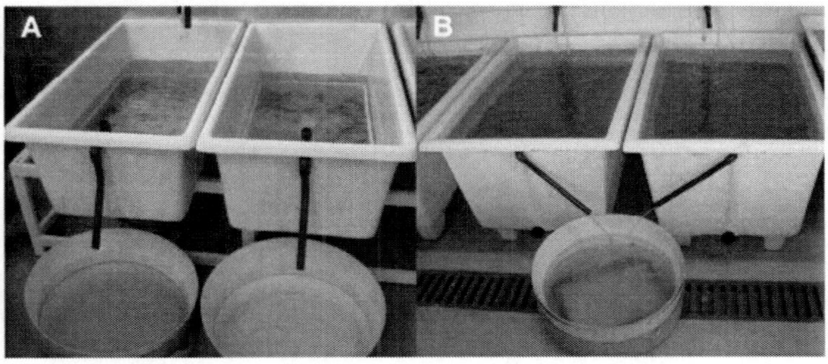

Figure 1. A. Clam maintenance system without substrate in plastic trays. B. Tanks with sand for razor clam burrowing.

Some authors use the unit "degree-days" (D°) in order to make comparisons between species and different experimental conditions in broodstock conditioning. This unit was defined by Mann (1979) and combines temperature and length of the conditioning period in bivalves. Herein, we present data of the investigations carried out by our research group and a review of published studies (Table 1). The effect of temperature on the conditioning of *Venerupis decussata* was investigated by our research group, testing the effect of a gradual increase of temperature from ambient to 20°C (GT) versus constant temperature (20°C, CT) at different periods of the year (autumn, winter and spring) (Ojea et al., 2008). It was observed that, in the three conditioning experiments, fewer degree-days were required to spawn clams successfully in tanks when a gradual increase in temperature was used, just 383 D° and 337.5 D° being necessary under this temperature regime in winter and spring conditioning, respectively. One aspect that is usually overlooked when data from conditioning trials is presented is that of the gametes released and the larval development (growth and survival) produced in the conditioned clams. In this study, a higher percentage of survival is reported in larvae from GT tanks (40, 27 and 38% for autumn, winter and spring conditionings, respectively) compared with CT tanks (32, 9 and 25% respectively, for the same conditioning experiments). These results support the assertion that the GT dietary regimen produces better-quality larvae than its CT counterpart. The effect of temperature and timing of broodstock collection on conditioning has also been tested in *V. decussata*, and investigated by Matias et al. (2009). In this latter study, these authors also investigated two different geographic origins in the Iberian Peninsula, in the North and the South. They reported different conditioning temperatures for the two different origins of the broodstock, that is, 20° C and 22°C for northern and southern populations, respectively. They provide two possible explanations for these findings, firstly that the two localities of origin have a different seasonal temperature regime and, secondly, that broodstock could have a different genetic adaptation in the populations. Moreover, winter conditioning performed better than autumn conditioning, Matias et al.'s results being congruent with those of Ojea et al. (2008). Delgado and Pérez-Camacho (2007a) investigated the effect of temperature (14 and 18°C) on gonadal development of *V. decussata* and *V. philippinarum*, finding that at 14°C both species needed over two months to mature, thus showing similar reproductive behavior. At 18°C, *V. philippinarum* showed a greater rate of gonadal development than *V. decussata*. Similarly, another study investigating the effect of temperature on *V. philippinarum* conditioning concluded that the temperature which yielded best results in promoting gametogenic development and survival of the broodstock was 18°C (Mann, 1979). At 12°C gametogenic development was low, spawning being only evident at 15, 18 and 21°C. Recently, the thermal threshold for gonadal maturation in *V. decussata* was established at 10°C, the temperature at which germinal cells proliferated but ripeness was not achieved (Blanco, 2010). This last author reported that the fastest rate of gonad development was at 22°C, although the values of the gonad occupation index were lower than at 18°C. In our laboratory, we investigated the effect of timing in broodstock collection (i.e. the effect of gonad developmental stage at the beginning of conditioning) on the razor clam *Ensis magnus*, *Ensis siliqua* and *Solen marginatus* conditioning, used as preliminary studies for investigating the effect of temperature (da Costa, 2009; da Costa et al., 2011a). We demonstrated, for *E. siliqua* and *S. marginatus*, that when conditioning was started at rest stage or early gametogenesis stage, the rate of gonadal maturation was not speeded up compared with the gametogenic cycle in wild populations, and thus ripeness was reached quite close to maturity in wild stocks.

Table 1. Conditioning experiments in clam species

Species	Season	Temperature	Feed type	Daily ration per animal	Best conditions and time necessary to obtain ripe broodstock (days, weeks or D°)	Reference
VENERIDAE						
Venerupis philippinarum	Spring	20°C	1. Dried *Tetraselmis suecica*; 2. Dried *T. suecica* (70%) + dried *Cyclotella cryptica* (30%); 3. Live *T. suecica*; 4. Live *Skeletonema costatum* and 5. Live *Dunaliella tertiolecta*	6% of meat dry wt. in dry wt. of algae	Live diets (3, 4 and 5) (6 weeks, 480 D°)	Laing and López-Alvarado (1994)
V. philippinarum	Autumn	1. 12°C; 2. 15°C; 3. 18°C and 4. 21°C	*S. costatum*	13.8% of meat dry wt. in dry wt. of algae	18°C (13 weeks)	Mann (1979)
V. philippinarum	nd	1. 14°C L; 2. 18°C L; 3. 18°C H and 4. 22°C H	*Isochrysis galbana* clone T-iso	L: low (750 μg) and H: high (1500 μg organic wt. of algae per g clam live wt.)	18°C H and 22°C H (70 days)	Delgado and Pérez Camacho (2007b); Fernández-Reiriz et al. (2007)
V. philippinarum and *V. decussata*	Winter	1. 14°C and 2. 18°C	*I. galbana* clone T-iso	0.5% dry wt. of the algae with regard to clam live wt.	18°C in both species (57 days in both species)	Delgado and Pérez-Camacho (2007a)
V. decussata	Autumn, winter and spring	1. 20°C constant (CT) and 2. gradual increase from 14-15°C up to 20°C (GT)	*Isochrysis* sp. clone T-iso + *Pavlova lutheri* + *S. costatum* + *Chaetoceros calcitrans* + *T. suecica*	4% of meat dry wt. in dry wt. of algae	Spring conditioning, GT (337.5 D°)	Ojea et al. (2008)
V. decussata	Autumn	1. 20°C (CT) L; 2. 20°C (CT) H; 3. 20°C (GT) L and 4. 20°C (GT) H. Starting temperature: 14-15°C	*Isochrysis* sp. clone T-iso + *P. lutheri* + *S. costatum* + *C. calcitrans* + *T. suecica*	L: 3% and H: 6% of meat dry wt. in dry wt. of algae	20°C (CT) and H ration (8 weeks)	Ojea (unpublished results)
V. decussata	Autumn and winter	1. 18°C; 2. 20°C and 3. 22°C	*Isochrysis* sp. clone T-iso + *C. calcitrans*	4% of meat dry wt. in dry wt. of algae	Winter conditioning, 20°C for northern and 22°C for southern population (11 weeks in both groups)	Matias et al. (2009)

Table 1. (Continued)

Species	Season	Temperature	Feed type	Daily ration per animal	Best conditions and time necessary to obtain ripe broodstock (days, weeks or D°)	Reference
V. decussata	nd	18ºC	*I. galbana* clone T-iso	1. 2.5%; 2. 5.0% and 3. 10.0%. Percentages correspond to the organic wt. of food supplied as a proportion of the live wt. of the clam	Rations 2 (5%) and 3 (10%) (35 days both rations)	Delgado et al. (2004)
V. decussata	nd	18ºC	*I. galbana* clone T-iso	1. 0.96%; 2. 0.48%; 3. 0.24%; 4. 0.05%; 5. 0.10% and 6. 0.025%. Percentages correspond to the organic wt. of food supplied as a proportion of the live wt. of the clam	Rations 1 (0.96%) and 2 (0.48%) (46 days both rations)	Delgado and Pérez-Camacho (2003); Pérez-Camacho et al. (2003)
V. decussata	Autumn and spring. Photoperiod 1. 8 h light: 16 h darkness) and 2. 16 h light: 8 h darkness	19ºC	*Isochrysis* sp. clone T-iso + *P. lutheri* + *S. costatum* + *C. calcitrans* + *Phaeodactylum tricornutum* + *T. suecica*	5% of meat dry wt. in dry wt. of algae	Spring, photoperiod 16 h light: 8 h darkness (less than 30 days)	Pazos et al. (2003)
V. decussata	Winter	19ºC	*Isochrysis* sp. clone T-iso + *P. lutheri* + *S. costatum* + *C. calcitrans* + *T. suecica*	1. 6% (8 h light: 16 h darkness); 2. 6% (16 h light: 8 h darkness) and 3. 9% of meat dry wt. in dry wt. of algae (16 h light: 8 h darkness)	9% ration and summer photoperiod (16 h light: 8 h darkness) (less than 45 days)	Martínez et al. (2005)
V. decussata	nd	1. 14ºC; 2. 18ºC and 3. 22ºC	*Isochrysis* sp. clone T-iso + + *T. chuii*	0.05 mg organic wt. of food supplied as a proportion of the live wt. of the clam per hour	22ºC (41 days)	Blanco (2010)
PHARIDAE						
Ensis magnus	Summer and autumn	18ºC	*I. galbana* + *P. lutheri* + *S. costatum* + *C. calcitrans* + *P. tricornutum* + *T. suecica*	4% of meat dry wt. in dry wt. of algae	Autumn, advanced gametogenesis stage (54 days)	da Costa (2009)

Species	Season	Temperature	Feed type	Daily ration per animal	Best conditions and time necessary to obtain ripe broodstock (days, weeks or D°)	Reference
E. magnus	Autumn	1. Environmental seawater temperature (12-15°C) and 2. gradient temperature of 20°C	*I. galbana + P. lutheri + S. costatum + C. calcitrans + P. tricornutum + T. suecica*	4% of meat dry wt. in dry wt. of algae	Environmental seawater temperature (12-15°C) (60 days)	da Costa et al. (2005)
Ensis siliqua	Autumn and winter	18°C	*I. galbana + P. lutheri + S. costatum + C. calcitrans + P. tricornutum + T. suecica*	4% of meat dry wt. in dry wt. of algae	Winter, advanced gametogenesis stage (60 days)	da Costa (2009)
E. siliqua	Winter	20°C	*I. galbana + P. lutheri + S. costatum + C. calcitrans + P. tricornutum + T. suecica*	1. 3%; 2. 6% and 3. 9% of meat dry wt. in dry wt. of algae	6% of meat dry wt. in dry wt. of algae (60 days)	da Costa et al. (2009)
E. siliqua	Autumn, photoperiod 1. 8 h light: 16 h darkness) and 2. 16 h light: 8 h darkness	nd	Unfiltered seawater	nd	16 h light: 8 h darkness (65 days)	Cromie et al. (2008)
SOLENIDAE						
Solen marginatus	Autumn and winter	18°C	*I. galbana + P. lutheri + S. costatum + C. calcitrans + P. tricornutum + T. suecica*	4% of meat dry wt. in dry wt. of algae	Winter, advanced gametogenesis stage (38 days)	da Costa (2009)
S. marginatus	Winter	1. 18°C constant (CT) and 2. gradual increase of 1°C/week from 12°C up to 18°C (GT)	*I. galbana + P. lutheri + S. costatum + C. calcitrans + P. tricornutum + T. suecica*	6% of meat dry wt. in dry wt. of algae	Gradual increase of temperature (50 days)	da Costa et al. (2011a)
S. marginatus	Spring	1. 17±1°C; 2. 20±1°C; 3. 23±1°C and 4. environmental temperature	*S. costatum + I. galbana + T. suecica + C. gracilis*	Continuous food supply of 340,000 cells mL^{-1} per tank	23±1°C (17 days)	Moreno et al. (2007)

D°: degree-days; nd: no data reported; Wt: weight.

Nevertheless, if broodstock conditioning in *E. siliqua* and *S. marginatus* was initiated with adults at advanced gametogenesis stage, maturity was promoted under experimental conditions, ripe individuals being found two months prior to maturity under natural conditions. These findings highlighted the importance of the timing of collection, in agreement with the results for *V. decussata* submitted by Ojea et al. (2008) and Matias et al. (2009). However, in *E. magnus*, the increase of temperature does not imply greater gonad maturation compared with natural populations. This can be related to the behavior of adults in the wild, as gametogenesis begins when seawater temperature decreases to a certain level in the environment (Darriba et al., 2004).

Consequently, we decided to study the effect of temperature during *E. magnus* artificial maturation in the hatchery by means of comparing the environmental temperature (12-15ºC) found in the wild beds where this species inhabits as against heated seawater from this temperature up to 20ºC, at a rate of 1ºC per week. We found a higher gonad condition index (CI=gonad fresh weight/valve dry weight) in those individuals subjected to environmental temperature; however, no statistically significant differences were observed. The percentage of individuals in ripe and spawning stage was also higher in the environmental temperature. Therefore, maturation can occur at higher temperatures than the environmental temperature, in spite of the fact that the gonad maturation rate is accelerated at lower temperatures. Moreover, this suggests a greater importance of nutrition in *E. magnus* conditioning. The usefulness of these results is that the cost of heating seawater in hatchery facilities during winter can be reduced. The opposite pattern was observed in our investigations with *S. marginatus* conditioning. We found greater gonad development being reached in the group subjected to a gradual increase in temperature up to 18ºC at the end of the conditioning trial (da Costa et al., 2011a). Moreno et al. (2007) studied the effect of temperature in gonadal development of *S. marginatus*, and observed that temperature influences gonadal maturation. They found a long delay in the emission of gametes and the reorganization of the gonad at 17ºC, whereas a reduction in spawning period and acceleration of emission of gametes was observed at 23ºC.

9.2.2. The Supply of Feed

Bivalve maturation in the laboratory and in wild populations are greatly influenced by food availability and quality, which subsequently influences the rate of gonad development and the quality of the gametes produced (Utting and Millican, 1997, 1998). In our hatchery, we investigated the effect of different dietary regimes (3, 6 and 9% rations) on *E. siliqua* conditioning, using a diet consisting of a mixture of *Tetraselmis suecica*, *Isochrysis galbana*, *Pavlova lutheri*, *Chaetoceros calcitrans*, *Phaeodactylum tricornutum* and *Skeletonema costatum* in equal proportions. We observed that the best ration at the end of the trial promoting gonad condition index (GCI) increase was 6% of the mean dry meat weight of the adults per day. Both the 3% and 9% diets reached a similar GCI at the end of the experiment. This shows that a 6% ration produces more gametes than a 9% ration.

A higher percentage of individuals at ripe and spawning stage were found in the 6% ration. Utting and Millican (1997) pointed out that a suitable ration for bivalve broodstocks is 6% of the dry meat weight in the dry weight of algae per day for most species reared at 20ºC, whilst for species reared at lower temperatures 3% may be sufficient.

The effect of food availability on conditioning is usually investigated from the perspective of the amount of food supplied. However, Delgado and Pérez-Camacho (2003) and Pérez-Camacho et al. (2003) have tackled this issue focusing on the energy balance of the conditioned *V. decussata* clam. These authors have tested several food rations on the basis of three energy balance states (positive, negative and zero).

Strikingly, they found that gonad development occurred under all energy balance states. Nevertheless, the rate of gonad maturation is reduced as the ration decreases, and consequently energy balance passes from a positive to a negative state. The mass-production of microalgae has been pinpointed by several authors as the main bottle-neck for hatchery and nursery culture of bivalves (Coutteau and Sorgeloos, 1992).

These authors estimated that microalgal production accounted for more than 30% of production costs in hatchery facilities.Therefore, many authors have investigated the use of algal substitutes as an alternative to live foods in broodstock, larvae and spat (Coutteau et al., 1994c; Knauer and Southgate, 1997; Caers et al., 1998, 1999a,b; Nóvoa et al., 2002; Badillo-Salas et al., 2009). Laing and López-Alvarado (1994), testing dried and live microalgae as food for *V. philippinarum* conditioning, found that despite dried algal diets being effective for conditioning and spawning Manila clam, conditioned clam fecundity and the proportion of parental lipids in the released eggs with dried algae were lower than for broodstock fed on live microalgae.

9.2.3. Photoperiod

Light is considered to play an important role in the regulation of gametogenesis in bivalves and is believed to be of particular significance in scallops, due to the presence of a cornea, two retinas and a crystalline lens within their eye structure (Devauchelle and Mingant, 1991). The literature dealing with the photoperiod effect on gonad maturation in bivalves is almost entirely restricted to scallops (Devauchelle and Mingant, 1991; Saout et al., 1999; Louro et al., 2006; Martínez et al., 2006; Desanctis et al., 2007; Mallet and Carver, 2009) and oysters (Chávez-Villalba et al., 2002; Fabioux et al., 2005). Cromie et al. (2008) reported on the effect of the photoperiod in *E. siliqua* conditioning. They conditioned broodstock under winter (8 h light: 16 h darkness) and summer (16 h light: 8 h darkness) photoperiod regimes. *E. siliqua* broodstock could be conditioned to start spawning in early January, a few months earlier than was observed in wild populations. Under the summer regime (16 h light: 8 h darkness) 78.6% of individuals reached spawning stage after 65 days, compared with just 33.3% in those individuals kept under the winter regime (8 h light: 16 h darkness).

Similarly, Pazos et al. (2003), investigating the effect of the photoperiod in *V. decussata* in autumn and spring, found promising results with the summer photoperiod (16 h light: 8 h darkness). In autumn conditioning they found males fully mature and females with well developed oocytes after 75 days under a summer photoperiod, whilst under a winter photoperiod (8 h light: 16 h darkness) broodstock were at the onset of gametogenesis.

In the spring experiment, these authors reported that gonad maturation was inhibited in individuals under the winter photoperiod and that those under summer photoperiod conditions spawned spontaneously after a month. Moreover, the combined effects of photoperiod and food rations were investigated in *V. decussata* (Martínez et al., 2005).

In this study, it was observed that the higher feeding ration (9%) and summer photoperiod (16 h light: 8 h darkness) conditions promoted partial spawning before day 45 of the experiment. Meanwhile, reabsorption of gametes was observed in those subjected to a 6% feeding ration and in the winter photoperiod.

9.3. Spawning Induction

Spawning induction allows production in hatchery facilities to be planned and also enables controlled fertilization to be carried out. However, the lack of efficient artificial stimuli to induce spawning still hinders the controlled development of several bivalve species (Mouëza et al., 1999). When a spawning method is chosen we must bear in mind several factors: effectiveness, ease, price, whether or not it causes mortalities in broodstock after spawning induction, possible effects on the viability of the gametes and subsequently during embryonic and larval development, etc. Thermal shock successfully induced spawning in several clam species (Table 2), it being a cheap, easy and widely-used method to induce spawning. In our laboratory, thermal shock has yielded satisfactory results in *Venerupis philippinarum*, *V. corrugata*, *Donax trunculus*, *Ensis siliqua*, *Solen marginatus* and *V. decussata* (Figure 2A). Thermal stimulation was also combined with the addition of gametes (preferably sperm) and microalgae, thus improving spawning yields. Nevertheless, it was not an efficient method for most species, even within the same family. Consequently, other methods have been developed to obtain gametes in clams, a review of which can be found in Table 2. An alternative method was developed in our laboratory for *E. magnus*, due to the failure of thermal shock as an effective stimulus for this species. This method consisted of changing the water level by simulating tides, together with brief dry periods (da Costa et al., 2008) (Figure 2B). Another method is that reported by Breese and Robinson (1981), consisting of immersing the razor clam *Siliqua patula* in seawater with a high concentration of the algae *Pseudoisochrysis paradoxa*. This method yielded better larval transformation than stripping gonads in *S. patula* (Breese and Robinson, 1981). Stripping gonads is a more invasive method and can cause mortality in the broodstock. Moreover, low fertilization success is generally observed, because during stripping there is no way to separate the immature oocytes from the mature eggs.

In addition to this, the technique of stripping gonads cannot be successfullly used with some clam species (eg. *V. philippinarum* and *V. decussata*) because in gonads the eggs are blocked at a first meiotic division prophase; the gametogenesis process ends only when oocytes pass through the oviduct (Utting and Spencer, 1991; Hamida, 1994). Other methods involve chemicals, such as serotonin injection in the gonad or in the anterior adductor muscle. This method was successfully tested in different species of clams (see Table 2), obtaining a response from the ripe individuals within a few minutes (Gibbons and Castagna, 1984; Pérez-Camacho and Román, 1987). The stimulus is very intense and frequently provokes the emission of immature oocytes and fragments of the gonad (Pérez-Camacho and Román, 1987).

9.4. Larval Rearing

In the larval rearing section data is presented in two main groups. Firstly, we present data describing standard larval development of the different species from studies that have not dealt with factors affecting larval culture in depth. Secondly, we will report on studies that investigate the effect of different factors (temperature, food quality and quantity, salinity, etc.) on larval production.

Figure 2. A. *V. decussata* male spawning. B. Release of eggs from the razor clam *E. magnus*.

9.4.1. Standard Larval Development

The general method used in our hatchery for clam larval rearing starts when the embryos are transferred to 500-L larval culture tanks with aerated water at 19±1°C and 32-35‰ salinity for incubation (Figure 3). After 1 or 2 days of incubation, depending on the species, newly developed D-shaped larvae are collected through nylon screens and then counted for transformation estimation. The water is changed every 2 days, using 1-μm sand-filtered, ultraviolet sterilized seawater. Larval density in each container at the beginning of the experiment is 5-10 individuals mL^{-1}. D-stage veliger larvae are fed daily with *T. suecica*, *I. galbana*, *P. lutheri* and *C. calcitrans* in equal proportions at 40 cells μL^{-1} as the initial ration. Larval culture ends when over 50% of the population are newly-settled postlarvae and they are transferred to containers for settlement. Currently, in our hatchery facilities we are rearing *Venerupis philippinarum*, *V. decussata*, *V. corrugata*, *Polititapes virgineus* (=*V. rhomboides*), *Donax trunculus*, *Ensis magnus*, *E. siliqua* and *Solen marginatus*. Data about their standard larval culture is provided in Table 3. Among these, the largest oocytes are found in *S. marginatus* (Figure 4B), due to the fact that its oocytes are surrounded by a chorionic envelope that stores large amounts of energy reserves, thus allowing a short larval development (8 days). The other species have oocytes ranging from 65 to 85 μm in diameter, with similar appearance across species. Larval morphology in these species is quite similar under the light microscope, as can be seen in Figure 4. Razor clams are the species showing the shortest larval development times, with 8, 14 and 20 days in *S. marginatus*, *E. siliqua* and *E. magnus*, respectively. Moreover, the clam *V. corrugata* also settles at day 20, whilst *V. philippinarum*, *V. decussata*, *P. virgineus* and *D. trunculus* need 30 days to undergo metamorphosis. Survival varies noticeably between these species, being high in *V. corrugata* and *S. marginatus* and low in *D. trunculus*. In our facilities we have found very variable larval performances in two of the clam species, *V. philippinarum* and *V. decussata*.

9.4.2. Larval Nutrition

Nutrition can be the dominant factor influencing larval bivalve growth and survival, being extensively reviewed in Marshall et al. (2010). The criteria for selecting a suitable algal diet for bivalve larvae must be based on form, mobility, size, ease of culture, absence of toxicity and the ability of the larvae to trap, ingest, digest and assimilate the algae.

Table 2. Spawning induction methods

Spawning method	Family	Species	Details of the method	Reference
Thermal shock	VENERIDAE	*Venerupis philippinarum*	Temperature rising from 13 to 23ºC	Martínez-Patiño (unpublished results)
		Venerupis decussata	Temperature rising from 13 to 23ºC	Martínez-Patiño (unpublished results)
		Venerupis corrugata	Temperature rising from 12 to 20ºC	Cerviño-Otero (2011)
	PHARIDAE	*Ensis siliqua*	Temperatures up to 25-27ºC for 1 h, decreasing to 10-12ºC for 30 min. 2 or 3 cycles.	da Costa et al. (2010c)
		Ensis directus	Temperature rising from 13 to 25ºC	Loosanoff and Davies (1963)
		Ensis macha	Temperatures up to 17-18ºC , decreasing to 11ºC, with 2 or 3 cycles for 30 min each	Lépez et al. (2011)
	SOLENIDAE	*Solen marginatus*	Temperatures up to 25-27ºC for 1 h, decreasing to 10-12ºC for 30 min. 2 or 3 cycles.	da Costa and Martínez-Patiño (2009)
Exposure to air	DONACIDAE	*Donax trunculus*	Period of dryness and temperature at 16ºC	Louzán (2008)
Injection of serotonin	MACTRIDAE	*Spisula solidissima*	Injection in the anterior adductor muscle of 0.4 mL of 2mM serotonin solution	Gibbons and Castagna (1984)
	VENERIDAE	*Mercenaria mercenaria*	Injection in the anterior adductor muscle of 0.4 mL of 2mM serotonin solution	Gibbons and Castagna (1984)
		V. philippinarum	Injection in the gonad of 0.2 to 0.4 mL of serotonin with concentrations of 0.08 and 0.8 mg cm^{-3}	Pérez-Camacho and Román (1987)
		V. decussata	Injection in the gonad of 0.2 to 0.4 mL of serotonin with concentrations of 0.08 and 0.8 mg cm^{-3}	Pérez-Camacho and Román (1987)
Stripping gametes	PHARIDAE	*Ensis directus*		Loosanoff and Davies (1963)
		Siliqua patula		Breese and Robinson (1981)
High concentration of microalgae	PHARIDAE	*Siliqua patula*	2-2.5 million cells *Pseudoisochrysis paradoxa* mL^{-1}	Breese and Robinson (1981)
Changing water levels	PHARIDAE	*Ensis magnus*	Changing water level by simulating tides, with brief dry periods	da Costa et al. (2008)

Figure 3. Larval rearing tanks.

Moreover, food value is determined principally by biochemical composition (lipids, carbohydrates and proteins). Lipids, mainly neutral lipids in the form of triacylglycerol (TAG), have been identified as a major source of energy for bivalve larvae, especially in periods of low food availability or starvation (Gallager et al., 1986). Essential fatty acids (EFAs), particularly omega-3 fatty acids, eicosapentaenoic acid (20:5n-3, EPA) and docosahexaenoic acid (22:6n-3, DHA), are important for growth and development (Langdon and Waldock, 1981) because they are major membrane components (Hendriks et al., 2003) and possible modulators of membrane function (Palacios et al., 2005).

Table 3. Embryonic and larval development

Species	Egg size (µm)	Larval transformation (%)	Settlement (day)	Larval survival	Temperature (ºC)	Reference
VENERIDAE						
Venerupis philippinarum	65-70	-	30	5-50%	18-22ºC	Martínez et al. (2008)
V. philippinarum	71.9	71-74%	13-14	9-23%	17-19ºC	Zhang and Yan (2006)
V. decussata	62-68	-	30	9-40%	20ºC	Ojea et al. (2008)
V. largillierti	63.8	-	16-19	<3%	20ºC	Kent et al. (1999)
V. corrugata	65-75	84%	22-25	24.0-87.0%	20ºC	Cerviño-Otero (2011)
Polititapes virgineus	70	10-35%	30-35	40%*2	19ºC	Cerviño-Otero et al. (2011)
Marcia opima	47.8	-	11	13.9-56.2%	28ºC	Muthiah et al. (2002)
Callista chione	-	-	25	31.9%*1	19ºC	Delgado et al. (2008)
Phapia undulata	40	-	13	2-3%	-	del Norte-Campos et al. (2010)
Cyclina sinensis	92.4	84-88%	12	57%	26ºC	Liu et al. (2002)
Katelysia rhytiphora	66.3	-	21	42%*2	20ºC	Nell et al. (1994)
DONACIDAE						
Donax trunculus	75-80	-	30	5-10%	21ºC	Louzán (2008); Martínez et al. (2008)

Table 3. (Continued)

Species	Egg size (μm)	Larval transformation (%)	Settlement (day)	Larval survival	Temperature (ºC)	Reference
PHARIDAE						
Ensis magnus	70-90	25-80%	20	4.8-24.8%	19ºC	da Costa (2009); da Costa et al. (2011b)
E. siliqua	75-90	70-80%	14	20.0-52.7%	19ºC	da Costa (2009); da Costa et al. (2010c)
E. directus	64-73	-	10	-	24ºC	Loosanoff and Davies (1963)
E. macha	-	-	17-20	25-50%[*2]	17ºC	Lépez et al. (2005, 2011)
Siliqua patula	-	60-38%	20-25	-	16.5ºC	Breese and Robinson (1981)
SOLENIDAE						
Solen marginatus	143-155	70-80%	8	28.0-81.0%	19ºC	da Costa (2009); da Costa and Martínez-Patiño (2009)

- Not reported. [*1]survival from D-shaped veliger to settlement. %[*2]survival from D-shaped veliger to pre-metamorphic larvae.

DHA is involved in maintaining a suitable membrane structure, while EPA fulfils a role both as an energy source and as a precursor of eicosanoids.

In addition, the omega-6 fatty acids docosapentaenoic acid (22:5n-6, DPA) and arachidonic acid (20:4n-6, AA) have been identified as affecting growth and survival during larval and postlarval stages (Pernet et al., 2005; Milke et al., 2006).

DPA is mainly accumulated in membranes, thus suggesting a specific role for this fatty acid in cell membranes (Delaunay et al., 1993). The importance of AA in invertebrate species is due to its role in eicosanoid production and stress response (Howard and Stanley, 1999). Moreover, AA is a major precursor of prostaglandins, which influence reproduction in bivalves (Osada et al., 1989). The ability to synthesise EFAs in bivalves is very limited, and is inadequate to meet their nutritional requirements; therefore, they must be supplied exogenously (Laing et al., 1990). The sterol composition of food has been linked to dietary success in supporting larval bivalve growth (Soudant et al., 1998). Sterols are ubiquitous lipid components of all eukaryotic cells, their main role being as structural components of cellular membranes. The ability to synthesize *de novo* or bioconvert sterols is generally low or absent; therefore, dietary sources are essential for bivalve growth and survival. Consequently, key molecules must be supplied by the diet to guarantee larval survival and growth. On the one hand, the fatty acid and biochemical composition of the diet could be varied using different microalgal species (live diets) in different proportions. On the other hand, inert diets (preserved phytoplankton and artificial feeds) have been tested both on their own and as supplements for bivalve feeding, and will be reviewed in this section.

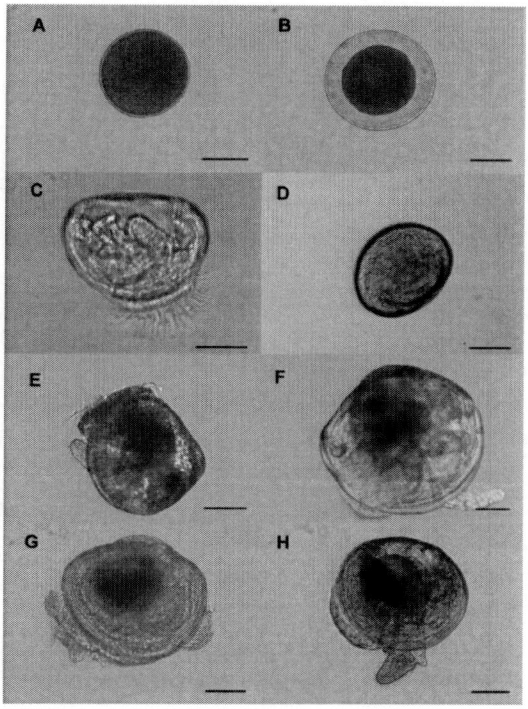

Figure 4. A. *E. magnus* egg. B. *S. marginatus* egg. C. Three-day-old larvae of *E. siliqua*. D. Umbonate larvae of *D. trunculus*. E. Pediveliger larvae of *V. decussata*. F. *V. corrugata* pediveliger. G. *E. magnus* pediveliger. H. Pediveliger of *S. marginatus*. Scale bar: 50 μm.

The nutritional value of four live diets for *E. magnus* larvae was evaluated at our facilities in a trial lasting 20 days (Table 4). Of these, the diet consisting of a mixture of *T. suecica* + *I. galbana* + *P. lutheri* + *C. calcitrans* was the one that resulted in the highest larval survival rate (25%), when compared with the other three diets, which showed a survival rate ranging between 5 and 10% (da Costa et al., 2011a). It may be that this diet displays a more balanced profile of essential fatty acids than the others. Helm and Laing (1987), observed that differences between treatments as to the resulting development may not be solely due to the presence or absence of essential components in the diet, but to the balance between them. This same diet, which is the standard one for all clam species in our hatchery, has been compared with the diet of reference in bivalve culture (*I. galbana*) in the larval development of *S. marginatus*, the multispecific diet giving rise to an improved larval performance. The fatty acid analysis of the diets and larvae suggested that AA deficiencies in *I. galbana* diet might be related to the lower growth and survival shown in larvae fed on this diet (da Costa, unpublished results), while the low values of EPA found in *I. galbana* may also explain these results. Conversely, Matias et al. (2011) investigated the effect of two mono-specific diets on the growth and survival of *V. decussata*, demonstrating that *Isochrysis aff. galbana* (T-iso) is a more suitable diet for this clam species, due to the higher energy content registered in larvae fed on T-iso immediately before metamorphosis, compared with larvae fed on *Chaetoceros calcitrans*. By contrast, the results of our examination of the effect of pluri-specific diets on *V. decussata* suggest that the presence of *Chaetoceros mulleri* in the diet increases the survival rate (Aranda et al., 2011). Gross biochemical analysis of the diets and larvae does not by itself explain the results observed, though it must be pointed out that fatty acid analysis could furnish support for these findings in the future. Similarly, Ruiz-Azcona et al. (1996) reported that the presence of the diatom *Chaetoceros gracilis* in the diet enhanced *D. trunculus* larval growth. These authors also found that a diet of T-iso on its own proved to be of poor food value for *D. trunculus* larvae in comparison with bispecific diets, as has also been reported for *V. philippinarum* (Helm and Laing, 1987). The n-6 fatty acids and DHA seem to be important in the diet of the larvae of *V. corrugata*, since it has been observed that the low content of n-6 fatty acids and the absence of DHA in *Tetraselmis suecica* might contribute to the massive mortality observed in larvae fed on this diet (Fernández-Reiriz et al., 2011). The vast majority of studies investigating the effect of inert diets on larval feeding have been performed on oysters (Chu et al., 1982, 1987; Robert et al., 2001; Brown and Robert, 2002; Ponis et al., 2003a), whilst research into clams is less commonly met with. A thoroughgoing review of the use of inert diets in clam species is to be found in the Nursery Culture section in this chapter, since the bulk of scientific literature focuses on this culture phase. Early work by Hidu and Ukeles (1962), investigating the use of dried unicellular algae for *M. mercenaria* larvae, demonstrated that clam larvae can be successfully reared to metamorphosis feeding on the dried algae *Dunaliella euchlora*, *Scenedesmus obliquus* and *I. galbana.* Spray-dried *Nannochlorosis* sp. and *T. suecica* fed to *V. philippinarum* larvae supported growth equal to, or greater than, their live algal counterparts, but less than the mixed diet made up of T-iso and *C. calcitrans*, used as a control diet (Laing et al., 1990). The importance of lipids to bivalve larval nutrition has led to the development of artificial diets able to provide specific lipid supplements during larval culture (Coutteau et al., 1994c). These authors demonstrated, with regard to *M. mercenaria* larvae, that supplying 50% (of algal dry weight) DHA-rich lipid supplement meant better growth and survival were found, compared with T-iso (control diet). Another way to deliver specific fatty acids to bivalve larvae is the use of olive oil gelatin-acacia microcapsules (GAMs) to supplement deuterium-labeled arachidonic acid (*AA) for *V. corrugata* larvae (Novoa et al., 2002). These microcapsules were ingested, digested and assimilated by the larvae, incorporating AA to neutral and polar lipids of larval tissues.

9.4.3. The Effects of Temperature

An increase in culture temperature shortens the larval period of many bivalves (e.g. *M. mercenaria*, in Loosanoff and Davis, 1963). Nonetheless, every species has an optimal temperature of larval culture, at which growth and survival are maximized. We studied the effects of temperature on *E. magnus* larval development, finding that as temperature increased growth was enhanced and the larval period reduced. Surprisingly, larval survival was higher at the lower temperature tested (14ºC), when compared with 18ºC and 24ºC (da Costa et al., 2011a). Larvae reared at 24ºC died after 19 days of culture, before settlement could take place. One explanation for the results observed could be that *E. magnus* adults spawn in wild populations during winter and early spring, when sea water temperatures are low (Darriba et al., 2004), and thus, larvae of this species may be well adapted to low temperatures. In contrast to these findings, higher temperatures maximizing larval growth without affecting survival were reported in other clam species of temperate climates, such as 26ºC in the case of *V. corrugata* (Pérez-Camacho et al., 1977), 28ºC in *V. decussata* (Beiras et al., 1994) and

30ºC in *M. mercenaria* (Loosanoff and Davies, 1963). The optimum temperature for *V. philippinarum* larvae, taking into account growth and survival, is 25ºC (Robinson and Bresse, 1984). Maximum growth was observed at 24ºC in *D. trunculus* larvae, although the highest survival rate was found at 20ºC (Márquez et al., 2009).

Table 4. Studies testing live diets on clam larval culture

Species	Algal species	Best conditions	Reference
VENERIDAE			
Venerupis philippinarum	1. *Isochrysis aff. galbana* (T-iso); 2. *Chaetoceros calcitrans* and 3. T-iso + *C. calcitrans* (1:1)	Diet 2: *C. calcitrans* and diet 3: T-iso + *C. calcitrans* (1:1)	Helm and Laing (1987)
V. philippinarum	1. *Isochrysis* spp.; 2. *Chlorella* spp. and 3. *Isochrysis* spp. + *Chlorella* spp. (1:1)	Diet 1: *Isochrysis* spp. and diet 3: *Isochrysis* spp. + *Chlorella* spp. (1:1)	Yan et al. (2006)
Mercenaria mercenaria	1. *I. aff. galbana* (T-iso); 2. *C. calcitrans* and 3. T-iso + *C. calcitrans* (1:1)	Similar growth in all the diets	Helm and Laing (1987)
V. decussata	1. *I. aff. galbana* (T-iso) and 2. *C. calcitrans*	Diet 2: *C. calcitrans*	Matias et al. (2011)
V. decussata	1. *I. galbana* + *Pavlova lutheri* + *C. muelleri* (1:1:1); 2. *I. galbana* + *P. lutheri* + *C. muelleri* (1:1:2); 3. *I. galbana* + *P. lutheri* (1:1) and 4. *C. muelleri*	Diet 2: *I. galbana* + *P. lutheri* + *C. muelleri* (1:1:2)	Aranda et al. (2011)
V. corrugata	1. *I. galbana*; 2. *Tetraselmis suecica* and 3. *I. galbana* + *T. suecica*	Diet 1: *I. galbana* and diet3: *I. galbana* + *T. suecica*	Fernández-Reiriz et al. (2011)
Meretrix meretrix	1. *I. galbana*; 2. *Dunaliella* sp.; 3. *Phaeodactylum tricornutum*; 4. *Platymonas subcordiformis*; 5. *Pavlova viridis*; 6. *I. galbana* + *Dunaliella* sp. (1:1); 7. *I. galbana* + *P. tricornutum* (1:1) and 8. *I. galbana* + *P. subcordiformis* (1:1)	Diet 1: *I. galbana*; diet 6: *I. galbana* + *Dunaliella* sp. (1:1); diet 7: *I. galbana* + *P. tricornutum* (1:1) and diet 8: *I. galbana* + *P. subcordiformis* (1:1)	Tang et al. (2006)
Paphia malabarica	1. *I. galbana*; 2. *Nannochloropsis salina* and 3. *I. galbana* + *N. salina*	Diet 1: *I. galbana* and diet 3: *I. galbana* + *N. salina*	Raghavan and Gopinathan (2008)
DONACIDAE			
Donax trunculus	1. *I. aff. galbana* (T-iso); 2. T-iso + *C. gracilis*; 3. T-iso + *T. suecica*; 4. T-iso + *Rhodomonas baltica*; 5. T-iso + *C. gracilis* + *T. suecica*; 6. T-iso + *C. gracilis* + *R. baltica* and 7. T-iso + *T. suecica* + *R. baltica*	Diet 5: T-iso + *C. gracilis* + *T. suecica* and diet 6: T-iso + *C. gracilis* + *R. baltica*	Ruiz-Azcona et al. (1996)
D. trunculus	1. *I. aff. galbana* (T-iso); 2. T-iso + *S. costatum*; 3. T-iso + *S. costatum* + *C. calcitrans* and 4. T-iso + *T. suecica*	Diet 3: T-iso + *S. costatum* + *C. calcitrans*	Sánchez-Lazo et al. (2009)
D. trunculus	1. *I. galbana* 1st week and then *C. muelleri* was added and 2. *I. galbana* + *P. lutheri* + *T. suecica* 1st week and then *C. muelleri* was added	Similar growth and survival in both diets	Louzán et al. (2010)
PHARIDAE			
Ensis magnus	1. *I. galbana* + *P. lutheri*; 2. *I. galbana* + *P. lutheri* + *C. calcitrans*; 3. *I. galbana* + *C. calcitrans* + *T. suecica* and 4. *T. suecica* + *I. galbana* + *P. lutheri* + *C. calcitrans*	Diet 4: *T. suecica* + *I. galbana* + *P. lutheri* + *C. calcitrans*	da Costa et al. (2011a)
SOLENIDAE			
Solen marginatus	1. *I. galbana* and 2. *T. suecica* + *I. galbana* + *P. lutheri* + *C. calcitrans*	Diet 2: *T. suecica* + *I. galbana* + *P. lutheri* + *C. calcitrans*	da Costa et al. (2011a)

9.4.4. Other Factors Affecting Larval Culture

One other environmental factor that has received attention in bivalve larval culture is salinity. Although the literature is scarce, some early works relating to the effect of salinity on growth and survival deserve mention here. Davis (1958) determined that the optimum salinity for growth of *M. mercenaria* larvae is 27‰ or higher. Later, it was ascertained that the effect of reduced salinity on *M. mercenaria* larvae was to reduce the range of temperatures they would tolerate (Davis and Calabrese, 1964). Larvae of the coot clam *Mulinia lateralis* grew satisfactorily within a 20 to 32.5‰ range of salinities, optimum salinity being 25‰ (Calabrese, 1969). A more recent study investigated short term (24h and 72h of exposure) effects of salinity on *V. philippinarum* larvae, finding that mortality was low, from 12‰ to 33‰ (Numaguchi, 1998). *V. philippinarum* larvae do not swim, mortality increasing at 8‰; while 2.5‰ salinity led to total mortality of the batch.

Stocking density of bivalve larvae is an important factor to be considered in hatchery conditions because it influences both the growth and survival of the larvae reared. In closed larval rearing systems, as stocking density increases the amount of metabolic wastes may increase in the water, giving rise to detrimental effects (Yan et al., 2006). In our hatchery facilities all bivalve species are reared at a density ranging between 5 to 10 larvae mL^{-1}, in accordance with the results for *V. corrugata*, *V. decussata*, and *V. philippinarum* larvae (Pérez-Camacho et al., 1977; Yan et al., 2006). A similar stocking density (10 larvae mL^{-1}) was found to be the best for promoting growth, survival and percentage of metamorphosis in *Spisula solidissima similis* larvae (Hurley and Walker, 1996). Nonetheless, in the clam *Meretrix meretrix* larval survival is independent of stocking density (Liu et al., 2006). These authors recommended a density of 10-20 larvae mL^{-1} in large-scale culture for cost-effective production. By contrast, a density of 1-3 larvae/mL^{-1} appeared to be optimal for normal growth of *Paphia malabarica* larvae (Raghavan and Gopinathan, 2008).

The geographic origin of broodstock may influence larval quality, since broodstock in wild populations is adapted to certain environmental conditions (i.e. temperature and quality and quantity of food) which in great measure determine the quality of the gametes produced. We investigated the effect of different geographic origins on larval performance of *V. corrugata*, using broodstock from five populations during the four seasons of the year (Cerviño-Otero, 2011). Our results show that spawning quality in *V. corrugata* is unrelated to the condition index, biochemical composition and maturity index of broodstock, since an examination of these parameters fails to provide information either about the number of eggs released or about the success of larval development.

9.5. Nursery Culture

The following section is organized in a similar way to that on Larval Rearing. Firstly, we will report on standard postlarval and spat culture. Secondly, we will analyze the different factors affecting growth and survival in this cultivation phase.

9.5.1. Standard Post-Larval and Seed Culture

Postlarvae were reared in sieves with inverted flow (down-wellers), kept in 500-L larval rearing tanks (Figure 5A). Water was fully renewed every two days and the temperature set at 18±1°C. A mixed diet was provided on a daily basis, consisting of a mixture of *P. lutheri*, *I. galbana*, *T. suecica* and *C. calcitrans* at an initial ration of 80 cells μl^{-1}. When seed reached a certain length (0.5 mm and 1 mm in clams and razor clams, respectively) postlarvae were transferred to 150 µm mesh sieves in a down-welling system in 1000-L rectangular tanks with closed circuit (Figure 5B). Sieves were changed when necessary to increase mesh size.

Table 5 shows data about the growth and survival of different clam species reared in our facilities; Figure 6 presents microphotographs of postlarvae of several clam species. Postlarval and spat culture of *Venerupis philippinarum* yielded better results in terms of survival than its larval culture. Manila clam exhibited high growth rates, similar to those found in *Venerupis corrugata*. The latter species shows high survival rates during nursery culture, and the majority of the cultures are successfully reared to an optimum size for their transfer to intermediate cultivation. *V. decussata* postlarvae and seed displayed a lower growth rate and lower survival than *V. philippinarum* and *V. corrugata*. Despite *Donax trunculus* larval culture showing low and variable survival rates, its spat culture showed high survival compared with other clam species reared in hatcheries. The highest mortality was recorded when postlarvae reached 0.5 mm long.

Postlarval survival from settlement until 1 mm long in *Ensis magnus* was reported to be as low as 5-10% (da Costa et al., 2011b). In 2011, nine larval batches were reared using low temperature during larval and postlarval culture (15-16°C), a higher survival rate being observed in early postlarval culture (25% on average, with values from 8% to 46%) than that previously reported.

Figure 5. A. System for settling metamorphosing larvae. B. Spat culture systems.

Table 5. Growth and survival of postlarvae and spat in our facilities. Survival is calculated using the number of postlarvae and spat at a given time, and refers to the number of settlers. The age of the seed is shown between brackets

Species	Early postlarvae		Spat	
	Length (mm)	Survival (%)	Length (mm)	Survival (%)
Venerupis philippinarum	0.8-1 (day 60)	70-80% (day 60)	2 (day 100)	13% (day 100)
V. decussata	0.3 (day 35)	10-30% (day 35)	0.8-1.7 (day 120)	3-15% (day 120)
V. corrugata	0.6-0.7 (day 40)	48% (day 40)	4-5 (day 150)	30-50% (day 150)
Donax trunculus	0.25-0.3 (day 30)	60% (day 30)	2 (day 120)	36% (day 120)
Ensis magnus	1 (day 30)	25% (day 30)	30 (day 120)	2-5% (day 120)
E. siliqua	1 (day 30)	5-10% (day 30)	30 (day 120)	2-5% (day 120)
Solen marginatus	1.5-2 (day 30)	63.5% (day 30)	20 (day 120)	2-5% (day 120)

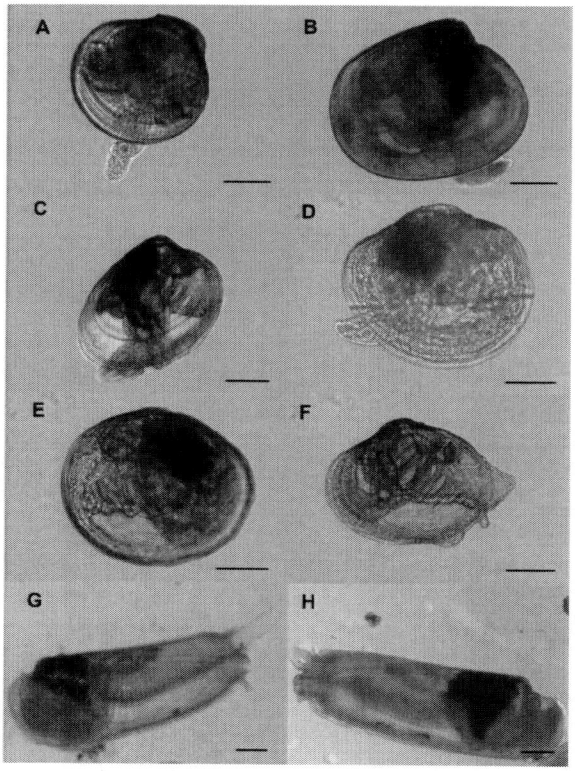

Figure 6. A. *V. decussata* postlarvae. B. *V. corrugata* newly-settled postlarvae at day 21 post-fertilization (pf). C. Thirty-day-old postlarvae of *D. trunculus*. D. *E. magnus* newly-settled postlarvae at day 20 pf. E. Eight-day-old postlarvae of *S. marginatus*. F. *E. siliqua* newly-settled postlarvae at day 14 pf. G. One-month-old postlarvae of *E. magnus*, showing the curvature of the shell. H. One-month-old postlarvae of *S. marginatus*. Scale bar: 100 μm.

Similarly, the low survival rates (8%) observed in the same culture phase of *Solen marginatus* (da Costa and Martínez-Patiño, 2009) were improved, reaching a mean survival of 63.5% (da Costa et al., 2011a). Conversely, the survival obtained in *Ensis siliqua* until postlarvae reached 1 mm long is still low (5-10%).

Moreover, from 1 mm long until seed were transferred to wild beds, survival was low (2-5%) in all three razor clam species.

Hence, more research is needed to ascertain the specific dietary and environmental requirements (in terms of rearing temperature) in razor clam nursery culture. The following sections dealing with the factors affecting nursery culture bring together some of the research studies carried out by our research group on razor clam seed culture.

9.5.2. Spat Nutrition

The optimal ration in bivalve juveniles depends on the species of bivalve reared and the culture conditions of algae making up the diet (Coutteau et al., 1994a). Moreover, stocking density and seed size must be considered in order to allow comparisons to be made between studies. Thus, the optimal daily ration for *M. mercenaria* seed (in size range 0.4 to 7 mg live weight), fed on a mixture of *C. gracilis* and *I. galbana* (clone T-iso), was established at 1.5 to 2% dry weight of food per wet weight of clams (DW WW^{-1} day^{-1}) (Coutteau et al., 1994b). Coutteau et al. (1994a) reported the optimal daily ration for *V. philipppinarum* to be 1% DW WW^{-1} day^{-1} when fed on *Chaetoceros neogracile* concentrates and 1.3% when fed on a live mixed diet of *C. neogracile* and T-iso. Laing and Millican (1991) stated that 0.2 g of dry weight of food per gram of live weight of biomass per week was the ration most efficiently used. On the other hand, these aforementioned rations are excessive for *V. decussata* seed of sizes between 4 and 7 mm, which displayed higher growth rates when 0.625% DW WW^{-1} day^{-1} of *T. suecica* and *P. tricornutum* mixture is provided (Enes and Borges, 2003).

Albentosa et al. (1996a) reported maximum growth for 1.6 mm-long *V. decussata* seed occurring at a concentration of 50 cells μL^{-1} of *I. galbana* (clone T-iso), whilst Beiras et al. (1993) using *V. corrugata* seed of a similar size (2 mm) found that 100 and 300 cells μL^{-1} of *I. galbana* were, respectively, optimum and maximum values for growth performance.

To a great extent, juvenile growth and survival are determined by diet quality (see Larval Nutrition section). In our laboratory we assayed different diets on spat of the razor clam *S. marginatus*, finding that a multispecific diet with *I. galbana*, *P. lutheri*, *C. calcitrans* and *T. suecica* gave the best yields (da Costa et al., 2010a) (Table 6). The high lipid and carbohydrate content of this diet could explain the optimum growth rates achieved by spat fed on a mixed diet. In the razor clam *E. magnus* we found that the algal diets which produced greatest growth were those that included *I. galbana* and *P. lutheri* (da Costa et al., 2010b). Early studies on clam seed nutrition suggested the importance of including *I. aff. galbana* (T-iso) in the diet of the *V. philippinarum*, *V. decussata* and *M. mercenaria* (Laing et al., 1987). In this study, it was also observed that a monospecific diet of *S. costatum* stimulated similar growth to that of *V. philippinarum* seed fed on T-iso. The large amounts of the essential fatty acids DHA (in T-iso) and EPA (in *S. costatum*) means that both are considered of good food value (Laing et al., 1987). *V. decussata* seed exhibited higher growth rates when fed with the microalga *T. suecica* (Albentosa et al., 1996c), which has low protein and lipid content, whilst carbohydrate content is high. These last authors pointed out that the latter biochemical substrate, that is, carbohydrate, could replace part of the protein and lipid content in the diet of *V. decussata* seed. A slightly lower growth was observed in clams fed with T-iso (Albentosa et al., 1996b,c). By contrast, in *V. corrugata* seed a higher growth rate was observed using a mixed diet consisting of T-iso and *T. suecica* than with a diet of T-iso alone (Albentosa et al., 1993). These results can be explained by the fact that *T. suecica* is considered an alga with a relatively low food value, due mainly to the limited amount of EPA and DHA (Webb and Chu, 1982). Albentosa et al. (1996b) showed that *V. decussata* seed growth was not limited by the absence of EPA in T-iso or DHA in *T. suecica*. Consequently, these authors suggested that it is questionable whether these fatty acids are essential.

There are several studies dealing with the substitution of live microalgae for spat feeding (Table 6), due to the demand for food in hatcheries during the cultivation period for clam spat, i.e. from metamorphosis until the spat is ready to be transferred to intermediate and on-growing culture. The pressure to reduce production costs of hatchery-produced spat has led to a total or partial replacement of live foods, thereby enabling a larger amount of spat to be produced.

Concentrated algae paste is one potential replacement for fresh microalgae that has been successfully tested in bivalve larvae and juveniles (Nell and O´Connor, 1991; Robert et al., 2001; Ponis et al., 2003a,b). However, its use in clams has only been reported in *V. philippinarum* by Bonaldo et al. (2005). One other alternative is dried microalgae, which can be divided into two groups: freeze dried algae and spray-dried microalgae. One drawback of freeze-dried algae is the loss of nutritional value, which may be due to some process that occurs during freeze-drying, affecting negatively the digestibility of the microalgae (Albentosa et al., 1997). Consequently, lower growth was observed in *V. decussata* when fed exclusively on freeze-dried diets, with freeze-dried *I. aff. galbana* (T-iso) being the diet that promoted highest clam seed growth rates (Albentosa et al., 1997). By contrast, spray-dried microalgae are heterotrophically produced, a process by which algae are grown in fermenters, using organic carbon instead of light as an energy source (Langdon and Ö-Nal, 1999).

The growth of *V. philippinarum* and *V. decussata* juveniles fed on spray-dried *T. suecica* was compared with growth with live diets, it being shown that there were no differences between clams fed dried and live *T. suecica* (Laing and Gil Verdugo, 1991). However, these same authors pointed out that far greater growth was achieved when live *T. suecica* was mixed with the diatom *C. calcitrans*. Laing and Millican (1991) increased live algal substitution with good results, using 70% spray-dried *T. suecica* combined with 30% of live *S. costatum*. The next step forward was a trial by Laing and Millican (1992), who replaced live microalgae by spray-dried *T. suecica* combined with the spray-dried diatom *Cyclotella cryptica*. This dried diet exhibited only slightly lower growth than a mixed diet of *T. suecica* and *S. costatum*, though a higher growth rate was observed using 90% spray-dried *T. suecica* combined with 10% of live *S. costatum*. Thus, all these studies highlighted the importance of including a diatom in clam juvenile nutrition. The high nutritional value of many diatom species widely used in juvenile bivalve cultivation is believed to be due to their relatively high content of essential long-chain polyunsaturated fatty acids, especially EPA (Langdon and Waldock, 1981). Moreover, Curatolo et al. (1993) achieved similar growth to that observed in the control diet by replacing up to 40% of the live diatom *C. calcitrans* with spray-dried *T. suecica*. More recently, it has been demonstrated that spray-dried *Schizochytrium* sp. can replace 80% of live algae in *V. philippinarum* juveniles when they are

Table 6. Nutritional studies on clam spat

Species	Type of diet	Diets tested	Best conditions	Reference
VENERIDAE				
Venerupis philippinarum	Live algae and microcapsules	Live *Chaetoceros calcitrans* and microcapsules	Live *C. calcitrans* and microcapsules + 15% *C. calcitrans*	Laing (1987)
V. philippinarum	Live algae	*Isochrysis aff. galbana* (T-iso), *Skeletonema costatum*, *C. calcitrans*, *Chroomonas salina*, *Thalasiosira pseudoenana*, *Tetraselmis suecica*, *Phaeodactylum tricornutum* and *Chlamydomonas coccoides*	Diet 1. *I. aff. galbana* (T-iso) and diet 2. *S. costatum*	Laing et al. (1987)
V. philippinarum	Live algae and yeast	Manipulated yeast, algal mixture (*I. galbana*, *T. pseudoenana*, *S. costatum*, *Chaetoceros gracilis* and *T. suecica*), manipulated yeast enriched with a growth promoting agent and single diets of *I. galbana*, *S. costatum*; *C. gracilis*, *T. suecica*	Diet 1. 80% manipulated yeast enriched with a growth promoting agent + 20% of algal mixture (*I. galbana*, *T. pseudoenana*, *S. costatum*, *C. gracilis* and *T. suecica*) and diet 2. 100% of algal mixture	Albentosa et al. (1989)
V. philippinarum	Live and spray-dried algae	Spray-dried *T. suecica* and live *C. calcitrans* and *T. suecica*	Live *C. calcitrans* + *T. suecica*	Laing and Gil Verdugo (1991)
V. philippinarum	Live and spray-dried algae	Spray-dried *T. suecica*, live *T. suecica* and live *S. costatum*	30% live *S. costatum* + 70% spray-dried *T. suecica*	Laing and Millican (1991)
V. philippinarum	Live and spray-dried algae	Dried *T. suecica*, live *S. costatum* and *T. suecica*	Diet 1. 90% dried *T. suecica* + 10% live *S. costatum* and diet 2. 70% live *T. suecica* + 30% live *S. costatum*	Laing and Millican (1992)
V. philippinarum	Live and spray-dried algae	Live *C. calcitrans* and spray-dried *T. suecica*	Diet 1. 100% live *C. calcitrans*; diet 2. 80% live *C. calcitrans* + 20% dried *T. suecica* and diet 3. 60% live *C. calcitrans* + 40% dried *T. suecica*	Curatolo et al. (1993)
V. philippinarum	Live and spray-dried algae	Spray-dried *Schizochytrium* sp., live *T. suecica* and *Chaetoceros* sp.	Live *T. suecica* + *Chaetoceros* sp.	Boeing (1997)
V. philippinarum	Live algae and lipid emulsions	Live *I. aff. galbana* (T-iso), *Chaetoceros neogracile*, *Dunaliella tertiolecta*, *T. suecica* and lipid emulsion	T-iso + *C. neogracile*	Caers et al. (1998)
V. philippinarum	Live algae and lipid emulsions	Live *Isochrysis aff. galbana* (T-iso), *T. suecica* and lipid emulsion	Diet 1. Live T-iso + *T. suecica*, 1% DW WW^{-1} day^{-1}; diet 2. *T. suecica*, ration 1.5% DW WW^{-1} day^{-1} and diet 3. *T. suecica* + emulsion, ration 1.5% DW WW^{-1} day^{-1}	Caers et al. (1999b)

Table 6. (Continued)

Species	Type of diet	Diets tested	Best conditions	Reference
V. philippinarum	Live algae and wheatgerm diets	Live *I. galbana* and wheatgerm	Diet 1. 100% *I. galbana*, 4 mg AFDW and diet 2. 50% *I. galbana* + 50% wheatgerm, 4 mg AFDW	Albentosa et al. (2002); Fernández-Reiriz et al. (2006)
V. philippinarum	Live algae and macroalgal meals	Live *I. aff. galbana* (T-iso) and two macroalgal meals (*Undaria* sporophytes and gametophytes)	Diet 1. 100% T-iso and diet 2. 50% T-iso + 50% *Undaria* sporophytes	Albentosa and Pérez-Camacho (2002)
V. philippinarum	Microalgae paste	Concentrated 100% *I. galbana*, concentrated 80% *I. galbana* + 20% *T. suecica* and concentrated 40% *I. galbana* + 20% *T. suecica* + 40% *C. muelleri*. 1% dry weight live weight of the clams^{-1}	Diet 1. Concentrated 40% *I. galbana* + 20% *T. suecica* + 40% *C. muelleri*	Bonaldo et al. (2005)
V. philippinarum	Live and spray-dried algae	Live *I. galbana* + *C. calcitrans* and spray-dried *Schizochytrium* sp. or *Haematococcus pluvialis*	Live *I. galbana* + *C. calcitrans* and spray-dried *H. pluvialis*	Ö-Nal et al. (2005)
V. philippinarum	Live algae and supplemented sugars	*C. calcitrans* and supplemented sugars (glucose, maltose, maltotriosa, maltopentaose and pullulan)	Diet 1. *C. calcitrans* + glucose 10 mg L^{-1} and diet 2. *C. calcitrans* + glucose 100 mg L^{-1}	Uchida et al. (2010)
V. decussata	Live algae	*I. aff. galbana* (T-iso), *C. calcitrans*, *T. pseudoenana* and *P. tricornutum*	*I. aff. galbana* (T-iso)	Laing et al. (1987)
V. decussata	Live and spray-dried algae	Spray-dried *T. suecica*, live *C. calcitrans* and *T. suecica*	Diet 1. 100 cells μL^{-1} *C. calcitrans* + 10 cells μL^{-1} *T. suecica* with no significant differences with the other diets	Laing and Gil Verdugo (1991)
V decussata	Live algae, flours and powdered fish	Live *T. suecica*, *Dunaliella bardawil* and *Nitzschia acicularis* grown under autotrophic or mixotrophic conditions; and inert diets (rye and soy flours and powdered fish)	Live microalgae grown in autotrophic conditions	Lamela et al. (1996)
V. decussata	Live diets	*I.aff. galbana* (T-iso), *T. suecica* and *P. tricornutum*	*T. suecica*	Albentosa et al. (1996b,c)
V. decussata	Live and freeze-dried algae	Live *I. aff. galbana* (T-iso) and freeze-dried *I. aff. galbana* (T-iso), *T. suecica* and *P. tricornutum*	Live *I. aff. galbana* (T-iso)	Albentosa et al. (1997)
V. decussata	Live algae and cornstarch	*I. galbana* and cornstarch	Live *I. galbana* 2% ration	Pérez-Camacho et al. (1998)
V. decussata	Live algae and cornmeal	*I. galbana* and cornmeal	75% *I. galbana* and 25% cornmeal	Pérez-Camacho et al. (1998)
V. decussata	Live and wheatgerm diets	Live *I. aff. galbana* (T-iso) and wheatgerm	Diet 1. T-iso, 2% of spat live weight and diet 2. T-iso (50%) + 50% wheatgerm, 2% ration	Albentosa et al. (1999)

Species	Type of diet	Diets tested	Best conditions	Reference
V. decussata	Live algae and single cell detritus (SCD)	Live *I. aff. galbana* (T-iso) and single cell detritus (SCD) from *Laminaria saccharina*	Diet 1. Live T-iso and diet 2. Single cell detritus (SCD) from *L. saccharina* (80-90%) + T-iso (10-20%)	Pérez-Camacho et al. (2002, 2007)
V. decussata	Live algae and cheese whey	Live *T. suecica* + *P. tricornutum* and cheese whey	25% (*T. suecica* + *P. tricornutum*) + 75% cheese whey	Enes and Borges (2003)
Mercenaria mercenaria	Live algae and yeasts	Live *Thalassiosira pseudonana* and yeast *Candida utilis*	Diet 1. 100% live *T. pseudonana*; diet 2: 75% live *T. pseudonana* + 25% yeast; and diet 3. 50% live *T. pseudonana* + 50% yeast	Epifanio (1979a)
M. mercenaria	Live algae	Live *Carteria chui*, *I. galbana*, *Platymonas suecica*, *T. pseudonana*	Diet 1. *I. galbana* + *P. suecica* + *T. pseudonana*; diet 2. *C. chui* + *I. galbana* + *P. suecica* + *T. pseudonana*; diet 3. *P. suecica* + *T. pseudonana*; and diet 4. *C. chui* + *P. suecica* + *T. pseudonana*	Epifanio (1979b)
M. mercenaria	Live algae and microcapsules	Live *C. calcitrans* and microcapsules	Live *C. calcitrans* and microcapsules + 15% *C. calcitrans*	Laing (1987)
M. mercenaria	Live algae	*I. aff. galbana* (T-iso), *C. calcitrans* and *T. pseudoenana*	*I. aff. Galbana* (T-iso)	Laing et al. (1987)
M. mercenaria	Live and spray-dried algae	Spray-dried *T. suecica*, live *C. calcitrans* and *T. suecica*	Live *C. calcitrans* + *T. suecica*	Laing and Gil Verdugo (1991)
M. mercenaria	Live algae and yeast	Live *Chaetoceros gracilis* + *I. aff. galbana* (T-iso) and five commercial yeasts	Live *C. gracilis* + *I. aff. galbana* (T-iso) at 2% ration	Coutteau et al. (1991)
M. mercenaria	Live algae and yeast	Live *Chaetoceros gracilis* + *I. aff. galbana* (T-iso) and baker´s yeast *Saccharomyces cerevisiae*	Live *C. gracilis* + *I. aff. galbana* (T-iso) at 2% ration	Coutteau et al. (1994b)
M. mercenaria	Live algae and commercially available diets	Live *I. galbana* and commercially available diets (preserved phytoplankton)	Diet 1. Live *I. galbana* and diet 2. DT´s Live Marine Phytoplankton	Espinosa and Allam (2006)
V. corrugata	Live diets	*I. aff. galbana* (T-iso) and *T. suecica*	Diet 1. T-iso + *T. suecica* and diet 2. T-iso	Albentosa et al. (1993, 1994b)
PHARIDAE				
Ensis magnus	Live diets	*I. galbana*, *P. lutheri*, *S. costatum* and *T. suecica*	Diet 1. *I. galbana* + *P. lutheri* + *S. costatum* + *T. suecica*; and diet 2. *I. galbana* + *P. lutheri* + *T. suecica*	da Costa et al. (2010b)
SOLENIDAE				
Solen marginatus	Live diets	*I. galbana*, *P. lutheri*, *C. calcitrans*, *S. costatum* and *T. suecica*	*I. galbana* + *P. lutheri* + *C. calcitrans* + *T. suecica*	da Costa et al. (2010a)

fed on live *T. suecica*. However, when the control diet was a mixture of *T. suecica* and *Chaetoceros* sp., only 40% of the live algae can be replaced, otherwise the outcome is lower growth (Boeing, 1997), in agreement with the aforementioned results. Similarly, Ö-Nal et al. (2005) found that up to 75% of live algae can be substituted by spray-dried *Haematococcus pluvialis* or *Schizochytrium* sp. without significantly reducing live (wet) growth of the clams.

Several studies have tested the effect of microcapsules and lipid emulsions as nutrient carriers in oyster and scallop spat (Langdon, 1989; Knauer and Southgate, 1997; Caers et al., 2000a,b), though they have rarely been used in clams. Only Laing (1987) reported on the use of microcapsules in *V. philippinarum* and *M. mercenaria* spat, although this study stated that up to 85% of the algae can be substituted by microcapsules. In spite of the promising results achieved by Laing (1987) with microcapsules, their use has not been generalized in nutritional studies with clam seed. This could be explained by the fact that microcapsules can severely pollute the water or are too complex to prepare on a regular basis in a hatchery (Coutteau et al., 1994c). More recently, the use of lipid emulsions as an artificial lipid supplement to live algae has emerged as an alternative. Lipid supplementation of *Dunaliella tertiolecta* produced a vast improvement in DHA levels and daily growth rates in the spat of *V. philippinarum* compared with individuals fed solely on *D. tertiolecta* (Caers et al., 1998). Even though DHA levels also increased in clams fed on *T. suecica* supplemented with lipids, no effect was observed on daily growth rates. Conversely, another study with lipid emulsion in *V. philippinarum* found a preferential accumulation for DHA, but not for EPA, in those clams that were thus fed, which indicated the greater importance of DHA as against EPA (Caers et al., 1999b).

Yeasts have also been tested as algal substitutes, due to the fact that their production is more economical and efficient than that of live algae (Coutteau et al., 1994b). Epifanio (1979a), using diets consisting of 50% dried yeast *Candida utilis* and 50% of the diatom *Thalassiosira pseudonana*, achieved growth rates in length and dry weight of soft tissue comparable to those of a 100% algal ration for *Mercenaria mercenaria*. A study with *V. philippinarum* compared single and mixed algal diets with partial and total algal substitution with yeasts, finding that algae can be partially replaced by manipulated yeasts, especially manipulated yeast enriched with a growth-promoting agent (Albentosa et al., 1989). Coutteau et al. (1994b) achieved 50% of substitution of the algal ration with manipulated yeast, without significantly decreasing growth rate relative to algal-fed controls.

Other inert diets that have been tested in bivalves include carbohydrate-rich cereal flours, given the importance of carbohydrates in bivalve nutrition. The use of wheatgerm as feed for *V. decussata* (Albentosa et al., 1999) and *V. philippinarum* spat (Albentosa et al., 2002) and the use of cornflour for *V. decussata* spat (Pérez-Camacho et al., 1998) allowed 50% of the maximum daily microalgal ration to be replaced without affecting growth.

Some studies have focused on macroalgal meals and macroalgal biotransformates as substitutes for live algae in nursery feeding trials (Albentosa and Pérez-Camacho, 2002; Pérez-Camacho et al., 2002, 2004, 2007). Preliminary results of the use of the macroalgae *Undaria* sporophytes and gametophytes as partial and total substitutes for live algae in *V. philippinarum* seed showed that 50% of replacement of the microalgae by sporophytes yielded growth comparable to that of the control diet (Albentosa and Pérez-Camacho, 2002). The same research group reported the method for preparation of a single cell detritus (SCD) from *Laminaria saccharina* as a hatchery diet for bivalves using enzymes and bacteria (Pérez-Camacho et al., 2004). The use of this SCD from *L. saccharina* showed promising results, achieving 80 to 90% replacement of the live phytoplankton content (Pérez-Camacho et al., 2002, 2007).

Other feeds, such as industrial by-products (for example cheese whey), have been tested as substitutes for microalgae (Enes and Borges, 2003). In this study with *V. decussata*, the best results in live weight and length were produced by a mixture of 25% of microalgae and 75% cheese whey.

A recent study supplemented the algal diet with sugars in juvenile *V. philippinarum* (Uchida et al., 2010). This study is based on the evidence that bivalves are able to absorb dissolved organic matter as nutrients. Several sugars were added to the rearing water, finding that only glucose significantly promoted soft body growth of the clams.

9.5.3. Other Factors Affecting Nursery Culture

Temperature is one of the most important factors regulating clam growth in hatcheries (Laing et al., 1987). Generally speaking, a temperature increase, within certain limits, speeds up feeding activity, metabolism and growth (Newell and Branch, 1980). *V. philippinarum* and *M. mercenaria* exhibited maximum organic growth at 25ºC; however, 20ºC is a more appropriate culture temperature due to the higher efficiency in the use of the diet supplied (Laing et al., 1987). The latter authors also reported a lower temperature for *V. decussata* seed culture, because in this species the most efficient growth occurred at 15-20ºC. Similarly, the optimum temperature for growth in *V. corrugata* was established at 20ºC, a temperature at which maximum values of ingestion rates, absorption efficiency and scope for growth were observed (Albentosa et al., 1994a).

Certain species have special requirements for seed culture, such as the razor clams. Razor clams have a weak adductor muscle to prevent shell gaping so their adults cannot live for more than one or two days without being buried. Moreover, razor clam seed are also very sensitive to substrate deprivation. Nonetheless, for large-scale seed production, it is important to reduce the volume of the sand needed or even to avoid the use of substratum. Thus, we have investigated the effect of seed transfer into substrate at different times during the nursery culture of *E. magnus*, *E. siliqua* and *S. marginatus*. Higher survival was observed when settlers are kept without substrate and also mortality outbreaks caused by proliferation of bacteria in the sand are prevented. When the seed reach 1-2 mm long an interspecific need for substrate was observed. Thus, *E. magnus* seed held with sand in cylinders in a down-welling system exhibited higher survival rates compared to the same containers without substratum (da Costa et al., 2011b), whilst a higher survival rate was observed in *S. marginatus* and *E. siliqua* seed kept without sand (da Costa and Martínez-Patiño, 2009; da Costa et al., 2010c). The latter two species can be kept without substratum until a length of 10 mm is reached. However, growth is always enhanced when razor clam seed are kept burrowed in substrate. From this size upwards, seed started to become deformed when it was deprived of substrate.

Acknowledgments

We are grateful to the staff of the Centro de Cultivos Marinos de Ribadeo. The studies of the Centro de Cultivos Marinos de Ribadeo reviewed in this chapter were supported in part by funds for marine investigation from the Xunta de Galicia and by the following grants: ALMEJAS (2005-2007), ALMEJAS (2008-2010), and HATCHERIES from the Junta Asesora Nacional de Cultivos Marinos (JACUMAR), SHARE-90 and TIMES from the European Initiative Interreg IIIB "Atlantic Area", PGIDITO6RMA50801PR (ALBA) and 10MMA103013PR from the Xunta de Galicia. Dr. Fiz da Costa's research work was partially funded by a Fundación Juana de Vega post-doctoral fellowship at IFREMER.

References

Albentosa, M., Naessens, E., Léger, P., Coutteau, P., Lavens, P. and Sorgeloos, P. (1989). Promising results in the seed culturing of the Manila clam *Tapes semidecussata* with a manipulated yeast product as a partial substitute for algae, *Aquaculture´89*. Bourdeaux, France. 10, 7-8.

Albentosa, M., Pérez-Camacho, A., Labarta, U., Beiras, R. and Fernández-Reiriz, M.J. (1993). Nutritional value of algal diets to clam spat *Venerupis pullastra*. *Mar. Ecol. Progr. Ser.* 97, 261-269.

Albentosa, M., Beiras, R. and Pérez-Camacho, A. (1994a). Determination of optimal thermal conditions for growth of clam (*Venerupis pullastra*) seed. *Aquaculture.* 126, 315-328.

Albentosa, M., Labarta, U., Pérez-Camacho, A., Fernández-Reiriz, M.J. and Beiras, R. (1994b). Fatty acid composition of *Venerupis pullastra* spat fed on different microalgae diets. *Comp. Biochem. Phys. A.* 108, 639-648.

Albentosa, M., Pérez Camacho, A. and Beiras, R. (1996a). The effect of food concentration on the scope for growth and growth performance of *Ruditapes decussatus* (L.) seed reared in an open-flow system. *Aquac. Nutr.* 2, 213-220.

Albentosa, M., Labarta, U., Fernández-Reiriz, M.J. and Pérez-Camacho, A. (1996b). Fatty acid composition of *Ruditapes decussatus* spat fed on different microalgae diets. *Comp. Biochem. Phys. A.* 113, 113-119.

Albentosa, M., Pérez-Camacho, A., Labarta, U. and Fernández-Reiriz, M.J. (1996c). Evaluation of live microalgal diets for the seed culture of *Ruditapes decussatus* using physiological and biochemical parameters. *Aquaculture.* 148, 11-23.

Albentosa, M., Pérez-Camacho, A., Labarta, U. and Fernández-Reiriz, M.J. (1997). Evaluation of freeze-dried microalgal diets for the seed culture of *Ruditapes decussatus* using physiological and biochemical parameters. *Aquaculture.* 154, 305-321.

Albentosa, M., Fernández-Reiriz, M.J., Pérez-Camacho, A. and Labarta, U. (1999). Growth performance and biochemical composition of *Ruditapes decussatus* (L.) spat fed on microalgal and wheatgerm flour diets. *J. Exp. Mar. Biol. Ecol.* 232, 23-37.

Albentosa, M. and Pérez-Camacho, A. (2002). Valor nutritivo de harinas de macroalgas para el cultivo de semilla de almeja japonesa *Ruditapes philippinarum* (Adams and Reeve, 1850): pruebas preliminares. *Bol. Inst. Esp. Oceanogr.* 18, 281-287.

Albentosa, M., Pérez-Camacho, A., Fernández-Reiriz, M.J. and Labarta, U. (2002). Wheatgerm flour in diets for Manila clam, *Ruditapes philippinarum*, spat. *Aquaculture.* 212, 335-345.

Aranda, J. A., da Costa, F., Cerviño-Otero, A., Louzán, A., Fernández-Pardo, A., Nóvoa, S., Ojea, J. and Martínez-Patiño, D. (2011). Efecto de diferentes dietas en crecimiento, supervivencia y composición bioquímica de larvas de almeja fina *Ruditapes decussatus* (Linné, 1758). *XIII Congreso Nacional de Acuicultura.* Casteldefels, Spain.

Badillo-Salas, C., Valenzuela-Espinoza, E., González-Gómez, M., Pares-Sierra, G., Ley-Lou, F. and García-Esquível, Z. (2009). Comparative growth of Pacific oyster (*Crassostrea gigas*) postlarvae with microfeed and microalgal diets. *Aquacult. Int.* 17, 173-186.

Beiras, R., Pérez-Camacho, A. and Albentosa, M. (1993). Influence of food concentration on energy balance and growth performance of *Venerupis pullastra* seed reared in an open-flow system. *Aquaculture.* 116, 353-365.

Beiras, R., Pérez-Camacho, A. and Albentosa, M. (1994). Influence of temperature on the physiology of growth in *Ruditapes decussatus* (L.) larvae. *J. Shellfish Res.* 13, 77-83.

Blanco, M.J. (2010). Influencia de la temperatura en el desarrollo gonadal y la movilización de reservas de la almeja fina (*Ruditapes decussatus* (L.)) *Ph.D Thesis. University of A Coruña.*

Boeing, P. (1997). Use of spray-dried *Schizochytrium* sp. as a partial algal replacement for juvenile bivalves. *J. Shellfish Res.* 16, 284.

Bonaldo, A., Badiani, A., Testi, S., Corso, G., Mordenti, A.L. and Gatta, P.P. (2005). Use of centrifuged and preserved microalgae for feeding juvenile Manila clam (*Tapes philippinarum*): effects on growth and fatty acid composition. *Ital. J. Anim. Sci.* 4, 375-384.

Breese, W.P. and Robinson, A. (1981). Razor clams, *Siliqua patula* (Dixon): Gonadal development, induced spawning and larval rearing. *Aquaculture.* 22, 27-33.

Brown, M R. and Robert, R. (2002). Preparation and assesment of microalgal concentrates as feeds for larval and juvenile Pacific oyster (*Crassostrea gigas*). *Aquaculture.* 207, 289-309.

Caers, M., Coutteau, P., Lombeida, P. and Sorgeloos, P. (1998). The effect of lipid supplementation on growth and fatty acid composition of *Tapes philippinarum* spat. *Aquaculture.* 162, 287-299.

Caers, M., Coutteau, P., Cure, K., Morales, V., Gajardo, G. and Sorgeloos, P. (1999a). The Chilean scallop *Argopecten purpuratus* (Lamarck, 1819): II. manipulation of the fatty acid composition and lipid content of the eggs via lipid supplementation of the broodstock diet. *Comp. Biochem. Phys. B.* 123, 97-103.

Caers, M., Coutteau, P. and Sorgeloos, P. (1999b). Dietary impact of algal and artificial diets, fed at different rations, on the growth and fatty acid composition of *Tapes phillipinarum* (L.) spat. *Aquaculture.* 170, 307-322.

Caers, M., Coutteau, P. and Sorgeloos, P. (2000a). Impact of starvation and of feeding algal and artificial diets on the lipid content and composition of juvenile oysters (*Crassostrea gigas*) and clams (*Tapes philippinarum*). *Mar. Biol.* 136, 891-899.

Caers, M., Coutteau, P. and Sorgeloos, P. (2000b). Incorporation of different fatty acids, supplied as emulsions or liposomes, in the polar and neutral lipids of *Crassostrea gigas* spat. *Aquaculture.* 186, 157-171.

Calabrese, A. (1969). Individual and combined effects of salinity and temperature on embryos and larvae of the coot clam, *Mulinia lateralis* (Say). *Biol. Bull.* 137, 417-428.

Castagna, M. (2001). Chapter 15. Aquaculture of the Hard Clam, *Mercenaria mercenaria*. In: J. Knauer, M. Castagna (Eds.), *Biology of the Hard Clam*. pp. 675-699, Elsevier, Amsterdam.

Cerviño-Otero, A. (2011). Ciclo reproductivo, cultivo en criadero y en el medio natural de la almeja babosa *Venerupis pullastra* (Montagu, 1803). *Ph.D Thesis. University of Santiago de Compostela.* 288 pp.

Cerviño-Otero, A., Aranda-Burgos, J. A., Nóvoa, S., Ojea, J. and Martínez-Patiño, D. (2011). Primeras experiencias de cultivo en criadero de la almeja rubia, *Venerupis rhomboides* (Pennant, 1777), en Galicia. *XIII Congreso Nacional de Acuicultura.* Castelldefels, Spain.

Chávez-Villalba, J., Barret, J., Mingant, C., Cochard, J.C. and Le Pennec, M. (2002). Autumn conditioning of the oyster *Crassostrea gigas*: a new approach. *Aquaculture.* 210, 171-186.

Chu, F.L., Webb, K.L., Hepworth, D. and Roberts, M. (1982). The acceptability and digestibility of microcapsules by larvae *Crassostrea virginica*. *J. Shellfish Res.* 2, 29-34.

Chu, F.L., Webb, K.L., Hepworth, D. and Casey, B.B. (1987). Metamorphosis of larvae of *Crassostrea virginica* fed microencapsulated diets. *Aquaculture.* 64, 185-197.

Coutteau, P., Hadley, N. H., Manzi, J., and Sorgeloos, P. (1991). Manipulated yeast diets as a partial algal substitute for the nursery culture of the hard clam *Mercenaria mercenaria*. *Spec. Publ. Eur. Aquacult. Soc.* 14, 77-78.

Coutteau, P. and Sorgeloos, P. (1992). The use of algal substitutes and the requiremen, for live algae in the hatchery and nursery rearing of bivalve molluscs: an international survey. *J. Shellfish Res.* 11, 467-476.

Coutteau, P., Curé, K. and Sorgeloos, P. (1994a). Effect of algal ration on feeding and growth of juvenile Manila clam *Tapes philippinarum* (Adams and Reeve). *J. Shellfish Res.* 13, 47-55.

Coutteau, P., Hadley, N.H., Manzi, J.J. and Sorgeloos, P. (1994b). Effect of algal ration and substitution of algae by manipulated yeast diets on the growth of juvenile *Mercenaria mercenaria*. *Aquaculture.* 120, 135-150.

Coutteau, P., Caers, M., Mallet, A., Moore, W., Manzi, J.J. and Sorgeloos, P. (1994c). Effect of lipid supplementation on growth, survival and fatty acid composition of bivalve larvae. In: P. Kestemont, J. Muir, F. Sevilla, P. Williot (Eds.), *Measures for Success.* pp. 213-218, Cemagref Editions.

Cromie, A., Fernández Tajes, J., Gaspar, M., da Costa, F., Martínez-Patiño, D., Méndez Felpeto, J., Nóvoa, S., Ojea, J. and Roberts, D. (2008). Final Report: Towards Integrated Management of *Ensis* Stocks. INTERREG IIIB-project 206-TIMES.

Curatolo, A., Ryan, M.J. and Mercer, J.P. (1993). An evaluation of the performance of Manila clam spat (*Tapes philippinarum*) fed on different rations of spray-dried algae (*Tetraselmis suecica*). *Aquaculture.* 112, 179-186.

da Costa, F., Nóvoa, S., Ojea, J., and Martínez, D. (2005). Acondicionamiento del solénido *Ensis arcuatus* en 2 regímenes de temperatura: relación con las proteínas y el glucógeno. *Actas del X Congreso Nacional de Acuicultura*. pp. 608-609, Gandía, Spain.

da Costa, F., Darriba, S. and Martínez-Patiño, D. (2008). Embryonic and larval development of *Ensis arcuatus* (Jeffreys, 1865) (Bivalvia: Pharidae). *J. Molluscan Stud.* 74, 103-109.

da Costa, F. (2009). Optimización del cultivo en criadero de navaja (*Ensis arcuatus* Jeffreys, 1865), longueirón (*E. siliqua* Linneo, 1758) y longueirón vello (*Solen marginatus* Pennánt, 1777): composición bioquímica y de ácidos grasos en los desarrollos larvarios. *Ph.D Thesis. University of Santiago de Compostela.* 336 pp.

da Costa, F. and Martínez-Patiño, D. (2009). Culture potential of the razor clam *Solen marginatus* (Pennant, 1777). *Aquaculture.* 288, 57-64.

da Costa, F., Nóvoa, S., Ojea, J. and Martínez-Patiño, D. (2009). Investigating the effect of feeding regimens on gametogenesis of broodstock of razor clam *Ensis siliqua* under hatchery conditions. *Aquaculture Europe 2009.* Thronheim, Norway.

da Costa, F., Ojea, J., Nóvoa, S. and Martínez-Patiño, D. (2010a). The effect of microalgal diets on growth and biochemical composition of the razor clam *Solen marginatus* spat. *Aquaculture Europe 2010.* Porto, Portugal.

da Costa, F., Nóvoa, S., Ojea, J. and Martínez-Patiño, D. (2010b). Effects of diets on growth and biochemical composition of spat of the razor clam *Ensis arcuatus* (Jeffreys, 1865). *Aquaculture Europe 2010.* Porto, Portugal.

da Costa, F., Martínez-Patiño, D., Ojea, J. and Nóvoa, S. (2010c). Larval rearing and spat production of the razor clam *Ensis siliqua* (Bivalvia: Pharidae). *J. Shellfish Res.* 29, 347-351.

da Costa, F., Nóvoa, S., Ojea, J. and Martínez-Patiño, D. (2011a). Razor clam culture in Galicia (NW Spain). In: A. Guerra, C. Lodeiros, M.B. Gaspar, F. da Costa (Eds.), *Razor clams: Biology, aquaculture and fisheries.* pp. 181-218, Consellería do Mar, Xunta de Galicia, Santiago de Compostela, Spain.

da Costa, F., Darriba, S., Martínez-Patiño, D. and Guerra, A. (2011b). Culture possibilities of the razor clam *Ensis arcuatus* (Pharidae: Bivalvia). *Aquac. Res.* 42, 1549-1557.

Darriba, S., San Juan, F. and Guerra, A. (2004). Reproductive cycle of the razor clam *Ensis arcuatus* (Jeffreys, 1865) in northwest Spain and its relation to environmental conditions. *J. Exp. Mar. Biol. Ecol.* 311, 101-115.

Davis, H.C. (1958). Survival and growth of clam and oyster larvae at different salinities. *Biol. Bull.* 114, 296-307.

Davis, H.C. and Calabrese, A. (1964). Combined effects of temperature and salinity on development of eggs and growth of larvae of *M. mercenaria* and *C. virginica*. *Fish. B.-NOAA.* 63, 643-655.

Del Norte-Campos, A., Nabuab, F.M., Palla, R.Q. and Burlas, M.R.M. (2010). The Early Development of the Short-Necked Clam, *Paphia undulata* (Born 1778) (Mollusca, Pelecypoda: Veneridae) in the Laboratory. *Science Diliman.* 22, 13-20.

Delaunay, F., Marty, Y., Moal, J. and Samain, J.F. (1993). The effect of monospecific algal diets on growth and fatty acid composition of *Pecten maximus* (L.) larvae. *J. Exp. Mar. Biol. Ecol.* 173, 163-179.

Delgado, M. and Pérez-Camacho, A. (2003). A study of gonadal development in *Ruditapes decussatus* (L.) (Mollusca, Bivalvia), using image analysis techniques: Influence of food ration and energy balance. *J. Shellfish Res.* 22, 435-441.

Delgado, M., Pérez Camacho, A., Labarta, U. and Fernández-Reiriz, M.J. (2004). The role of lipids in the gonadal development of the clam *Ruditapes decussatus* (L.). *Aquaculture.* 241, 395-411.

Delgado, M. and Perez-Camacho, A. (2007a). Comparative study of gonadal development of *Ruditapes philippinarum* (Adams and Reeve) and *Ruditapes decussatus* (L.) (Mollusca: Bivalvia): Influence of temperature. *Sci. Mar.* 71, 471-484.

Delgado, M. and Pérez-Camacho, A. (2007b). Influence of temperature on gonadal development of *Ruditapes philippinarum* (Adams and Reeve, 1850) with special reference to ingested food and energy balance. *Aquaculture.* 264, 398-407.

Delgado, M., Pérez-Larruscain, J. and Gairín, J.I. (2008). First data on larval and spat growth of bivalve *Callista chione* (L. 1758) under culture conditions. *European Aquaculture Society, Special Publication Nº 37.* pp. 187-188, Krakov, Poland.

Desanctis, P.N., Tetrault, K. and Tettelbach, S.T. (2007). Effects of photoperiod manipulation on reproductive condition of the Northern Bay Scallop, *Argopecten irradians irradians* (Lamarck, 1819). *Veliger.* 49, 15-18.

Devauchelle, N. and Mingant, C. (1991). Review of the reproductive physiology of the scallop, *Pecten maximus*, applicable to intensive aquaculture. *Aquat. Living Resour.* 4, 41-51.

Enes, P. and Borges, M.T. (2003). Evaluation of microalgae and industrial cheese whey as diets for *Tapes decussatus* (L.) seed: effects on water quality, growth, survival, condition and filtration rate. *Aquac. Res.* 34, 299-309.

Epifanio, C.E. (1979a). Comparison of yeast and algal diets for bivalve molluscs. *Aquaculture.* 16, 187-192.

Epifanio, C.E. (1979b). Growth in bivalve molluscs: Nutritional effects of two or more species of algae in diets fed to the American oyster *Crassostrea virginica* (Gmelin) and the hard clam *Mercenaria mercenaria* (L.). *Aquaculture.* 18, 1-12.

Espinosa, E.P. and Allam, B. (2006). Comparative growth and survival of juvenile hard clams, *Mercenaria mercenaria*, fed commercially available diets. *Zoo Biol.* 25, 513-525.

Fabioux, C., Huvet, A., Le Souchu, P., Le Pennec, M. and Pouvreau, S. (2005). Temperature and photopeiod drive *Crassostrea gigas* reproductive internal clock. *Aquaculture.* 250, 458-470.

Fernández-Reiriz, M.J., Labarta, U., Albentosa, M. and Pérez-Camacho, A. (2006). Lipid composition of *Ruditapes philippinarum* spat: Effect of ration and diet quality. *Comp. Biochem. Phys. B.* 144, 229-237.

Fernández-Reiriz, M.J., Pérez-Camacho, A., Delgado, M. and Labarta, U. (2007). Dynamics of biochemical components, lipid classes and energy values on gonadal development of *R. philippinarum* associated with the temperature and ingestion rate. *Comp. Biochem. Phys. A.* 147, 1053-1059.

Fernández-Reiriz, M. J., Pérez-Camacho, A., Peteiro, L. G., and Labarta, U. (2011). Growth and kinetics of lipids and fatty acids of the clam *Venerupis pullastra* during larval development and postlarvae. *Aquac. Nutr.* 17, 13-23.

Gallager, S.M., Mann, R. and Sasaki, G.C. (1986). Lipid as an index of growth and viability in three species of bivalve larvae. *Aquaculture.* 56, 81-103.

Gibbons, M.C. and Castagna, M. (1984). Serotonin as an inducer of spawning in six bivalve species. *Aquaculture.* 40, 189-191.

Guo, X.M., Ford, S.E. and Zhang, F.S. (1999). Molluscan aquaculture in China. *J. Shellfish Res.* 18, 19-31.

Hamida, L. (1994). Reproduction de la palourde *Ruditapes decussatus* en milieu natural (sud Tunisie) et en milieu contrôle (écloserie expérimentale): relation avec le système immunitaire. *Ph.D Thesis. Université de Bretagne Occidentale.* 96 pp.

Helm, M.M. and Laing, I. (1987). Preliminary observations on the nutritional value of "Tahiti *Isochrysis*" to bivalve larvae. *Aquaculture.* 62, 281-288.

Helm, M.M., Bourne, N. and Lovatelli, A. (Comp./Ed.) (2004). *Hatchery culture of bivalves. A practical manual.* FAO Fisheries Technical Paper. No. 471. Rome, FAO. 177 pp.

Hendriks, I.E., van Duren, L.A. and Herman, P. M. J. (2003). Effect of dietary polyunsaturated fatty acids on reproductive output and larval growth of bivalves. *J. Exp. Mar. Biol. Ecol.* 296, 199-213.

Hidu, H. and Ukeles, R. (1962). Dried unicellular algae as food for larvae of the hard-shell clam, *Mercenaria mercenaria. Proc. Natl. Shellfish. Ass.* 53, 85-101.

Howard, R.W. and Stanley, D.W. (1999). The tie that binds: Eicosanoids in invertebrate biology. *Ann. Entomol. Soc. Am.* 92, 880-890.

Hurley, D.H. and Walker, R.L. (1996). The effects of larval stocking density on growth, survival, and development of laboratory-reared *Spisula solidissima similis* (Say, 1822). *J. Shellfish Res.* 15, 715-718.

Kent, G.N., Maguire, G.B., Pugh, R. and Duthie, I. (1999). Spawning, settlement, and growth of the New Zealand Venerid *Ruditapes largillierti* (Philippi 1849) in culture. *New Zeal. J. Mar. Fresh. Res.* 33, 55-62.

Knauer, J. and Southgate, P.C. (1997). Growth and fatty acid composition of Pacific oyster (*Crassostrea gigas*) spat fed a microalga and microcapsules containing varying amounts of eicosapentaenoic and docosahexaenoic acid. *J. Shellfish Res.* 16, 447-453.

Laing, I. (1987). The use of artificial diets in rearing bivalve spat. *Aquaculture.* 65, 243-249.

Laing, I., Utting, S.D. and Kilada, R. W. S. (1987). Interactive effect of diet and temperature on the growth of juvenile clams. *J. Exp. Mar. Biol. Ecol.* 113, 23-38.

Laing, I., Child, A.R. and Janke, A. (1990). Nutritional value of dried algae diets for larvae of Manila clam (*Tapes philippinarum*). *J. Mar. Biol. Assoc. U.K.* 70, 1-12.

Laing, I. and Gil Verdugo, C. (1991). Nutritional value of spray-dried *Tetraselmis suecica* for juvenile bivalves. *Aquaculture.* 92, 207-218.

Laing, I. and Millican, P.F. (1991). Dried-algae diets and indoor nursery cultivation of Manila clam juveniles. *Aquaculture.* 95, 75-87.

Laing, I. and Millican, P.F. (1992). Indoor nursery cultivation of juvenile bivalve molluscs using diets of dried algae. *Aquaculture.* 102, 231-243.

Laing, I. and Lopez-Alvarado, J. (1994). Effect of dried algae diets on conditioning and fecundity of Manila clam, *Tapes philippinarum* (Adams and Reeve). *Aquac. Res.* 25, 157-166.

Lamela, T., Otero, A., Arredondo-Vega, B.O., Patiño, M. and Fabregas, J. (1996). Artificial diets as substitutes of microalgae for the culture of clam spat (*Tapes decussatus*). *J. Mar. Biotechnol.* 3, 278-282.

Langdon, C.J. (1989). Preparation and evaluation of protein microcapsules for a marine suspension-feeder, the Pacific oyster *Crassostrea gigas. Mar. Biol.* 102, 217-224.

Langdon, C.J. and Waldock, M.J. (1981). The effect of algal and artificial diets on the growth and fatty acid composition of *Crassostrea gigas* spat. *J. Mar. Biol. Assoc. U.K.* 61, 431-448.

Langdon, C. and Ö-Nal, E. (1999). Replacement of living microalgae with spray-dried diets for the marine mussel *Mytilus galloprovincialis*. *Aquaculture*. 180, 283-294.

Lépez, I., Arriagada, D., Padget, C., Tarifeño, E., Vega, A., Riquelme, R. and Araya, J.M. (2005). Cultivo larval y post larval de *Ensis macha* en ambiente controlado. *Workshop Internacional Estado Actual de la Biología y Cultivo de Bivalvos Enterradores*. Concepción, Chile.

Lépez, I., Véjar, F.E. and Arriagada, D. (2011). Razor clam (*Ensis macha*) culture in Chile. In: D. Guerra, C. Lodeiros, M.B. Gaspar, F. da Costa (Eds.), *Razor clams: Biology, aquaculture and fisheries*. pp. 219-226, Consellería do Mar, Xunta de Galicia, Santiago de Compostela, Spain.

Liu, W., Ma, Y., Hu, S., Miao, G. and Li, J. (2002). Rearing Venus clam seeds, *Cyclina sinensis* (Gmelin), on a commercial scale. *Aquaculture*. 211, 109-114.

Liu, B., Dong, B., Tang, B., Zhang, T. and Xiang, J. (2006). Effect of stocking density on growth, settlement and survival of clam larvae, *Meretrix meretrix*. *Aquaculture*. 258, 344-349.

Loosanoff, V. and Davies, H.C. (1963). Rearing of bivalve larvae. *Adv. Mar. Biol.* 1, 1-136.

Louro, A., De la Roche, J.P., Sánchez, J.L., Silva, A., Martínez, P., Pérez-Paralle, M.L., Martínez, I. and Román, G. (2006). The effect of photoperiod on the conditioning of the black scallop, *Chlamys varia*. I - Gonadal development. *J. Shellfish Res.* 25, 292-293.

Louzán, A. (2008). Ciclo reproductor, composición bioquímica y patología de los progenitores de *Donax trunculus* (Linné, 1758), de un banco natural de Valdoviño (NO Galicia), para la obtención de semilla en criadero. *Msc Thesis. University of Santiago de Compostela*. 67 pp.

Louzán, A., da Costa, F., Nóvoa, S., Ojea, J. and Martínez-Patiño, D. (2010). Efecto de dos tipos de dietas en el desarrollo larvario de la coquina *Donax trunculus* (Linné, 1758). *XIII Foro dos Recursos Mariños e da Acuicultura das Rías Galegas*. O Grove, Spain.

Mallet, A.L. and Carver, C.E. (2009). The effect of photoperiod on the reproductive development of the northern bay scallop, *Argopecten irradians irradians*. *J. Exp. Mar. Biol. Ecol.* 374, 128-133.

Mann, R. (1979). Effect of temperature on growth, physiology, and gametogenesis in the Manila clam *Tapes philippinarum* (Adams and Reeve, 1850). *J. Exp. Mar. Biol. Ecol.* 38, 121-133.

Márquez, L., Ruiz-Azcona, P. and Moreno, O. (2009). Crecimiento y supervivencia de larvas de *Donax trunculus* (Mollusca, Bivalvia) cultivadas a distintas temperaturas. *XII Congreso Nacional de Acuicultura*. Madrid, Spain.

Marshall, R., McKinley, S. and Pearce, C. M. (2010). Effects of nutrition on larval growth and survival in bivalves. *Rev. Aquacult.* 2, 33-55.

Martínez, D., Cerviño-Otero, A., Louzán, A., da Costa, F., Ojea, J. and Nóvoa, S. (2008). Producción de especies de interés marisquero en la planta de cultivos marinos de Ribadeo (CIMA). *II Foro Iberoamericano de los recursos marinos y la acuicultura. FIRMA 2008*. Cumaná, Venezuela.

Martínez, I., Silva, A., Domínguez, L., Álvarez-Fariña, P. and Sánchez, J.L. (2005). Efecto combinado del fotoperíodo y la ración de alimento sobre el acondicionamiento de *Ruditapes decussatus*, Linné 1758, en criadero. *X Congreso Nacional de Acuicultura*. pp. 520-521, Gandía, Spain.

Martínez, I., Pazos, A.J., Louro, A., de la Roche, J.P., Roman, G., Abad, M. and Sanchez, J.L. (2006). The effect of photoperiod on the conditioning of the black scallop, *Chlamys varia* II - Histological study. *J. Shellfish Res.* 25, 294-294.

Matias, D., Joaquim, S., Leitão, A. and Massapina, C. (2009). Effect of geographic origin, temperature and timing of broodstock collection on conditioning, spawning success and larval viability of *Ruditapes decussatus* (Linné, 1758). *Aquacult. Int.* 17, 257-271.

Matias, D., Joaquim, S., Ramos, M., Sobral, P. and Leitão, A. (2011). Biochemical compounds' dynamics during larval development of the carpet-shell clam *Ruditapes decussatus* (Linnaeus, 1758): effects of mono-specific diets and starvation. *Helgoland Mar. Res.* 65, 369-379.

Milke, L.M., Brijelj, V.M. and Parrish, C.C. (2006). Comparison of early life history stages of the bay scallop, *Argopecten irradians*: Effects of microalgal diets on growth and biochemical composition. *Aquacult. Int.* 260, 272-289.

Moreno, O., Palanco, I., Ruiz, P. and Torres, M.A. (2007). Influencia de la temperatura en la evolución del desarrollo gonadal de reproductores de longueirón *Solen marginatus* (Pulteney, 1799). *Actas del XI Congreso Nacional de Acuicultura.* pp. 591-594, Vigo, Spain.

Mouëza, M., Gros, O. and Frenkiel, L. (1999). Embryonic, larval and postlarval development of the tropical clam, *Anomalocardia brasiliana* (Bivalvia, Veneridae). *J. Molluscan Stud.* 65, 73-88.

Muthiah, P., Rodrigo, J.X. and Suja, N. (2002). Larval rearing and spat production of *Marcia opima* (Gmelin). *Aquaculture.* 211, 393-401.

Nell, J.A. and O´Connor, W.A. (1991). The evaluation of fresh algae and stored algal concentrates as a food source for Sydney rock oyster, *Saccostrea comercialis* (Iredale and Roughley), larvae. *Aquaculture.* 99, 277-284.

Nell, J.A., O´Connor, W.A., Heasman, M.P. and Goard, L. . (1994). Hatchery production for the venerid clam *Katelysia rhytiphora* (Lame) and the Sydney cockle *Anadara trapezia* (Deshayes). *Aquaculture.* 119, 149-156.

Newell, R.C. and Branch, G.M. (1980). The influence of temperature on the maintenance of metabolic energy balance in marine invertebrates. *Adv. Mar. Biol.* 17, 329-396.

Nóvoa, S., Martínez, D., Ojea, J., Soudant, P., Samain, J.F., Moal, J. and Rodríguez, J.L. (2002). Ingestion, digestion, and assimilation of gelatin-acacia micrapsules incorporating deuterium-labeled arachidonic acid by larvae of the clam *Venerupis pullastra*. *J. Shellfish Res.* 21, 649-658.

Numaguchi, K. (1998). Preliminary experiments on the influence of water temperature, salinity and air exposure on the mortality of Manila clam larvae. *Aquacult. Int.* 6, 77-81.

Ö-Nal, E., Langdon, C. and Ö-Nal, U. (2005). The evaluation of spray-dried microalgae in diets for juvenileess Manila clam, *Tapes philippinarum*. *J. Shellfish Res.* 24, 1061-1065.

Ojea, J., Pazos, A.J., Martínez, D., Nóvoa, S., García-Martínez, P., Sánchez, J.L. and Abad, M. (2008). Effects of temperature regime on broodstock conditioning of *Ruditapes decussatus*. *J. Shellfish Res.* 27, 1093-1100.

Osada, M., Nishikawa, M. and Nomura, T. (1989). Involvement of prostaglandins in the spawning of the scallop, *Patinopecten yessoensis*. *Comp. Biochem. Physiol. C.* 94, 595-601.

Palacios, E., Racotta, I.S., Kraffe, E., Marty, Y., Moal, J. and Samain, J.F. (2005). Lipid composition of the giant lion's-paw scallop (*Nodipecten subnodosus*) in relation to gametogenesis: I. Fatty acids. *Aquaculture*. 250, 270-282.

Pazos, A.J., Silva, A., Sordo, A., Abad, M. and Sánchez, J.L. (2003). Influencia del acondicionamiento sobre el acondicionamiento de la almeja fina (*Ruditapes decussatus*), Linné 1787. *IX Congreso Nacional de Acuicultura*. pp. 293-295, Cádiz, Spain.

Pérez-Camacho, A. and Román, G. (1987). La reproducción en los moluscos bivalvos. In: J. Espinosa, U. Labarta (Eds.), *Reproducción en la Acuicultura*. pp. 133-184, Plan de Formación de Técnicos Superiores en Acuicultura (FEUGA), Madrid, Spain.

Pérez-Camacho, A., Román Cabello, G. and Torre, M. (1977). Experiencias de cultivo de larvas de tres especies de moluscos bivalvos: *Venerupis pullastra* (Montagu), *Venerupis decussata* (Linnaeus) y *Ostrea edulis* (Linnaeus). *Bol. Inst. Esp. Oceanogr.* 235, 10-61.

Pérez-Camacho, A., Albentosa, M., Fernández-Reiriz, M.J. and Labarta, U. (1998). Effect of microalgal and inert (cornmeal and cornstarch) diets on growth performance and biochemical composition of *Ruditapes decussatus* seed. *Aquaculture*. 160, 89-102.

Pérez-Camacho, A., Salinas, J.M., Fuertes, C. and Delgado, M. (2002). Utilización de biotransformados de *Laminaria saccharina* (L.) Lamouroux en la alimentación de la semilla de la almeja *Ruditapes decussatus* (L., 1758). *Bol. Inst. Esp. Oceanogr.* 18, 321-328.

Pérez-Camacho, A., Delgado, M., Fernández-Reiriz, M.J. and Labarta, U. (2003). Energy balance, gonad development and biochemical composition in the clam *Ruditapes decussatus*. *Mar. Ecol. Progr. Ser.* 258, 133-145.

Pérez-Camacho, A., Salinas, J.M., Fuertes, C. and Delgado, M. (2004). Preparation of Single Cell Detritus from *Laminaria saccharina*; as a Hatchery Diet for Bivalve Mollusks. *Mar. Biotechnol.* 6, 642-649.

Pérez-Camacho, A., Salinas, J.M., Delgado, M. and Fuertes, C. (2007). Use of single cell detritus (SCD) produced from *Laminaria saccharina* in the feeding of the clam *Ruditapes decussatus* (Linnaeus, 1758). *Aquaculture*. 266, 211-218.

Pernet, F., Bricelj, V.M. and Parrish, C.C. (2005). Effect of varying dietary levels of omega 6 polyunsaturated fatty acids during the early ontogeny of the sea scallop, *Placopecten magellanicus*. *J. Exp. Mar. Biol. Ecol.* 327, 115-133.

Ponis, E., Robert, R. and Parisi, G. (2003a). Nutritional value of fresh and concentrated algal diets for larval and juvenile Pacific oysters (*Crassostrea gigas*). *Aquaculture*. 221, 491-505.

Ponis, E., Robert, R., Parisi, G. and Tredici, M. (2003b). Assessment of the performance of Pacific oyster (*Crassostrea gigas*) larvae fed with fresh and preserved *Pavlova lutheri* concentrates. *Aquacult. Int.* 11, 69-79.

Raghavan, G. and Gopinathan, C.P. (2008). Effects of diet, stocking density and environmental factors on growth, survival and metamorphosis of clam, *Paphia malabarica* (Chemnitz) larvae. *Aquac. Res.* 39, 928-933.

Robert, R., Parisi, G., Rodolfi, L., Poli, B.M. and Tredici, M. R. (2001). Use of fresh and preserved *Tetraselmis suecica* for feeding *Crassostrea gigas* larvae. *Aquaculture*. 192, 333-346.

Robinson, A.M. and Bresse, W.P. (1984). Gonadal development and hatchery rearing techniques for the Manila clam *Tapes philippinarum* (Adams and Reeve). *J. Shellfish Res.* 4, 161-163.

Ruiz-Azcona, P., Rodríguez-Sierra, R. and Martín, J.B. (1996). Culture of coquina clam, *Donax trunculus*, larvae. *Aquaculture.* 139, 151-155.

Sánchez-Lazo, C., Martínez-Pita, I. and Moreno, O. (2009). Ensayo de dietas para la alimentación larvaria de la coquina (*Donax trunculus* Linne, 1758). *XII Foro dos Recursos Mariños e da Acuicultura das Rías Galegas.* O Grove, Spain.

Saout, C., Quere, C., Donval, A., Paulet, Y.M. and Samain, J. F. (1999). An experimental study of the combined effects of temperature and photoperiod on reproductive physiology of *Pecten maximus* from the Bay of Brest (France). *Aquaculture.* 172, 301-314.

Soudant, P., Le Coz, J.-R., Marty, Y., Moal, J., Robert, R. and Samain, J.-F. (1998). Incorporation of microalgae sterols by scallop *Pecten maximus* (L.) larvae. *Comp. Biochem. Physiol. A.* 119, 451-457.

Tang, B., Baozhong, L., Wang, G., Zhang, T. and Xiang, J. (2006). Effects of various algal diets and starvation on larval growth and survival of *Meretrix meretrix. Aquaculture.* 254, 526-533.

Uchida, M., Kanematsu, M. and Miyoshi, T. (2010). Growth promotion of the juvenile clam, *Ruditapes philippinarum*, on sugars supplemented to the rearing water. *Aquaculture.* 302, 243-247.

Utting, S.D. and Spencer, B.E. (1991). *The hatchery culture of bivalve mollusc larvae and juveniles.* Ministry of Agriculture, Fisheries and Food (UK). Laboratory Leaftlet No. 68. 32 pp.

Utting, S.D. and Millican, P.F. (1997). Techniques for the hatchery conditioning of bivalve broodstock and the subsequent effect on egg quality and larval viability. *Aquaculture.* 155, 45-54.

Utting, S.D. and Millican, P.F. (1998). The role of diet in hatchery conditioning of *Pecten maximus* L.: a review. *Aquaculture.* 165, 167-178.

Webb, K.I. and Chu, F.E. (1982). Phytoplankton as a food source for bivalve larvae. In: G. D. Pruder, C. Langdon, D. Conklin (Eds.), *Proceedings of the second international conference on aquaculture nutrition; biochemical and physiological approaches to shellfish nutrition.* pp. 272-290, Louisiana State University, Baton Rouge, Louisiana, US.

Yan, X., Zhang, G. and Yang, F. (2006). Effects of diet, stocking density, and environmental factors on growth, survival, and metamorphosis of Manila clam *Ruditapes philippinarum* larvae. *Aquaculture.* 253, 350-358.

Zhang, G. and Yan, X. (2006). A new three-phase culture method for Manila clam, *Ruditapes philippinarum*, farming in northern China. *Aquaculture.* 258, 452-461.

In: Clam Fisheries and Aquaculture
Editor: Fiz da Costa González
ISBN: 978-1-62257-518-3

Chapter 10

CLAM PRODUCTION AND CULTIVATION IN GALICIA (NW SPAIN): THE ROLE OF HATCHERIES

A. Guerra Díaz[*]
Centro de Investigacións Mariñas. Consellería do Mar.
Vilanova de Arousa (Pontevedra), Spain

ABSTRACT

Over 90% of clams produced commercially in Spain are harvested on the Galician coast. This production is based mainly on recruitment in the natural environment and, to a lesser extent, clam seed cultivated in hatcheries. The main species produced are: the Manila clam *Venerupis philippinarum*, the pullet carpet shell *Venerupis corrugata* (=*V. pullastra*) and the grooved carpet shell (*Venerupis decussata*). It is estimated that 30,000 tonnes/year of clams are consumed in Spain, of which about 5,000 tonnes/year are produced in Galicia, the rest being imported from other countries, mainly European ones. Basing ourselves on data from production areas on the Galician coastline, potential commercial clam production is assessed. The development and status of mollusc hatcheries in Galicia, methods used for intermediate culture, nurseries as ancillary facilities, essential to achieve a seed size suitable for seeding, are all analysed. We also assess the use of effluent from fish farms on land as a model of multi-trophic aquaculture for intermediate culture of clam spat.

10.1. INTRODUCTION

Bivalve molluscs dominate aquaculture production in Spain, ranging over the years between 80% and 85% of all species of marine and continental origin. Mussels accounts for 97% of the production of shellfish in Galicia and represent the bulk of the volume of Spanish aquaculture.

[*] Corresponding author: A. Guerra Díaz. E-mail address: aguerradiaz@gmail.com.

The rest of the production (0.02% of the total in aquaculture, and 3% of molluscs) is of clams. 90% of the clams produced in Spain are harvested on the Galician coast. Current production is in the following order of species: the Manila clam *Venerupis philippinarum*, the pullet carpet shell *Venerupis corrugata*, the grooved carpet shell *Venerupis decussata,* and, less so, the locally produced *Polititapes virgineus* (=*Venerupis rhomboides*).

Approximately 2-3% of shellfish production comes from spat obtained from hatcheries. This indicates that there is a huge potential to gradually replace shortfalls in seed recruited in the wild, so shellfish farms must compete and offer improvements in the seed, bettering in quality and performance that obtained in the natural environment.

However, although recruitment of most species of commercial bivalves is in decline, the price of the seed has remained constant over the years (Gutiérrez, 2010).

The demand for and commercial success of this seed are linked in with its proper cultivation up to the moment of sale, a stage usually in the hands of the traditional farmer, who is often ignorant of the characteristics of the spat proceeding from the hatchery or fails to handle it properly, in some cases blaming the origin of the seed for failures that are actually the result of poor culture practices. Hatcheries are designed to produce large quantities of small spat at low costs, but the management of this seed is not easy for professional growers. Nurseries ensure the transition from tightly controlled production facilities (hatcheries), and the natural environment, and must meet the right biological and economic requirements to allow costs to be reduced and production increased. There is a notorious lack of seed clam of the right size for seeding, to recover currently unproductive areas, or to enhance production in natural beds. Despite the economic and social importance of shellfish culture and production in Galicia, over recent years seed farms have not developed and expanded in accordance with existing market potential.

10.2. Commercial Clam Production

Clam production in Spain, according to Eurostat statistics, has shown a steady decline in recent years (1997-2008), with production down from 9,129 tonnes in the first of these years, to 3,158 tonnes, reported in 2007. There was a sharp decline during the period 2002 to 2005, with values close to 2,000 tonnes (according to JACUMAR), or even about 1,500 tonnes (according to Eurostat (Pérez-Camacho, 2007). The dispersal of mussel populations, adverse natural phenomena causing mass mortality, oscillations in quantity, as well as the existence of unofficial marketing channels, and so on, have meant that the bulk of shellfish harvested has not followed the established channels for registration and control.

There seems to be no doubt that catches are higher than those actually recorded; although statistics allow us to observe developments and trends in production, and it is on the Galician coast where practically all catches take place.

10.3. Production Model

In Galicia there are sixty-two fishermen's associations that bring together groups of professional shellfish gatherers, who jointly manage most areas of operations. The work of clam cultivation and harvesting is done on foot in the intertidal zone at low tide, or alternatively in the subtidal zone, working from small boats with rakes (Figure 1).

The harvesting of clams and exploitation of natural beds in Galicia is regulated by the administration by means of several norms and, in general, contemplates three: free shellfish-gathering, authorization and/or shellfish farming, and growing facilities The most traditional shellfishing is carried out in intertidal zones -*shellfish harvesting on foot*- and is performed by around 4,000 people, of whom about 85% are women. Extraction in subtidal areas is performed from fishing vessels –*shellfish-harvesting afloat*- and involves some 3,500 people working from around 3,000 vessels (De Coo, 2008). Chapter 11 contains a detailed analysis of the management system of shellfish gathering in Galicia.

Official data on the collection and sale of clam at first sale, its importance and distribution, depending on the mode of harvesting (shellfish-gathering on foot or from vessels), reveal the origin and area of distribution of each species.

*Polititapes virgineus*and *V*are subtidal species, most of their collection being carried out by fishermen from boats, whereas *Venerupis* and *Venerupis* are predominant in shellfish-gathering performed on foot (Table 1).

Overall, catches of the species have remained stable over the last five years, except for *V.* whose production almost tripled during the period 2005 to 2011, rising from 848 tonnes in 2005 to more than 2,000 tonnes in 2011.

10.4. Production Potential of the Galician Coast

Arnaiz (2005), in his study of shellfish harvesting in Galicia, assesses the productive potential of commercial shellfish exploited in natural beds on the Galician coast. asing himself on a series of actual hard facts (sales at auction, bed surface area, year of optimum production, maximum values of mean densities extracted in terms of units per m^2, etc.), Arnaiz makes an estimate of the potential of production expected and of the bio-mass that is not produced. 510 natural beds, occupying more than 81 million m^2 of littoral area, were analyzed. On this basis, and in relation to mussel species with the highest levels of production (*V. philippinarum, V. decussatus* and *V. corrugata*), data are extrapolated calculating actual first sales (percentage harvested) over the total potential expected (t), it being inferred that there is a huge untapped potential in the production of clams in Galicia (Table 2).

Figure 1. Common clam-fishing methods on the Galician coast. Image A, from aboard vessels, and B, from land at low tide.

Table 1. Summary of data recording clam catches and first sales in Galicia during the period 2005 to 2011. This is broken down into the two forms of capture (shellfish-gathering on foot and from boats). The total also includes catches where the mode of collection is unknown. We indicate the value for each species, this being the sum of the annual totals in the series (2005-2011), and the values of the first (2005) and last (2011) yearly totals within this series

Species	Years	Total		Shellfishing on foot		Shellfishing from vessels	
		Amount (kg)	Value (€)	Amount (kg)	Value (€)	Amount (kg)	Value (€)
V. corrugata	2005-2011	7,402,737	98,623,745	835,454	10,247,395	5,324,672	71,906,322
	2005	1,097,165	15,255,031	154,670	1,983,529	633,641	8,959,472
	2011	1,059,861	13,013,892	123,796	1,427,583	821,094	10,351,470
V. decussata	2005-2011	4,895,723	111,181,259	2,748,010	63,227,131	1,586,875	35,168,972
	2005	553,031	16,410,406	342,007	9,594,228	138,241	4,441,208
	2011	705,267	13,479,644	382,797	7,803,440	241,123	4,189,072
P. virgineus	2005-2011	4,439,990	33,170,395	7,931	55,201	3,842,779	28,720,764
	2005	364,129	3,101,016	856,856	5,786	283,625	2,387,239
	2011	420,285	3,477,678	12	75	412,552	3,414,450
V. philippinarum	2005-2011	12,039,856	85,653,571	6,720,142	50,026,896	2,725,607	18,727,646
	2005	848,377	6,363,576	340,271	2,862,968	221,396	1,642,634
	2011	2,018,969	12,136,555	1,097,646	6,828,412	504,154	2,918,993

Source: www.pescadegalicia.com.

Table 2. Data taken from Arnaiz (2005) which, based on extraction data, calculates clam production potential in natural banks on the Galician coast

Species	Density of sample (indiv./m 2)	Untapped potencial (t)	Potential extracted (%)
V. decussata	26.03	11,087	6.9
V. corrugata	34.58	9,698	21.1
V. philippinarum	11.58	3,372	16.5

The data from that study (1998-2004) are based on first sales in fishmarkets, which are usually lower than actual sales. Nowadays, first sales of *V.* and *V.* are comparable to those cited in the study, and, in the case of *V.*, increased to over 2,000 tonnes (www.pescagalicia.com).

10.5. A MODEL OF CLAM CULTURE: THE "CARRIL CULTIVATION PARKS" (RIA DE AROUSA)

In Galicia there are a total of 1,134 "cultivation parks" leased officially, mainly of *V.*, *V. V.* of which 90% are the so-called "Carril parks", located at the mouth of the river Ulla, at the end of the Ría de Arousa. Farmers sow their parks at high densities, compared with those in open shellfishing areas. Exploitation of these areas is private and exclusive.

About 200 years ago these areas were devoted to the cultivation of oysters and mussels, but in time they were adapted to the cultivation of *V. V.* In the nineteen-nineties *V.* (Manila clam) was introduced, as it is resistant to changes in salinity due to the influence of the river during rainy periods. The parks occupy about one million m^2, of which area 90% is in operation, and the bulk of the individual cultivation plots are no larger than 500 m^2 each. Seed for culture is obtained by natural spatfalls or from nurseries, the amount, in either case, depending on the intensity of natural recruitment.

Capacity and production vary considerably, depending on location and management: in a subtidal park these can reach 2,000 units/m^2, while in intertidal areas they can range from 700 to 1000 units m^2. A properly managed subtidal park produces 15-20 kg/m^2/year, including all species produced simultaneously. Average production of all parks, regardless of their location, ranges between 5-8 kg/m^2 (Harguindey, 2010). Drawing on recorded sales data from the Carril Fishermen's Association, total production is about 400 tonnes, providing an income of 22 million euros, and affording direct employment to about 1,000 people.

10.6. MANAGEMENT OF CLAM PRODUCTION AREAS

roduction data outlined in the previous sections can be improved not only by extraction-oriented techniques, improvement of substratum, cleaning, and reseeding, but also by improving management and particularly by restocking with seed obtained from nurseries. This means encouraging shellfish collectors to become growers, thereby increasing both their production and professionalism.Arnaiz (2005) calls for the promotion of firms engaged in commercial bivalve mollusc seed production.

Figure 2. The "Carril cultivation parks" (Ría de Arousa), at high tide (left) and low tide (right), respectively.

10.7. Hatcheries, a Complement to Natural Production

Guerrero (1995, 2010) analyses the evolution of hatcheries and their role in the development of shellfish production. This began as a commercial activity in Galicia in the nineteen-seventies, spreading to other areas of Spanish territory. Although the basic technology of hatchery production has been maintained over the years, fresh conceptions of hatcheries have emerged, mainly affecting the automation of systems.

Guerrero (2010) points out the most outstanding innovations in recent years, which have been developed particularly by Seasalter Shellfish Ltd., and which can be summarized in the following three aspects: the use of pasteurizers for microalgae production, an up-flow system in hatcheries and seedbeds, and the use of open circuits in all sections and production lines (microalgae, larvae and post-larvae). In Chapter 9 of this book a comprehensive review is made of both the current status and degree of development of commercial clam species achieved in hatcheries. Hatcheries are designed to produce large quantities of small seed at low cost, but management of this seed (1-2 mm) is not easy for professional growers. Only after an additional period of cultivation in the nursery will the seed reach a suitable size for planting out in a natural environment. The success of the nursery phase is a prerequisite for the controlled development of reproduction in bivalve molluscs. However, the future of nurseries depends on their ability to produce seed from 10 to 15 mm at competitive prices, that is, lower than those of today. This requires mechanised nurseries to be available which are easy to operate and able to regularly produce 20-30 tonnes/year of seed ready for sowing (Gutierrez, 2010).

10.8. Mini-Hatcheries for Molluscs, a Model for Technology Transfer

Mini-hatcheries are modular units for the production of commercial clam seed for planting and cultivation, or restocking in cultivation areas managed by associations of shellfish-gatherers on the coasts of Galicia. aim is to encourage shellfish-collectors to become shellfish-cultivators, and they provide some of the seed to encourage the transformation of production and management models. The purpose and structure of these facilities, the specific processes and operational techniques followed are all describedin Guerra (2002a, b, 2010a), Pouso et al. and Lastres et al.. The operational staff in these hatcheries are basically advanced aquaculture technicians from the Galician Aquaculture Training Institute (IGAFA), who, armed with a two-year theoretical and training course, are able to wholly operate and manage a small-scale shellfish hatchery (Figure 3). These units follow the general guidelines and operational protocols concerning hatcheries. They were originally seen in Wilson (1981), and are based on versatile, modular facilities (greenhouses), under lightweight covers, making up the different sections of traditional hatcheries (Figure 4). Low-cost, and simple to install, the structures operate primarily during molluscs' natural spawning periods. They house the basic conventional components of traditional hatcheries: tanks for keeping progenitors for short periods of conditioning, in order to cultivate the larvae, and systems for continuous microalgal production, which make use of natural daylight, with a capacity for daily production of 800 litres and a density of 8,000,000 cells/ml (based on the system designed by Seasalter Shellfish Ltd., UK).

The activity of these facilities is seasonal, comprising molluscs' natural spawning periods. The essential feature of their operations is that they take advantage of climatic conditions in the area where they are located. Environmental conditions regulate the activity of the hatcheries to a large extent, so they do not usually operate in periods of maximum and minimum temperatures (some summer and winter months).

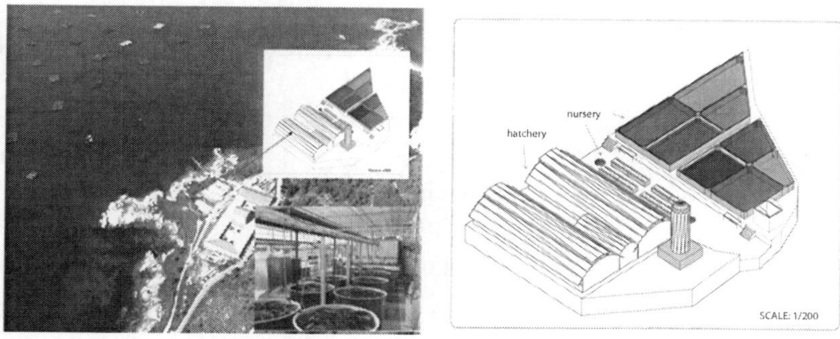

Figure 3. Left, floor plan of the Galician Aquaculture Training Institute (IGAFA) on the island of Arousa. Right, diagram of a mini-hatchery, according to the initial design of its exterior.

Figure 4. Interior view of mini-hatchery facilities.

Table 3. Clam seed supplied for sowing, in the years of hatchery and associated nursery activity in Camariñas(Source: CIMA activity reports, www.cimacoron.org)

Year	Clam species (units)			Size
	V. corrugata	*V. decussata*	*V. philippinarum*	
2011	4,664,760	184,000	6,050,840	>5 mm <12 mm
2010	4,446,957	116,252	4,036,054	>5 mm <12 mm
2009	1,426,582	-	5,878,331	>10 mm
2008	-	543,120	1,839,463	>10 mm
2007	-	7,023,290	2,018,507	>2 mm <5 mm
2006	-	355,793	1,788,982	>2 mm <4 mm

Operational protocols are based on the use of semi-continuous open circuits. In general, cultures grown in these circumstances have both a better growth rate and improved viability in the larval period. This will both balance and stabilize changes caused by the outside temperature. In these circumstances, both thermal efficiency and economy of use will improve, while at the same time bacteriological problems will be reduced (Bayes, 1987).

They are designed for an annual production of 40 million units/year of clam seed with 1 mm mesh sieves (equivalent to 40 kg of seed biomass). This facility has a floating nursery with forced reverse flow -upwelling or "flupsy"- (described in a section below), with an annual production capacity of 12-15 million units, reaching the target size of 10-12 mm, which is appropriate for the concept of a "mini-hatchery" (Lastres et al., 2011b).

The seed is shared out among the shellfishing associations managing the production areas (No Couto, 2010).The aim is that seed will have a size of >10mm, so that shellfish-collectors who receive it can plant and grow it without running any great risks. Seed produced in a mini-hatchery over several years is shown in table 3. It is estimated that over 50% of the seed sown will reach commercial size, given proper management.

10.9. Nurseries

Hatcheries are prepared to produce large quantities of small seeds (1-2 mm), but their capacity is very limited, and operating and maintenance costs, as well as risks, gradually increase if the seed remains on site in order to reach a larger size. As the seed grows, the biomass multiplies exponentially and needs a great deal of nutrients. Therefore it is necessary to find systems within a natural environment, efficient and low-cost, which will allow us to care for this seed until it reaches the right size for planting in substratum (Figure 5).

Nurseries associated with or complementary to hatcheries are the essential basis for success; the availability of hatchery seed, of sufficient quantity and quality, at the time necessary for each fattening strategy, is a fundamental prerequisite for the future of stable production of commercial clams (Lucas, 1981; Guerra, 1984, Bayes, 1987, Perez-Camacho, 2007; Gutiérrez, 2010). There are no fixed rules as to the form, type and capacity of nurseries.

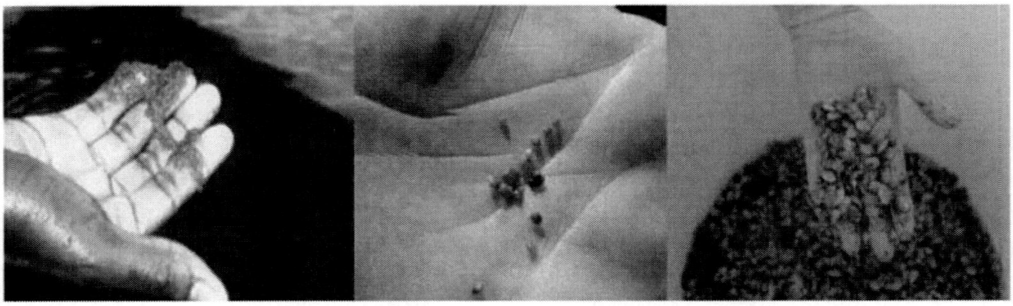

Figure 5. The sizes of clam seed that can be kept in nurseries. Left: 1 mm sieve (T1, 1.5 to 2.0 mm, 0.6 mg/u), suitable for systems of reversed forced flow. Centre: 2 mm sieve (T2, 3 to 5 mm, 200 mg/u), more suitable for passive tidal flow. Right: 8 mm sieve (T8, 10 to 12 mm, 200 mg/u) which is the minimum recommended size for sowing.

The basic models and infrastructures of this type of facilities are described in detail in Claus et al. Guerra (2010 b, c) also discusses those facilities currently operating in Galicia as well as others whose main market is the sale of seed to Galician mollusc-farmers (fishermen's associations and privately owned seed beds).

Seed producers have opted for 0.7 or 1 mm size sieves. A 2 mm sieve, which limits production, is too large for a hatchery and too small for cultivators that have to grow to market size.

To overcome this limitation and to achieve seed of the above-mentioned sowing size of 10 mm in length, in Spain and especially Galicia, two methods of intermediate culture are practised: with shellfishing equipment in passive tidal flow (raised in the intertidal zone or suspended from floating rafts or "bateas"), and nurseries based on forced circulation systems, in marshy areas, or floating. The advantages and disadvantages of different types of seed beds are described below.

10.10. Equipment for Intermediate Culture in Passive Tidal Flow

In the equipment for the pre-fattening (intermediate culture) of seed, hanging from rafts or "pochones" in the intertidal zone, the supply of suspended nutrition depends on tidal currents; the nursery receives the nutrition produced within a much greater area than that of the nursery itself, primary production being imported from adjacent areas (Figure 6). Cervino-Otero (2011) describes in detail the variations of culture and the equipment used in Galicia in the intermediate culture phase of seed clams; in particular, Cerviño-Eiroa (2010) and Cervino-Eiroa et al. analysed different gear, hanging on rafts (bateas), used for intermediate culture.

These authors are positive about a system based on a steel cage structure, suspended from the rafts, which includes bags of intermediate culture substances (40x19 cm, 2 and 3 mm mesh), stacked in layers and varying in number. Intermediate culture starts with 150 g/bag, and seed of >2 mm and displays clam survival close to 95% at the end of the intermediate culture stage. Many authors have analysed the biological efficiency and economic performance of a wide variety of intermediate culture gear types, in seeds of various commercial species, in different growing areas.

Results vary greatly, although most authors agree on the high operational costs of linear flow systems, and the difficulty of handling seed smaller than 2-3 mm. Cervino-Otero (2011) points out that in numerous studies carried out on intertidal culture in the intermediate culture phase of clam seed (*V. corrugata*), growth rates are lower and mortality higher compared with suspension culture from rafts. She also notes that suspended pre-fattened seeds present less parasitism than their counterparts in intertidal areas.

The author analyses in detail the factors involved, which intervene in each type of culture, affecting the stability of the environmental conditions in culture in rafts as a key factor in obtaining optimum outcomes. In this sense, she indicates that, while the total cultivation cycle of *V.* from larval stage, nursery and growing stages until commercial size is achieved in 26 months in the intertidal zone, this is achieved in just 18 months in suspended culture.

Figure 6. Some of the gear used in Galicia for intermediate culture of seed in passive linear flow caused by tidal currents. Suspended from rafts: 1A and 1B, multi-tiered trays; 2A and 2B in bags stacked in metal frames.In an intertidal zone: 3, in mesh bags on a metal frame "pochones".

Figure 7. A: High productivity catamaran nursery within port facilities by Remagro SA (Pontevedra, Galicia). B: Land-based nursery by TINAMENOR, SA (Santander), having two 6- and 10-hectare marine lagoon areas, mutually interchangeable via the nursery and able to house around 300 million units.

10.11. Nurseries with Forced Water Circulation

acilities are generally floating ones, in which the seed is arranged inside large containers with mesh bottoms, and in which a continuous upward flow of water is generated through the mass of seed. This flow, which provides feed in suspension and carries away faeces, is generated by pumping, propellers, paddle wheels, etc. (Figure 7).

With these systems we seek to pre-fatten, outdoors, seed smaller than a 2 mm sieve size, in large containers, until it grows larger than 10-12 mm, a size at which it can be planted in cultivation areas and natural beds without any great risks of predation or of its being dragged away by marine currents.

Figure 8. Floating semi-industrial nursery system located in Camariñas port area (A Coruña, Spain). From left to right: Platform, details of the containers, and detail of the seed at the beginning and end of the intermediate culture process.

These facilities are designed for outputs of over 100 million units/year of seed, located in calm areas, protected, easily accessible and able to receive electric power (sea-marshes, marine lagoons, harbor areas, etc.).

Semi-industrial in nature, the nursery associated with the Camariñas mini-hatchery (Figure 8) is based on the exploitation of the abundant phytoplankton that exist naturally in Galician estuaries, particularly in port areas. It is a floating platform (7x6 m) that holds up to 40 containers with seeds of varying sizes, with air-powered reversed forced flow, as in the system described by Williams (1981). The seed starts at the size it leaves the hatchery, that is, 1.5 mm, being distributed for sowing and cultivation in production areas when it measures up to 12-15 mm.

The management of this system is simple, and operational costs are low, it being designed to handle about 12-15 million clams/year, that is, approximately 3,000 kgs of seed for sowing.

Lopez et al. indicate a survival rate in *V.* seed of the order of between 70 and 78% in these intermediate culture systems. Similarly, Guerra (2010b, c) suggests the possibility of initiating the forced flow intermediate culture in seed measuring 1 mm and upwards, indicating recapture rates averaging above 60% in *V.* and over 80% in *V.* for seed sizes between 1 mm and 9 mm mesh. The period the seed remains in the nursery to reach the target planting size (12 mm), varies, depending on the species and on when the intermediate culture process begins. In *V.* intermediate culture usually takes between 50-60 days between February and July, and in *R.* between 40 and50 days when intermediate culture takes place between May and October, when seawater temperatures are at their highest.

10.12. Intermediate Culture of Clam Seed Using Land-Based Fish Farm Effluent

Multitrophic aquaculture is based on recycling the waste of one species so that it becomes food for another, thus integrating fish-farming and shellfish cultivation. This is one alternative aimed at obtaining mollusc biomass from effluents of these fish farms. By means of the filtration of the finest particles, the impact caused by the emission of organic matter to the marine environment is decreased, promoting environmental bioremediation. In a recent publication, Guerrero and Cremades (in press), reproduce the observations of many researchers and managers on the beneficial aspects of integrated aquaculture, in order to promote the development of multi-trophic aquaculture in Galicia.

The productive process in fish farming creates great amounts of mud, most of which remains settled in settling basins. Its composition is determined according to several parameters, such as the non-consumed scraps of feed and excrement, or other organic droppings from fish. In Galicia, these land hatcheries produce around 7,000 tonnes of turbot, which produce a stable stock of 5,000 tonnes of fish biomassThis production of turbot, which accounts for 84% of the total in Spain, requires about 8,000 tonnes/year of feed, that is, about 22 tonnes/day, of which about 3-5 tonnes/day are poured out as effluent. Most hatcheries with open circuits pour out around 75,000 m^3/h, with an amount of organic particulate matter ranging between 2 and 3 ppm and mainly made up of fodder scraps and fish excrement, which are protein-based. nitiatives involving integrated bivalve mollusc-fish culture, put into practice in the grow-out of clam seeds obtained in hatcheries, are an alternative which make it possible for seed farms to work together with fish producers. This makes it possible to reduce the amount of organic effluent and its charge, as clam biomass will act as bio-filters. Previous tests (Guerrero and Gonzalez, 1991; Jara-Jara, 1995, Miranda et al., 2006, Guerra et al., 2007, 2010; Guerrero et al., 2011) show the suitability of seeds for the sowing of clam obtained with this practice. With reference to the tests carried out over several years in a turbot fish farm located in O Grove in the Ría de Arousa (Insuiña SL), which is the oldest farm producing turbot in Galicia (about 150 tonnes/year, with an effluent flow rate of between 1,000 and 1,500 m^3/h; part of this flow (60 m^3/h) is filtered with a rotary filter (<200 microns) and is diverted to longitudinal pools (7x10x0.5 m), in which the water flows inverted (up-flow) through bivalve mollusc seed containers (Figure 9).

Figure 9. Intermediate clam seed culture unit, deriving part of the effluent from a turbot farm (Insuiña SL). A: Intermediate culture area, with channels and containers holding the seed, to which comes water diverted and pre-filtered from the effluent. B: Detail of the containers with the seed. C: Effluent discharged outside.

V. corrugata is the species which presents the best growth throughout the year; in second place *V.* from 18° C water temperature of the effluent, with risks of specific mortalities when the temperature regularly exceeds 22°C. The species *V.* responds irregularly to pre-fattening in the effluent, with occasional mortalities and size dispersion throughout the process. At certain times of the year, when there is an intense load of fish culture on the farm together with heavy rainfall, pH values drop, and from 7.2-7.4 and up, the calcium carbonate of the shells dissolves, episodes of mortality over 30% in some cases appearing, higher in *V.* than in other species.

In short, with the right measures and fine adjustments an efficient system for intermediate seed clam culture, which is at the same time user-friendly and efficient, is validated. inimum seed size should not be less than 3 mm. The work protocol of the fish-farm (cleaning, treatment, changes, etc.) should follow work procedures in the nursery.

10.13. Culture and Management in the Natural Environment: Restocking with Hatchery Seed

The fattening phase begins with the sowing of seed, of an appropriate size to maximize the result of the growing when legal market size is reached: 38 mm for *V.* and 40 mm for *V.* and *V. philippinarum.*

Protocols and manuals of good practice for growing clams in substratum are well-known and widely used, although each area and species possess some specific requirements essential to the commercial success of the crop. As a general rule, and as indicated by Cerviño-Otero (2011), substrate in growing areas should be conditioned prior to planting in order to ease the planting of the seed and eliminate predators. To optimize the growth and survival of seed, it is important to follow the usual procedures of culture management: removal of macro-algae, placing traps for predators and removing their shelters, replacement of protection nets, removal of dead individuals, seed-thinning, etc.

Santiago (2010) analyzes the factors that affect commercial clam cultivation on the Galician coast, particularly the biological and economic yield after sowing with seed from hatcheries. He notes that with a recapture rate of 30% of commercial clam, from hatchery clam seed of 10-12 mm, the economic return far surpasses investment. He also assesses each of the factors limiting current clam production, citing the following issues: social, environmental, economic, administrative and seed supply. Lastres et al. (2011b), taking mean commercial clam production values as a yardstick, find, from planting hatchery seed 2-3 mm final recapture rates of around 40% in *V.* and 30% of *V.* , and a mean period of between 18 and 22 months. Parada (2007) analyzes economic and material viability on an industrial scale, of the fattening of *V.* and *V.* from seed ranging from 8-12mm up to commercial size, in suspended systems with substrate. This author finds that *V.* of 8 mm reach commercial size after 339 days, culture of *V.* with initial mean size of 9.56 mm reached commercial size at 706 days. These rates of growths, in *V.* are greater than those reported by Santamaría et al. (2009) planting seeds of average size 12.76 mm on a plot in the "Carril cultivation parks", 20% of the population exceeding legal commercial size (38mm) at 500 days.

Based on biological results reported from numerous clam farming activities carried out over several years, an estimated one million units of seed of 10 to 12 mm (0.1 g/unit), which is the size of planting and that at which they leave the nursery, after 14 or 18 months of culture, and taking into account a loss of 50% of the seed sown, 6-8 tonnes reach commercial size clam (20 g/unit). That is, for an increase of 1,000 tonnes of clams (about 50 million adults, accounting for 20% of current clam production in Spain) it was necessary to plant and properly cultivate around 100 million seeds. In particular, and taking as a benchmark for Galician first sales data the year 2008 (www.pescadegalicia.com) it is estimated that about 32 million units of *V. philipinarum* and 16 million units of *V.* must be planted (size> 12 mm) and grown, in order to increase the production of these species by 10%.

Solis et al. (2009) and De Coo (2010) indicate that in natural beds commercial species of bivalves present mortalities ranging from 60 to 80%. To improve the yields of clam sown and cultivated, they suggest intervention strategies that address factors affecting predation, suggesting that prevention and protection systems (Figure 10), increase the survival of individuals planted. These authors note that a reduction of 15-20% of the effects of predators means doubling the commercial stock available.

Based on FAO data regarding production and considering 100 units/kg, it is found that about 6,000 million units of clams are consumed in Europe each year. bout 300 million seeds are produced in European hatcheries, of which about 50% are lost before they reach commercial size. That is, hatcheries provide about 5% of the clam that reaches the European market (Gutiérrez, 2010). This reveals the huge development potential of clam production activity, whether seed recruits come from the wild or from hatcheries.

note that the cultivation and production of clam are generally performed from seeds obtained from populations that spawn in the wild or from local broodstock, which are taken to the hatchery to spawn. The offspring (seeds) are planted in intertidal or subtidal areas and feed on microalgae produced naturally in the waters surrounding natural beds and areas of exploitation.

Figure 10. A: Preparation of planting areas at low tide. B: Overview of the activity of shellfish-harvesting on foot. Below: Seed of a size appropriate for outdoor growing. Right: Planting in progress at low tide.

Thus, the activity is associated with an environment of good-quality seawater and affects sectors of society that have traditionally lived on the coast, carrying out shellfish-gathering as a profession. Shellfish culture in general, and especially that of clams, represents the most clear-cut example of sustainable aquaculture, as shellfish feed on particles in the environment, and companies and professionals devoted to the management and exploitation of this resource are the main promoters of and beneficiaries of environmental sustainability, since the future of their businesses and, ultimately, their livelihood, all depend on it.

REFERENCES

Arnaiz, I. R. (2005). *Georreferenciación y cartografiado de los bancos marisqueros de Galicia. Evaluación del potencial marisquero.* Los recursos marinos de Galicia. Serie técnica nº 4. Consellería de Pesca e Asuntos Marítimos, Xunta de Galicia. Santiago de Compostela, Spain. 281 pp.

Bayes, J. (1987). Acondicionamiento y engorde de ostra. *Cuad. Marisq. Public. Téc.* 10, 35-42.

Cerviño-Eiroa, A., García Fernández, A. and De Coo Martín, A. (2006). Sistema de Bolsas para preengorde de almejas en Batea. *VIII Foro dos Recursos Mariños e da Acuicultura das Rías Galegas.* pp. 341-346, O Grove, Spain.

Cerviño-Eiroa, A. (2010). Preengorde de semilla de almejas. In: A. Guerra (Ed.), *Jornadas sobre criaderos, semilleros y cultivo de almejas.* pp. 43-51, Consellería do Mar, Xunta de Galicia, Santiago de Compostela, Spain.

Cerviño-Otero, A. (2011). Ciclo reproductivo, cultivo en criadero y en el medio natural de la almeja babosa *Venerupis pullastra* (Montagu, 1803). *Ph.D Thesis. University of Santiago de Compostela.* 288 pp.

Claus, C., Pauw, N. and Jaspers, E. (Eds.) (1981). Nursery culturing of bivalve Mollusc. *Proccedings of the International Workshop. Special publication nº 7.* Belgium. 393 pp.

De Santiago, J. A., Andrés, M. C. and Guerra, A. (2007). Preengorde de almeja babosa *Venerupis pullastra* (Montagu, 1830), mediante un sistema de flujo invertido forzado por air-lift, en la Ría de Camariñas (A Coruña). *XI Congreso Nacional de Acuicultura.* pp. 423-426, Vigo, Spain.

De Santiago, J. A., Fernandez, A., Ruiz, M., and Guerra, A. (2008). Preengorde de almeja babosa (*Venerupis pullastra*), almeja fina (*Ruditapes decussatus*) y almeja japonesa (*Ruditapes philippinarum*) en tres sistemas de preengorde. *X Foro dos Recursos Mariños e da Acuicultura das Rías Galegas.* pp. 381-388, O Grove, Spain.

De Santiago, J. A. (2010). Aspectos del cultivo de infaunales en Galicia. In: A. Guerra (Ed.), *Jornadas sobre criaderos, semilleros y cultivo de almejas.* pp. 133-138, Consellería do Mar, Xunta de Galicia, Santiago de Compostela, Spain.

De Coo, A. (2008). Revisión a dez anos de marisqueo. *X Foro dos Recursos Mariños e da Acuicultura das Rías Galegas.* pp. 117-128, O Grove, Spain.

De Coo, A. (2010). Importancia dos depredadores nos bivalvos comerciais. Sistemas de prevención e protección. En: A. Guerra (Ed.), *Jornadas sobre criaderos, semilleros y cultivo de almejas.* pp. 139-145, Consellería do Mar, Xunta de Galicia, Santiago de Compostela, Spain.

Guerra, A. (1984). Desarrollo y alternativas de los cultivos de ostra en Galicia. *Alimentaria.* 21, 89-91.

Guerra, A. (2002a). Los minicriaderos: propuesta de un modelo de instalación para la obtención de semilla de moluscos bivalvos comerciales. *IV Foro dos Recursos Mariños e da Acuicultura das Rías Galegas.* pp. 67-77, O Grove, Spain.

Guerra, A. (2002b). La ostricultura. Técnica de producción y cultivo. In: Fundación Alfonso Martín Escudero (Ed.), *Impulso, desarrollo y potenciación de la ostricultura en España.* pp. 37-64, Madrid, Spain.

Guerra, A., Novoa, S., Besada, M., Búa, I., Lastres, M., Fernandez, J., and Asela, R. (2007). Crecimiento y composición bioquímica de semilla de almeja japonesa (*Tapes philippinarum*) y almeja babosa (*Venerupis pullastra*), obtenida en criadero y cultivada en diferentes sistemas de preengorde y en parque de cultivo. *XI Congreso Nacional de Acuicultura.* pp. 467-470, Vigo, Spain.

Guerra, A. (Ed.) (2010a). *Jornadas sobre criaderos, semilleros y cultivo de almejas.* Consellería do Mar, Xunta de Galicia, Santiago de Compostela, Spain. 165 pp.

Guerra, A. (2010b). Valoración de un modelo de transferencia tecnológica para la producción de semilla de bivalvos: los minicriaderos. *XII Foro dos Recursos Mariños e da Acuicultura das Rías Galegas.* pp. 39-46, O Grove, Spain.

Guerra, A. (2010c). Proyecto demostración de criaderos y semilleros para almeja en recintos de alta productividad. In: A. Guerra (Ed.), *Jornadas sobre criaderos, semilleros y cultivo de almejas.* pp. 23-32, Consellería do Mar, Xunta de Galicia, Santiago de Compostela, Spain.

Guerra, A., Marcet, P., Otero, M., Fernández, J., and González, S. (2010). Use and treatments of organic matter of turbot hatcheries effluents in Galicia (NW Spain). *Aquaculture Europe 2010.* pp. 540-541, Porto, Portugal.

Guerrero, S. and Gonzalez, X. O. (1991). Clam nursery (*Tapes decussatus*) in the efluente of a fish farm in Ría de Arousa, Spain. *EAS Special publication.* 14, 132-133.

Guerrero, S. (1995). Análisis de producción de ostra plana (*Ostrea edulis* L.) en las fases de cría, semilla y engorde. Estrategias para el cultivo comercial en Galicia. *Ph.D Thesis. University of Santiago de Compostela.*

Guerrero, S. (2010). Breve historia de los criaderos gallegos. In: A. Guerra (Ed.), *Jornadas sobre criaderos, semilleros y cultivo de almejas.* pp. 13-17, Consellería do Mar, Xunta de Galicia, Santiago de Compostela, Spain.

Guerrero, S., Cremades, J., Ancosmede, C., Dominguez, J. C., and Muñiz, V. (2011). Preengorde en parque de cultivo de semilla de almeja fina y japonesa procedente de semillero multitrófico. *XIII Foro dos Recursos Mariños e da Acuicultura das Rías Galegas.* pp. 429-434, O Grove, Spain.

Guerrero, S. and Cremades, J. (Eds.) (in press). *Acuicultura multitrófica integrada. Una alternativa sostenible y de futuro para los cultivos marinos de Galicia.* Consellería do Mar, Xunta de Galicia, Santiago de Compostela, Spain.

Gutiérrez, F. (2010). El futuro de os criaderos de bivalvos. In: A. Guerra (Ed.), *Jornadas sobre criaderos, semilleros y cultivo de almejas*, pp. 9-12. Consellería do Mar, Xunta de Galicia, Santiago de Compostela, Spain.

Harguindey, B. (2010). Cultivo diferenciado de bivalvos nos parques de Carril. In: A. Guerra (Ed.), *Jornadas sobre criaderos, semilleros y cultivo de almejas.* pp. 147-152, Consellería do Mar, Xunta de Galicia, Santiago de Compostela, Spain.

Jara Jara, R. (1995). Crecimiento de semilla de *Venerupis pullastra* M. y de *Ruditapes decussatus* L. en agua residual procedente del cultivo de rodaballo (*Psetta maxima* L.). *Ph.D Thesis. University of Santiago de Compostela.* 263 pp.

Lastres, M. A., Andrés, C., Santamaría, I., and Guerra, A. (2009). Proyecto formativo para la especialización de técnicos de acuicultura en procedimientos de cultivo de moluscos en criaderos. *Foro Iberoam. Rec. Mar. Acui.* II, 89-117.

Lastres, M. A., Andrés, C., Santamaría, I., and Guerra, A. (2011a). Descripción de las instalaciones generales de un minicriadero de moluscos. *XIII Foro dos Recursos Mariños e da Acuicultura das Rías Galegas.* pp. 275-281, O Grove, Spain.

Lastres, M. A., Andrés, C., Santamaría, I., and Guerra, A. (2011b). Resultados de la aplicación de técnicas de cultivo de moluscos bivalvos en criaderos bajo el modelo de empresa tutelada. *Foro Iberoam. Rec. Mar. Acui.* III, 265-269

López, J., Carrasco, J. F. and Rodríguez, C. (2011). Cultivo y repoblación de almeja fina (*Ruditapes decussatus*) en el Principado de Asturias: 2009-2010. *XIII Foro dos Recursos Mariños e da Acuicultura das Rías Galegas.* pp. 259-265, O Grove, Spain.

Lucas, A. (1981). Le rôle du naissain d´écloserie dans la culture des bivalves en 1980. *Tiré à part de La Peche Maritime.* 4 pp.

Miranda, M., Santamaría, I., Casal, J., Ruiz, A., and Guerra, A. (2006). Aprovechamiento de aguas residuales de una piscifactoría de rodaballo en Galicia para el engorde de moluscos bivalvos. *VII Foro dos Recursos Mariños e da Acuicultura das Rías Galegas.* pp. 181-185, O Grove, Spain.

No Couto, E. (2010). Minicriaderos 2009: valoración e características de preengorde e sementeiras. In: A. Guerra (Ed.), *Jornadas sobre criaderos, semilleros y cultivo de almejas.* pp. 127-132, Consellería do Mar, Xunta de Galicia, Santiago de Compostela, Spain.

Parada, J. M. (2007). Engorde a escala industrial de almeja babosa (*Venerupis senegalensis*) y almeja fina (*Tapes decussatus*) en batea hasta talla comercial. *Congreso Nacional de Acuicultura.* pp. 373-378, Vigo, Spain.

Pardo Vuelta, M. J. (2007). Sistema para el preengorde de semilla de moluscos comerciales. *VIII Foro dos Recursos Mariños e da Acuicultura das Rías Galegas.* pp. 127-132, O Grove, Spain.

Perez-Camacho, A. (2007). El cultivo de bivalvos en Galicia: entre el paradigma y la entelequia. *XI Congreso Nacional de Acuicultura.* pp. 373-378, Vigo, Spain.

Pérez Corvacho, E. and Pardo Vuelta, M. J. (2007). O cultivo de ameixa en batea: a experiencia da Cofradía de Moaña. *VIII Foro dos Recursos Mariños e da Acuicultura das Rías Galegas.* pp. 263-269, O Grove, Spain.

Palanco, I. and Royo, A. (2008). Aplicación de diferentes sistemas de preengorde en la obtención de semilla de almeja fina (*Ruditapes decussatus* Linneo, 1758). *X Foro dos Recursos Mariños e da Acuicultura das Rías Galegas.* pp. 531-543, O Grove, Spain.

Pouso, O., Poza, G., Longa, E., and Santamaría, I. (2004). Formación en réxime de empresa tutelada para alumnos do Instituto Galego de Formación en Acuicultura. *VI Foro dos Recursos Mariños e da Acuicultura das Rías Galegas.* pp. 265-268, O Grove, Spain.

Rama Villar, A. (2011). Preengorde de moluscos bivalvos en batea. Sistema mixto (caja-bolsa). *XIII Foro dos Recursos Mariños e da Acuicultura das Rías Galegas.* pp. 421-428, O Grove, Spain.

Royo, A. (2007). Optimización del preengorde de la semilla de moluscos bivalvos (almejas). *XI Congreso Nacional de Acuicultura.* pp. 379-384, Vigo, Spain.

Santamaría, I., Otero, M., Fernandez, M. ,and Guerra, A. (2009). Engorde de la almeja babosa *Venerupis pullastra* (Montagu, 1803), en los parques de Carril, Ría de Arousa (Pontevedra). *XI Foro dos Recursos Mariños e da Acuicultura das Rías Galegas.* pp. 371-376, O Grove, Spain.

Solis, L., Alcalde, A., Conde, M. L., De Coo, A., Fariña, J., García, A., García, J., Poza, G., and Santos, I. (2009). Potencial de afectación das especies depredadoras de moluscos bivalvos comerciais en Galicia. Primeiros mostraxes de poboacións e conclusións preliminares. *XI Foro dos Recursos Mariños e da Acuicultura das Rías Galegas.* pp. 549-554, O Grove, Spain.

Wilson, J. (1981). A greenhouse oyster hatchery in Ireland. *Fish Farming International.* 5, 20-23.

Williams, P. (1981).Offshore nursery-culture using upwelling principle. In: Claus, C., Pauw, N. and Jaspers, E. (Eds.). Nursery culturing of bivalve Mollusc. *Proccedings of the International Workshop. Special publication nº 7.* pp. 311-315, Ghent, Belgium.

In: Clam Fisheries and Aquaculture
Editor: Fiz da Costa González
ISBN: 978-1-62257-518-3

Chapter 11

ARTISANAL EXPLOITATION OF NATURAL CLAM BEDS: ORGANIZATION AND MANAGEMENT TOOLS

J. M. Parada[1*] and J. Molares[2]
[1]Centro de Investigacións Mariñas. Consellería do Mar. Vilanova de Arousa (Pontevedra), Spain
[2]Dirección Xeral de Ordenación e Xestión dos Recursos Mariños. Consellería do Mar. Galicia, Spain

ABSTRACT

Most of the world production of clams from natural beds comes from mechanized fisheries that generally use dredges. However, other ways of exploitation are employed. Artisanal harvesting afloat involves a large number of people who use rakes to extract the clams onboard small vessels manned by crews of 1 to 3 shellfishers. These molluscs are also hand-gathered by clam diggers working in the intertidal zone with small rakes and tools similar to those used in traditional agriculture.

Artisanal clam fisheries are of great social importance, since the wealth obtained is distributed among a large number of people. Galicia, located in the NW Iberian peninsula, is one of the places where clams are harvested only from natural beds by artisanal fishers. Over 9,000 people, registered as professionals and grouped into approximately 70 local organizations, are involved in this activity.

Shellfishing in Galicia is not limited to extraction. Management incorporates tools typical of mollusc culture. When so many people are involved, the implementation of non extractive and communal activities into a non privative extraction zones, requires a high level of organization. Furthermore, efficient management must be supported by suitable technical assessment. This chapter reviews institutional arrangements and management tools in the artisanal clam fishery from Galicia.

[*] Corresponding author: J. M. Parada. E-mail address: paradajm@gmail.com.

11.1. Introduction

From a technical standpoint, the methods used to harvest clams and cockles have evolved in different directions depending on whether the harvested area is accessible to hand gatherers on foot or located underwater. In the intertidal beds accessible on foot, diggers use tools inspired by agriculture, such as hoes and rakes, adapted in various ways. While extraction from submerged areas may be carried out by diving, most of this activity is performed afloat with two techniques: dredging and manually operated rakes. In the latter, similar to the ones used by on-foot shellfishers, the vessel is merely a means to access the submerged beds.

From an administrative point of view, the exploitation of these resources has been organized differently according to the specific characteristics of the resource. While in some places extraction is no more than a purely recreational activity, in others, harvesting is carried out by professionals who use highly mechanized vessels. The large cockle beds in Europe are harvested by approximately 30-40 vessels manned by 3-7 crewmembers and equipped with hydraulic dredges (Gosling, 2003). On the contrary, in many countries clams fisheries involves many little boats with a low technological development.

The largest clam and cockle beds in Galicia, covering an area of roughly 500 ha with mean densities of commercial-sized individuals ranging from 300 to 400 ind/m², are artisanally harvested by 300-400 vessels with 1-3 crewmembers and 200-300 on-foot shellfishers (Ríos, 2001; UTPB, 2005; Parada et al., 2006). Therefore, the harvesting of these bivalves in Galicia, along with the economic importance entailed, is of great social interest owing to the many direct jobs generated. Landings of cockles and clams accounted for a total of 65 million euros in 2010 (Plataforma Tecnolóxica da Pesca http://www.pescadegalicia.com/default.htm). Over 8000 people grouped into some 70 locally-based organizations, exercise this activity professionally in an area of approximately 80,000 ha (UTPB, 2005). The main species harvested include cockles (*Cerastoderma edule*), pullet carpet shell (*Venerupis corrugata*); grooved carpet shell (*Venerupis decussata*), manila clam (*Venerupis philippinarum*) and banded carpet shell (*Polititapes virgineus*).

Management of this fishery requires a major regulatory effort by the authority and organization of the shellfishers associations. As of 1993, the extraction of bivalves from the sand banks began to be considered a year-round profession, subject to regulations backed by technical support (Marugán, 2004; Mahou, 2008). This transformation was driven by i) changes in the regulatory framework, ii) improved scientific and technical support, iii) ongoing communication between the authority and shellfishers, and iv) extensive training. In addition to harvesting, professional shellfishing involves practices typical of extensive aquaculture or "semiculture" (habitat enhancement and re-stocking), surveillance of the beds and organizational activities.

Professionalization led to informative co-management (Sen et al., 1996), in which the fisheries authority devolves decision-making to the shellfishers associations, which, in turn, report back to the authority. Decisions are overseen by the authority from a technical standpoint, and must comply with the regulatory framework. This type of management system has proved economically, socially and in biologically sustainable in many different contexts (Caddy and Defeo, 2003).

This chapter analyzes three aspects of the management of the harvesting of clams and cockles in Galicia: i) organization, ii) technical support, and iii) management tools.

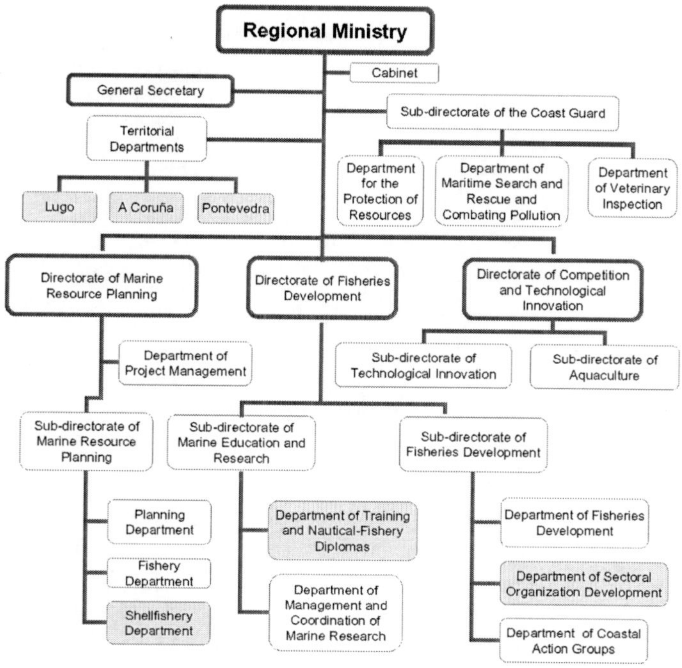

Source: http://conselleriamar.xunta.es/web/pesca/organigrama.

Figure 1. Organization of the "Consellería do Mar". The boxes shaded in grey show the services directly involved in shellfishery management.

11.2. Organizational Aspects

Within the "Consellería do Mar" (Regional Ministry of the Sea) (Figure 1), the "Dirección Xeral de Ordenación e Xestión dos Recursos Mariños" (Directorate of Marine Resource Planning and Management) is in charge of drawing up the regulations related to granting licences and to marine resource management. The Territorial Departments located in coastal towns are in charge of processing the documentation related to the two types of licences mentioned earlier. Training and organization of the shellfishers are carried out by the "Dirección Xeral de Desenvolvemento Pesqueiro" (Directorate of Fisheries Development).

In Galicia, "marisqueo" (shellfishing) is defined as the practice of extracting invertebrates on foot along the seashore or on board of vessels, using selective and/or specific gear to target one or more species of marine invertebrates for the purpose of commercialization. Since 1994 shellfishing in both authorized (natural beds whose exploitation is granted under an exclusive access system) and open access zones it is required to hold a harvesting permit (PERMEX). In the case of on-foot shellfishers a PERMEX is granted to a legal person and must be renewed every year, while the permit to shellfish from a fishing boat is granted to the vessel and said permit will be valid during the useful life of the vessel. PERMEXs issued to vessels may include up to five fishing methods, including shellfishing with a rake. The latter may be used alternately provided that the vessel owner has registered the vessel for the fishery selected. Simultaneous use of more than one fishing method on a given day is not allowed.

Starting in 1993 shellfishing has been regulated through annual "exploitation plans", designed by associative entities of the sector. They may include limits on catches, capacity and effort, and other measures adapted to particular circumstances of time and place.

Development and approval of exploitation plans (Figure 2) begins in September, with surveys of the most important shellfish beds, and ends in December with the publication of the approval for the following year.

Agents involved in the design of these plans include managers (directors of associative agencies) and technical experts contracted by each organization. The technical staff, considered in recent years to be a key part of the whole shellfish management system in Galicia, is largely made up of biologists or people with degrees in marine sciences who have the knowledge needed to conduct sampling surveys established in the procedures proposed by the fisheries authority, and to provide appropriate advice to the managers.

Decisions made in the associative entities do not follow a single path, but may vary depending on the nature and degree of organization of each association (Figure 3).

If the organization has no technical expert or Shellfishers Association, the procedure is simplified, with the board of directors generally taking over the design and monitoring of the exploitation plans. The next step consists of the assessment of the plans by the area biologists, the technical experts at the Territorial Departments.

Once all the exploitation plans have been presented and approved, the 'Xefatura de Servizo de Marisqueo' (Department of Shellfishery Services) prepares an 'Order of the General Plan for the Exploitation of Shellfish Stocks' which includes information from each specific plan.

The number of associative entities currently submitting plans for shellfish management totals 70, more than half of which have their own technical experts. Some 200 management plans are approved annually. Presently the administration enables shellfishing by 4,123 on-foot women shellfishers and approximately 3,900 shellfishers signed up with 2,589 authorized vessels. Most of the associative entities are fisher associations (62), non-profit public law corporations with administrative duties that act as bodies for consultancy and cooperation with the fisheries authority.

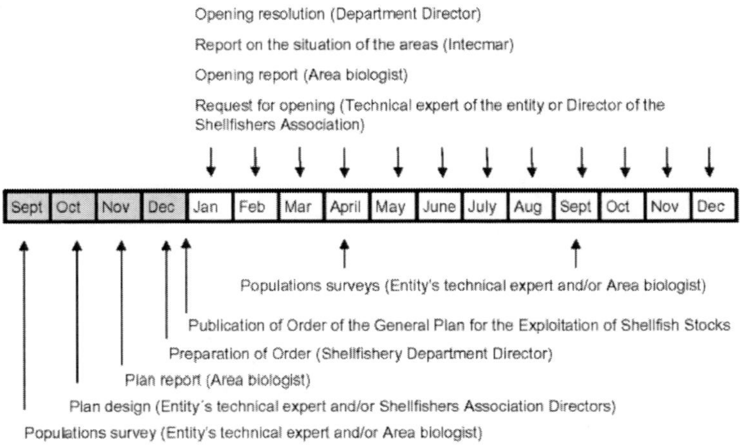

Figure 2. Schedule of shellfishery management in Galicia, including all the procedures and the people in charge of each stage.

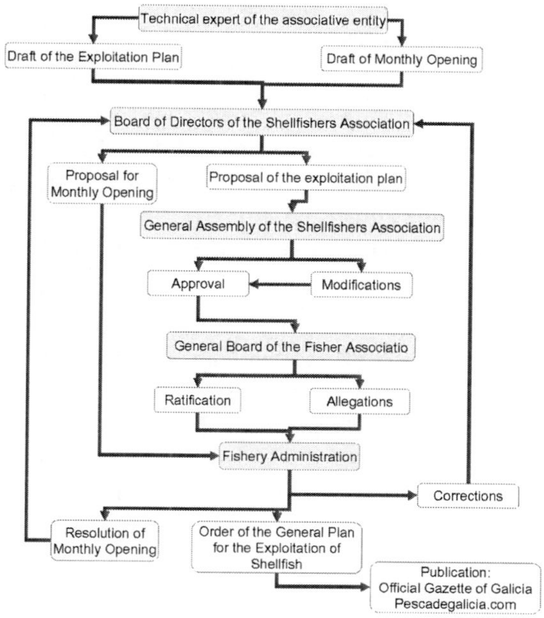

Figure 3. General diagram reflecting the decision making process in shellfishery management within the fisher associations.

At the same time they represent the economic and corporative interests of the sector's professionals and their associates, and can be active in organization and commercialization issues. The rest of the organizations are associations (5) or cooperatives (3).

Over the year, as monitoring proceeds, modification of the plans may be required. Examples are closures due to biotoxins, drops in prices causing a gap between supply and demand, occasional mortalities produced by events of high precipitation or temperature, etc. To organize the daily harvesting activity of each plan, another procedure exists, i.e. the Resolution of Monthly Opening. Fifteen days before the activity is scheduled to start, the organizations must apply for permission to open the zone where they plan to extract shellfish, specifying the exact dates and maximum amounts per person, which may be lower than the maxima established in the plan depending on demand. To proceed with the approval of this application, the "Xefatura de Servizo de Recursos Mariños" (Department of Marine Resources Services) of the "Delegacións Territoriais" (Territorial Delegations) must have a favourable report issued by the area biologist and by Intecmar, which is in charge of monitoring seawater quality.

Control points are established places in the bed itself or in the vicinity thereof, used during the work day of the activity carried out on foot or on board of vessels. Control points are used to verify that the shellfishers abide by the approved quotas in the opening resolution and with size regulations, and to control of the duration of the workday. Some organizations use the control points to classify the catch by size and to transport the product to the market. In some cases each shellfisher takes charge of the transport individually. Specimens included in groups that exceed the allotted catch quota are withheld at the control point and later returned to the sea in the same area where they were extracted, or in areas where stocks are being enhanced. In same cases, people in charge of the control points are guards contracted by the organizations. However, the directors of the shellfishers association are usually the ones in charge of controlling these points.

Catches from shellfish beds must go to authorized points of sale. At markets shellfish is sold in a lowest bid auction system, although other types of commercial transactions may be made. Weight of the catch from each harvesting permit holder and the amounts paid for each group are automatically recorded in a centralized database. Entries by marketplace, species and day may be checked by consulting the website of the Fishery Technological Platform of the "Consellería do Mar" (http://www. pescadegalicia.com/default.htm).

By consulting the information contained in the website of the platform since 1997, it is possible to carry out comparative analyses and to follow the evolution of the production of the different associative bodies. However, without the information from control points, the conclusions may be erroneous because a marketplace can receive products from different organizations.

To carry out extraction on foot, shellfishers must have a certificate of qualification (Shellfisher's Certificate). Applicants to a new PERMEX must attend training courses offered by the fisheries authority and pass the appropriate tests. Technical experts of the associative entities have a wealth of information available to them over the Internet regarding procedures and management of computer applications developed to standardize sampling surveys of the populations, and the control of the extractive activity (http://sites.google.com/site/arousa09). Area biologists and the other agents involved in shellfishery management are offered courses designed to update their knowledge of the fishery legislation.

11.3. Technical Support

11.3.1. Annual Surveys

Each shellfish bed is surveyed at least once between September and October. Sampling design depends on the target species, the characteristics of the surveyed bed, and the logistics available to the shellfishers organization. The general design consists of periodic stratified sampling surveys; sampling units are collected using different techniques, ranging from quadrats and standardized dredges hauls to raket samples of an estimated surface area. The most comprehensive surveys provide estimates of weight and density (individuals per square meter) of the stock's commercial fraction and recruitment.

Other data obtained are size-frequency distributions which make it possible to estimate the available stock of specified size ranges. On the basis of this information abundance is mapped, making it possible to investigate spatial patterns of distribution and identify settlement areas (Figure 4).

Technical experts of the shellfishers associations may also collect physicochemical data to characterize the environmental conditions prevailing in the shellfish beds. The most commonly studied parameters are sediment composition and bathymetry, as well as, on occasion, compaction and organic matter content of the sediment. These parameters are also analyzed spatially (Figure 5). Habitat maps are complemented with information on freshwater inputs, location of runoffs, seagrass meadows, zones prone to accumulation of seaweed, or areas with high abundance of predators.

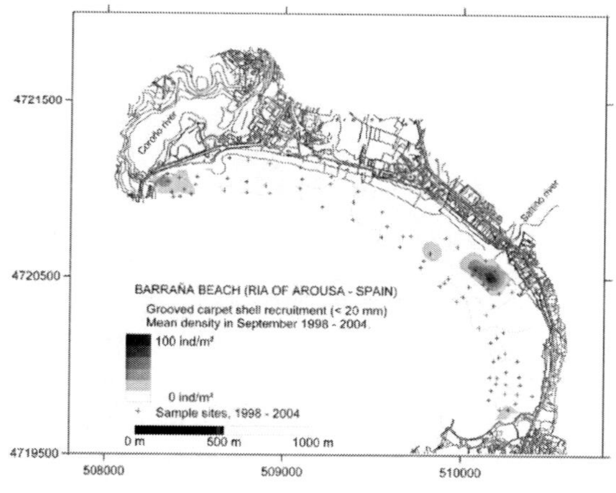

Figure 4. Spatial distribution of the recruitment of the grooved carpet shell averaged for the years from 1998 to 2004 on Barraña Beach (Ria of Arousa).

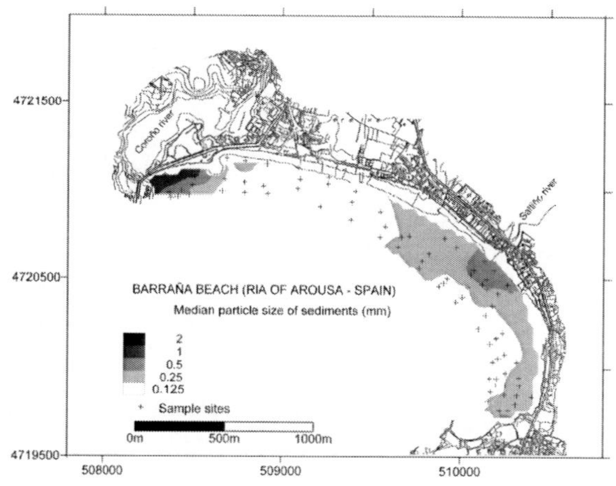

Figure 5. Spatial distribution of the median particle size of sediments on Barraña Beach (Ria de Arousa).

11.3.2. Dynamic Information

This includes information on population dynamics (growth and mortality rates), monitoring of the fishing process, and environmental data. Data on the populations originate from periodic sampling surveys or experiments in which tagged individuals are kept in boxes to monitor growth and survival on a monthly or bimonthly basis (Figure 6a). Monitoring of the fishing process includes data about effort, catches and size structure of the catches related with the extraction zone. Environmental data include high- and low-frequency time series. The first category corresponds to temperature and salinity recorded with data loggers with a ten-minute frequency (Figure 6b), and daily records of precipitations or freshwater runoff. Low-frequency series record changes in more stable factors, like sediment composition, tidal level, extension of seagrass meadows, etc. In these series spatially-explicit information is examined over periods ranging from 5 to 10 years.

11.3.3. Integration of Technical Information

By integrating the sampling surveys and dynamic information, it is possible to investigate relations between species and their habitat, and catches and environmental characteristics, which are of interest to management. The use of different statistical tools or superimposed layers of information in a geographic information system allows the identification of zones of interest for management, such as high growth or recruitment areas, or risky zones because of freshwater runoff or the accumulation of macroalgae (Figure 7). Integration that information, models can be developed (Parada and Molares, 2008; Parada et al., 2011) and population responses can be analyzed (Parada and Molares, 2009). This is helpful for diagnostics, projections of stock size, and evaluation of alternative management options.

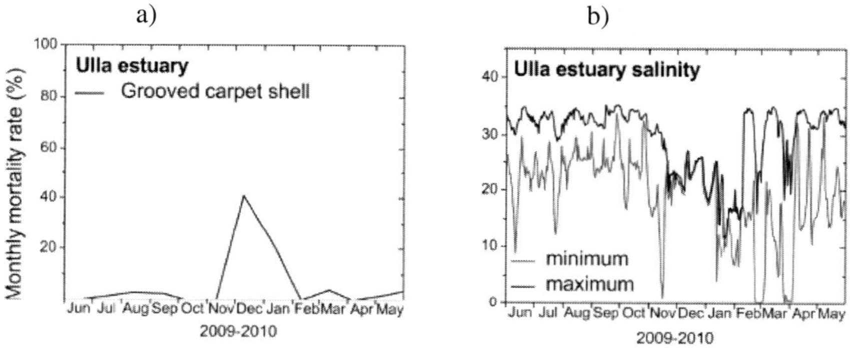

Figure 6. Monitoring of the monthly mortality of the grooved carpet shell (a) and salinity fluctuation (b) in the estuary of the Ulla River.

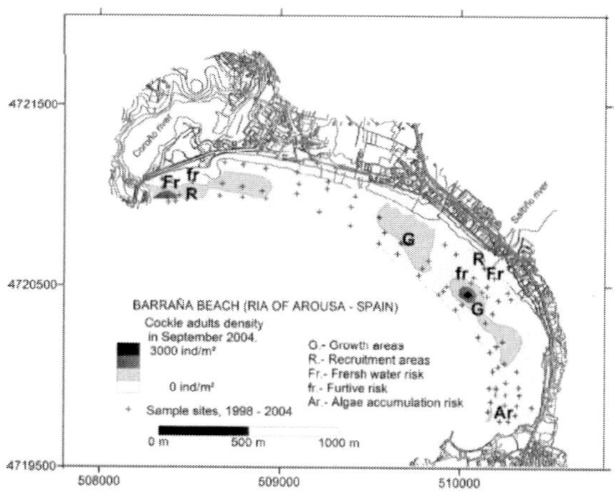

Figure 7. Identification of areas of interest for the management of the shellfish bed.

One of the main problems confronting the Galician fisheries administration and technical experts of the organizations was the diversity of techniques and methodologies used to carry out the assessments. Heterogeneity makes it difficult to compare the results obtained by the different organizations, and, within the same organization, the results found by the different technical experts contracted over the years.

Starting in 2000 several different initiatives have been launched to standardize the sampling techniques to be used in stock assessments. In 2009 these initiatives provided the technical experts with a computer application consisting of a set of tools for Excel (Microsoft Office) developed in the Visual Basic Application. The ARouSA application (available at https://sites.google.com/site/arousa09) contains standardized tools for the calculation of stocks with different techniques, description of the environmental characteristics, size-weight ratios, the calculation of mortalities, population analyses, and the control of catch size, among other features.

11.4. Management Tools

The management tools used in the exploitation of clams and cockles can be grouped into three categories: i) exploitation plans, ii) indicators of the evolution of the extraction campaign, and iii) semiculture techniques. An exploitation plan is a document through which every shellfisher association must program its activities in terms of organization, extraction and economic issues. The indicators of the evolution of the extraction campaign comprise a series of parameters that provide managers with information regarding the evolution of the stock and its market price. This information makes it possible to change the exploitation strategy within the limits established by the exploitation plan. Semiculture techniques make use of survey and dynamic data on stocks and habitat to increase the production of the shellfish beds.

11.4.1. Exploitation Plan

Each shellfisher's organization is required to submit an annual exploitation plan to the fisheries authority. The plan regulates activity in the fishing zones which have been authorized for exploitation. By virtue of this document, and bound by the legislation, shellfishers formulate their internal administrative regulations and technical measures. These pertain to local governance, harvesting and technical support. In addition to establishing a personalized harvester's permit (PERMEX), starting in 1993 the requirements of the exploitation plan marked a turning point in the professionalization of the shellfishery sector (Mahou, 2008).

One of the most significant consequences of the implementation of the exploitation plans was that it opened up the possibility of individualized management of each shellfish bed. Prior to the existence of these plans, the exploitation of clams and cockles on the entire Galician coast was organized into six-month-long harvesting campaigns (October 1st to March 31st), with identical regulations regardless of the specific characteristics of each shellfish bed or of the markets that they would be able to serve. Besides the lack of control over access to the resource, the structure of the harvesting activity was disorganized and most of the stock was removed in a few days. During the 1979-1980 harvesting campaign, before the exploitation plans went into effect, in Redondela (Spain) 107 t of cockles were extracted. However 75% of the entire stock was extracted in less than 10 days by 900 people, while the campaign lasted 30 days (Figure 8a). In Galicia the product is sold through a lowest-bid auction system, which led to market saturation, a rapid reduction of prices, and as a result, reduced income for the shellfishers.

A similar extraction (104 t) during 2006 provided work for 396 professionalized shellfishers (Figure 8b). Under the exploitation plans, in keeping with the limits set according to the regulations of the Administration, each organization can establish a harvesting calendar over a yearly period which takes into account the biological situation of the stock of each bed as well as the evolution of the demand and, therefore, prices. This, along with the implementation of the personalized harvester's permit granted to a limited number of shellfishers in each organization, improved the management efficiency of the beds and their economic profitability (Figure 8b).

The exploitation plans make it possible to plan many different aspects of the activity:

- Harvesting calendar
- Catch quotas
- Forecasting of catches
- Sales organization
- Organization of semiculture activities
- Annual economic plan

The harvesting calendar establishes the working days and timetables during which extraction will be carried out, within the limits laid down by the Administration. Some organizations may choose to maintain short-term campaign systems, while others establish all working days along the year in keeping with the time of the neap tides. The former strategy facilitates coordination with the large canning companies, which program their activity to can cockles during this season.

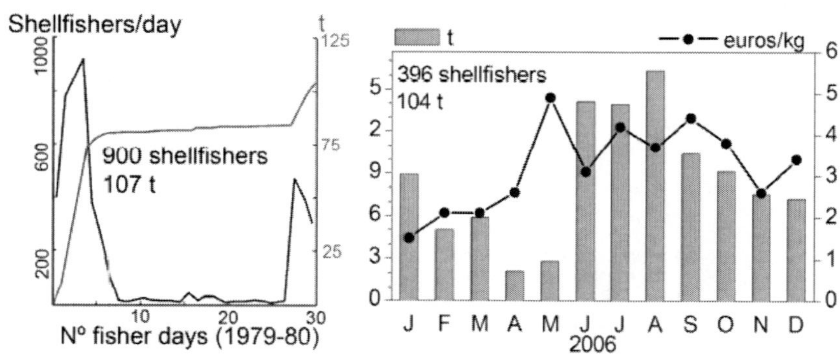

Figure 8. Cockle extraction during the harvesting campaigns of 1979-1980 (a; adapted from Fernández et al., 1987) and 2006 (b), in the shellfish beds of Redondela (Ría de Vigo) according to Plataforma Tecnolóxica da Pesca.

The latter strategy allows the organizations to supply material to the canning companies as well as permitting them to adapt their supply to the fresh fish market. The price evolution of this market is linked to the different seasons of the year such as Christmas or other periods when demand is high (for example the summer).

Figure 9 presents a comparison of harvesting strategies between two groups of shellfishers with similar catches (989 and 1031 t/year in 2005). The shellfishers of Noia (Ria of Muros and Noia) harvested during 5 or 6 months (one third of the product was extracted in only one month) to serve the canning companies, whereas the shellfishers of Rianxo and Cabo de Cruz (Ria of Arousa) carried out their activity over the whole year to supply the market with fresh product, whose prices increase in summer.

Although the two strategies differ, the final economic outcome is similar. The gross income generated by cockle extraction in 2005 to around 1100 shellfishers from Noia and 850 from Rianxo and Cabo de Cruz ranged between 5000 and 6000 euros per person.

Catch quotas contemplated in the exploitation plan establish a maximum number of kilos per person and day that shellfishers are allowed to extract of each species.

However, shellfishers may lower these quotas, provided that they inform the fisheries authority in advance. This makes it possible for shellfishers to adapt their supply to the demand in order to sustain prices or preserve the stock during periods of low demand. At the same time, it allows them to regulate extraction, keeping track of available stock in the bed or in the area of the bed planned to be harvested each day.

Forecasting the catches for the year in which the exploitation plan is in force is programmed according to stock assessments made immediately before the exploitation plan is drawn up. This forecast is also dependent upon the daily extraction quotas. There are different ways of estimating the catches foreseen to be taken, but the most reliable ones use the data series that associate the stock assessments with the catches yielded (Figure 10). The Administration requires that cockles and clams be sold exclusively at fish markets or authorized points of sale. Under an exploitation plan, shellfishers organizations may set up more detailed sales arrangements.

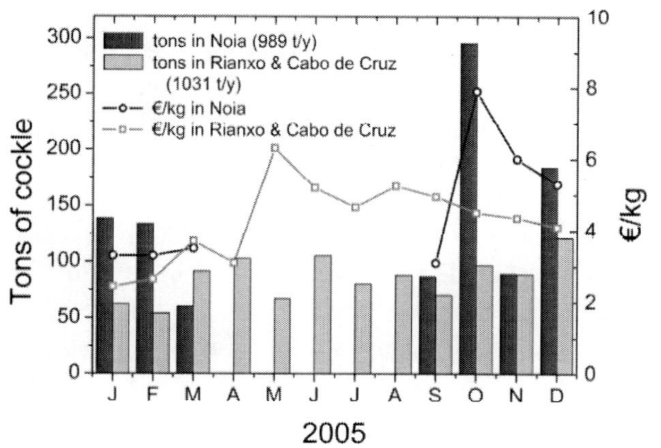

Figure 9. Two shellfish harvesting strategies. In Noia harvesting took place during only a few months of the year to cater to the demand of the canning industry. In Rianxo and Cabo de Cruz harvesting was spread out over the whole year to take advantage of the rise in prices associated with the consumption of the fresh product in summer (data from Plataforma Tecnolóxica da Pesca).

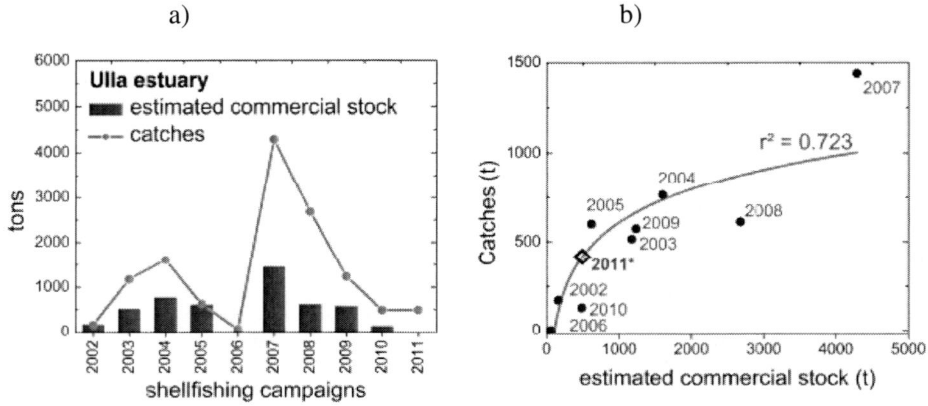

Figure 10. Relationship between estimated commercial stock prior to the start of each shellfishing campaign and catches obtained from the corresponding campaign (a) and model for the estimation of the catches forecasted during a campaign according to previous stock estimation (b). The dates shown in b indicate the ratio of estimated stock – catches obtained during the campaigns of each year. 2011*; catches forecasted for 2011. r^2=0.723; model fit.

Some organizations select catches by size and then sell them classified. Other organizations allow each shellfisher to sell his or her catch individually and it is the shellfisher who decides whether or not to classify the product by size. Some markets that cater to only one or two buyers come to a price agreement with them prior to extraction, while in others; the most common system is governed by the lowest-bid auction system.

Many exploitation plans include the possibility of setting a minimum price at which the product may be withdrawn from the auction. When the auction reaches this minimum price, the product can be taken to other markets and at times the organization may even decide to halt extraction until the demand increases. This allows shellfishers a certain degree of control over the prices. In addition to price control, they may also choose to manipulate demand by regulating the amount of product offered each day, after adjusting the harvesting calendar and catch quotas.

Most organizations program their harvesting calendars to include periods during which no extraction takes place depending on the evolution of the demand, the state of the stock, or in order to carry out semiculture activities. The latter, discussed later, are carried out by the shellfishers who organize their work into shifts, with working teams varying in number according to the requirements of each task. The calendar and organization of these activities - including timetables, points of action, group make-up, shift systems, warning systems, attendance control, etc.- are specified in the exploitation plan; all members of the organization are familiar with them.

Some specific tasks such as the control of catch sizes and quotas or guarding against poachers involve collecting bivalves that will later be returned to the sea. The mechanisms through which the collection of these bivalves is managed are also determined in the exploitation plan.

The annual economic plan, which is part of the exploitation plan, includes the forecasting of expenditures and income resulting from the activity. This forecast covers a wide range of areas from administrative aspects to projects aimed at the improvement and protection of the resources and their habitat.

11.4.2. In-Season Indicators

In-season indicators are based on effort and catch data from the control points and from market sales.

The capture effort is measured as the daily number of persons or vessels that carry out their activity on a daily basis in the bed being monitored (Figure 11a). A decrease in the number of shellfishers or vessels that are shellfishing may indicate a drop in the commercial stock, but it also may be related to dwindling market demand. In sectors that combine the harvesting of cockles and clams with other fishing activities it may be linked to the harvests of other resources. Evolution of daily accumulated catch provides information on the state of the stock. When the curve representing the accumulation of catches shows a declining slope, (Figure 11b) it may be an indication of dwindling effort or of decreased stock abundance. Catch per unit effort (CPUE) is a classic index used in fisheries studies. On the basis of sales information it is possible to examine the daily evolution of this index. However, under exploitation plans in which the quotas established are low, there is no expected relation between CPUE and stock abundance. Shellfishers usually compensate for a drop in abundance by spending more time on the job or by being less careful in the selection of the harvested product.

Some organizations require shellfishers to go through the control point before beginning their daily activity and again at the end of the said activity. In these cases the start and finish of the harvesting activity of each shellfisher or vessel is recorded. This yields a catch index per unit of effort in terms of kilos harvested per person and hour each day.

In both cases, a decrease in CPUE reflects the gradual dwindling of the stock. However, when it is expressed in terms of catch per hour, CPUE is more sensitive (Figure 12). The difference between the two indices is clearly seen at the beginning of the campaigns when the stock is still abundant. Sudden increases in CPUE are related to the location of areas with a high density of bivalves, changes in harvesting areas or to increases in the maximum established quota.

Market sale price is generally linked to size (Figure 13), implying that shellfishers will not only try to catch the maximum allotted weight according to the daily quota, but they also target the largest-sized specimens. Hence, as the harvesting campaign proceeds, both the stock volume and mean catch length decrease.

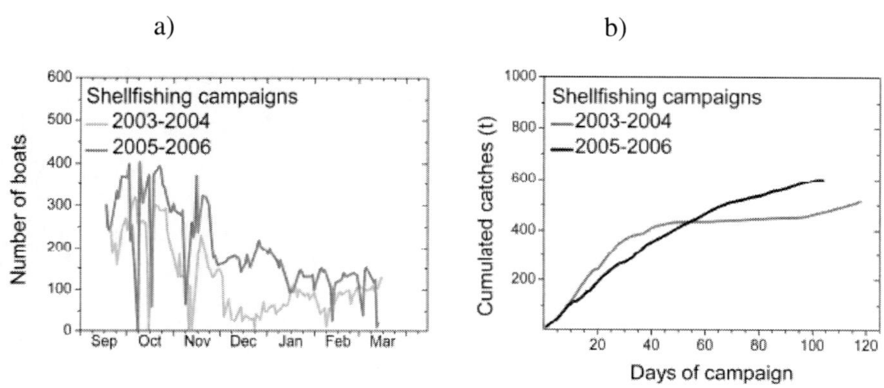

Figure 11. Evolution of daily effort (a) and cumulative catch (b) during several shellfishing campaigns.

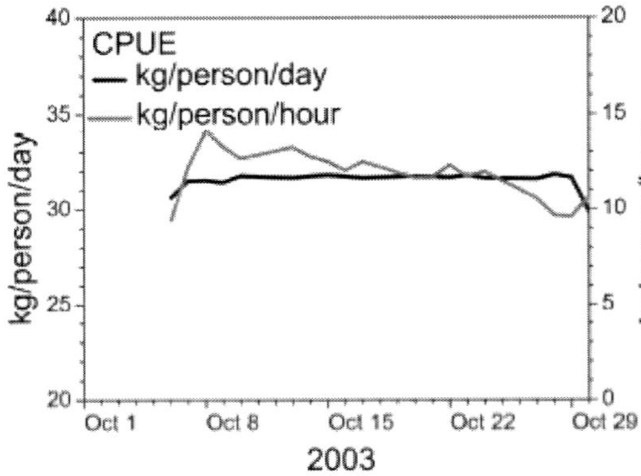

Figure 12. Evolution of catch per unit effort (CPUE) in terms of kg per person and day and kg per person and hour at the start of the 2003 shellfishing campaign in the Ulla Estuary.

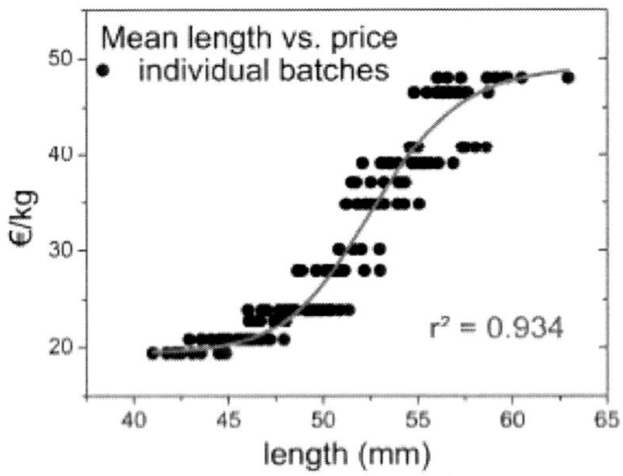

Figure 13. Relation between the mean length of individual catches of the pullet carpet shell sold at the marketplace in December, 2010, and auction price; r^2=0.934.

a) b)

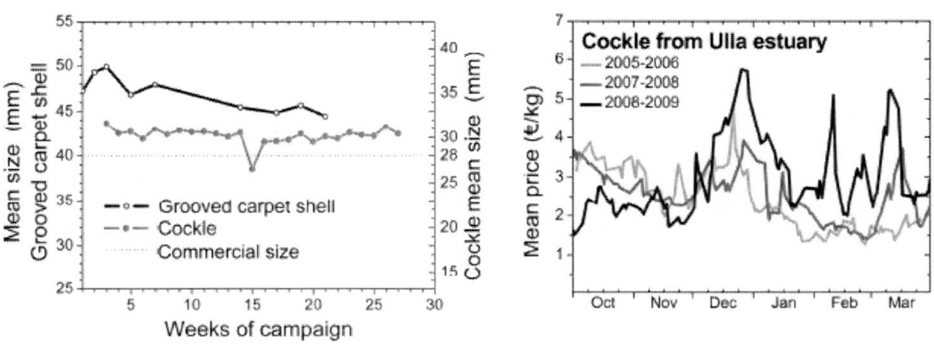

Figure 14. Evolution of the mean size in the catch (a) and the prices reached during different seasons (b).

However, this decrease may also be related to less strictly enforced surveillance and control. In this way, an in-season increase in mean length may be due to the shellfishers' changing zones or increased surveillance of size control (Figure 14a). All these indices must be interpretated jointly because of their interrelationships. There is a pattern of price increase at certain times of the year (Figure 14b), so that a decrease in the mean size of the catch does not necessarily correspond to a drop in price.

Similarly, a drop in the accumulation of catches is not important if it coincides with dwindling effort. It even could be misleading if it coincides with a decrease in mean size of catch, because could be related with a lack of interest of shellfishers more than with a decreased stock abundance. Managers may use this information to adjust quotas, change harvesting zones, increase monitoring of size, apply the minimum price established into the plan to withdraw the product from the auction, and cease harvesting activity on certain days or seasons.

11.4.3. Semiculture Techniques

In the exploitation of clams and cockles (unlike in crustaceans or fish fisheries) it is possible to make use of semiculture techniques to enhance productivity. Although specimens cultured in hatcheries are some times introduced into the natural beds, the most important activities are related with the natural bed itself, so that the main goal focuses on increasing just the natural stock and improving its survival rate. This can be considered as extensive culture, although in most cases only "wild" specimens that were born in the natural bed are involved (Figure 15). Techniques consist mainly of transplants and seeding. Transplanted specimens are collected in recruitment zones or high density areas and transferred to low-density or high-growth areas. The main goal of this practice is to reduce the incidence of density-dependent phenomena that increase mortality or negatively affect settlement and growth (Montaudouin and Bachelet, 1996). This technique is also used for repopulation. Transplants are sometimes directed at zones with better access conditions during neap tides, or areas that are more sheltered form storms, in order to increase the number of working days available for extraction. Oftentimes, the areas having the greatest density coincide with highly accessible areas, so it is easy to collect large quantities in a short period of time. In these cases the purpose of the transplant is to shelter the clams from intruders.

Figure 15. Transplants (a) and beach cleaning (b) carried out by a group of shellfishers.

Seeding differs from transplants in that the specimens planted are reared in hatcheries. While it is hard to assess the efficiency of this form of stock enhancement in economic terms, some of the studies carried out suggest that viability is questionable in the case of clams (Burton et al., 2001). However, the objective is not so much to step up production for the sale of the cultivated spat itself, but to repopulate and increase the reproductive potential of the bed (Arnold, 2001).Like transplants, seeding requires information about the characteristics of the bed and stock distribution, being more likely to succeed when knowledge is available regarding sources and sinks in the population. Major drawbacks of seeding are availability of seed, high mortality and labor requirements. Habitat enhancement is oriented towards the increase of the net useable bed area, restoring degraded areas, and maintaining optimum conditions for the development of the target species. Practices include removal of seaweed and objects piled up on the beds, breaking up the substrate in areas that tend to become compacted or accumulate muddy sediments, and adding gravel to balance sediment texture in the muddy zones. The drawback of these practices is that they usually require continued attention.

11.5. Future Needs

Despite the economic importance of this activity, standardized indices to draw cartographic information showing the suitability of Galician coasts for the development of shellfishing have not yet been developed. This information should help to integrate this activity into coastal management projects. Increased demand for RandD in shellfishing in Galicia focusing on habitat enhancement and development of diagnostic procedures for environmental issues in relation with falling productivity. Low success when implementing habitat enhancement techniques is often linked to diagnostic errors. Changing configuration of coastline and increasing urban pressure seem related to increasing need for actions keeping the environmental and health conditions. Studies of basic biology involved in stock management, as density dependent factors affecting pathology or growth, relationships of coastal dynamics and larval biology, or the bloom dynamic of toxic phytoplankton, should not be neglected. Similarly, knowledge about the effects of acidification, increasing temperature, and the introduction of exotic species must be improved. More applied studies such as developing low-cost devices for algae cleaning, industrial use of this subproduct and developing of traps and baits for predator control, could contribute to increased production.

Acknowledgments

Referees' sugestions have improved the final manuscript.

References

Arnold, W. S. (2001). Bivalve enhancement and restoration strategies in Florida, US. *Hydrobiologia*. 465, 7-19.

Burton, C. A., MacMillan, J. T. and Learmouth, M. M. (2001). Shellfish ranching in the UK. *Hydrobiologia.* 465, 1-5.

Caddy, J. F. and Defeo, O. (2003). Enhancing or restoring the productivity of natural populations of shellfish and other marine invertebrate resources. *FAO. Fisheries technical paper.* 448. 159 pp.

Fernández Cortés, F., Moscoso, E. R. and Pazó, J. P. (1987). Análisis de la explotación de moluscos bivalvos en la ría de Vigo. II. Discusión sobre la producción comercial de moluscos. *Cuad. Marisq. Publ. Téc.* 9, 121-136.

Gosling, E. (2003). *Bivalve molluscs. Biology, ecology and culture.* Galway. Blackwell Publishing.

Mahou Lago, X. M. (2008). Implementación y gobernanza: La política de marisqueo en Galicia. Escola Galega de Administración Pública. *Monografías.* 13. 310 pp.

Marugán Pintos, B. (2004). *E colleron ese tren. Profesionalización das mariscadoras galegas.* Santiago de Compostela. Xunta de Galicia. Consellería de Pesca e Asuntos Marítimos.

Montaudouin, X. and Bachelet, G. (1996). Experimental evidence of complex interactions between biotic and abiotic factors in the dynamics of an intertidal population of the bivalve *Cerastoderma edule. Oceanol. Acta.* 19, 449-463.

Parada, J. M., Molares, J., Sánchez-Mata, A., Martínez, G., Darriba, C., and Mariño, J. (2006). Plan de actuación para la recuperación del banco "Lombos do Ulla": Campañas marisqueras desde 2002 a 2005. *Revista Galega dos Recursos Mariños (Artículos e Informes Técnicos).* 1, 1-37.

Parada, J. M. and Molares, J. (2008). Natural mortality of the cockle *Cerastoderma edule* (L.) from the Ria of Arousa (NW Spain) intertidal zone. *Revista de Biología Marina y Oceanografía.* 43, 501-511.

Parada, J. M. and Molares, J. (2009). *Unha ferramenta para a avaliación de recursos marisqueiros. ARouSA.* Consellería do Mar; Xunta de Galicia. Santiago de Compostela.

Parada, J. M., Molares, J. and Otero, X. (2011). Multispecies mortality patterns of commercial bivalves in relation to estuarine salinity fluctuation. *Estuar. Coast.* doi: 10.1007/s12237-011-9426-2.

Plataforma Tecnolóxica da Pesca. Consellería do Mar. Xunta de Galicia. [Online] Available on www.pescadegalicia.com. [Acceded on 11/10/2011].

Ríos, P. (2001). Bancos marisqueros de moluscos bivalvos de sustratos blandos explotados en las costas de Galicia. *Ph.D Thesis. University of Santiago de Compostela.* 509 pp.

Sen, S. and Nielsen, J. R. (1996). Fisheries co-management: a comparative analysis. *Marine Policy.* 20(5), 405-418.

Unidade Técnica de Pesca de Baixura (UTPB) (2005). *Georreferenciación y cartografiado de los bancos marisqueros de Galicia. Evaluación del potencial marisquero.* Xunta de Galicia (Ed.). Los recursos marinos de Galicia. Serie técnica N°4. 280 pp.

In: Clam Fisheries and Aquaculture
Editor: Fiz da Costa González
ISBN: 978-1-62257-518-3

Chapter 12

CLAM FISHERIES WORLDWIDE: MAIN SPECIES, HARVESTING METHODS AND FISHING IMPACTS

M. B. Gaspar,[1*] I. Barracha,[2#] S. Carvalho[1#] and P. Vasconcelos[1#]
[1]Instituto Português do Mar e da Atmosfera (IPMA.) / IPIMAR, Olhão, Portugal
[2]Inês Barracha: arte | design | multimédia, Quinta das Oliveiras, Portimão, Portugal

ABSTRACT

This chapter presents a comprehensive compilation of information available on fisheries targeting bivalves belonging to the family Veneridae worldwide. Based on an extensive search of data, from both bibliographic references and web-sites, the chapter provides information on the most important clam fisheries, the main harvesting methods and fishing gears employed to target venerids, and the respective impacts on the marine environment. Overall, it was gathered information concerning the exploitation of 36 genera and at least 89 species of venerids, which are caught worldwide using quite different harvesting methods and fishing gears. Depending on where the bivalves occur (intertidal flats vs. subtidal areas) and on the type of fishery (subsistence, recreational or commercial), harvesting can be performed through highly diverse methods, from very simple and rudimentary techniques (e.g. hand picking) to sophisticated gears (e.g. hydraulic dredge). Accordingly, the extent and magnitude of bivalve fishing impacts on the marine ecosystems depend on the combined influences of local hydrodynamic and environmental conditions (currents, tidal strength, water depth, nature of the substrata and

[*] Corresponding author: M. B. Gaspar. Instituto Português do Mar e da Atmosfera (IPMA) / IPIMAR, Avenida 5 de Outubro s/n, 8700-305, Olhão, Portugal. E-mail address: mbgaspar@ipimar.pt.

[#] I. Barracha: Inês Barracha: arte | design | multimédia, Quinta das Oliveiras, Lote 29, Loja 18, 8500-818, Portimão, Portugal. S. Carvalho, P. Vasconcelos: Português do Mar e da Atmosfera (IPMA) / IPIMAR, Avenida 5 de Outubro s/n, 8700-305, Olhão, Portugal.

structure of the benthic communities) and the type and intensity of the fishery (type of gear, fishing frequency, towing speed and gear penetration into the sediment).

12.1. Introduction

There are approximately 9,200 living species of bivalves worldwide, belonging to 1,260 genera and 106 families (Huber, 2010), a diversity that makes the class Bivalvia the second largest within the phylum Mollusca (the biggest phylum in the marine environment). Among these, marine bivalves (including brackish water and estuarine species) represent about 8,000 species, distributed within 1,100 genera and 99 families (Huber, 2010). Besides this massive diversity, bivalves also play key ecological roles in marine ecosystems.

Additionally, several bivalve species are used for human consumption and constitute important natural resources in many parts of the world. Most bivalve species are exploited because they are highly productive, low in the food chain, abundant in shallow coastal waters and are tasty and nutritious.

Bivalve fisheries play a significant part for the social, economic and cultural well-being of many coastal communities. Although bivalves contribute with a small percentage to global fishery landings, in general their commercial value (high unit price) compensates for the smaller landed weight compared to other categories of marine organisms (fishes, crustaceans and other molluscs) (Gosling, 2003). However, it is virtually impossible to determine the actual contribution of bivalve fisheries to the world fishery catches, because at local, regional or national levels, statistics frequently group together the fishery / harvesting landings and the aquaculture production.

Moreover, in some cases, different *taxa* are grouped under generic, vague and indefinite terms such as "marine molluscs". Likewise, whenever different bivalves are not discriminated in fishermen logbooks or by the authorities, several species are included under the generic denomination of "bivalves". For these reasons, some records of bivalve landings are ambiguous and inaccurate, especially in the case of small scales fisheries.

According to the latest statistics available on the world capture production by groups of species, in 2009 clams' fisheries (together with cockles and arkshells) amounted almost 740 thousand tonnes (FAO, 2010a), which corresponded to less than 1% of the world total capture production of marine organisms (FAO, 2010b). Concerning the overall economic value, those catches were estimated to account around 725 million US$, which was also under 1% of the total economic value of fisheries worldwide (FAO, 2010c).

Many of the most important edible bivalve species are commonly known simply as "clams". The present chapter, devoted to the clams' fisheries worldwide, focuses on the exploited species belonging to the family Veneridae (hard-shell clams or Venus clams). Veneridae is the largest recent family of marine bivalves, including over 680 living species worldwide (Huber, 2010) and thus constituting the most speciose family of heterodont bivalves (Chen et al., 2011). Veneridae is a cosmopolitan and ubiquitous family, distributed across all world oceans, from intertidal flats to deep-sea areas, colonising all types of soft bottoms. In general, venerids burrow in muddy or sandy habitats, but vary considerably in lifestyles, thus inhabiting mangrove zones, coastal lagoons, estuaries, bays, surf zones and the deep sea (below 200 metres depth) (e.g. Tebble, 1966; Poppe and Goto, 1993; Macedo et al., 1999).

Veneridae includes from minuscule (4 mm in shell length) to large species (>10 cm in shell length) (Canapa et al., 2003), but most venerids are edible and many are exploited as a food source for human consumption.

Some venerid species are among the favourite shellfish of people everywhere, reaching high economic value in the seafood market. Many of those commercially important venerids are subjected to intensive exploitation and make up a significant proportion of the world fishery of edible bivalves. Besides fishing / harvesting for human consumption, some venerid species are also gathered for using their meat as fishing bait and their shells for producing lime, for ornamental purposes and for malacological collections.

12.2. Collection of Information/Data

The present chapter presents a comprehensive compilation of information on fisheries targeting venerid species worldwide. In general, data was collected through a bibliographic search of scientific articles, books, reports and web-sites containing relevant information on this topic. Although the extensive amount of information gathered, it is virtually impossible to include all venerid fisheries worldwide, therefore the present data should be interpreted as an effort to encompass the most relevant cases.

Despite the existence of quite different types of venerid fisheries (some are commercial, while others are recreational or for subsistence), the information presented did not consider such distinction. Moreover, this kind of data is very scarce for some geographical regions (e.g. African and Middle East countries) or even lacking for several countries worldwide.

Therefore, the information compiled in the text and illustrated in the maps certainly does not cover all fisheries of a certain species, instead meaning that for other regions / countries data was not available or was not found.

The species nomenclature was revised following two well-known and freely available databases: WoRMS (Appeltans et al., 2011) and SeaLifeBase (Palomares and Pauly, 2011). Priority was given to WoRMS but SeaLifeBase was consulted whenever there was no information available in the former. Nomenclature remained unchanged (i.e. species names were kept as cited in the original source of data) whenever there was no information available in those databases. Each time species nomenclature changed, the currently valid designation was adopted (and accordingly all information was grouped under the current species name). In every case, the original nomenclature was kept unchanged in the respective sources of information (bibliographic references or web-sites cited below).

The present chapter focuses exclusively on clam fisheries targeting species belonging to the family Veneridae, particularly in what concerns the main exploited species, most important fisheries, harvesting methods and fishing gears.

However, several bivalve fisheries are multi-specific, catching both venerid and non-venerid species when using the same harvesting / fishing techniques, therefore their impacts are expected to be similar.

For this reason, the sub-chapter summarising the main fishing impacts of these activities reports information available for bivalve fisheries *sensu lato* and not exclusively for clam fisheries targeting venerid species.

12.3. Most Important Clam Fisheries

The geographic distribution of the most relevant venerid fisheries worldwide (either commercial, recreational or for subsistence) is presented in figures 1 to 9. In this context, it is worth mentioning that the position of the symbols refers only to the existence of the fishery in the country, not necessarily to its precise location within the country.

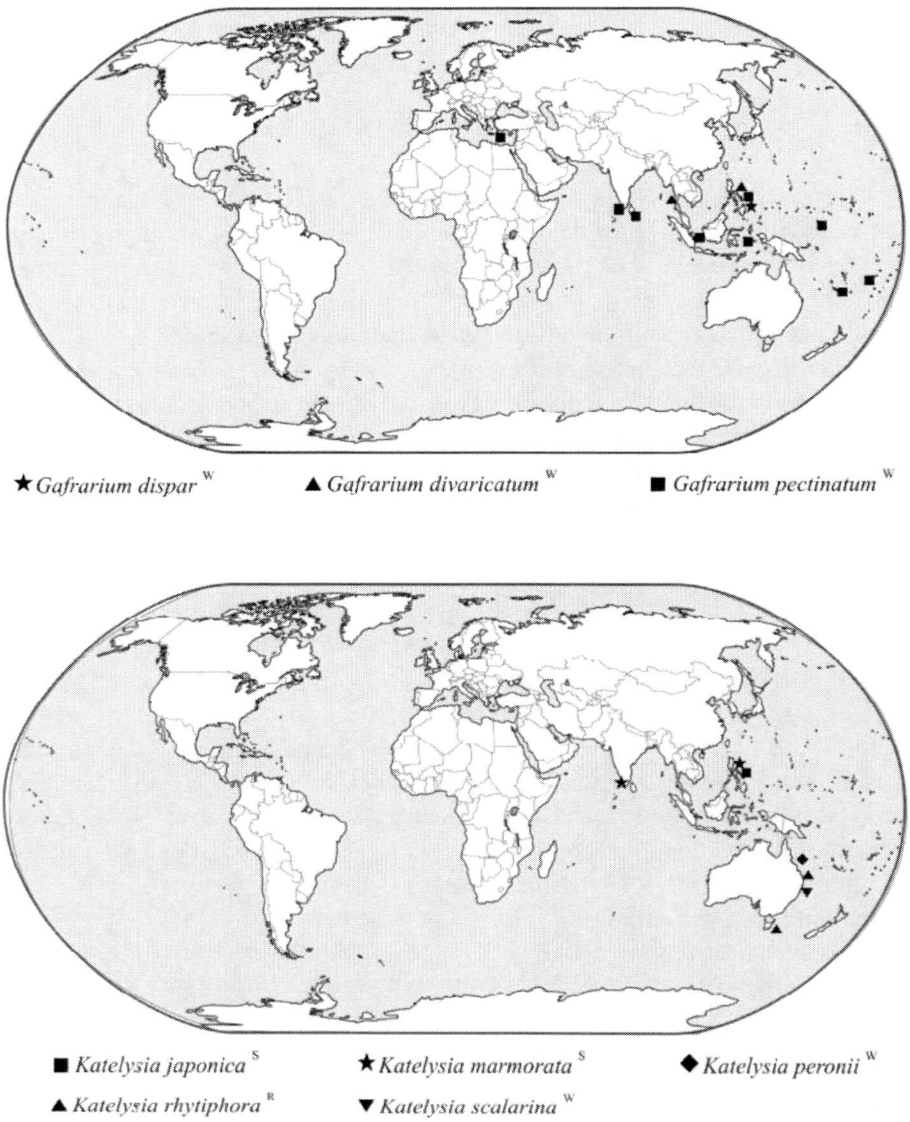

Figure 1. Venerid fisheries of the genera *Gafrarium* (top) and *Katelysia* (bottom). Nomenclature adopted according to W - WoRMS, S - SeaLifeBase or R - bibliographic reference.

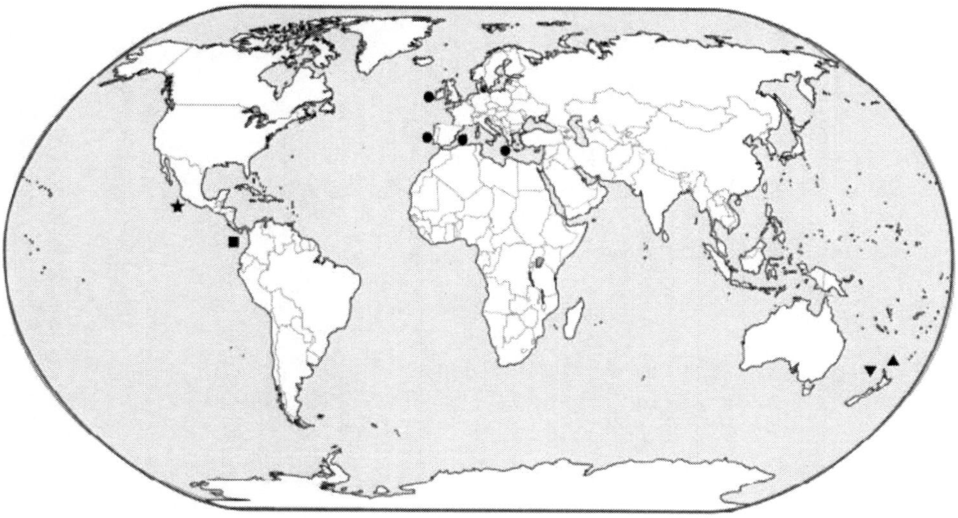

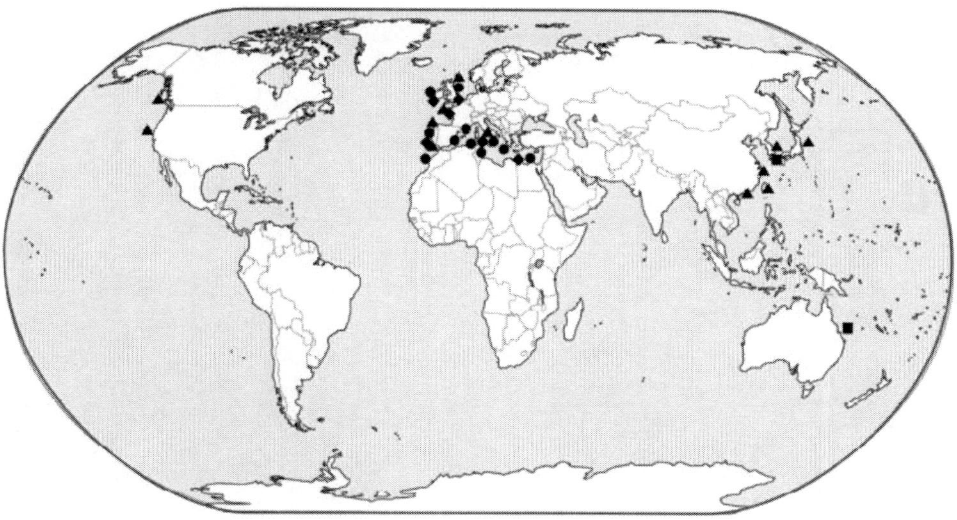

Figure 2. Venerid fisheries of the genera *Dosinia* (top), *Ruditapes* and *Venerupis* (bottom). Nomenclature adopted according to W - WoRMS or S - SeaLifeBase.

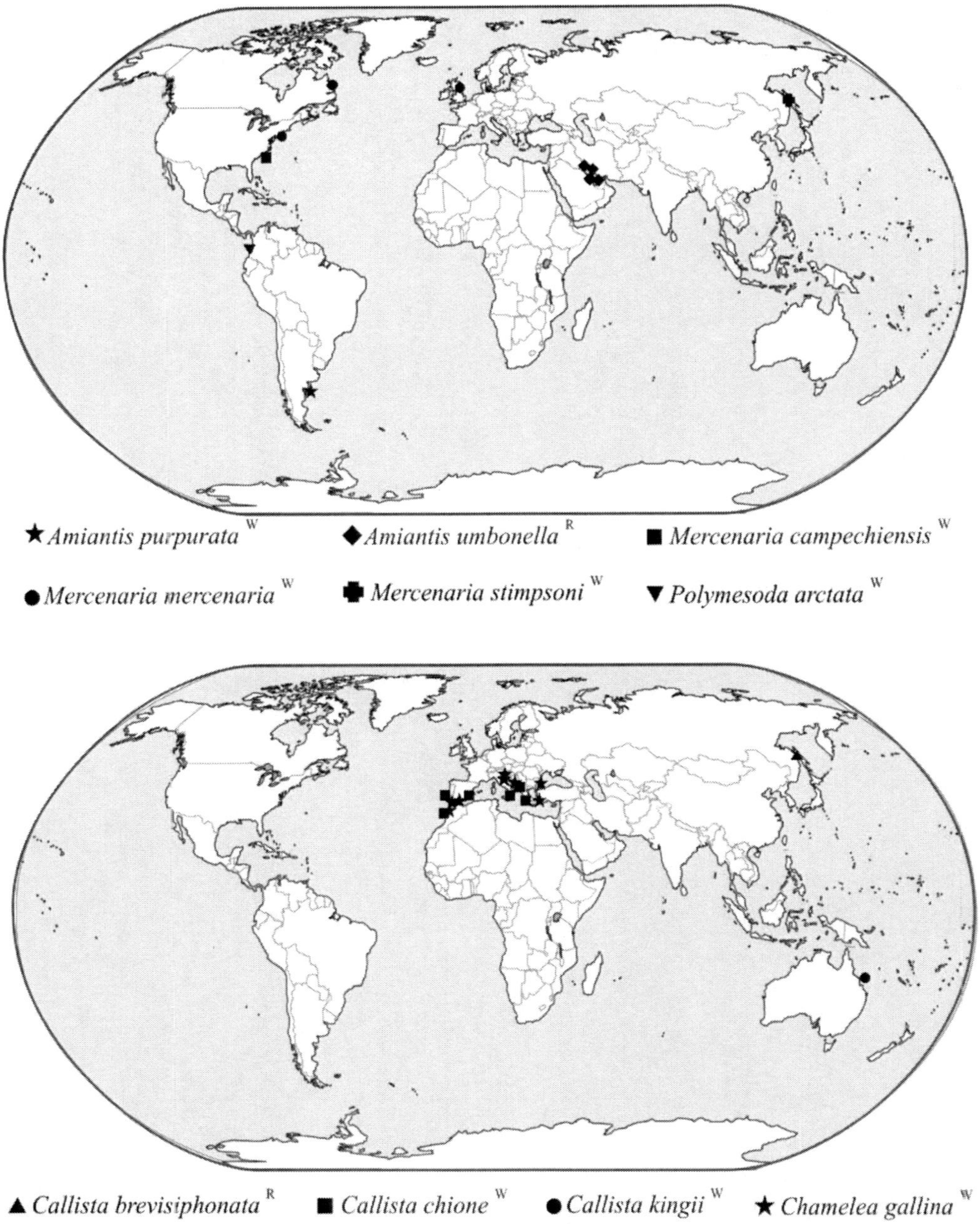

Figure 3. Venerid fisheries of the genera *Amiantis, Mercenaria, Polymesoda* (top), *Callista and Chamelea* (bottom). Nomenclature adopted according to [W] - WoRMS or [R] - bibliographic reference.

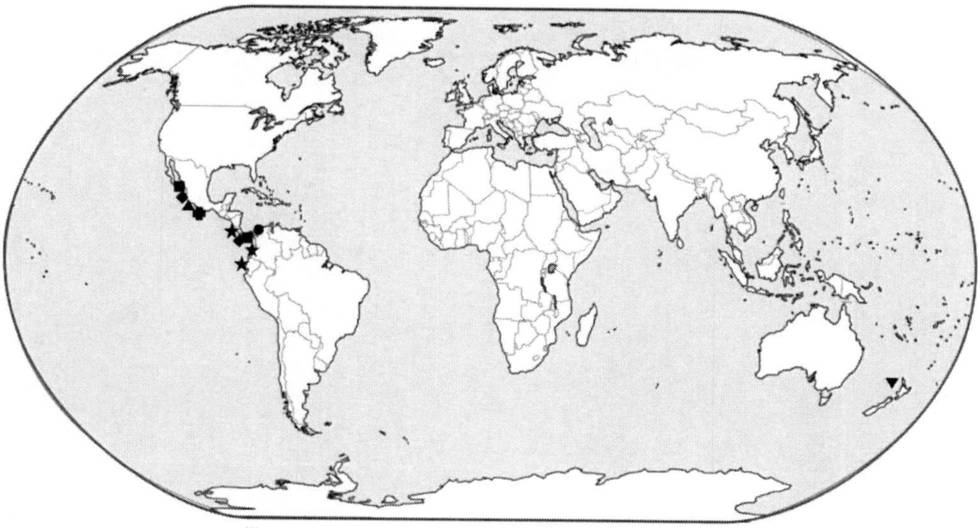

■ *Chione californiensis* [W] ● *Chione cancellata* [W] ✚ *Chione gnidia* [R] ▼ *Chione stutchburyi* [S]
★ *Chione subrugosa* [S] ◆ *Chione undatella* [W] ▲ *Chionista fluctifraga* [W]

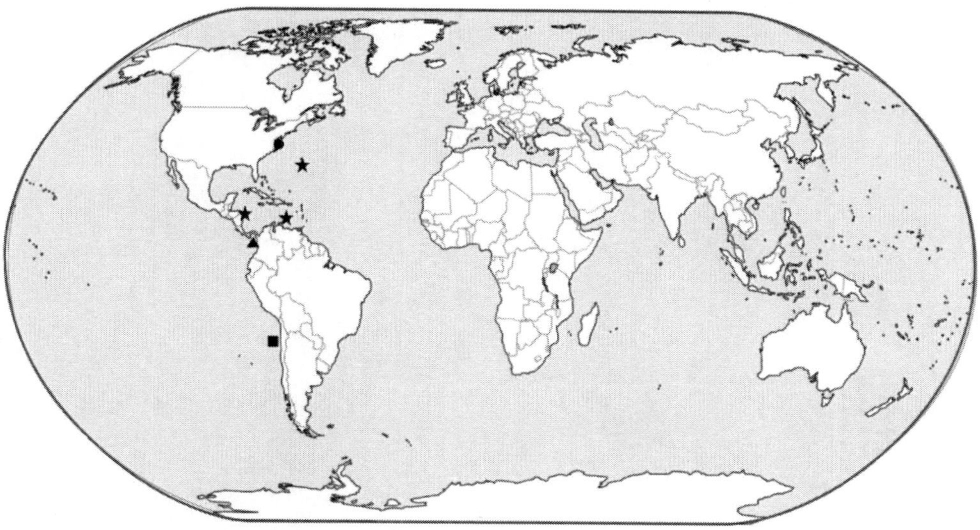

■ *Eurhomalea exalbida* [W] ▲ *Macrocallista aurantiaca* [R]
★ *Macrocallista maculata* [W] ● *Macrocallista nimbosa* [W]

Figure 4. Venerid fisheries of the genera *Chione*, *Chionista* (top), *Eurhomalea* and *Macrocallista* (bottom). Nomenclature adopted according to [W] - WoRMS, [S] - SeaLifeBase or [R] - bibliographic reference.

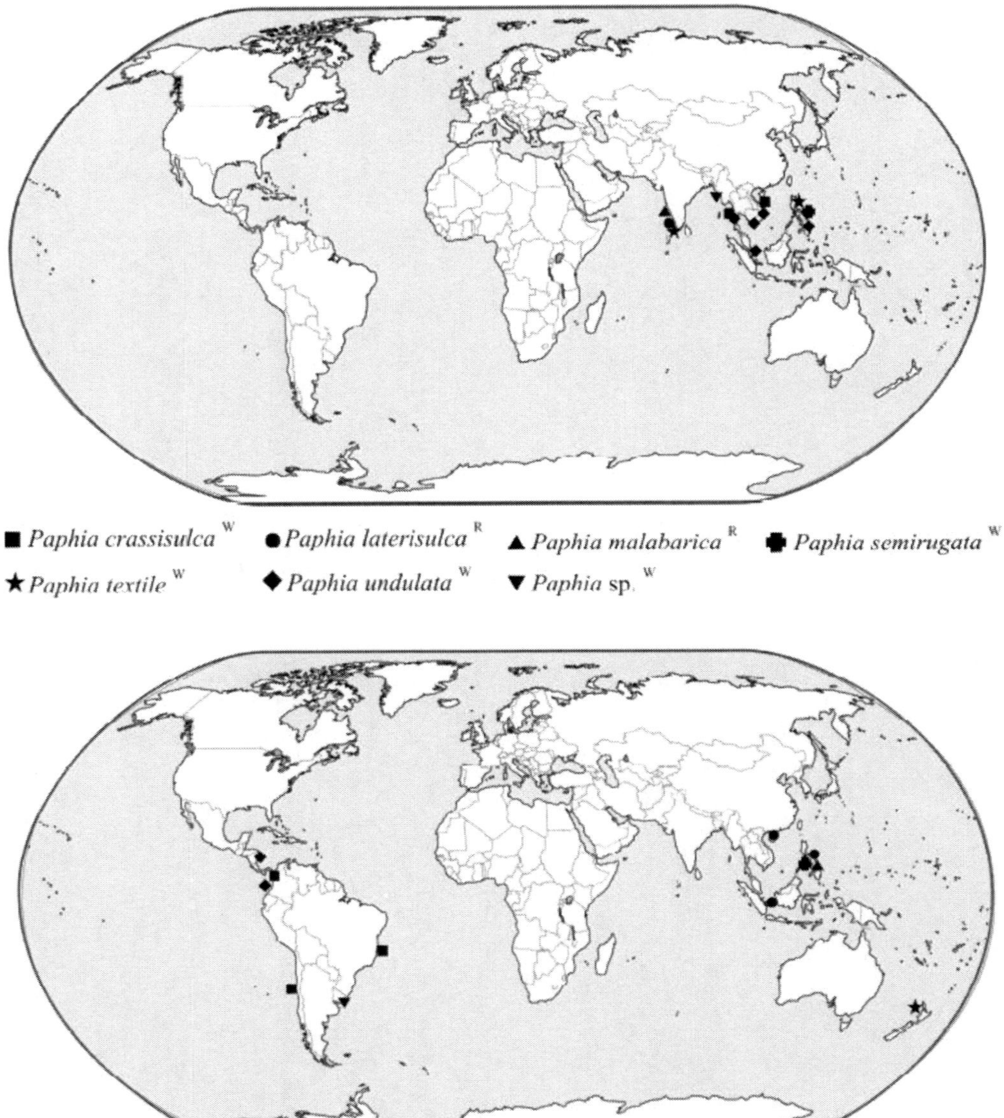

Figure 5. Venerid fisheries of the genera *Paphia* (top), *Anomalocardia*, *Circomphalus*, *Pitar* and *Placamen* (bottom). Nomenclature adopted according to [W] - WoRMS, [S] - SeaLifeBase or [R] - bibliographic reference.

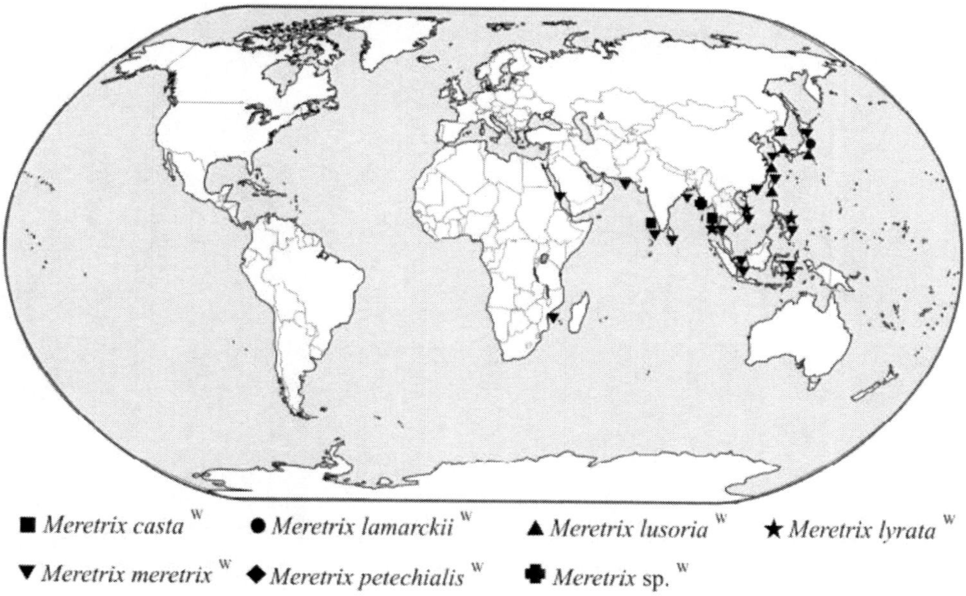

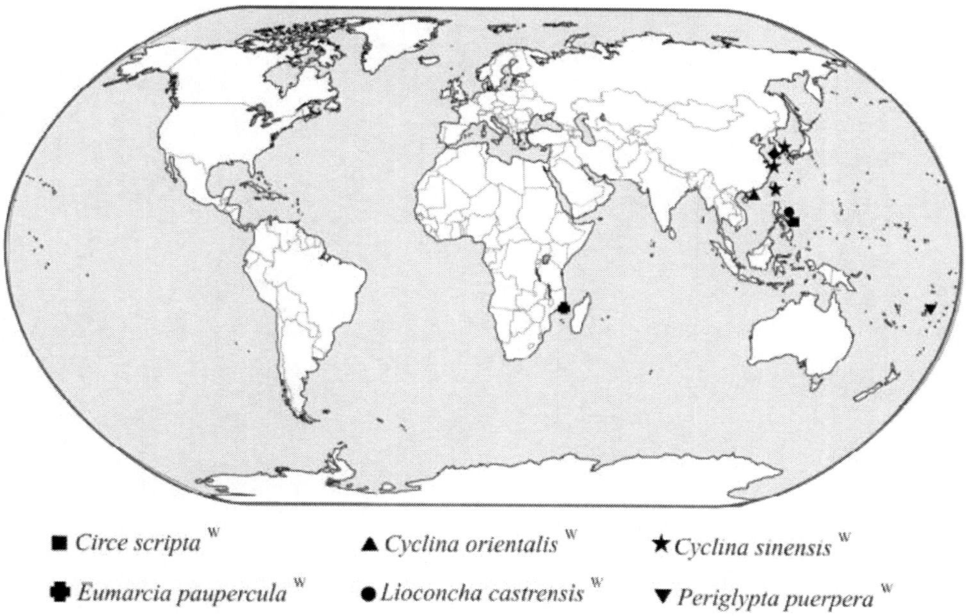

Figure 6. Venerid fisheries of the genera *Meretrix* (top), *Circe*, *Cyclina*, *Eumarcia*, *Lioconcha* and *Periglypta* (bottom). Nomenclature adopted according to W - WoRMS.

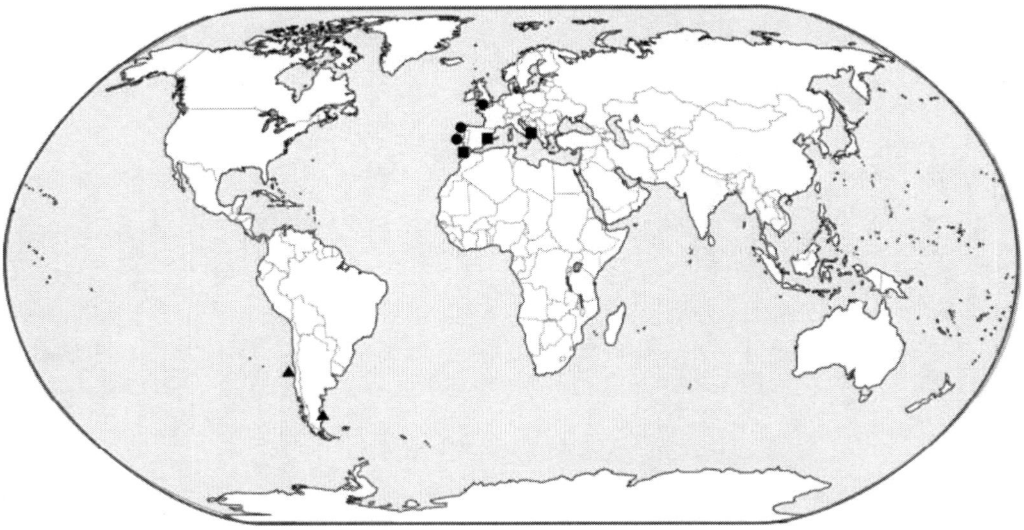

■*Polititapes aureus* [W] ●*Polititapes virgineus* [W] ▲*Tawera gayi* [S]

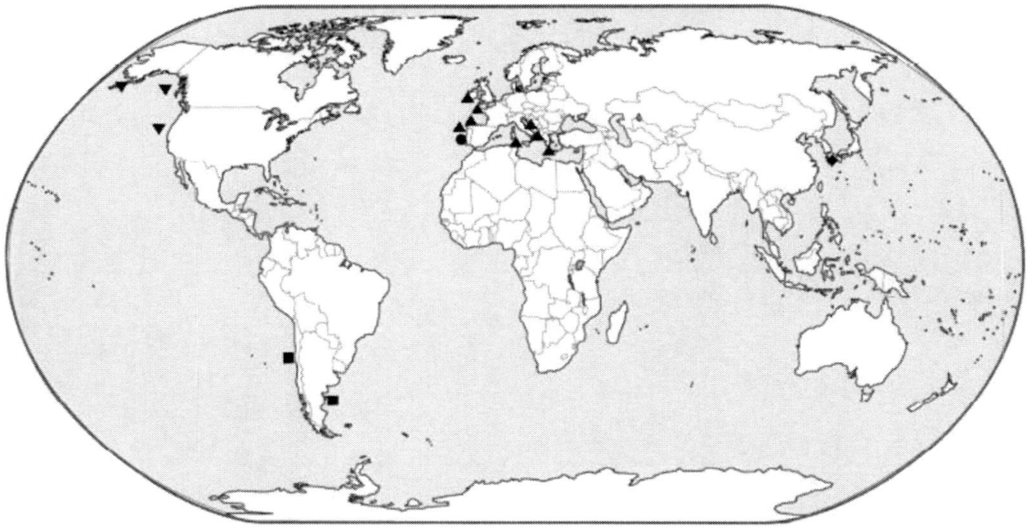

■ *Ameghinomya antiqua* [W] ▼ *Saxidomus gigantea* [W] ◆*Saxidomus purpurata* [W]

●*Venus casina* [W] ▲*Venus verrucosa* [W]

Figure 7. Venerid fisheries of the genera *Polititapes*, *Tawera* (top), *Ameghinomya*, *Saxidomus* and *Venus* (bottom). Nomenclature adopted according to [W] - WoRMS or [S] - SeaLifeBase.

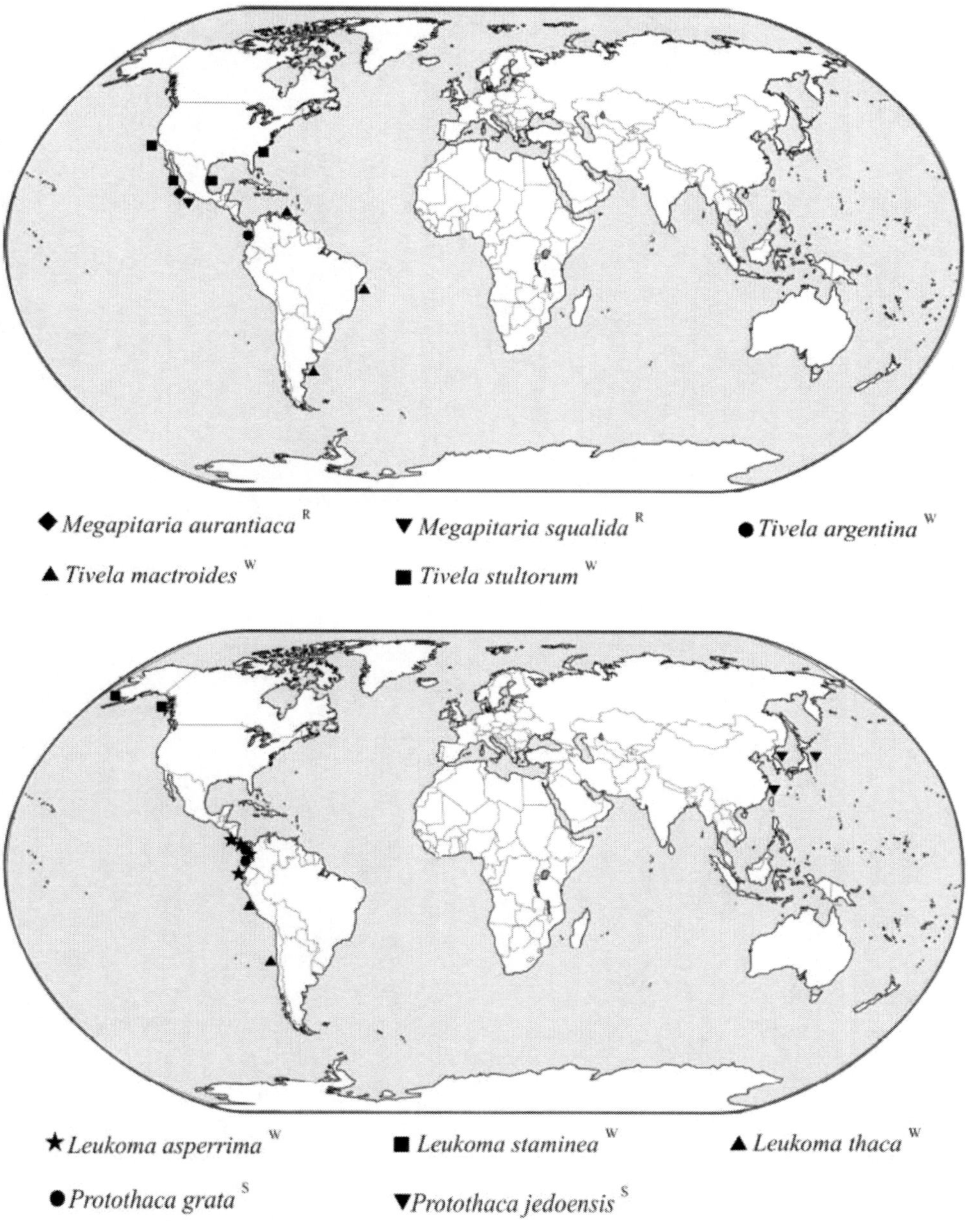

Figure 8. Venerid fisheries of the genera *Megapitaria*, *Tivela* (top), *Leukoma* and *Protothaca* (bottom). Nomenclature adopted according to W - WoRMS, S - SeaLifeBase or R - bibliographic reference.

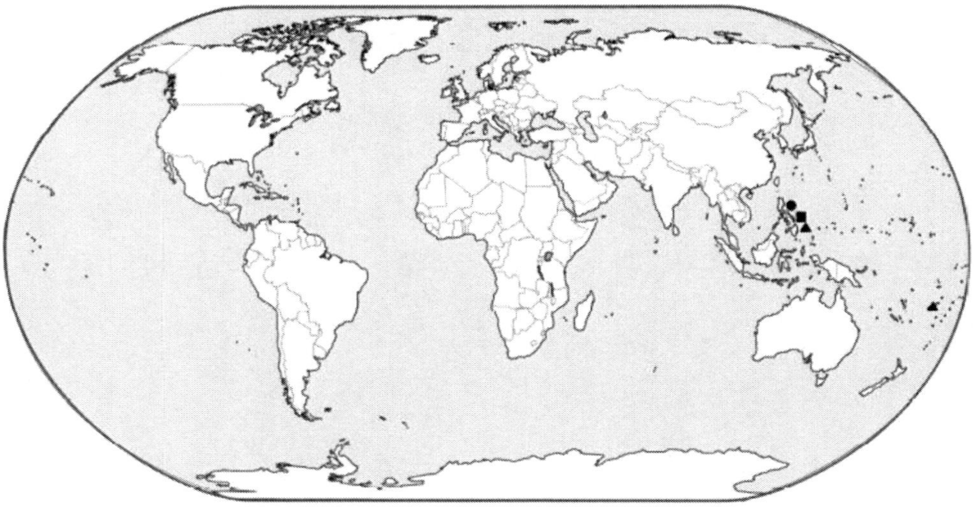

■ *Tapes belcheri* [W] ● *Tapes dorsatus* [S] ▲ *Tapes literatus* [W]

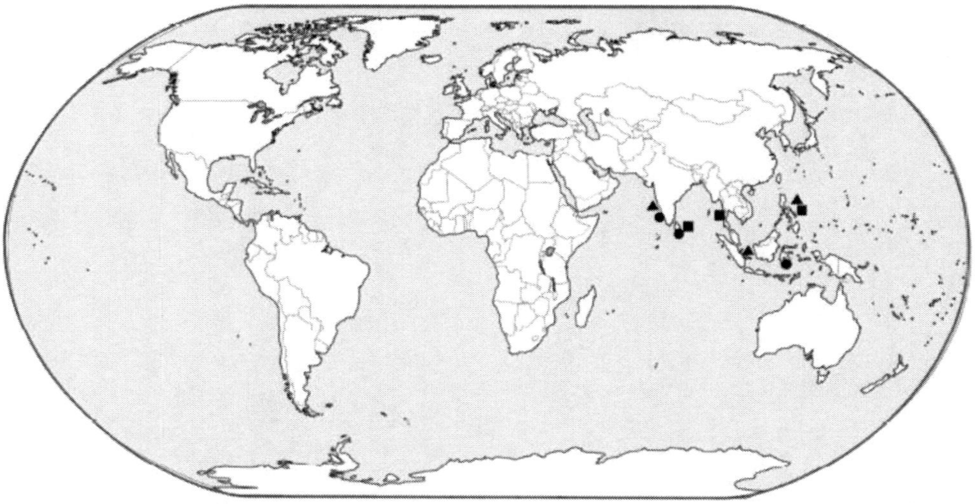

■ *Marcia hiantina* [W] ▲ *Marcia marmorata* [R] ● *Marcia opima* [W]

Figure 9. Venerid fisheries of the genera *Tapes* (top) and *Marcia* (bottom). Nomenclature adopted according to [W] - WoRMS, [S] - SeaLifeBase or [R] - bibliographic reference.

Because of their worldwide distribution, venerids are target species of bivalve fisheries throughout the world's oceans. In general, a greater diversity of venerid species is harvested in the Pacific and Indian oceans, namely compared to the Atlantic Ocean and Mediterranean Sea. Several species are exploited in different geographical areas because of their broad distributional range, whereas others with more restricted distribution are only harvested in more confined areas.

For instance, species belonging to the genera *Callista, Dosinia* and *Venerupis* are fished in the Mediterranean Sea, Atlantic and Pacific oceans. Venerids of the genera *Gafrarium, Katelysia, Marcia, Meretrix* and *Paphia* are harvested mainly in the Indian and Pacific oceans. Fisheries targeting species belonging to the genera *Ameghinomya, Anomalocardia, Chione, Macrocallista, Mercenaria, Pitar, Tawera* and *Tivela* occur both in the Atlantic and Pacific oceans. The genera *Chamelea, Polititapes* and *Venus* are fished both in the Atlantic and Mediterranean. Several venerids are only exploited in a single geographical area, namely the genus *Amiantis* in the Atlantic, the genus *Eumarcia* in the Indian, and the genera *Chionista, Circe, Circomphalus, Cyclina, Eurhomalea, Leukoma, Lioconcha, Megapitaria, Periglypta, Placamen, Polymesoda, Protothaca, Ruditapes, Saxidomus* and *Tapes* in the Pacific.

Finally, it is worth mentioning that this extensive list of exploited venerids includes an autochthonous species from the Pacific Ocean, the Manila clam (*Venerupis philippinarum*), that became a common invasive species in the Atlantic Ocean and Mediterranean Sea, being subjected to commercial exploitation in some European and North American countries nowadays.

12.4. Main Harvesting Methods and Fishing Gears

Although some venerids fisheries occur in the subtidal zone of coastal areas, the majority takes place in intertidal flats and shallow subtidal areas of estuaries, coastal lagoons, sheltered bays and mangroves. In these areas, the most common harvesting methods are hand picking, signing, digging or raking (Figure 10). In hand picking, harvesters get down on their hands and knees or squat down and then sweep their hands through the sediment to dislodge the clams. Signing and digging consists in walking along the shore looking for siphon holes. When a hole is found, the harvester picks the clam from the sediment with his hands or digs the clam out of the sediment using simple tools such as spoons, knives, rakes, forks, shovels or grubber hoes. Some harvesters use a clam gun to collect clams that are buried deeper into the sediment. This tool comprises a 10-15 cm diameter tube with a handle and a small air vent at the closed upper end. When a hole is found, the tube is placed over it and pushed down with an up-and-down, rocking or twisting movement. When the clam is enclosed, the air vent is blocked with a finger and the core of sand pulled up and dropped on the beach to collect the clam. Raking is a technique that involves dragging a rake through the bottom until a scraping is felt. By that time, the harvester pushes the rake into the sediment, pulls it towards him and upwards to collect the clam.

In shallow subtidal areas, treading is a widely used method that consists in probing the bottom with the foot until a clam is felt. When a clam is found, the harvester simply bends down and picks it up. Hand dredging and bullraking are also two common techniques used in shallow areas (ranging between 0.3 and 1.5 m depth). Hand dredges (Figure 10) have a metallic frame with a digging blade or a toothed lower bar. Clams are retained in a mesh bag (made of metal rings or netting material) or in a grid cage (constructed with metal, bamboo or wood) that is tied or welded to the dredge frame. The mouth of the gear is tied, screwed or riveted to a wood or metal handle. In Italy, hydraulic rakes are also used in some areas. This type of rake has at the mouth a blade to cut the sediment and multiple jets that expel seawater under high pressure to fluidize the sediment. Seawater is supplied to the jets through a hose connected to a pump onboard the fishing vessel. Bullrakes are very similar to hand dredges, but instead of having a collecting bag they have a metallic basket for holding the clams as they are dug up. In this gear, the pool ends in a T-shaped handle to help its operation during harvesting. In hand dredging, while walking backwards, the fisherman pushes and pulls the handle repeatedly to facilitate towing the gear over the sediment surface. In bullraking, the rake is positioned on the sediment surface with the tines pointing down. Then, the fisherman pushes the teeth into the sediment and pulls the rake towards him with short, quick jerks, forcing the entrance of clams into the basket. Sometimes, before starting hand dredging or bullraking, fishermen seek for clam beds using the treading technique.

Figure 10. Harvesting techniques employed in intertidal areas: hand picking using rudimentary tools and hand dredging.

Hand dredges and bullrakes (Figure 11) can also be operated from boats in order to exploit deeper clam beds. In this case, the pool used is much longer and might attain 5 m length. Clam tongs are also operated from boats, consisting of two long wooden or metal handles attached together like scissors and with a rake-like basket. Clam tonging (Figure 11) is usually carried in clear waters up to 10 m depth. In some areas, bigger tongs have a hydraulic system to raise and lower the tongs, as well as to close them on the bottom.

Clam kicking (Figure 12) is another harvesting method employed in the low intertidal. In this method, the fisherman uses the propeller backwash to dislodge clams out of the substrate, throwing them on the sediment surface. Then, clams are picked up by hand, or using hand dredges or bullrakes. Some fishermen also use a dredge or a small trawl that is towed from the stern, in order to catch the clams that were kicked out by the propeller backwash.

Figure 11. Manual fishing gears operated from boats: bullrake and clam tong.

Dredging is a technique employed to exploit subtidal clam beds and involves towing a gear using the vessel propeller or a winch (Figs. 12-14). In sheltered areas, clams can be gathered using pump-scoop dredges, hydraulic escalator dredges and hand winch dredges ("sarilho"). These fishing gears are used in the U.K., North America and Portugal, respectively. The pump-scoop dredge comprises a metallic grid cage with a toothed lower bar and a set of water jets attached to the entrance of the scoop to clear sediment from the dredge. Hydraulic escalator dredges (Figure 12) are towed from the side of the vessel and employ jet streams of seawater to cut through the sediment. The head of the dredge, comprising the manifold and a cutting blade, is connected to a steel-mesh conveyor belt. Clams, bycatch and debris that were loosened by the high-pressure seawater jets go through a scoop and onto the conveyor belt in a steady stream, where a fisherman culls the legally-sized clams. Unwanted catch and debris continue along the belt until reaching the top and falling back into the water. The "sarilho" fishery uses a hand metal winch screwed to the vessel hull to tow a dredge (Figure 13). This dredge has a semi-circular metal frame and a lower bar equipped with 10 cm length teeth. Clams are retained in a net bag attached to the metal frame of the gear. The fisherman hauls the dredge from the stern of the boat, pays out around 100 m of cable and anchors the vessel from the bow, and then starts dragging the dredge across the bottom by using the hand winch. A similar technique is employed in Morocco, where the only difference is that the winch is made of wood and operated using the hands and feet (Figure 13). In general, Moroccan fishermen drag three dredges simultaneously.

Figure 12. Dredges towed by boats: hydraulic escalator dredge and clam kicking method.

Figure 13. Dredges operated from boats: hand winch dredges.

In coastal areas, the gears more frequently used to catch clams are mechanical and hydraulic dredges. Mechanical clam dredges (Figure 14) have a metallic frame fitted with a toothed lower bar or a digging blade, and a mesh net bag or a rectangular metallic grid cage made of steel bars to retain the catch. A diving vane can also be introduced at the back of the cage, to maintain bottom contact when the digging blade encounters resistance. The maximum burrowing depth of the target species dictates the length of the dredge teeth, but it never exceeds 20 cm length. Small boats can operate up to six dredges, whereas large vessels can operate up to 24 dredges simultaneously. Dredges are usually towed with a cable at 3:1 warp:depth ratio. The towing speed varies between one and four knots and each tow lasts 10 to 30 min, depending on the density of the target species.

Figure 14. Dredges operated from boats: Mechanical and hydraulic dredges.

Hydraulic dredges (Figure 14) comprise a large metal cage made of steel bars to retain the catch, a cutting blade, two skids along both sides of the collecting cage and a system for delivering pressurised seawater through jets. A large pump on the boat pumps high-pressure seawater through a hose, which is delivered from a series of pressure jets at the entrance of the dredge and to the dredge blade. The manifold jets seawater into the sand fluidizing it, allowing the blade to cut into the sediment easily and thus to extract the clams. While some vessels tow hydraulic dredges from the stern, others haul the gear from the bow. In this case, the dredge is towed astern, either by warping on a big anchor using a winch or by moving backwards using the propeller. Usually, the towing speed does not exceed two knots.

Finally, in sheltered areas and coastal zones ranging from 1 to 50 m depth, clams can be harvested by divers. Diving can be carried out in apnea, or using scuba or "hookah" equipment (Figure 15). In "hookah" fishing, divers are supplied with air through long hoses connected to an air compressor onboard the fishing boat.

Divers can collect clams using solely their hands, rudimentary tools, water jets (clam kicking) or sucking devices, depending on the burrowing depth of the target species. When a siphon hole is found or a siphon projecting above the sediment surface is observed, the diver quickly pushes the thumb and forefinger into the sediment to grasp and pull off the clam, or rapidly deploys a tool such as tong pichers and knives to take the animal out of the bottom. Water jets are used to clear sediment away, helping to dig out larger clams that are buried deeper into the substrata. Suction devices that act like a vacuum pump are also used to dig out clams that are deeply buried into the sediment. Suction consists in pumping high pressure seawater through a water jet that creates a suction effect in a flexible tube. Then, the material sucked off the sea bottom is collected in a net bag that is dumped in the vessel deck from time to time for sorting the catch.

Figure 15. Harvesting techniques employed in subtidal areas: apnea, "hookah" and SCUBA diving.

12.5. Main Fishing Impacts

Fishing is one of the main anthropogenic drivers of ecosystems alterations by inducing changes in biological communities, in the pathways of energy transfer and in the sea-floor habitats (e.g. Jennings et al., 2001; Zhang et al., 2009). Ecosystem changes are mainly associated with mobile bottom gears, especially dredges, which impact the benthic habitat and associated assemblages of species. In order to catch target species, fishing gears are dragged over the seafloor imposing several damages that can be reflected either in sediment, water column or in the associated species (e.g. Hall and Harding, 1997; Pranovi et al., 1998; Hiddink, 2003; Masero et al., 2008). The extent and magnitude of these impacts depend on the combined influences of factors such as the fishing frequency, towing speed, gear type, gear penetration into the sediment, time of year, local environmental conditions (such as water depth, tidal strength and currents), nature of the substrata and the structure of the benthic communities affected (de Groot, 1984; Churchill, 1989; Mayer et al., 1991). In recent decades, extensive literature has been produced concerning the impacts of bivalve fishing on the ecosystems (e.g. Eleftheriou and Robertson, 1992; Dare et al., 1993; Hall et al., 1993; Jennings and Kaiser, 1998; Hall-Spencer and Moore, 2000), which will be briefly reviewed in this sub-chapter. Change of seabed topography is an immediate effect resulting from the passage of the mobile fishing gears through the sediment, which may cause habitat fragmentation and loss (Wilcove et al., 1998; Sih et al., 2000; Thrush and Dayton, 2002; Tudela, 2004). When fishing is undertaken in areas of structurally complex habitats such as natural reefs, mangroves, maerl beds, seagrass and kelp meadows, their physical integrity is generally affected (Norse and Watling, 1999; Hall-Spencer and Moore, 2000; Valiela et al., 2001; Duarte, 2002; Fossa et al., 2002; Hall-Spencer et al., 2002; Hauton et al., 2003a; Tudela, 2004; Hiscock et al., 2005; Wolff, 2005; Rabaut et al., 2008), and their ecological relevance (e.g. source of food resources, nursery areas, spawning grounds, refuge from predators) may be impaired. Indeed, the loss of habitat heterogeneity may lead to significant changes on fish communities and ecosystems (Thrush et al., 1995; Sainsbury et al., 1997; Barbera et al., 2003; Hall-Spencer et al., 2003), to reduced biodiversity, decreased benthic production, loss of slow-growing long-lived species, and may also affect predator-prey interactions and the energetic needs of individuals (McConnaughey et al., 2000; Jennings et al., 2001; Sewell and Hiscock, 2005). On the other hand, those impacts are expected to be much lower in less complex habitats such as sandy bottoms (e.g. Constantino et al., 2009).

With the passage of the fishing gear, several furrows of variable dimensions (depending on the specifications of the gear being used: gear width, tooth length, cutting depth of the blade and/or the water jets pressure) are created along the seabed. Furrows can last for a few hours (Gaspar et al., 2003), few days (DeAlteris et al., 1999; Tuck et al., 2000; Gaspar et al., 2003), several months (Pickett, 1973; DeAlteris et al., 1999) or few years (Gilkinson et al., 2003), as a function of the cohesive characteristics of the sediments, depth and the local hydrodynamic regime of the area where the fishery takes place (Caddy, 1973; Dernie et al., 2003). Besides, while mobile fishing gears are being towed across the sea floor, sediment is re-suspended and sediment plumes are produced in the water column, whose residence time also depends on the local hydrodynamics, but usually disappear within a short-period of time (Medcof and Caddy, 1971; Caddy, 1973; Butcher et al., 1981; Meyer et al., 1981; Mayer et al., 1991; Gaspar, 1996; Gaspar et al., 2003; Pranovi et al., 2004). In low-energy areas, the prevalence of turbidity for long periods may result in increased mortality of invertebrates, particularly suspension-feeders (Currie and Parry, 1996), as well as reduced light availability for photosynthetic organisms. In shallow waters, because of natural disturbance, turbidity is very common, and only minor effects are expected to be observed. Nevertheless, the dispersal of sediment particles after fishing will be intensified if currents are above the critical threshold for deposition (Falcão et al., 2003). The persistence of these impacts may affect primary productivity, as phytoplankton growth is limited by light intensity (Barnes et al., 1991).

The loss of the sediment fine fraction from the upper sediment layers is common, since these particles once in suspension will travel further than coarser particles and will leave the fished area (Mayer et al., 1991; Tuck et al., 2000; Watling et al., 2001). The restoration of the sediment composition may vary between a few days (Tuck et al., 2000) to up to six months (Watling et al., 2001). Nevertheless, a permanent change in the sediment's composition may occur when repeated fishing operations are undertaken within a certain area (Langton and Robinson, 1990; Pranovi and Giovanardi, 1994; Schwinghamer et al., 1996, 1998; Watling et al., 2001). All these changes will have repercussions in benthic communities as they have strong, but narrow, affinities to certain biotic and abiotic conditions, and the disruption of the existing characteristics may lead to changes in the distribution and abundance of benthic species (Dayton, 1971; Dayton and Hessler, 1972; Thistle, 1981; Lissner et al., 1991; Gibson and Robb, 1992).

Fishing activities not only directly affect target species, but also benthic communities as a whole (Pranovi et al., 1998), and potentially all other species in the system with which they interact (Dayton et al., 1995). Moreover, they can also affect non-benthic species, as they are food items for several coastal fish and bird species (Norse and Watling, 1999; Masero et al., 2008). Fishing impacts depend on the size of benthic animals (meiofauna and macrofauna), their life stage and phase of reproductive cycle, the position of the individuals in the sediment (infauna or epifauna), as well as on the nature of the sediment (soft or hard), fishing effort, resilience and recovery potential of the ecosystem and other local environmental conditions (for a review see Gaspar and Chícharo, 2007).

The short-term environmental impacts of fishing are traduced in the reduction of benthic biomass, diversity, and secondary production (Kaiser, 1998; Collie et al., 2000; Kaiser et al., 2000; Jennings et al., 2001), with major impacts being observed mainly on the non-target species (Ardizzone et al., 2000; Ramsay et al., 2000), as the design of the fishing gear is directed to the commercial species. The reduction in density of non-target fauna in disturbed areas may result from direct mortality caused by fishing, redistribution from disturbed to undisturbed areas or both (Hiddink, 2003), but also from discarding of by-catch specimens, which are dredged, subjected to aerial exposure on the deck, sorted and then returned back to the sea. During any of these procedures, animals may suffer physical damage that may impair their survival potential (Hauton et al., 2003b). However, it is worth mentioning that the survival of discarded individuals seems to be species-specific, being related to their ecology and life histories (Gaspar and Monteiro, 1998) and to their ability to rebury into the sediment and escape from scavengers and predators that move to the fished area (e.g. Fonds, 1994; Ramsay et al., 1996; Fonds et al., 1998; Ramsay and Kaiser, 1998; Coffen-Smout and Rees, 1999; Jenkins and Brand, 2001). For example, species inhabiting sandy beaches and subjected to periodic aerial exposure have a natural resistance to higher temperatures that favours their survival rates (Gaspar and Monteiro, 1998, 1999). On the other hand, after long-term impacts, both biota and abiotic characteristics of the ecosystem can be permanently affected. Nevertheless, studies on long-term changes in benthic ecosystems are seriously constrained by the interactions between fishing and natural disturbance (Currie and Parry, 1996; Kaiser et al., 1998). An alternative is the analysis of time-series data from before and after the beginning of large-scale commercial fisheries (e.g. Reise and Schubert, 1987; Greenstreet and Hall, 1996; De Vooys and Van der Meer, 1998; Hill et al., 1999), although the interpretation of the results may be biased by factors such as climate change (Southward et al., 1995). In general, decrease of benthic diversity, abundance and biomass (e.g. Norse and Watling, 1999;

Veale et al., 2000; Schratzberger and Jennings, 2002), changes in dominant trophic groups (Veale et al., 2000), namely through the selective removal of sensitive taxa (Eleftheriou and Robertson, 1992), the increase of scavengers (Britton and Morton, 1994; Kaiser and Spencer, 1994, 1996; Evans et al., 1996; Collie et al., 1997) and changes in community structure (Lindeboom and de Groot, 1998) have been reported.

As a selective process, fishing has also potential to affect the genetic diversity and genetic structure of a species (Jennings and Kaiser, 1998; Law and Stokes, 2005). Genetic diversity is an essential component of biodiversity and encompasses all the genetically determined differences that occur within and among populations of a species (Bagley et al., 2002) and changes at this level may impair the sustainability of natural populations. Indeed, species with greater genetic diversity are more able to cope with environmental change and therefore have higher chance for long term survival (Reed and Frankham, 2003).

Conclusion

Although with a small contribution to global fishery landings, bivalve fisheries play a significant role in the social, economic and cultural well-being of several coastal communities worldwide. The present chapter confirmed the high diversity of venerid species targeted by subsistence, recreational or commercial fisheries. Similarly, and depending on where those species occur and on the type and development degree of the fishery, venerids are caught using highly diverse harvesting methods and fishing gears, from rudimentary techniques and tools to sophisticated gears. Bivalves harvesting inevitably impacts the benthic habitat and associated assemblages of species, especially when the fishery is performed using mobile bottom gears, namely dredges. The extent and magnitude of those impacts depend upon a number of factors, including the local hydrodynamic and environmental conditions (currents, tidal strength, water depth, nature of the substrata and structure of the benthic communities), combined with the type and intensity of the fishery (type of gear, fishing frequency, towing speed and gear penetration into the sediment).

Acknowledgments

The authors would like to thank the colleagues that kindly provided information on venerid fisheries in their countries, namely Óscar Moreno (Spain), Dai Roberts, Gavin Burnell and Ian Lawler (Ireland), Nikolay Selin (Russia) and César Lodeiros (Venezuela). Thanks are also due to the book editor (Dr. Fiz da Costa) and two anonymous reviewers for valuable comments and suggestions that helped improving the revised manuscript.

Sources of Information/Data

Abubakr, M. M. (2004). *The Republic of Yemen Marine Biotic Ecosystem (Resources, Habitats and Species).* Ministry of Water and Environment, Environment Protection Authority, 128 pp.

Anon. (2010). Giao Thuy Association of Mollusc Farmers. *Vietfish Int.* 7, 26-27.

Anon. (2011). *Guide des Engins de Pêche Artisanale Utilisées à Ghannouch et Akarit (Golfe de Gabès, Tunisie).* FAO CopeMed II - ArtFiMed, Málaga, 29 pp.

Al-Khayat, J. A. and Al-Mohannadi, M. S. (2006). Ecology and biology of the benthic bivalve *Amiantis umbonella* (Lamark) in Khor Al-Adaid, Qatar. *Egypt. J. Aquat. Res.* 32, 419-430.

Alyahya, H., El-Gendy, A. H., Al Farraj, S., and El-Hedeny, M. (2011). Evaluation of heavy metal pollution in the Arabian Gulf using the clam *Meretrix meretrix* Linnaeus, 1758. *Water Air Soil Pollut.* 214, 499-507.

Ardila, N., Navas, G. R. and Reyes, J. (2002). *Libro Rojo de Invertebrados Marinos de Colombia.* Ministerio de Medio Ambiente, INVEMAR, Bogotá, 177 pp.

Arneri, E., Giannetti, G. and Antolini, B. (1998). Age determination and growth of *Venus verrucosa* L. (Bivalvia: Veneridae) in the southern Adriatic and the Aegean Sea. *Fish. Res.* 38, 193-198.

Babaei, M. M., Sahafi, H. H., Ardalan, A. A., Ghaffari, H., and Abdollahi, R. (2010). Morphometric relationship of weight and size of clam *Amiantis umbonella* L., 1818 (Bivalvia: Veneridae) in the eastern coasts of Bandar Abbas, Persian Gulf. *Adv. Environ. Biol.* 4, 376-382.

Bagher, N. S. M., Negar, G., Preetha, K., and Simin, D. M. (2007). Population growth of the venerid bivalve *Circenita callipyga* in the Hendijan coast, Persian Gulf. *Pakistan J. Biol. Sci.* 10, 3185-3189.

Baron, J. and Clavier, J. (1992a). Effects of environmental factors on the distribution of the edible bivalves *Atactodea striata*, *Gafrarium tumidum* and *Anadara scapha* on the coast of New Caledonia (SW Pacific). *Aquat. Living Resour.* 5, 107-114.

Baron, J. and Clavier, J. (1992b). *Etude des Populations de Bivalves Intertidaux sur le Littoral Sud-Ouest de Nouvelle Calédonie.* Convention ORSTOM / FIDES. Conventions Sciences de la Mer: Biologie Marine, No. 5, 76 pp.

Bartosh, G. and Hrynyshyn, J. (Eds.) (2001). *Fish Stocks of the Pacific Coast.* Fisheries and Oceans Canada, 152 pp.

Bechtol, W. R. and Gustafson, R. L. (1998). Abundance, recruitment, and mortality of Pacific littleneck clams *Protothaca staminea* at Chugachik Island, Alaska. *J. Shellfish Res.* 17, 1003-1008.

Bello-Garzón, D., López, J. A. and Diaz, J. (2003). Captura artesanal de almejas con el arte de pesca tellinaro en la zona costera del santuario de flora y fauna Los Flamencos La Guajira - Caribe Colombiano. *X Congreso Latinoamericano de Ciencias del Mar*, 22-26 de Setiembre 2003, San José, 4 pp.

Berthou, P., Poutiers, J. M., Goulletquer, P., and Dao, J. C. (2009). Shelled Molluscs. In: R. Radke (Ed.). *Fisheries and Aquaculture, Volume II, Marine Fish and Invertebrates: Biology and Harvesting Technology*. Encyclopedia of Life Support Systems (EOLSS), developed under the auspices of the UNESCO. pp. 135-158, EOLSS Publishers, Oxford (available at http://www.eolss.net).

Boominathan, M., Subash Chandran, M. D. and Ramachandra, T. V. (2008). Economic Valuation of Bivalves in the Aghanashini Estuary, West Coast, Karnataka. ENVIS Technical Report No. 30. Sahyadri Conservation Series - 9. Centre for Ecological Sciences, Bangalore, 33 pp.

Borzone, C. A., Vargas, K. M., Pezzuto, P. R., and Tavares, Y. A. G. (2001). Aspectos da reprodução e dinâmica populacional de *Chione pubera* (Bory Saint-Vicent) (Bivalvia, Veneridae) no sul do Brasil. *Rev. Bras. Zool.* 18, 333-349.

Brandão, J., de Paula, R., Ferreira, P., Lavander, H., Rodrigues, S., Oliveira, L., Souza, A., and Gálvez, A. (2010). Identificação do aparelho de pesca na coleta da *Anomalocardia brasiliana* (Gmelin, 1971), praia de Mangue Seco, Pernambuco - Brasil. *X Jornada de Ensino, Pesquisa e Extensão - JEPEX 2010.* 18 a 22 de Outubro, UFRPE, Recife, 3 pp.

Caddy, J. F. and Defeo, O. (2003). *Enhancing or restoring the productivity of natural populations of shellfish and other marine invertebrate resources.* FAO Fisheries Technical Paper No. 448. FAO, Rome, 159 pp.

Cárdenas, E. B. and Aranda, D. A. (2000). A review of reproductive patterns of bivalve mollusks from Mexico. *Bull. Mar. Sci.* 66, 13-27.

Carpenter, K. E. and Niem, V. H. (Eds.) (1998). *The Living Marine Resources of the Western Central Pacific. Volume 1. Seaweeds, Corals, Bivalves and Gastropods.* FAO Species Identification Guide for Fishery Purposes. FAO, Rome, 686 pp.

Castilla, J. C. and Gelcich, S. (2008). Management of the Loco (*Concholepas concholepas*) as a Driver for Self-Governance of Small-Scale Benthic Fisheries in Chile. In: R. Townsend, R. Shotton, H. Uchida (Eds.), *Case Studies in Fisheries Self-Governance.* pp. 441-451, FAO Fisheries Technical Paper No. 504, FAO, Rome.

Chung, E.-Y. (2007). Oogenesis and sexual maturation in *Meretrix lusoria* in western Korea. *J. Shellfisheries Res.* April 2007. (available at http://findarticles.com/p/articles/mi_m0QPU/is_1_26/ai_n27245006/?tag= content;col1).

Ciocco, N. F., Lasta, M. L. and Bremec, C. (1998). Pesquerías de Bivalvos: Mejillón, Vieiras (Tehuelche y Patagónica) y Otras Especies. In: E. E. Boschi (Ed.), *El Mar Argentino y sus Recursos Pesqueros. Tomo 2 - Los Moluscos de Interés Pesquero.* pp. 143-166, Instituto Nacional de Investigación y Desarrollo Pesquero. Secretaria de Agricultura, Ganadería, Pesca y Alimentación, Mar del Plata.

Coen, L. D. (1995). *A Review of the Potential Impacts of Mechanical Harvesting on Subtidal and Intertidal Shellfish Resources.* South Carolina Department of Natural Resources, Marine Resources Research Institute, 46 pp.

Coen, L., Hadley, N. and Anderson, B. (2005). Clam update: State of South Carolina's coastal resources. South Carolina Department of Natural Resources, Columbia, 2 pp. (available at http://www.dnr.sc.gov/marine/mrri/pubs/yr2005/update%20of%20 clams.pdf).

Corte, G. N., Yokoyama, L. Q. and Amaral, A. C. Z. (2011). Dinâmica reprodutiva e taxa de renovação do berbigão *Anomalocardia brasiliana* em áreas de captura do litoral paulista. *V Simpósio Brasileiro de Oceanografia. Oceanografia e Políticas Públicas,* Santos, 5 pp.

Cranfield, H. J., Michael, K. P., Stotter, D., and Doonan, I. J. (1994). *Distribution, biomass and yield estimates of surf clams off New Zealand beaches.* New Zealand Fisheries Assessment Research Document 94/1. N. Z. Ministry of Agriculture and Fisheries, Wellington, 17 pp.

Cranfield, H. J. and Michael, K. P. (2001). *The surf clam fishery in New Zealand: description of the fishery, its management, and the biology of surf clams.* New Zealand Fisheries Assessment Report 2001/62. Ministry of Fisheries, Wellington, 24 pp.

Cruz, R. A. (1986). Gametogenesis y desove de *Chione subrugosa* (Wood, 1928) (Bivalvia: Veneridae) en el Golfo de Nicoya, Costa Rica. *Uniciencia.* 3, 31-34.

Davy, F. B. and Graham, M. (1984). Cultivo de Bivalvos en Asia y el Pacifico: Trabajos Presentados en un Seminario Celebrado en Singapur del 16 al 19 de Febrero de 1982. Centro Internacional de Investigaciones para el Desarrollo (CIID), Ottawa, 94 pp.

De Boer, W. F. and Prins, H. H. T. (2002). Human exploitation and benthic community structure on a tropical intertidal flat. *J. Sea Res.* 48, 225-240.

Dobbinson, S. J., Barker, M. F. and Jillett, J. B. (1989). Experimental shore level transplantation of the New Zealand cockle *Chione stutchburyi. J. Shellfish Res.* 8, 197-212.

Domingo, A., Forselledo, R., Miller, P., and Passadore, C. (2008). Plan de Acción Nacional para la Conservación de Condrictios en las Pesquerìas Uruguayas. DINARA, Montevideo, 88 pp.

Donrung, P., Tunkijjanukij, S., Jarayabhand, P., and Poompuang, S. (2011). Spatial genetic structure of the surf clam *Paphia undulata* in Thailand waters. *Zool. Stud.* 50, 211-219.

Gab-Alla, A. A.-F.A., Mohamed, S. Z., Mahmoud, M. A. M. and Soliman, B.A., 2007. Ecological and biological studies on some economic bivalves in Suez Bay, Gulf of Suez, Red Sea, Egypt. *J. Fish. Aquat. Sci.* 2, 178-194.

Gabr, H. R. and Gab-Alla, A. A.-F. A. (2008). Effect of transplantation on heavy metal concentrations in commercial clams of Lake Timsah, Suez Canal, Egypt. *Oceanologia.* 50, 83-93.

Gaspar, M. B. and Monteiro, C. C. (1998). Reproductive cycles of the razor clam *Ensis siliqua* and the clam *Venus striatula* off Faro, southern Portugal. *J. Mar. Biol. Assoc. U.K.* 78, 1247-1258.

Gaspar, M. B., Castro, M. and Monteiro, C. C. (1999). Effect of tooth spacing and mesh size on the catch of the Portuguese clam and razor clam dredge. *ICES J. Mar. Sci.* 56, 103-110.

Gaspar, M. B., Dias, M. D., Campos, A., Monteiro, C. C., Santos, M. N., Chicharo, A., and Chicharo, L. (2001a). The influence of dredge design on the catch of *Callista chione* (L. 1758). *Hydrobiologia.* 465, 153-167.

Gaspar, M. B., Leitão, F., Santos, M. N., Sobral, M., Chícharo, L., Chícharo, A., and Monteiro, C. C. (2002a). Influence of mesh size and tooth spacing on the proportion of damaged organisms in the catches of the Portuguese clam dredge fishery. *ICES J. Mar. Sci.* 59, 1228-1236.

Gaspar, M. B., Leitão, F., Santos, M. N., Chícharo, L., Damásia, D., Chícharo, A., and Monteiro, C. C. (2003). A comparison of direct macrofaunal mortality using three types of Portuguese clam dredge. *ICES J. Mar. Sci.* 60, 733-742.

Gaspar, M. B., Pereira, A. M., Vasconcelos, P., and Monteiro, C. C. (2004). Age and growth of *Chamelea gallina* from the Algarve coast (southern Portugal): influence of seawater temperature and gametogenic cycle on growth rate. *J. Molluscan Stud.* 70, 371-377.

Gibson, F. A. and Duggan, C. B. (1970). Experiments with the American hard-shelled clam (*Mercenaria mercenaria*). Fishery Leaflet No. 24. Fisheries Division, Dublin, 6 pp.

Gillett, R. (2004). *The marine fisheries of Cambodia.* FAO / FishCode Review No. 4. FAO, Rome, 57 p.

Glover, E. and Beech, M. J. (2005). The environment and economy of an Ubaid-related settlement on Dalma Island, United Arab Emirates. *Paléorient.* 31, 97-107.

González, J. and Barahona, N. (2007). Situación de la pesquería del recurso almeja (*Venus antiqua*) en la X Región. Seguimiento del Estado de Situación de las Principales Pesquerías Nacionales. Investigación Situación Pesquerías Bentónicas 2006 - Informe Técnico No. 2. Instituto de Fomento Pesquero, Valparaíso, 23 pp.

Goulletquer, P. (1997). A Bibliography of the Manila Clam *Tapes philippinarum*. IFREMER, RIDRV-97.02/RA/ La Tremblade, 122 pp.

Hadley, N. and Coen, L. (n/d). Hard clams (*Mercenaria mercenaria* and *M. campechiensis*). South Carolina Department of Natural Resources (available at http://www.dnr.sc.gov/cwcs/pdf/Hardclam.pdf).

Hamida, L., Medhiouband, M.-N., Cochard, J. C., Romdhame, M. S., and Le Pennec, M. (2004). Étude comparative du cycle de reproduction de la palourde *Ruditapes decussatus* en milieu naturel (sud Tunisie) et contrôlé (écloserie). *Cah. Biol. Mar.* 45, 291-303.

Huaquin, L. G. (2002). *Guía de Características e Identificación de Moluscos de Importancia Económica para Chile.* Universidad de Chile, Facultad de Ciencias Veterinarias y Pecuarias, Departamento de Ciencias Biológicas Animales, Santiago, 10 pp.

James, P. S. B. R. (Ed.) (1988). National seminar on shellfish resources and farming. Session I. 19-21 January 1987, Tuticorin. *CMFRI Bulletin* No. 42, Part One. Central Marine Fisheries Research Institute, Cochin, 238 pp.

Jamieson, G. S., Chew, L., Gillespie, G., Robinson, A., Bendell-Young, L., Heath, W., Bravender, B., Tompkins, A., Nishimura, D., and Doucette, P. (2001). Review of the Environmental Impacts of Intertidal Shellfish Aquaculture in Baynes Sound. Research Document 2001/125. Canadian Science Advisory Secretariat, 103 p. (available at http://www.dfo-mpo.gc.ca/csas/).

Joaquim, S., Matias, D., Matias, A. M., Moura, P., Arnold, W. S., Chícharo, L., and Gaspar, M. B. (2011). Reproductive activity and biochemical composition of the pullet carpet shell *Venerupis senegalensis* (Gmelin, 1791) (Mollusca: Bivalvia) from Ria de Aveiro (northwestern coast of Portugal). *Sci. Mar.* 75, 217-226.

Jung, H. T., Kim, J. and Choi, S. D. (2004). Phylogenetic relationship of the five Korean Veneridae clams (Bivalvia, Veneroida) according to morphological characters. *J. Aquaculture.* 17, 197-208.

Kang, H.-W., Choi, K.-H., Jun, J.-C., Lee, K.-Y., and Park, K.-H. (2010). Ultrastructural studies of vitellogenesis in oocytes and follicle cells during oogenesis in female *Protothaca (Notochione) jedoensis* (Bivalvia: Veneridae). *Anim. Cells System.* 14, 343-349.

Kashorte, M. (2003). Moving subsistence fisheries to commercial fisheries in South Africa. Fisheries Training Programme, Final Project 2003. The United Nations University, Reykjavik, 46 pp.

Khebbeb, M. E. H., Nadji, S. and Amrani, A. (2010). The effect of cadmium exposure on malonedialdehyde and reduced glutathione concentrations in several tissues of a bivalve mollusc (*Ruditapes decussatus*) fished from Mellah lagoon (North East of Algeria). *Ann. Biol. Res.* 1, 166-173.

Kithsiri, H. M. P., Wijeyaratne M. J. S. and Amarasinghe, U. S. (2000). Influence of some environmental factors on the abundance of three commercially important bivalve species (Family: Veneridae) in the Puttalam lagoon and Dutch bay, Sri Lanka. *Sri Lanka J. Aquat. Sci.* 5, 27-38.

Kloff, S., Trebaol, L. and Lacroix, E. (2007). Pêche aux Bivalves and Environnement: Panorama Mondial, Études de Cas et Application à l'Exploitation des Praires en Mauritanie. *Atelier Scientifique et Technique:"Pour une Pêche Responsable des Praires en Mauritanie"*, 15-16 Février 2007, Nouadhibou. Fondation Internationale du Banc d'Arguin, 191 pp.

Kotchawong, J., Ajjimangkul, S., Meksumpun, C., and Srisomwong, J. (2010). Hard clam (*Meretrix* spp.) and ridged venus clam (*Paphia crassisulca*) fishery in Kapor Bay. *World Small-Scale Fisheries Congress 2010*. 17-22 October 2010, Bangkok, 98 pp.

Koutsoubas, D., Galinou-Mitsoudi, S., Katsanevakis, S., Leontarakis, P., Metaxatos, A., and Zenetos A. (2007). Bivalve and Gastropod Molluscs of Commercial Interest for Human Consumption in the Hellenic Seas. In: C. Papaconstantinou, A. Zenetos, V. Vassilopoulou, G. Tserpes (Eds.), *State of Hellenic Fisheries*. pp. 70-84, Hellenic Centre for Marine Research, Athens.

Kraeuter, J. N., Klinck, J. M., Powell, E. N., Hofmann, E. E., Buckner, S. C., Grizzle, R. E., and Bricelj V. M. (2008). Effects of the fishery on the northern quahog population in Great South Bay, New York: a modeling study. *J. Shellfisheries Res.* August 2008. (available at http://findarticles. com/p/articles/mi_m0QPU/is_4_ 27/ai_n28047450/ ?tag=content;col1).

Lasta, M. L., Ciocco, N. F., Bremec, C., and Roux, A. (1998). Moluscos Bivalvos y Gasterópodos. In: E. E. Boschi (Ed.), *El Mar Argentino y sus Recursos Pesqueros. Tomo 2 - Los Moluscos de Interés Pesquero*. pp. 115-142. Instituto Nacional de Investigación y Desarrollo Pesquero. Secretaria de Agricultura, Ganadería, Pesca y Alimentación, Mar del Plata.

Lavander, H. D., Cardoso Júnior, L. O., Oliveira, R. L., Silva Neto, S. R., Galvez, A. O., and Peixoto, S. R. M. (2011). Biologia reprodutiva da *Anomalocardia brasiliana* (Gmelin, 1791) no litoral norte de Pernambuco, Brasil. *Rev. Bras. Cienc. Agr.* 6, 344-350.

Laxmilatha, P. and Appukuttan, K. K. (2002). A review of the black clam (*Villorita cyprinoides*) fishery of the Vembanad Lake. *Indian J. Fish.* 49, 85-91.

Leal, J. H. (2002). Bivalves. In: K. E. Carpenter (Ed.), *The Living Marine Resources of the Western Central Atlantic: Volume 1. Introduction, Molluscs, Crustaceans, Hagfishes, Sharks, Batoid Fishes, and Chimaeras*. pp. 25-98, FAO Species Identification Guide for Fishery purposes. FAO, Rome.

Leblanc, K., Ouellette, M., Chouinard, G. A., and Landry, T. (2005). Commercial harvest and population structure of a northern quahog (*Mercenaria mercenaria* Linnaeus 1758) population in St. Mary's Bay, Nova Scotia, Canada. *J. Shellfish Res.* 24, 47-54.

Lee, S. Y. (1996). Distribution pattern and interaction of two infaunal bivalves, *Tapes philippinarum* (Adams and Reeve) and *Anomalocardia squamosa* (Linnaeus) (Bivalvia: Veneridae). *J. Exp. Mar. Biol. Ecol.* 201, 253-273.

Leung, K. M. Y. (2004). Sand elimination by the Asiatic hard clam *Meretrix meretrix* (L.): influences of temperature, salinity and season. *J. Shellfish Res.* 23, 421-428.

López, A. and Urcuyo, J. (2008). *Moluscos de Nicaragua. I - Bivalvos.* MARENA, Managua, 133 pp.

Lovatelli, A., Farías, A. and Uriarte, I. (Eds). (2008). Estado actual del cultivo y manejo de moluscos bivalvos y su proyección futura: factores que afectan su sustentabilidad en América Latina. *Taller Técnico Regional de la FAO*. 20-24 de Agosto de 2007, Puerto Montt. FAO Actas de Pesca y Acuicultura No. 12. FAO, Roma, 359 pp.

Morriconi, E., Lomovasky, B. J., Calvo, J., and Brey, T. (2002). The reproductive cycle of *Eurhomalea exalbida* (Chemnitz, 1795) (Bivalvia: Veneridae) in Ushuaia Bay (54° 50' S), Beagle Channel (Argentina). *Invertebr. Reprod. Dev.* 42, 61-68.

Morriconi, E., Lomovasky, B. J. and Calvo, J. (2007). Reproductive cycle and energy content of *Tawera gayi* (Hupe 1854) (Bivalvia: Veneridae) at the southernmost limit of their distribution range. *J. Shellfish Res.* 26, 81-88.

Morsan, E. (2007). Spatial pattern, harvesting and management of the artisanal fishery for purple clam (*Amiantis purpurata*) in Patagonia (Argentina). *Ocean Coast. Manage.* 50, 481-497.

Mouëza, M., Gros, O. and Frenkiel, L. (2006). Embryonic development and shell differentiation in *Chione cancellata* (Bivalvia, Veneridae): an ultrastructural analysis. *Invertebr. Biol.* 125, 21-33.

Moura, P., Gaspar, M. B. and Monteiro, C. C. (2009). Age determination and growth rate of a *Callista chione* population from the south-western coast of Portugal. *Aquat. Biol.* 5, 97-106.

Murray-Jones, S. and Steffe, A. S. (2000). A comparison between the commercial and recreational fisheries of the surf clam, *Donax deltoides*. *Fish. Res.* 44, 219-233.

Narasimham, K. A. (1991). Present status of clam fisheries of India. *J. Mar. Biol. Assoc. India*. 33, 76-88.

Nayar, K. N. and Mahadevan, S. (1974). Edible Bivalves: Clams and Others. In: R. V. Nair, K. S. Rao (Eds.), *The Commercial Molluscs of India*. pp. 40-53, Bulletin of the Central Marine Fisheries Research Institute, No. 25. ICAR, Cochin.

Ng, P. K. L. and Sivasothi, N. (Eds.) (2001). A Guide to Mangroves of Singapore. 2001 Raffles Museum of Biodiversity Research, The National University of Singapore and The Singapore Science Centre. Accessed at http://mangrove.nus.edu.sg/guidebooks/text/ fauna6.htm on March 2012.

Norte-Campos, A. and Villarta, K. A. (2010). Use of population parameters in examining changes in the status of the short-necked clam *Paphia undulata* Born, 1778 (Mollusca, Pelecypoda: Veneridae) in coastal waters of southern Negros Occidental. *Sci. Diliman* 22, 53-60.

NZ Fisheries InfoSite (2011). Ministry of Fisheries. New Zealand Government. Accessed at http://www. fish.govt.nz/en-nz/default.htm on March 2012.

Olguín A. and Jerez, G. (2003). *Chile. Especies Bentónicas de Importancia Comercial.* Serie Chile: Recursos Pesqueros N° 1. IFOP, Valparaíso, 30 pp.

Oliveira, I., Amorim, A., Lavander, H., Peixoto, S., and Gálvez, A. O. (2011). Spatial and temporal distribution of the shellfish *Anomalocardia brasiliana* (Gmelin, 1791) on Mangue Seco beach, Pernambuco, Brazil. *Int. J. Aquat. Sci.* 2, 68-79.

Oliveira, M. M., Camanho, A. S. and Gaspar, M. B. (2010). Technical and economic efficiency analysis of the Portuguese artisanal dredge fleet. *ICES J. Mar. Sci.* 67, 1811-1821.

Olympia, E. G. (1983). Larval development of *Marcia marmorata* Lamarck 1818. *MSc Thesis in Marine Biology. University of the Philippines,* Quezon City, 39 pp.

Panjarat, S. (2008). *Sustainable Fisheries in the Andaman Sea Coast of Thailand.* Division for Ocean Affairs and the Law of the Sea. Office of Legal Affairs. The United Nations, New York, 107 pp.

Park, K.-I., Ngo, T. T. T., Choi, S.-D., Cho, M., and Choi, K.-S. (2006). Occurrence of *Perkinsus olseni* in the Venus clam *Protothaca jedoensis* in Korean waters. *J. Invertebr. Pathol.* 93, 81-87.

Park, K.-I., Tsutsumi, H., Hong, J.-S., Choi, K.-S. (2008). Pathology survey of the short-neck clam *Ruditapes philippinarum* occurring on sandy tidal flats along the coast of Ariake Bay, Kyushu, Japan. *J. Invertebr. Pathol.* 99, 212-219.

Pattison, C. A. and Lampson, K. M. (2008). Pismo Clam. Annual Status of the Fisheries Report. In: K. Barsky (Ed.), *Status of the Fisheries Report - An Update through 2006.* pp. 49-59, California Department of Fish and Game, Sacramento.

Paulay, G. (2000). *Benthic ecology and biota of Tarawa Atoll lagoon: influence of equatorial upwelling, circulation, and human harvest.* Atoll Research Bulletin No. 487. National Museum of Natural History, Washington, 41 pp.

Peharda, M., Ezgeta-Balić, D., Vrgoč, N., Isajlović, I., and Bogner, D. (2010). Description of bivalve community structure in the Croatian part of the Adriatic Sea - hydraulic dredge survey. *Acta Adriat.* 51, 141-158.

Pereira, A. M., Palanco, I., Rufino, M. M., Moreno, Ó., and Gaspar, M. B. (2007). *La Pesca de Bivalvos en el Litoral Oceánico de la Costa Suroccidental de la Península Ibérica: Descripción de la Pesquería y Artes de Pesca, Medidas de Gestión, Biología de las Especies Comerciales y Catálogo de Especies de Bivalvos.* Junta de Andalucía, Consejería de Agricultura y Pesca, Huelva, 141 pp.

Pereira, M. A. M. and Gonçalves, P. M. B. (2000). Influence of human exploitation of intertidal mollusk resources on the selection and utilisation of gastropod shells by the hermit crab *Clibanarius longitarsus* (de Haan) in Costa do Sol Mangrove, Maputo. *2nd National Conference on Coastal Zones Research*, 27-29 September 2000, Maputo, 13 pp.

Pitel, M., Berthou, P. and Fifas, S. (2001). Dredge Designs and Fisheries. Report of the Programme Ecodredge (1999-2001). IFREMER - Laboratoire Côtier Brest: Direction des Ressources Vivantes / Ressources Halieutiques, 95 pp.

Ponurovsky, S. K. and Yakovlev, Y. M. (1992). The reproductive biology of the Japanese littleneck, *Tapes philippinarum* (A. Adams and Reeve, 1850) (Bivalvia: Veneridae). *J. Shellfish Res.* 11, 265-277.

Poutiers, J. M. (1998). Bivalves (Acephala, Lamellibranchia, Pelecypoda). In: K. E. Carpenter, V. H. Niem (Eds.), *The Living Marine Resources of the Western Central Pacific: Volume 1. Seaweeds, Corals, Bivalves and Gastropods.* pp. 123-362. FAO Species Identification Guide for Fishery Purposes. FAO, Rome.

Ra, K., Kim, K. T., Bang, J. H., Lee, J. M., Kim, E. S., and Cho, S. R. (2011). A preliminary study of environmental impact assessment of tidal power plant in Shihwa Lake, Korea: Heavy metal accumulation in the transplanted Manila clam (*Ruditapes philippinarum*). *J. Coast. Res.* Spec. Iss. 64, 932-936.

Romanelli, M., Cordisco, C. A. and Giovanardi, O. (2009). The long-term decline of the *Chamelea gallina* L. (Bivalvia: Veneridae) clam fishery in the Adriatic Sea: is a synthesis possible? *Acta Adriat.* 50, 171-205.

Ross, D. J., Johnson, C. R., Hewitt, C. L., and Ruiz, G. M. (2004). Interaction and impacts of two introduced species on a soft-sediment marine assemblage in SE Tasmania. *Mar. Biol.* 144, 747-756.

Rowling, K., Hegarty, A.-M. and Ives, M. (Eds.) (2010). Cockles (Arcoida and Veneroida). In: K. Rowling, A.-M. Hegarty, M. Ives (Eds.), *Status of Fisheries Resources in NSW in 2008/09*. pp. 73-75, Industry and Investment NSW, Cronulla (available at http://www.dpi.nsw.gov.au/research/areas/systems research/wildfisheries/outputs/ 2010/1797/ StatusOfFisheriesResourcesNSW2008-09.pdf).

Rufino, M. M., Gaspar, M. B., Pereira, A. M., and Vasconcelos, P. (2006). Shape recognition of *Chamelea gallina* and *Chamelea striatula* (Bivalvia: Veneridae) based on linear and geometric morphometric methods. *J. Morphol.* 267, 1433-1440.

Salas, S., Chuenpagdee, R., Charles, A., and Seijo, J. C. (Eds.) (2011). *Coastal Fisheries of Latin America and the Caribbean.* FAO Fisheries and Aquaculture Technical Paper No. 544, FAO, Rome, 430 pp.

Sartori, A. F. and Printrakoon, C. (2008). Siphonal structure in the Veneridae (Bivalvia: Heterodonta) with an assessment of its phylogenetic application and a review of venerids of the Gulf of Thailand. *Raffles Bull. Zool.* Supplement 18, 103-125.

Savina, M. and Ménesguen, A. (2007). The growth of the banded carpet shell (*Paphia rhomboïdes*) in a contrasted region, the English Channel (Western Europe): A modelling study. *Ecol. Model.* 205, 39-51.

Savina, M. and Ménesguen, A. (2008). A deterministic population dynamics model to study the distribution of a benthic bivalve with planktonic larvae (*Paphia rhomboïdes*) in the English Channel (NW Europe). *J. Mar. Syst.* 70, 63-76.

Scarabino, F., Zaffaroni, J. C., Clavijo, C., Carranza, A., and Nin, M. (2006). Bivalvos Marinos y Estuarinos de la Costa Uruguaya: Faunística, Distribución, Taxonomía y Conservación. In: R. Menafra, L. Rodríguez-Gallego, F. Scarabino, D. Conde (Eds.), *Bases para la Conservación y el Manejo de la Costa Uruguaya*. pp. 157-170, Vida Silvestre Uruguay, Montevideo.

Scarlet, M. P. J. (2005). Clams as a resource in Maputo Bay - Mozambique. *MSc Thesis in Marine Ecology,* Göteborg University, Göteborg, 31 pp.

Schweers, T., Wolff, M., Koch, V., and Duarte, F. S. (2006). Population dynamics of *Megapitaria squalida* (Bivalvia: Veneridae) at Magdalena Bay, Baja California Sur, Mexico. *Int. J. Trop. Biol.* 54, 1003-1017.

SEAFDC (2011). Fishing Gear and Method in Southeast Asia: Dredges. Southeast Asian Fisheries Development Center. Accessed at http://map.seafdec.org/Monograph/index_gear.php?val=06 on March 2012.

Shafee, M. S. (1999). *Pêche des Bivalves sur la Côte Méditerranéenne Marocaine: Catalogue d'Espèces Exploitées et des Engins de Pêche.* FAO-COPEMED, Alicante, 58 pp.

Shin, S.-H. and Jong-Geel, J. E. (2008). Biological assessment of ecologically important areas for the coastal mollusks taxonomic group of the Yellow Sea ecoregion - Korea part. pp. 157-177, Biological Assessment Report of the Yellow Sea Ecoregion.

Silberfeld, T. and Gros, O. (2006). Embryonic development of the tropical bivalve *Tivela mactroides* (Born, 1778) (Veneridae: subfamily Meretricinae): a SEM study. *Cah. Biol. Mar.* 47, 243-251.

Solaiman, M., Huq, K. A., Faruk, M. A. R. and Hossain, M. A. R. (n/d). Mollusc Marketing System through Value Chain Analysis with Special Emphasis on Livelihood of the Poor Fishermen. In: *Value Chain and Market Assessment of Coastal and Marine Aquatic Products of Bangladesh.* pp. 56-85, Bangladesh Fisheries Research Forum (BFRF).

Tata, A. and Prieto, A. S. (1991). Producción secundaria en una población del bivalvo tropical *Tivela mactroides* (Veneridae) en el Oriente de Venezuela. *Carib. J. Sci.* 27, 28-34.

Thomas, F. (2007). The behavioral ecology of shellfish gathering in Western Kiribati, Micronesia. 2: Patch choice, patch sampling, and risk. *Hum. Ecol.* 35, 515-526.

Thomas, F. R. (2001). Mollusk habitats and fisheries in Kiribati: An assessment from the Gilbert Islands. *Pacific Sci.* 55, 77-97.

Thorpe, A., Ibarra, A. A. and Reid, C. (2000). The new economic model and marine fisheries development in Latin America. *World Develop.* 28, 1689-1702.

Tirado, C., Salas, C. and Márquez, I. (2003). Reproduction of *Venus verrucosa* L., 1758 (Bivalvia: Veneridae) in the littoral of Málaga (southern Spain). *Fish. Res.* 63, 437-445.

Toba, M. (2004). The decline of Manila clam stock in Tokyo Bay. *Bull. Fish. Res. Agen* Suppl. 1, 13-18.

Trigui El Menif, N., Guezzi, Y., Le Pennec, M., Boumaiza, M., and Le Pennec, G. (2005). Infestation of the clam *Venus verrucosa* by Sipunculoidea and the lithophagus bivalve, *Gastrochaena dubia*. *Acta Adriat.* 46, 83-90.

Trimble, M., Ríos, M., Passadore, C., Szephegyi, M., Nin, M., Garcia Olaso, F., Fagúndez, C., and Laporta, P. (2010). *Ecosistemas Costeros Uruguayos: Una Guía para su Conocimiento*. Editorial Imprenta Monteverde, Montevideo, 335 pp.

Tunçer, S. and Erdemir, C. Ç. (2002). A preliminary study on some properties for *Chamelea gallina* (L.) (Bivalvia: Verenidae) from Karabiga-Çanakkale. *Turk. J. Fish. Aquat. Sci.* 2, 117-120.

Uda, T. (2010). Impacts on sandy beach and habitat of Japanese hard clams due to construction of port breakwater. *International Symposium on Integrated Coastal Management for Marine Biodiversity in Asia*. 14-15 January, Kyoto, 28-34 pp.

Vargas-Zamora, J. A. and Sibaja-Cordero, J. A. (2011). Molluscan assemblage from a tropical intertidal estuarine sand-mud flat, Gulf of Nicoya, Pacific, Costa Rica (1984-1987). *Int. J. Trop. Biol.* 59, 1135-1148.

Vasconcelos, P., Morgado-André, A., Morgado-André, C., and Gaspar, M. B. (2011). Shell strength and fishing damage to the smooth clam (*Callista chione*): simulating impacts caused by bivalve dredging. *ICES J. Mar. Sci.* 68, 32-42.

Vélez, C. E. T., Díaz, L. J. O., Suárez, L. A. C., Franco, F. A., Morales, A. D., Solano, J. E., and Camelo, A. L. F. (2003). *Aportes al Entendimiento de la Bahía de Tumaco: Entorno Oceanográfico, Costero y de Riesgos*. Serie Publicaciones Especiales - Volumen 2. Centro Control Contaminación del Pacífico (CCCP), Tumaco, 168 pp.

Villalba, A., Carballal, M. J., López, C., Cabada, A., Corral, L., and Azevedo, C. (1999). Branchial rickettsia-like infection associated with clam *Venerupis rhomboides* mortality. *Dis. Aquat. Org.* 36, 53-60.

Villarta, K. A. and Norte-Campos, A. G. C. (2010). Fishery of the short-necked clam *Paphia undulata* in southern Negros Occidental, Central Philippines. *Sci. Diliman* 22, 43-51.

Voultsiadou, E., Koutsoubas, D. and Achparaki, M. (2010). Bivalve mollusc exploitation in Mediterranean coastal communities: an historical approach. *J. Biol. Res. Thessaloniki.* 13, 35-45.

Wood, J. B. and Jackson, K. J. (2005). Bermuda. In: P. Miloslavich, E. Klein (Eds.), *Caribbean Marine Biodiversity: the Known and the Unknown*. Census of Marine Life, Caribbean Regional Committee.

Xie, Q. and Burnell, G. M. (1994). A comparative study of the gametogenic cycles of the clams *Tapes philippinarum* (A. Adams and Reeve 1850) and *Tapes decussatus* (Linnaeus) on the south coast of Ireland. *J. Shellfish Res.* 13, 467-472.

Yamasaki, S., Higano, J. and Watanabe, T. (2002). Foot damages of the clam *Meretrix lamarckii* caused by clam dredging. *Nippon Suisan Gakk.* 68, 368-373.

Yan, X., Zhang, G. and Yang, F. (2006). Effects of diet, stocking density, and environmental factors on growth, survival, and metamorphosis of Manila clam *Ruditapes philippinarum* larvae. *Aquaculture.* 253, 350-358.

Yan, H., Li, Q., Yu, R., and Kong, L. (2010). Seasonal variations in biochemical composition and reproductive activity of Venus clam *Cyclina sinensis* (Gmelin) from the Yellow River Delta in northern China in relation to environmental factors. *J. Shellfish Res.* 29, 91-99.

CITED REFERENCES

Appeltans, W., Bouchet, P., Boxshall, G. A., Fauchald, K., Gordon, D. P., Hoeksema, B. W., Poore, G. C. B., van Soest, R. W. M., Stöhr, S., Walter, T. C., and Costello, M. J. (Eds.) (2011). World Register of Marine Species - WoRMS. Accessed at http://www.marinespecies.org on March 2012.

Ardizzone, G. D., Tucci, P., Somaschini, A., and Belluscio, A. (2000). Is bottom trawling partly responsible for the regression of *Posidonia oceanica* meadows in the Mediterranean Sea? In: M. J. Kaiser, S. J. de Groot (Eds.), *Effects of Fishing on Non-Target Species and Habitats*. pp. 37-46, Blackwell Science, Oxford.

Bagley, M. J., Franson, S. E., Christ, S. A., Waits, E. R., and Toth, G. P. (2002). Genetic Diversity as an Indicator of Ecosystem Condition and Sustainability: Utility for Regional Assessments of Stream Condition in the Eastern United States. US Environmental Protection Agency, Cincinnati, 88 pp.

Barbera, C., Bordehore, C., Borg, J. A., Glémarec, M., Grall, J., Hall-Spencer, J. M., de la Huz, Ch., Lanfranco, E., Lastra, M., Moore, P. G., Mora, J., Pita, M. E., Ramos-Esplá, A. A., Rizzo, M., Sánchez-Mata, A., Seva, A., Schembri, P. J., and Valle, C. (2003). Conservation and management of northeast Atlantic and Mediterranean maerl beds. *Aquat. Cons. Mar. Freshw. Ecosyst.* 13, 65-76.

Barnes, D., Chytalo, K. and Hendrickson, S. (1991). Final Policy and Generic Environmental Impact Statement on Management of Shellfish in Uncertified Areas Program. New York Department of Environment and Conservation, New York, 79 pp.

Britton, J. C. and Morton, B. (1994). Marine carrion and scavengers. *Oceanogr. Mar. Biol. Ann. Rev.* 32, 369-434.

Butcher, T., Matthews, J., Glaister, J., and Hamer, G. (1981). Study suggests scallop dredges causing few problems in Jervis Bay. *Aust. Fish.* 40, 9-12.

Caddy, J. F. (1973). Underwater observations on tracks of dredges and trawls and some effects of dredging on scallop ground. *J. Fish. Res. Board Can.* 30, 173-180.

Canapa, A., Schiaparelli, S., Marota, I., and Barucca, M. (2003). Molecular data from the 16S rRNA gene for the phylogeny of Veneridae (Mollusca: Bivalvia). *Mar. Biol.* 142, 1125-1130.

Chen, J., Li, Q., Kong, L., and Zheng, X. (2011). Molecular phylogeny of venus clams (Mollusca, Bivalvia, Veneridae) with emphasis on the systematic position of taxa along the coast of mainland China. *Zool. Scripta.* 40, 260-271.

Churchill, J. H. (1989). The effect of commercial trawling on sediment ressuspension and transport over the Middle Atlantic Bight continental shelf. *Cont. Shelf Res.* 9, 841-864.

Coffen-Smout, S. S. and Rees, E. I. S. (1999). Burrowing behaviour and dispersion of cockles *Cerastoderma edule* L. following simulated fishing disturbance. *Fish. Res.* 40, 65-72.

Collie, J. S., Escanero, G. A. and Valentine, P. C. (1997). Effects of bottom fishing on the benthic megafauna of Georges Bank. *Mar. Ecol. Prog. Ser.* 155, 159-172.

Collie, J. S., Hall, S. J., Kaiser, M. J., and Poiner, I. R. (2000). A quantitative analysis of fishing impacts on shelf-sea benthos. *J. Anim. Ecol.* 69, 785-798.

Constantino, R., Gaspar, M. B., Tata-Regala, J., Carvalho, S., Cúrdia, J., Drago, T., Taborda, R., and Monteiro, C. C. (2009). Clam dredging effects and subsequent recovery of benthic communities at different depth ranges. *Mar. Environ. Res.* 67, 89-99.

Currie, D. R. and Parry, G. D. (1996). Effects of scallop dredging on a soft sediment community: a large-scale experimental study. *Mar. Ecol. Prog. Ser.* 134, 131-150.

Dare, P. J., Key, D. and Connor, P. M. (1993). The efficiency of springloaded dredges used in the Western English Channel fishery for scallops, *Pecten maximus* (L.). ICES Fish Capture Committee CM1993/B-15, 8 pp.

Dayton, P. K. (1971). Competition, disturbances and community organization: the provision and subsequent utilization of space in a rocky intertidal community. *Ecol. Monogr.* 41, 351-389.

Dayton, P. K. and Hessler, R. R. (1972). Role of biological disturbance in maintaining diversity in the deep sea. *Deep Sea Res.* 19, 199-208.

Dayton, P. K., Thrush, S. F., Agardy, M. T., and Hofman, R. J. (1995). Environmental effects of marine fishing. *Aquat. Cons. Mar. Freshw. Ecosyst.* 5, 205-232.

DeAlteris, J., Skrobe, L. and Lipsky, C. (1999). The Significance of Seabed Disturbance by Mobile Fishing Gear Relative to Natural Processes: A Case Study in Narragansett Bay, Rhode Island. In: L. Benaka (Ed.), *Fish Habitat: Essential Fish Habitat and Rehabilitation*. pp. 224-237, American Fisheries Society, Symposium 22, Bethesda, Maryland.

De Groot, S. J. (1984). The impact of bottom trawling on benthic fauna of North Sea. *Ocean Manage.* 9, 177-190.

Dernie, K. M., Kaiser, M. J., Richardson, E. A., and Warwick, R. M. (2003). Recovery of soft sediment communities and habitats following physical disturbance. *J. Exp. Mar. Biol. Ecol.* 285/286, 415-434.

De Vooys, C. G. N. and Van der Meer, J. (1998). Changes between 1931 and 1990 in by-catches of 27 animal species from the southern North Sea. *Neth. J. Sea Res.* 39, 291-298.

Duarte, C. M. (2002). The future of seagrass meadows. *Environ. Cons.* 29, 192-206.

Eleftheriou, A. and Robertson, M. R. (1992). The effects of experimental scallop dredging on the fauna and physical environment of a shallow sandy community. *Neth. J. Sea Res.* 30, 289-299.

Evans, P. L., Kaiser, M. J. and Hughes, R. N. (1996). Behaviour and energetics of whelks, *Buccinum undatum* (L.), feeding on animals killed by beam trawling. *J. Exp. Mar. Biol. Ecol.* 197, 51-62.

Falcão, M. M., Gaspar, M. B., Caetano, M., Santos, M. N., and Vale, C. (2003). Short-term environmental impact of clam dredging in coastal waters (south of Portugal): chemical disturbance and subsequent recovery of seabed. *Mar. Environ. Res.* 56, 649-664.

FAO (2010a). Capture Production by Groups of Species - 2009. Yearbooks of Fishery Statistics - Summary Tables. FAO, Rome, 2 pp. (available at: ftp://ftp.fao.org/FI/STAT/summary/a1d.pdf).

FAO (2010b). World Capture Production - 2009. Yearbooks of Fishery Statistics - Summary Tables. FAO, Rome, 2 pp. (available at: ftp://ftp.fao.org/FI/STAT/summary/a1a.pdf).

FAO (2010c). Estimated Value by Groups of Species - 2009. Yearbooks of Fishery Statistics - Summary Tables. FAO, Rome, 5 pp. (available at: ftp://ftp.fao.org/FI/STAT/summary/ appIIybc.pdf).

Fonds, M. (1994). Mortality of Fish and Invertebrates in Beam Trawl Catches and the Survival Chances of Discards. In: S. J. de Groot, H. J. Lindeboom (Eds.), *Environmental Impact of Bottom Gears on Benthic Fauna in Relation to Natural Resources Management and Protection of the North Sea*. pp. 131-146. NIOZ Report No. 12.

Fonds, M., Groenewold, S., Hoppe, I., Kaiser, M. J., Munday, B. W., and Ramsay, K. (1998). Scavenger Responses to Trawling. In: H. J. Lindeboom, S. J. de Groot (Eds.), *The Effects of Different Types of Fisheries on the North Sea and Irish Sea Benthic Ecosystems*. pp. 85-244, NIOZ Report 1998-1, RIVO-DLO Report C003/98.

Fossa, J. H., Mortensen, P. B. and Furevik, D. M. (2002). The deep-water coral *Lophelia pertusa* in Norwegian waters: distribution and fishery impacts. *Hydrobiologia.* 471, 1-12.

Gaspar, M. B. (1996). Bivalves do litoral oceânico algarvio. Aspectos da biologia, ecologia e das pescarias dos mananciais de interesse económico: aplicação à gestão dos recursos. *Ph.D Thesis, Universidade do Algarve, Faro,* 282 pp.

Gaspar, M. B. and Monteiro, C. C. (1998). Indirect mortality caused by dredges on the survival of juveniles of *Spisula solida* and *Donax trunculus*. *Proceedings of the ICES Symposium - Marine Benthos Dynamics: Environmental and Fisheries Impacts*. Crete.

Gaspar, M. B. and Monteiro, C. C. (1999). Indirect mortality caused to juveniles of *Spisula solida* due to deck exposure. *J. Mar. Biol. Assoc. U.K.* 79, 566-568.

Gaspar, M. B. and Chícharo, L. M. (2007). Modifying Dredges to Reduce By-Catch and Impacts on the Benthos. In: S. J. Kennelly (Ed.), *By-Catch Reduction in the World's Fisheries*. pp. 95-140, Springer, Dordrecht.

Gaspar, M. B., Santos, M. N., Leitão, F., Chícharo, L., Chícharo, A., and Monteiro, C. C. (2003). Recovery of substrates and macro-benthos after fishing trials with a new Portuguese clam dredge. *J. Mar. Biol. Assoc. U.K.* 83, 713-717.

Gibson, R. N. and Robb, L. (1992). The relationship between body size, sediment grain size and the burying ability of juvenile plaice *Pleuronectes platessa* L. *J. Fish Biol.* 40, 771-778.

Gilkinson, K. D., Fader, G. B. J., Gordon, Jr D. C., Charron, R., McKeown, D., Roddick, D., Kenchington, E. L. R., MacIsaac, K., Bourbonnais, C., Vass, P., and Liu, Q. (2003). Immediate and longer-term impacts of hydraulic clam dredging on an offshore sandy seabed: effects on physical habitat and processes of recovery. *Cont. Shelf Res.* 23, 1315-1336.

Gosling, E. M. (2003). *Bivalve Molluscs: Biology, Ecology and Culture.* Fishing News Books, Blackwell Publishing Ltd, Oxford, 443 pp.

Greenstreet, S. P. R. and Hall, S. J. (1996). Fishing and ground-fish assemblage structure in the north-western North Sea: an analysis of long-term and spatial trends. *J. Anim. Ecol.* 65, 577-598.

Hall, S. J. and Harding, M. J. C. (1997). Physical disturbance and marine benthic communities: the effects of mechanical harvesting of cockles on non-target benthic infauna. *J. Appl. Ecol.* 34, 497-517.

Hall, S. J., Robertson, M. R., Basford, D. J., and Heaney, S. D. (1993). The possible effects of fishing disturbance in the northern North Sea: An analysis of spatial patterns in community structure around a wreck. *Neth. J. Sea Res.* 31, 201-208.

Hall-Spencer, J. M. and Moore, P. G. (2000). Impact of Scallop Dredging on Maerl Grounds. In: M. J. Kaiser, S. J. de Groot (Eds.), *Effects of Fishing on Non-Target Species and Habitats*. pp. 105-107, Blackwell Science, Oxford.

Hall-Spencer, J. M., Allain, V. and Fossa, J. H. (2002). Trawling damage to Northeast Atlantic ancient coral reefs. *Proc. R. Soc. London* 269, 507-511.

Hall-Spencer, J. M., Grall, J., Moore, P. G., Atkinson, R. J. A. (2003). Bivalve fishing and maerl-bed conservation in France and the U.K. - retrospect and prospect. *Aquat. Cons. Mar. Freshw. Ecosyst.* 13, 33-41.

Hauton, C., Hall-Spencer, J. M. and Moore, P. G. (2003a). An experimental study of the ecological impacts of hydraulic bivalve dredging on maerl. *ICES J. Mar. Sci.* 60, 381-392.

Hauton, C., Atkinson, R. J. A. and Moore, P. G. (2003b). The impact of hydraulic blade dredging on a benthic megafaunal community in the Clyde Sea area, Scotland. *J. Sea Res.* 50, 45-56.

Hiddink, J. G. (2003). Effects of suction-dredging for cockles on non-target fauna in the Wadden Sea. *J. Sea Res.* 50, 315-323.

Hill, A. S., Veale, L. O., Pennington, D., Whyte, S. G., Brand, A. R., and Hartnoll, R. G. (1999). Changes in Irish Sea benthos: possible effects of 40 years of dredging. *Estuar. Coast. Shelf Sci.* 48, 739-750.

Hiscock, K., Sewell, J. and Oakley, J. (2005). Marine Health Check 2005. A Report to Gauge the Health of the U.K.'s Sea-Life. WWF-UK, Godalming, 79 pp.

Huber, M. (2010). *Compendium of Bivalves. A Full-Color Guide to 3'300 of the World's Marine Bivalves. A Status on Bivalvia after 250 Years of Research.* ConchBooks, Hackenheim, 901 pp.

Jenkins, S. R. and Brand, A. R. (2001). The effect of dredge capture on the escape response of the great scallop, *Pecten maximus* (L.): implications for the survival of undersized discards. *J. Exp. Mar. Biol. Ecol.* 266, 33-50.

Jennings, S. and Kaiser, M. J. (1998). The effect of fishing on marine ecosystems. *Adv. Mar. Biol.* 34, 201-252.

Jennings, S., Dinmore, T. A., Duplisea, D. E., Warr, K. J., and Lancaster, J. E. (2001). Trawling disturbance can modify benthic production processes. *J. Anim. Ecol.* 70, 459-475.

Kaiser, M. J. (1998). Significance of bottom-fishing disturbance. *Cons. Biol.* 12, 1230-1235.

Kaiser, M. J. and Spencer, B. E. (1994). Fish scavenging behavior in recently trawled areas. *Mar. Ecol. Progr. Ser.* 112, 41-49.

Kaiser, M. J. and Spencer, B. E. (1996). The effects of beam-trawl disturbance on infaunal communities in different habitats. *J. Anim. Ecol.* 65, 348-358.

Kaiser, M. J., Edwards, D. B., Armstrong, P. J., Radford, K., Lough, N. E. L., Flat, R. P., and Jones, H. D. (1998). Changes in megafauna benthic communities in different habitats after trawling disturbance. *Neth. J. Sea Res.* 55, 353-361.

Kaiser, M. J., Ramsay, K., Richardson, C. A., Spence, F. E., Brand, A. R. (2000). Chronic fishing disturbance has changed shelf sea benthic community structure. *J. Anim. Ecol.* 69, 494-503.

Langton, R. W. and Robinson, W. E. (1990). Faunal associations on scallop grounds in the western Gulf of Maine. *J. Exp. Mar. Biol. Ecol.* 144, 157-171.

Law, R. and Stokes, K. (2005). Evolutionary Impacts of Fishing on Target Populations. In: E. A. Norse, L. B. Crowder (Eds.), *Marine Conservation Biology: The Science of Maintaining the Sea's Biodiversity*. pp. 232-246, Island Press, Washington.

Lindeboom, H. J. and de Groot, S. J. (Eds.) (1998). The Effects of Different Types of Fisheries on the North Sea and Irish Sea Benthic Ecosystems. NIOZ Report 1998-1, RIVO-DLO Report C003/98, 404 pp.

Lissner, A. L., Taghon, G. L., Diener, D. R., Schroeter, S. C., and Dixon, J. D. (1991). Recolonization of deep-water hard substrate communities: potential impacts from oil and gas development. *Ecol. Appl.* 1, 258-267.

Macedo, M. C. C., Macedo, M. I. C. and Borges, J. P. (1999). *Conchas Marinhas de Portugal (Seashells of Portugal).* Editorial Verbo, Lisboa, 516 pp.

Masero, J. A., Castro, M., Estrella, S. M., and Pérez-Hurtado, A. (2008). Evaluating impacts of shellfish and baitworm digging on bird populations: short-term negative effects on the availability of the mudsnail *Hydrobia ulvae* to shorebirds. *Biodiv. Cons.* 17, 691-701.

Mayer, L. M., Schick, D. F., Findlay, R. H., and Rice, D. L. (1991). Effects of commercial dragging on sedimentary organic matter. *Mar. Environ. Res.* 31, 249-261.

McConnaughey, R. A., Mier, K. L. and Dew, C. B. (2000). An examination of chronic trawling effects on soft-bottom benthos of the eastern Bering Sea. *ICES J. Mar. Sci.* 57, 1377-1388.

Medcof, J. C. and Caddy, J. F. (1971). Underwater observations on the performance of clam dredges of three types. ICES CM 1971/B:10, 7 pp.

Meyer, T. L., Cooper, R. A. and Pecci, K. J. (1981). The performance and environmental effects of a hydraulic clam dredge. *Mar. Fish. Rev.* 43, 14-22.

Norse, E. A. and Watling, L. (1999). Impacts of Mobile Fishing Gear: The Biodiversity Perspective. In: L. Benaka (Ed.), *Fish Habitat: Essential Fish Habitat and Rehabilitation*. pp. 31-40, American Fisheries Society, Symposium 22, Bethesda, Maryland.

Palomares, M. L. D. and Pauly, D. (Eds.) (2011). SeaLifeBase (version 10/2011). Accessed at www. sealifebase.org on March 2012.

Pickett, G. (1973). The Impact of Mechanical Harvesting on the Thames Estuary Cockle Fishery. MAFF Laboratory Leaflet (New Series) No. 29, 22 pp.

Poppe, G. T. and Goto, Y. (1993). *European Seashells. Volume II (Scaphopoda, Bivalvia, Cephalopoda).* Verlag Christa Hemmen, Wiesbaden, 221 pp.

Pranovi, F. and Giovanardi, O. (1994). The impact of hydraulic dredging for short-necked clams, *Tapes* spp., on an infaunal community in the lagoon of Venice. *Sci. Mar.* 58, 345-353.

Pranovi, F., Giovanardi, O. and Franceschini, G. (1998). Recolonization dynamics in areas disturbed by bottom fishing gears. *Hydrobiologia.* 375/376, 125-135.

Pranovi, F., Da Ponte, F., Raicevich, S., and Giovanardi, O. (2004). A multidisciplinary study of the immediate effects of mechanical clam harvesting in the Venice Lagoon. *ICES J. Mar. Sci.* 61, 43-52.

Rabaut, M., Braeckman, U., Hendrickx, F., Vincx, M., and Degraer, S. (2008). Experimental beam-trawling in *Lanice conchilega* reefs: Impact on the associated fauna. *Fish. Res.* 90, 209-216.

Ramsay, K. and Kaiser, M. J. (1998). Demersal fishing disturbance increases predation risk for whelks (*Buccinum undatum* L.). *J. Sea Res.* 39, 299-304.

Ramsay, K., Kaiser, M. and Hughes, R. (1996). Changes in hermit crab feeding patterns in response to trawling disturbance. *Mar. Ecol. Progr. Ser.* 144, 63-72.

Ramsay, K., Kaiser, M. J., Rijnsdorp, A. D., Craeymeersch, J. A., and Ellis, J. (2000). Impact of Trawling on Populations of the Invertebrate Scavenger *Asteria rubens*. In: M.J. Kaiser, S.J. de Groot (Eds.), *Effects of Fishing on Non-Target Species and Habitats*. pp. 151-162, Blackwell Science, Oxford.

Reed, D. H. and Frankham, R. (2003). Correlation between fitness and genetic diversity. *Cons. Biol.* 17, 230-237.

Reise, K. and Schubert, A. (1987). Macrobenthic turnover in the subtidal Wadden Sea: the Norderaue revisited after 60 years. *Helgol. Meeresunters.* 41, 69-82.

Sainsbury, K. J., Campbell, R. A., Lindholm, R., and Whitelaw, A. W. (1997). Experimental Management of an Australian Multispecies Fishery: Examining the Possibility of Trawl-Induced Habitat Modification. In: E. K. Pikitch, D. D. Huppert, M. P. Sissenwine (Eds.), *Global Trends: Fisheries Management*. pp. 107-112, American Fisheries Society, Bethesda, Maryland.

Schratzberger, M. and Jennings, S. (2002). Impacts of chronic trawling disturbance on meiofaunal communities. *Mar. Biol.* 141, 991-1000.

Schwinghamer, P., Guigne, J. Y. and Siu, W. C. (1996). Quantifying the impact of trawling on benthic habitat using high resolution acoustics and chaos theory. *Can. J. Fish. Aquat. Sci.* 53, 288-296.

Schwinghamer, P., Gordon, D. C., Rowell, T. W., Prena, J., McKeown, D. L., Sonnichsen, G., and Guigne, J. Y. (1998). Effects of experimental otter trawling on surficial sediment properties of a sandy-bottom ecosystem on the Grand Banks of Newfoundland. *Cons. Biol.* 12, 1215-1222.

Sewell, J. and Hiscock, K. (2005). Effects of Fishing within UK European Marine Sites: Guidance for Nature Conservation Agencies. Report to the Countryside Council for Wales, English Nature and Scottish Natural Heritage from the Marine Biological Association. CCW Contract FC 73-03-14A. Marine Biological Association, Plymouth, 195 pp.

Sih, A., Jonsson, B. G. and Luikart, G. (2000). Habitat loss: ecological, evolutionary and genetic consequences. *Trends Ecol. Evol.* 15, 132-134.

Southward, A. J., Hawkins, S. J. and Burrows, M. T. (1995). 70 years observations of changes in distribution and abundance of zooplankton and intertidal organisms in the western English Channel in relation to rising sea temperature. *J. Therm. Biol.* 20, 127-155.

Tebble, N. (1966). *British Bivalve Seashells. A Handbook for Identification.* The British Museum (Natural History), London, 212 pp.

Thistle, D. (1981). Natural physical disturbances and communities of soft bottoms. *Mar. Ecol. Progr. Ser.* 6, 223-228.

Thrush, S. F. and Dayton, P. K. (2002). Disturbance to marine benthic habitats by trawling and dredging: implications for marine biodiversity. *Ann. Rev. Ecol. Syst.* 33, 449-473.

Thrush, S. F., Hewitt, J. E., Cummings, V. J., and Dayton, P. K. (1995). The impact of habitat disturbance by scallop dredging on marine benthic communities - What can be predicted from the results of experiments. *Mar. Ecol. Progr. Ser.* 129, 141-150.

Tuck, I. D., Bailey, N., Harding, M., Sangster, G., Howell, T., Graham, N., and Breen, M. (2000). The impact of water jet dredging for razor clams, *Ensis* spp., in a shallow sandy subtidal environment. *J. Sea Res.* 43, 65-81.

Tudela, S. (2004). *Ecosystem effects of fishing in the Mediterranean: an analysis of the major threats of fishing gear and practices to biodiversity and marine habitats.* General Fisheries Commission for the Mediterranean, Studies and Reviews No. 74, FAO, Rome, 58 pp.

Valiela, I., Bowen, J. L. and York, J. K. (2001). Mangrove forests: one of the world's threatened major tropical environments. *BioScience.* 51, 807-815.

Veale, L. O., Hill, A. S., Hawkins, S. J., and Brand, A. R. (2000). Effects of long-term physical disturbance by commercial scallop fishing on subtidal epifaunal assemblages and habitats. *Mar. Biol.* 137, 325-337.

Watling, L., Findlay, R. H., Mayer, L. M., and Schick, D. F. (2001). Impact of a scallop drag on the sediment chemistry, microbiota, and faunal assemblages of a shallow subtidal marine benthic community. *J. Sea Res.* 46, 309-324.

Wilcove, D. S., Rothstein, D., Dobow, J., Phillips, A., and Losos, E. (1998). Quantifying threats to imperiled species in the United States. *BioScience.* 48, 607-615.

Wolff, W. J. (2005). The exploitation of living resources in the Dutch Wadden Sea: a historical overview. *Helgol. Mar. Res.* 59, 31-38.

Zhang, C. I., Kim, S., Gunderson, D., Marasco, R., Lee, J. B., Park, H. W., and Lee, J.H. (2009). Ecosystem-based fisheries assessment approach for Korean fisheries. *Fish. Res.* 100, 26-41.

In: Clam Fisheries and Aquaculture
Editor: Fiz da Costa González
ISBN: 978-1-62257-518-3

Chapter 13

THE HABITAT, FISHERIES AND AQUACULTURE OF THE VOLTA CLAM, GALATEA PARADOXA IN THE LOWER VOLTA RIVER IN GHANA: AN EXAMPLE OF THE WORLDWIDE IMPORTANCE OF BRACKISH WATER CLAMS

***C. Amoah*[1*] *and P. K. Ofori-Danson*[2#]**
[1]Volta Basin Research Program, University of Ghana, Legon, Ghana
[2]Department of Marine Science and Fisheries,
University of Ghana, Legon, Ghana

ABSTRACT

The Volta Clam, *Galatea paradoxa* (=*Egeria radiata*) occurs in the lower reaches of certain large West African rivers including the Volta River. The industry surrounding the Volta clam supports a very large riparian population and provides a substantial protein source for people in the lower Volta area of Ghana. The species is ideal for aquaculture because it is a sedentary filter feeder requiring no feeding apart from the natural algae content of the surrounding water and generally minimum husbandry. Although described as being brackish water species, it appears to be excluded from the lowermost 10 kilometres of the course of the Volta river. The lowermost limit of the range of *Galatea* appears to be the approximate limit of penetration of saline water from the sea. The factors which prevent *Galatea* from colonizing the whole length of the river is not clearly understood but could include the optimal grain size and organic matter content of the substratum.

The monthly catch assessment surveys indicate an average catch of about 5,400 kg of clam per month per aqualung diving fisher. At an estimated cost of US$0.12 per kg, an aqualung diver could therefore make an estimated income of about US$648.0 per canoe per month. This suggests that the clam fishery is a multi-million business for supporting

* Corresponding author: C. Amoah. Volta Basin Research Program, University of Ghana, Legon, Ghana, P.O.Box LG55, Legon, Ghana. E-mail address: capaloo@gmail.com.
# Department of Marine Science and Fisheries, University of Ghana, P.O.Box LG99, Legon, Ghana.

the livelihood of the riparian communities and urgently needs to be salvaged and developed to contribute sustainably to the income of the local fishing communities.

The closure of the clam fishery by traditional authorities from (December to March) coincides with the period when the clams have lower Condition Index emphasizing the importance of local knowledge in the development of the clam fishery and culture practices.

Observations from bottom grab samples indicate that the distribution of the clam in the sandy substratum is patchy. The gastropod, *Pachymelania aurita* which inhabits similar grounds as the clam could possess competitive advantage over the clam for habitable space due to their relatively higher abundance a situation which may not augur well for settlement of the larval stages of the clams.

Two size frequency distributions were found for the clam's population, which is attributed to possible display of protandric hermaphroditism-a typical phenomenon in bivalve populations. Protandric hermaphroditism has not been confirmed in *G. paradoxa*. Further studies on the sexuality of the species are needed to confirm this phenomenon, although it has been found in other bivalve that inhabit similar habitat.

Recent studies have shown by molecular technique that the endosymbiotic bacteria, Wolbachia, which could influence sex change in the clam, have been found in 59% of the clam mantle samples. The smaller size frequency distribution with average shell length <20 mm could represent males. These findings may influence the future of the industry in Ghana.

Preliminary experiments on clam culturing did not show appreciable growth in shell length and fresh weight per month.

13.1. INTRODUCTION

Clams form a significant part of the world's fisheries production. In 2000, landings of clams from capture fisheries and aquaculture operations totaled 14,204,152 tons (Helm et al., 2004). During the decade from 1991 to 2000,there was a continuing increase in production of clams, and landings more than doubled from 6.3 million tons in 1991 to 14 million tons in 2000 (Helm et al., 2004).

The global trend in the growth of human consumption of clams will undoubtedly continue. In 2000, landings of clams from capture fisheries and aquaculture operations totaled 14,204, 152 tones (Michael and Niel, 2004). During the decade from 1991 to 2000, there was a continuing increase in production of clams, and landings more than doubled from 6.3 million tones in 1991 to 14 million tones in 2000.

Clams are ideal animals for aquaculture; they are filter feeders requiring no additional feeding apart from the natural algae content of the water body and generally minimum husbandry. Although they have been cultured for hundreds of years, advances in culture technology and continued improvements in culture methodology and technology will be required to meet increasing demand and to make clam culture more economically attractive to both investors and people who wish to become shellfish farmers.

The clam, *Galatea paradoxa*, commonly called the Volta clam in Ghana, is reported to be distributed in estuaries from Guinea to Congo in Africa (Fischer et al., 1981). In West Africa they are accessible in the lower reaches of many large rivers including the Cross river in Nigeria (Ekanem and Achinewhu, 2006), the Sanaga and the Volta river estuaries in the Cameroons and Ghana respectively (Pople, 1966).

Of the various clams identified in Ghana, the Volta clam *G. paradoxa* is of paramount interest. Prior to the damming of the river at Akosombo, clam picking was mainly the predominant activity in the lower Volta estuary (Gyasi, 1999). This is because *G. paradoxa* is found in the lower reaches of the Volta River in Ghana. It has been the main source of animal protein, and therefore, played a vital role in the socio-economic life of the people over the years. They have been harvested for food and sold to earn income for living (Yankson, 2004; Gordon and Amatekpor, 1999). The clam is an important food source especially in the lower Volta area of the Volta basin.

Before to the construction of the two dams (Akosombo and Kpong), on the Volta river in 1964 and 1982 respectively, the clam fishery industry yielded 8,000 tons of *Galatea* per annum and employed about 2,000 women. Presently the clam industry supports about 80% of riparian population and the mantle provides a substantial protein source for the people of Ghana, while the shells are used as a source of calcium in the formulation of animal feed in the poultry industry. It is therefore imperative that everything be done to salvage and develop the clam industry. In the absence of a monitoring and control system, pressures from artisanal fishermen on the Volta Clam in the Lower Volta are on the increase. There is therefore the need to establish present habitat range and basic data collection on the species.

This Chapter outlines the following:

1 Attempt to define the pre and post impoundment limit of *Galatea paradoxa* in the Lower Volta.
2 Provide information on the new habitat of *G. paradoxa* after the creation of the dams at Akosombo and Kpong.
3 Give an overview of observations on the clam catch from 2004-2006 to provide scientific support for sustainable exploitation of the species as alternative livelihood approach for the communities in the Lower Volta area of Ghana.
4 Provide preliminary information on the detection, for the first time of the presence of the endosymbiotic bacteria *Wolbachia* species in the Volta Clam that is known to change sex of male into female in invertebrates.
5 Provide pertinent growth information on *G. paradoxa* culturing for clam fishery in the Lower Volta basin.

13.2. Pre and Post Impoundment Limit of G. Paradoxa in the Lower Volta

The area in which the clam occurred prior to the construction of the dams was between Torgome and Tefle (Figure 1 and table 1) the southernmost boundary was about 32 kilometres from the estuary. After the closure of the dam in 1964, the flow rate of the river which used to reach a peak of 10,000 m^3/s stabilized to about 850 m^3/s (Ennin and De Graft, 1977). This regulated flow of fresh water from the Volta caused a shift in saline water penetration limit downstream from Tefle to Agortaga (Figure 1).

This phenomenon also resulted in drifting of the habitat of the clams downstream from Torgome to Asutuare and Volivo towards the sea (Ennin and De Graft, 1977). This implies apparent reduction in the habitat range of the clam due to the effet of damming. The present southernmost boundary of the Volta clam is about 5 km upstream from the sea. Its habitat in the sediment of the Volta river extends upstream from this area for about 40 km to Agorkpo (Figure 1).

Observations from bottom grab samples indicate that the distribution of the clam in the sediment is patchy with areas of high abundance and areas of low abundance. This patchiness may be attributed to deeper areas where there is more likely to be less scour or hardiness (rocky) of the bottom.

Table 1. Summary of major trends of the G. paradoxa industry and differences in production from 1954-2005

Year	Location of clam beds (distance from estuary in Km)	Estimated yield in Kg per annum	Major events
1954-1963	Torgome (85 Km)-Tefle (32 Km)	8,000,000	Annual seasonal floods (August-October). Clams grown in 'farms' upstream and harvested just before the floods. $1^0/_{00}$ dry season salinity at about 30 km from mouth of river.
May 1964-1966	Torgome (85 Km)-Sogakope (30 Km)	8,000,000	Akosombo dam closed. Compensation flow through tunnel constant at 20 m^3s^{-1} until September 1965 when it was increased approximately to 100 m^3s^{-1} as the first turbines started working. Annual heavy flooding ceased.
1967-1970	-	-	Relatively constant flow of river. Dry season salinity zone of $1^0/_{00}$ is pushed towards the estuary. Development of submerged aquatic plants.
1970-1980	Asutuare (60 Km) - Big Ada (5 Km)	-	River almost fresh water (less than $0.07^0/_{00}$ salinity) at the estuary. Due to the formation of sand bar at mouth of river. Proliferation of aquatic weeds, harboring bilharzia vector snail and causing itching of the skin of the clam divers. Decrease in number of people harvesting clams because of bilharzia infection.
1980-1990	Agordomi(15 km)-Big Ada (5 km)	1,248,000	Sandbar forced open, causing increase in water salinity and decrease of aquatic weeds. Increase in the number of divers because of decrease in bilharzia risk.
1990-1995	-	-	No activity
1995-2005	Agorkpo (30 Km)-Big Ada (5 Km)	7,500,000	Sandbar forced open, salinity restored in the river estuary. Increase in the activity of clam divers.

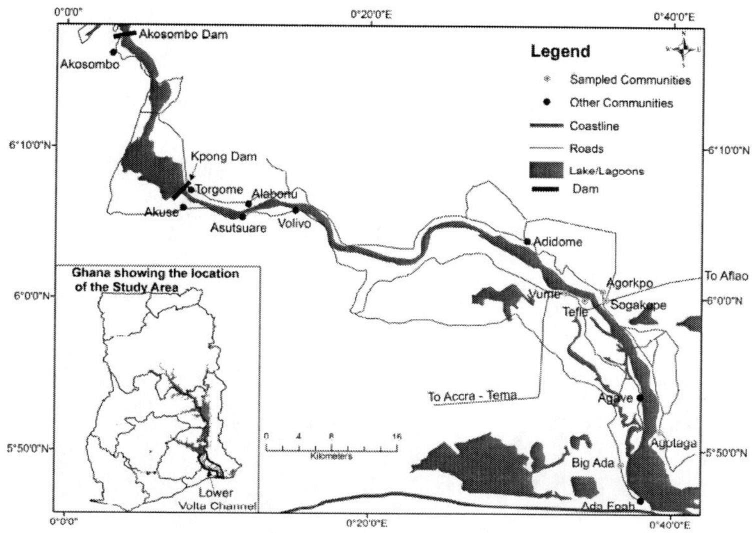

Figure 1. Lower Volta Basin showing some clam harvesting sites.

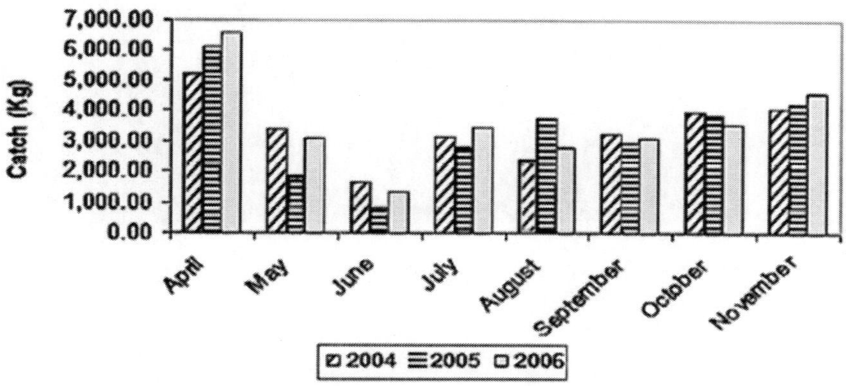

Figure 2.

In Big Ada, and its environs, the clam's population is over 400 individuals within a square metre whilst at Agorkpo, Tefle and Vume the population decreases to about 10 for the same area.

13.3. Overview of Clam Catch during 2004-2006

13.3.1. The Mean Annual Catch of Clams Harvested Per Unit Effort

The mean annual catch of clams harvested per unit effort (CPUE) (5 man/h) was recorded from April to November every year from 2004-2006. There was no harvesting from December to March because this period is regarded as closed season when no harvesting is allowed. The closed season allows for readjustment of the ecological balance in the clam's environment.

The CPUE per month was highest in April immediately after lifting the ban on harvesting during the period 2004-2006 (Figure 2). The clam catch ranged between 37.0-300 kg/day and a monthly catch of 962-7,800 kg/day of clams per diver (Figure 2).

Experimental recording of catch harvested from December to March was low with a mean of 2.8 kg/day. Figure 3 presents a picture of the aqualung diver in a canoe and figure 4 the collected clams. An estimated 120-150 people dive for the clams everyday.

Data on the body measurements are shown in table 2. The maximum body weight was 98 g per individual with an average of 48.95±1.15 g per individual. The maximum length was 99 mm with an average shell length of 36.15±0.25 mm.

Figure 3. Picture of aqualung divers in a local canoe.

Figure 4. A close up shot of the clams in a container.

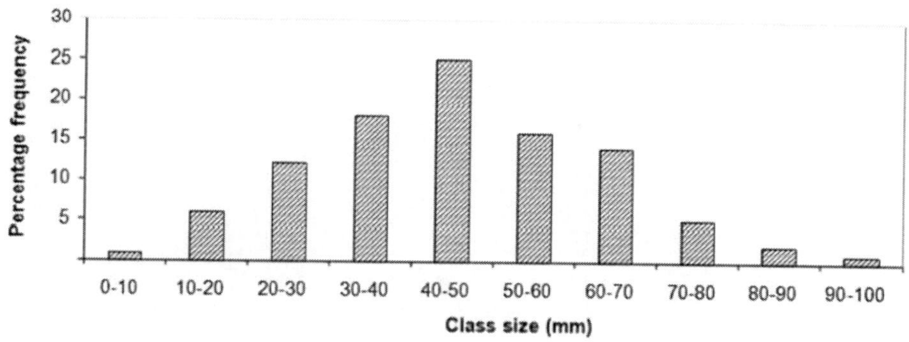

Figure 5. Shell size frequency distribution of clams in the Volta river estuary during 2004-2006.

Table 2. Measurement of the listed parameters of the Volta clam harvested from the lower Volta area (n =1,798)

Average shell length (mm)	36.15±0.25
Average total body weight (g)	48.95±1.15
Maximum length (mm)	98.40±2.2
Maximum total body weight (g)	98.00±0.5
Minimum shell length (mm)	17.0±0.15
Minimum total body weight (g)	1.60±0.13

13.3.3. Reproduction of *G. Paradoxa*

Using condition indices and histological methods, the reproductive cycle of *G. paradoxa* has been reported to follow an annual cycle with a single-spawning event between June and October (Adjei-Boateng and Wilson, 2011).

13.3.4. Shell Size-Frequency and Sex

Figure 6 shows two clearly separate curves of size groups for the shell length-weight relationship with relatively few juvenile clams. It has been reported that juvenile bivalves are typically males (Rice, 1992). In successive years they may change sex and produce eggs. This phenomenon is called protandric hermaphroditism. It is possible that this characteristic is being displayed by the clam and could offer explanation for the two size-frequency curves (Figure 6). If this is accepted then the smaller size group may represent males whilst the larger group represents the females. This implies that specimens >20 mm shell length could largely be females and those <20 mm were largely males. The gap between the two curves (ranging between 12-20 mm shell lengths) could be the transition length range which remains to be clarified.

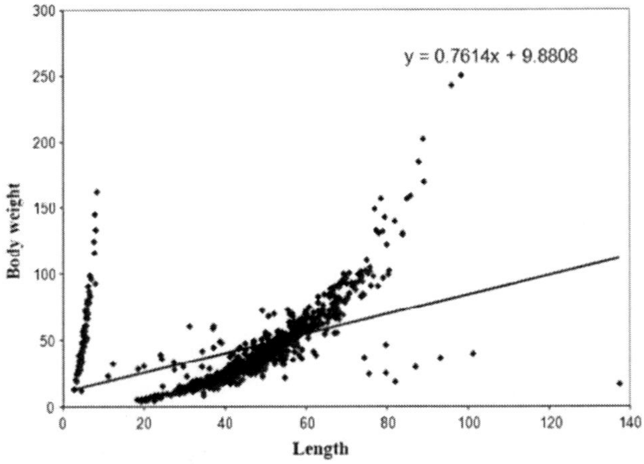

Source: Ofori Danson and Amoah (2005).

Figure 6. Body weight against length of *Galatea paradoxa* from April 2004 to April 2006.

On this basis, the two curves in figure 6 were treated separately as males and females to obtain the length-weight relationship equation respectively as males and females as shown in figure 7a and b. There was a curvilinear relationship between the shell length and total body weight with an exponential value of 2.3882 and 2.1674 for males and females respectively (Figure 8). The value was close to 3.0 indicating isometric growth.

a)

b)

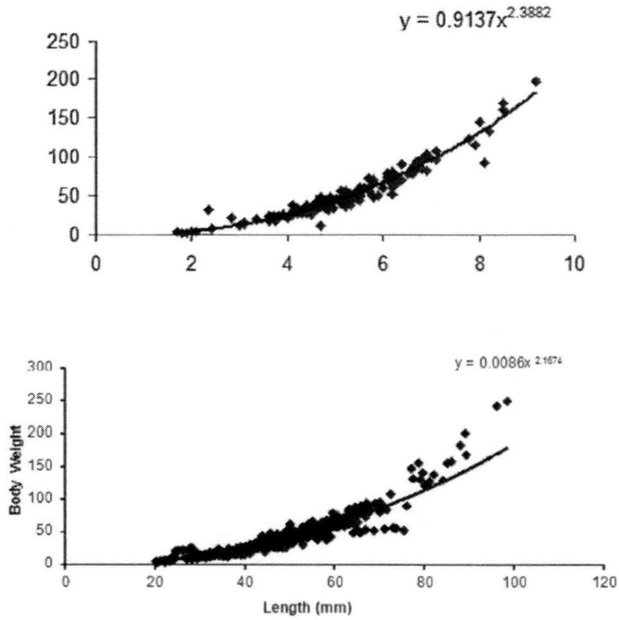

Figure 7. Length-weight relationship of male (A) and female specimens (<20 mm in length) (B) of *Galatea paradoxa*.

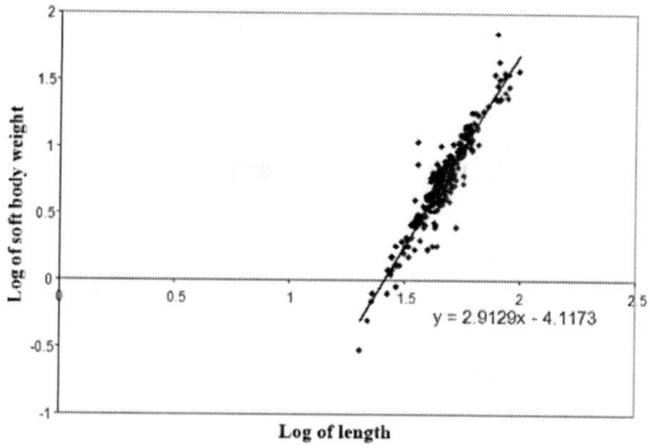

Source: Ofori Danson and Amoah (2005).

Figure 8. Log of body weight against log of length of *Galatea paradoxa* from April 2004 to April 2006.

In a recent study by Amoah (2009), she recognized the role the phenomenon of protandric hemaphroditism play in the reproduction and population of adult clams produced in the Volta river estuary. She found *Wolbachia* in the mantle of Volta Clam.

It was suggested that molecular biology methods could be used to distinguish between the sexes of the juvenile clams. Protandric hemaphroditism, may gain credence if *Wolbachia* strains are found in the life cycle of other clam in addition to *Galatea paradoxa* in West Africa.

13.3.5. Ecological Factors Influencing Larval Settlements and Juvenile Survival

One of the apparent factors that may determine the eventual distribution of the adult clam is the environmental (physical-chemical) conditions in the water for success of larval settlement and metamorphosis. It is possible that the period of settlement and metamorphosis may therefore be one of the most critical periods in the life cycle of the clam. It is also possible that a large number of larvae would not survive the transition. These observations may be clearer from continuous future analysis of physio-chemical data in relation to the data on the catches of the clam and this should be pursued to elucidate annual cycles. Recent studies showed that the temperature recordings of the Volta river at the water mud interface where the clams are found was 24.6-29.5°C, dissolved oxygen varied form 3.9-6.8 mg/L and pH 6.3-7.4. Present salinity values ranged between 0.01-0.60 PSU.

13.3.6. Another Vital Ecological Factor Affecting the Clam Population is the Gastropod *Pachymelania aurita*

The gastropod *Pachymelania aurita* were reported to occur in lagoons in Ivory Coast, Western Region of Ghana and some other places eastwards (Edmunds, 1978). Preliminary investigations of the sand bottom at the Ada shoreline near the study area indicate that this species occur in relatively large quantities than the clam in the ratio of about 64:1 for a given volume of sand. This suggests possible competition between these two mollusks for either food or space in the suitable sandy substratum.

From the examination of food item in 114 gut samples of *G. paradoxa* observed, the feeding of *G. paradoxa* can be described as selective because the bulk of the food items were phytoplankton. The phytoplankton *Ceratium sp.* contributed the greatest frequency of occurrence (64.04%), followed by *Anabaena sp.*

For comparison, 50 gut samples of *P. aurita* also examined were found to contain mainly detritus and miscellaneous materials. This implies that the two molluscs belong to two different feeding groups. *G. paradoxa* is a filter feeder while the *P. aurita* is a deposit feeder and so they may not compete directly for food.

13.3.7. Planktonic Algae

Planktons in the gut of the clam have recently been studied by Amoah et al. (2010). The microscopic algae encountered in the gut of the clam belonged to the Chlorophyta, Cyanophyta and Bacillariophyta. Members of the Chlorophyta were the most predominant, constituting 63% of the total population of phytoplanktons encountered. Thirteen (13) genera of Chlorophyta species isolated in the gut of the clams belonged to *Chlamydomonas*, *Chlorella*, *Closterium*, *Coelastum*, *Euglena*, *Oocytis*, *Palmella*, *Phacus*, *Phytoconis*, *Scenedesmus*, *Staurastrum*, *Ulotrix* and *Volvox* species. The diatoms (Bacillariophyta) found were *Cymbella tumida, Navicula* sp., *Pinnularia* sp. and *Syndra* sp. Out of the members of the Cyanophyta (Blue-green), *Merismopedia elegans*, *Microcystis aeroginosa*, *Oscillatoria princeps*, *Lyngbya limnetica* and *Planktothrix agardhii* were the blue-green algae (Cyanophyta) resident in the gut of the clams.

13.3.8. Culture Potential of the Clam

The clam fishery currently targets the adult population favouring recruitment of juveniles to replenish exploited adults. This suggests successful recruitment which could make a viable and sustainable fishery if the present fishing effort measured by the numbers of fishers does not increase tremendously.

Preliminary measurements on growth increments in weight and length of clam samples were cultured in "hapas". Hapas are specially sewn mosquito-proof net cages held in position by bamboo sticks and laden with sand. The experiments were carried out in ponds at the University of Ghana Aquaculture Research Centre near Ada, between the months of July to September 2006 (figures 9 & 10).

The growth in the mean length and weight during this early culture period was not very appreciable because the mean growth rate was 1.8% shell length and 3.2% body weight per month. This is in support of the report from Adjei-Boateng et al. (2010) that the pond environment is not suitable for the culture of *G. paradoxa* as the species is adapted to life in a river with its filter-feeding activity dependent on the water current.

Similar structures were erected in the 'wild' at the shoreline of the river near the VBRP Aquaculture Center. The "hapas" were lined with sand collected from the study area. Selected specimens of the clam were marked or 'tagged" after measurement of length and weight and implanted in the "hapas". The 'tagged' specimens were monitored for growth in length, weight and other growth parameters such as the length-weight relationships.

The data gathered were used in the assessment of the growth performance of the cultured clam. Preliminary observations of the clam kept in aquaria for 3 months (July to September, 2006) indicated that the clam can survive in glass aquaria while burrowed in suitable sand and covered with aerated water.

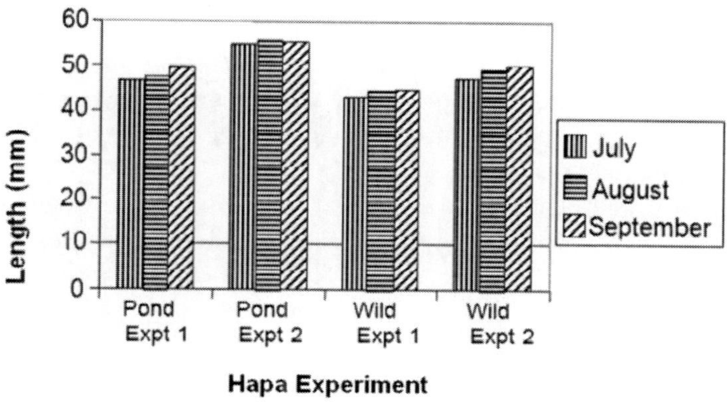

Figure 9. Growth increments by length of *G. paradoxa* cultured in hapas located in ponds and 'wild' from July to September, 2006.

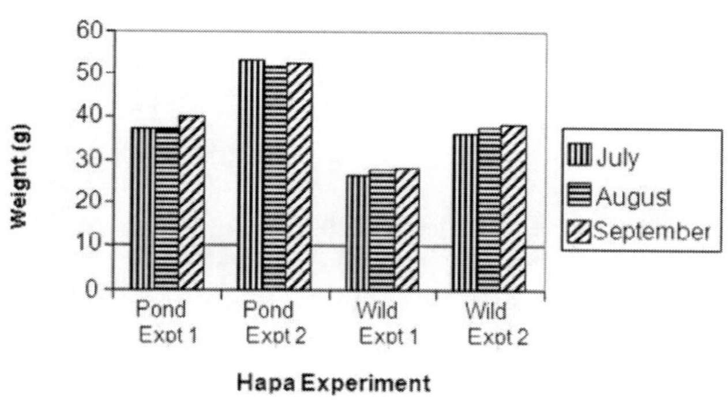

Figure 10. Growth increments by weight of G. paradoxa cultured in hapas located in ponds and 'wild' from July to September, 2006.

It was not clear what the animal fed on during this period (because they were not fed) but being a filter feeder, it is possible that the specimens relied on detritus within or lining the sand particles. This has been reported by Adjei-Boateng et al. (2010) that the substrate type affect survival, with sandy substrates yielding better survival than muddy ones.

It was also apparent that the character of the bottom sediment may influence the growth and settlement of the clam in the field. This was supported by the survival of the clam in "hapas" lined with sand in some ponds at the University of Ghana Aquaculture Research Center near Ada. In these studies, the clam had the tendency to grow and survive in the sand as opposed to the silt/clay bottom of the ponds.

Conclusion

The present habitat of the clams have drifted about 40 km downstream from Torgome to Agorkpo in the lower Volta River

The monthly catch assessment survey undertaken in 2004 to 2006 indicated that on the average, catch of about 5400 kg is taken per canoe of aqualung diver per month. At an estimated cost of US$0. 12 per kg, a canoe could make projected revenue of about US$646.20 per month. This suggests that the clam fishery could be a multi-million business supporting the livelihood of the riparian communities.

On the basis of numbers the gastropod, *Pachymelania aurita* was found to possess competitive advantage over the clam for habitable space in the sandy bottom. This situation is considered unfavorable for settlement of larval stages of the clam and hence poses threat for the development of the clam fishery.

The physiological well-being of the clams was assessed from the Condition Index (C.I.). The monthly mean Condition Index (C.I) of the clam was relatively high and stable between April and July with peak value in November. It is interesting to note that the closure of the clam fishery by traditional authorities from (November to March) coincided with the season of low C.I which emphasizes the importance of local knowledge in the development of the clam fishery and culture practices.

The detection of *Wolbachia sp* on the clam mantle requires further studies to show their role in protandric hermaphroditic phenomenon observed in the clam.

Culture trials in "hapas" located in ponds showed gradual increment in growth sustainable for the first three months only. Further investigations are needed to enhance higher growth performance through feeding experiments and artificial reproduction through induced spawning techniques.

References

Adjei-Boateng, D., Owusu-Appiah, B. and Amisah, S. (2010). Effect of sandy and muddy substrates on the growth and survival of the freshwater clam *Galatea paradoxa* (Born, 1778). *Aquac. Res.* 41, 84-88.

Amoah, C. M. (2009). Studies on selected physio-chemical and bacteriological quality of the water clam (*Galatea paradoxa* Born 1778) from Volta River and some of the mineral content of the mantle. *Ph.D Thesis, University of Ghana.* 260 pp.

Amoah, C. M., Odamtten, G. T. and Ofori-Danson, P. K. (2010). Occurrence of phytoplankton in the water and gut of the Volta Clam (*Galatea paradoxa* Born 1778) from the lower Volta estuary, Ghana. *J. Ghana Sci. Assoc.* 11, 103-112.

Edmunds, J. (1978). *Seashells and other Molluscs found on West African shores and estuaries.* Ghana University Press, Accra, 148 pp.

Ekanem, E. O. and Achinewhu, S. C. (2006). Mortality and quality indices of live West African Hard -Shell clams (*Galatea paradoxa* Born) during wet and dry post harvest storage. *J. Food Process. Pres.* 30, 247-257.

Ennin, M. A. and De Graft-Johnson, K. A. A. (1977). Studies on the ecology of *Egeria radiate* Larck in the Lower Volta Estuary. *Institute of Aquatic Biology (CSIR) Publication.* No. IAB 76.

Fischer, W. Bianchi, G. and Scott, W. B. (Eds.). (1981). *FAO Species identification sheets for fishery purposes. Eastern Central Atlantic; fishing area 34,37 (in part).* Canada Funds-in Trust. Ottawa, Department of Fisheries and Oceans Canada, by arrangement with Food and Agriculture Organization of the United Nations, vol. 6.

Gordon, C. and Amatekpor, J. A. (1999). The sustainable Integrated Development of the Volta Basin in Ghana. *Volta Basin Research Project, Accra,* 159 pp.

Gyasi, E. A. (1999). Foreword address. In: C. Gordon, J. A. Ametekpor (Eds.). The Sustainable Integrated Development of the Volta Basin in Ghana, *Volta Basin Research Project. Accra*, 159 pp.

Helm, M. M., Bourne, N. and Lovatelli, A. (comp./ed.) (2004). *Hatchery culture of bivalves. A practical manual.* FAO Fisheries Technical Paper, No. 471. Rome, FAO, 177 pp.

Pople, W. (1996). Comparison of the *Egeria* fishery of the Sanaga River, Federal Republic of Cameroun with that in the Volta River, Ghana. University of Ghana, *Volta Basin Research Project, Technical Report* XII.

Rice, M. A. (1992). *The Northern Quahog: The biology of Mercenaria mercenaria.* C. Jaworski and M. Schwartz (Eds.), Rhode Island Sea Grant, University of Rhode Island Bay Campus, Narragansett, US Department of Agriculture.

Yankson, K. (2004). *Fish from the shell: its potential in the quest for adequate protein in Ghana.* Media graphics and Press Ltd. Kokomlemle, Accra.

In: Clam Fisheries and Aquaculture
Editor: Fiz da Costa González
ISBN: 978-1-62257-518-3

Chapter 14

CLAMS AS BIOLOGICAL TOOLS IN MARINE ECOTOXICOLOGY

R. Beiras*
ECIMAT, Universidade de Vigo, Vigo, Galicia, Spain

ABSTRACT

Wild clam populations provide an opportunity to assess coastal pollution in a more ecologically relevant manner than sediment chemistry, by measuring the levels of chemicals in their tissues. Contaminants accumulated in the clams respond to the bioavailable fraction only, and do not correlate with the total metal concentrations in the sediments. However clams tend to accumulate less chemicals than mussels and oysters, likely due to physiological differences in feeding, metabolization and excretion rates.

Behavioural responses at a whole individual level are among the most sensitive indicators of environmental stress. Infaunal bivalves such as clams are suitable candidates for whole sediment toxicity testing. The burrowing activity of many marine clam species (e.g. genus *Abra*, *Macoma*, *Mya*, *Protothaca*, *Ruditapes*, *Venerupis*) has provided the endpoint for standard sediment toxicity testing. The ET_{50} value -time needed for complete burial of a half of the population- allows an objective estimate of burrowing speed. The ET_{50} has been proved to increase in chemically polluted sediments, and the test detects the bioavailable fraction of the metal only. However, several biotic and abiotic factors, including clam size and sediment texture, interfere with the measured response and must be taken into account.

Early life stages of clams provide a sensitive and ecologically relevant biological model for microscale toxicity testing. Assessments of sublethal toxic effects of pollutants and quality of environmental samples are based on the percentage of normal D-shaped larvae obtained after 48-h incubation of fertilised eggs in the test samples. Clam embryo-larval bioassays are sensitive to trace metals, pesticides, surfactants and certain hydrocarbons at concentrations as low as ppb. The embryos and larvae of American (*Mercenaria*, *Mulinia*) and European (*Ruditapes*, *Venerupis*) Veneridae and Mactridae clams have proved to be equally suitable for bioassays that other more frequent

* Corresponding author: R. Beiras. ECIMAT, Universidade de Vigo, E-36331,Vigo, Galicia, Spain. E-mail address: rbeiras@uvigo.es.

ecotoxicological models such as sea-urchins, mussels and oysters, and the combination of different species allows conducting the bioassay with natural stocks all year round.

14.1. Introduction

Marine ecotoxicology takes advantage of biological tools to provide, in addition to analytical chemistry, an integrative approach to coastal pollution assessment. These biological tools may consist of autochthonous or transplanted organisms that accumulate pollutants in their tissues (bioaccumulators), biochemical or physiological responses recorded in local wild populations (biomarkers), or sensitive lethal or sublethal responses to environmental matrices in standardised laboratory organisms (toxicity bioassays). With these aims, sessile, abundant, long-lived, pollution resistant organisms are particularly useful. Adults and early life stages of clams have been used in a number of ecotoxicology and pollution assessment studies. The aim of this chapter is to review this topic.

Adult clams provide information on the bioavailability of sediment pollutants, and a series of field studies have measured the bioaccumulation of persistent pollutants in native populations of different clam species. Since juvenile and adult clams are infaunal organisms, these organisms are also suitable biological models for laboratory testing of sediment toxicity (both natural and spiked sediments), by measuring the reburial activity and calculating bioassay endpoints such as the median reburial time and disturbed surface area. This approach has been used to assess the pollution of sand affected by oil spills.

Finally, clam embryos and larvae are suitable models, because of their extremely high sensitivity to water quality, ecological relevance and commercial importance, for the standard bivalve embryogenesis bioassay. Tens thousands of synthetic substances are in existence today, and hundreds of novel compounds are being introduced every year. Because of the complexity of the physic-chemical interactions between pollutants and the marine environment, the potential toxicity of contaminants can be assessed adequately only by means of bioassays with living organisms. From a practical point of view, a bioassay needs to be sensitive and scientifically valid, yield rapid results at moderate cost, and the organism in question must be readily available. Ecotoxicological bioassays with bivalve embryos, larvae and juveniles fulfil these criteria better than most other tests. They have increasingly come into use during the past four decades and are now commonly employed to ascertain the biological effects of pure chemicals, as well as to determine the quality of effluents, coastal waters and sediments sampled in the field. There do not appear to be very great differences between bivalve species with regard to larval sensitivity to toxicants. The bivalve embryo test has been originally developed using oysters and mussels, but the combined use of different clam species, with similar sensitivity to the other bivalves, allows conducting the bioassay all-year round.

For example, clam embryo tests contributed to the assessment of the impact of the *Prestige* fuel oil spill, that took place in winter, a period during which mussels and wild oysters are refractory to spawning. The 24-h clam reburial test, preferentially conducted with juveniles, is the ideal complement to the liquid phase embryo test, since it is suitable for whole sediment toxicity testing, and it is a more promising bioassay, in terms of sensitivity and feasibility, that the 10-d amphipod mortality test. Due to their sensitivity, rapid response, cost-effectiveness and ecological relevance, these biological tools bear potential to be applied in the assessment of the ecological status of surface water bodies, within the current context of the implementation of the European Water Framework Directive.

14.2. Bioaccumulation and Effects of Persistent Pollutants in Wild Clam Populations

14.2.1. Bioaccumulation

Adult clams from natural stocks are used to monitor chemical pollution in many parts of the world, including the Galician Rias (NW Iberian Peninsula). Pollutant concentrations found are summarised in tables 1 and 2. For essential trace metals such as Cu and Zn background levels are ppm units, while for xenobiotic metals such as Hg, Cd and Pb they are below 0.01 ppm. Enrichment factors in highly polluted areas may reach x100 for essential metals and x1,000 or higher for xenobiotics.

Other elements such as As, Co, Cr, Mn, Ni or Fe showed a lower bioaccumulation potential. Levels of hydrocarbons in clams from polluted sites are normally 1-100 ppm, and 1-10 ppb for other organic pollutants. Branched and cyclic aliphatics are preferentially accumulated over straight-chain hydrocarbons, and higher molecular weight aromatics are also selectively accumulated (Clement et al., 1980).

From the studies summarised in tables 1 and 2 we can also infer that: i. higher metal and organic concentrations are found in the inner parts of estuaries, ii. clams have a lower metal bioaccumulation potential compared to other bivalves such as mussels, oysters and scallops, iii. seasonal fluctuations in metal content have been described, so monitoring should be conducted in the same seasons.

Strong (x3 approx.) seasonal fluctuations in clam metal content have been described (Abaychi and Mustafa, 1988; Bordin et al., 1996). Bordin et al. (1996) coupled these field data with laboratory observations were metal uptake was enhanced in winter compared to summer, in parallel with the gametogenic activity. If metal is preferentially accumulated in the digestive gland compared to mantle and foot, such as the case of Hg in *Scrobicularia plana* (Langston, 1982), then spawning will increase the metal concentration in the total soft tissues. In laboratory studies it was demonstrated that metal uptake increases as temperature increases and salinity decreases, and competition among divalent cations stand for the later result (Jackim et al., 1977; Wright and Zamuda, 1987; Chong and Wang, 2001).

In other laboratory studies, clams were also useful for more theoretical investigations devoted to provide the basis for modelling metal uptake and accumulation. Absil et al. (1994) working with a copper radioisotope proved that Cu uptake through food in *Macoma balthica* was at least as efficient as uptake of waterborne metal. *Venerupis philippinarum* accumulates Ag and Cd but not Zn, and metallothioneins play a role in the detoxification of the accumulated metals (Ng and Wang, 2004).

After reviewing the scientific literature, no clear pattern emerges on the effect of clam size on metal accumulation, and positive, negative and no effect of size have all been reported (e.g. Strong and Luoma, 1981). In the freshwater clam *Corbicula fluminea*, Abaychi and Mustafa (1988) reported that in laboratory exposures larger individuals accumulated significantly more Cd, but the field data from Bilos et al. (1998) do not support that hypothesis and rather suggests that larger clams accumulate more copper and less zinc. Langston (1982) did not find a relation between Hg concentration and size in *S. plana*, and Lee et al. (1998) found higher metal concentrations in smaller clams, perhaps as a result of a relatively larger gill surface area. Usero et al. (1996) found an inverse effect of size on bioaccumulation for essential metals (Cu, Fe, Zn) but the opposite trend for non-essentials (Cd, Hg, As and Pb). Exceptions were also reported. Sánchez-Marín and Beiras (2008) found a strong and clear dependence of Pb accumulation in *Dosinia exoleta* on body size, with lead concentrations increasing exponentially with shell length. Wallace et al. (2003) proved that the compartmentalization of metal accumulation is size dependent, with larger clams tending to preferentially accumulate metals in metal-rich granules innocuous for the clam metabolism. This progressive retention of metals as non-excretable granules with age may lead to the size-dependency of metal concentrations found in some populations of *M. balthica* (Wallace et al., 2003).

Despite clams being infaunal organisms, total metal content in the sediment does not correlate with metal accumulated in clams (e.g. De Gregori et al., 1996). The metal content in *Venerupis corrugata* (=*V. pullastra*) clams from different sites of Ria de Vigo did not correlate with the total metal content of the sediment (except for mercury, where r=0.78, n=8) (V. Besada and R. Beiras, unpublished data). Interestingly, *C. fluminea* metal contents correlated better with particulate matter than with the dissolved form (Abaychi and Mustafa, 1988). In our Ria de Vigo study the Cd and Cu content of total particulate matter (TPM) were better predictors of mussel metal concentrations than the sediment, but we did not find significant correlations between the clam metal content and the metal content of TPM either (V. Besada, P. Sánchez-Marín and R. Beiras, unpublished data). For both sediment and suspended particulate matter, it is the easily extractable fraction of Cd, Cu and Pb (associated with oxides of iron and manganese), and not the total metal concentration, that better correlates with the metal content in clams (Thomas and Bendell-Young, 1998; Pierre Stecko and Bendell-Young, 2000).

This evidences the importance of chemical speciation on the bioavailability of metals, also for the particulate fraction. Predictions of metal bioaccumulation from sediments can be further improved by taking into account different sediment traits that influence uptake by clams. Luoma and Bryan (1978) stated that the biological availability of lead in sediments is controlled mainly by the concentration of iron, and the Pb concentration in *S. plana* may be predicted from the Pb/Fe ratio in weak acid extracts of surface sediment.

Langston (1982) takes into account the organic content to obtain a model that predicts 83.4% of the variability of mercury accumulation in *S. plana*. Once organic content was taken into account, Hg concentration in the clam was from 1.9 to 2.7 fold higher than in the sediment. Bourgoin et al. (1991) added that by normalising the weak-acid extracted lead with the total sulphur content in the sediments significantly improved the predictability of lead uptake in *M. balthica*.

The implications of the previous findings for biological monitoring of pollution are: i. for monitoring water column mussels and oysters are more useful than clams because their bioconcentration factors are higher (Jackim et al., 1977; Chong and Wang, 2001), probably due to physiological differences in feeding, metabolization and excretion; ii. chemical speciation affects bioavailability to clams, and monitoring strategies based on total metal analysis in sediments are less useful from an ecological standpoint than those supported on metal analysis in bivalves.

14.2.2. Molecular, Cellular, Histological and Physiological Biomarkers

Some studies take a step further and record biological effects at molecular, cellular, histological and physiological levels as a consequence of the accumulation of pollutants in clams. We do not aim to thoroughly review this topic here, but just to provide some illustrative examples of the use of clams as biological models for studies of biochemical and physiological responses to chemical pollution. These responses may just indicate exposure to chemical pollution (exposure biomarkers) or, in addition, evidence deleterious effects on the exposed individuals (effect biomarkers).

Exposure biomarkers include the following. Metal uptake in clams induces the synthesis of metallothioneins (Ng and Wang, 2004; Gagné et al., 2007), low molecular weight, cysteine-rich proteins with a role in metal sequestration and thus detoxification. Increased benzo(a)pyrene hydroxylase and catalase activities together with a significant reduction in lysosome membrane stability was described in *Mercenaria mercenaria* clams transplanted to a polluted estuary (Nasci et al., 1999), and increased glutathione S-transferase activity was recorded in *Mya arenaria* specimens native from polluted sites (Gagné et al., 2007). The lipid peroxidation was found to increase in the digestive gland tissues of clams exposed to dietary PAHs, but no differences were observed in the catalase activity (Frouin et al., 2007).

Effect biomarkers in clams were also reported. The accumulation of certain persistent organic pollutants, such as PCBs, has been associated to increased prevalence of neoplasias in *M. arenaria* (Strandberg et al., 1998). Chronic exposure to oil increased respiration and reduced growth and condition index in *M. balthica* (Stekoll et al., 1980). Exposure of *Venerupis decussata* to 10 μg/L copper increased respiration and decreased feeding rates, with the concomitant reduction in the energy budget (Sobral and Widdows, 1997). TBT concentrations down to 12 ng Sn/L decreased growth rates in juvenile *S. plana* (Ruiz et al., 1994). TBT at concentrations below 1 μg/L affected the cytochrome P450 monoxygenase system of *V. decussata* and induced the formation of hydroxylated steroid metabolites that may contribute to the masculinization of clam physiology (Morcillo et al., 1998). Finally, vitellogenin-like proteins in *M. arenaria* clams were significantly increased at sites influenced by domestic wastewaters (Gagné et al., 2007).

14.3. Clam Reburial Activity As an Endpoint in Sediment Toxicity Testing

We have just briefly reviewed molecular, cellular and physiological responses to pollutants. Behavioural responses at the whole individual level are among the most sensitive indicators of environmental stress (Eisler, 1979; Olla et al., 1980). They can provide real time responses to changes in water quality and are thus ideal candidates for biomonitoring water quality even in open flow conditions. In epibenthic bivalves such as mussels and oysters, shell closure can provide a continuous response suitable for biomonitoring the water column. Infaunal bivalves such as clams live inside the sediment and thus are suitable candidates for whole sediment toxicity testing.

Table 1. Accumulation of trace metals in clams collected from field studies (values are ranges). For dry-weight conversion an average water content of 80% in fresh tissues can be used

Test species	Location		Metal (μg g^{-1} wet weight except when otherwise stated)												Reference and notes
			Hg	Pb	Cu	Zn	Ni	Cd	As	Ag	Cr	Co	Mn	Fe	
Chamelea gallina	Atlantic coast (S Spain)	*a*	0.028-0.141	0.65-2.1	7.2-47	12-27	0.28-0.39	0.24-0.79	1.1-2.4		0.17-0.56		1.5-2.9	60-148	Uscro ct al. (1996) Samples: *a* <19 mm *b* 19-24 mm *c* 24-29 mm *d* >29 mm
		b	0.033-0.158	0.72-2.2	7.1-41	12-25	0.28-0.43	0.26-0.81	1.3-2.6		0.12-0.51		1.3-2.8	50-166	
		c	0.036-0.169	0.81-2.0	6.2-38	11-22	0.27-0.40	0.32-0.82	1.4-2.8		0.11-0.60		1.3-2.8	48-173	
		d	0.039-0.224	0.82-2.4	4.8-27	10-20	0.30-0.40	0.40-0.83	1.6-3.0		0.09-0.57		1.4-2.8	38-194	
Cerasto-derma glaucum	Etang de Tau, France		0.384-0.735	0.08-3.36	7.2-13.8	50.4-96.6	43.2-105.6	0.6-2.76		2.4-0.36	1.28-3.52	6.72-21.12	5.6-24.08	480-1056	Szefer et al. (1999)
Corbicula fluminea	Shat al-Arab river, Iraq			0.15-5.1*	40-1065*	30-83*	0.2-4.7*	2.2-60.4*					1.8-3*	0.4-15.5*	Abaychi and Mustafa (1988) *Dry weight
C. fluminea	Río de la Plata estuary (Argentina)				28-89*	118-316*	1.3-6.4*	0.5-1.7*			1.3-11*		15-81*		Bilos et al. (1998) *Dry weight
Macoma balthica	Dalhouse harbour (Canada)			32.6± 10.7-124± 21.7*											Bourgoin et al. (1991) *Dry weight
M. balthica	Westerschelde estuary (Netherlands)				13.8-37.1*	284-868*		0.11-1.27*							Bordin et al. (1996) *Dry weight

Test species	Location	Metal ($\mu g\ g^{-1}$ wet weight except when otherwise stated)												Reference and notes
		Hg	Pb	Cu	Zn	Ni	Cd	As	Ag	Cr	Co	Mn	Fe	
M. balthica	The Fraser River estuary (Canada)	0.148-265	0.5-13.5	9.5-308.4	86-527	4.2-26.9	0.15-1.5							Thomas et al. (1998)
Mercenaria mercenaria	Tampa Bay (Florida)		0.16	1.63	15.2	0.41	0.20			0.08	0.08	17.5	27.2	Nasci et al. (1999)
	St. Petersburg (Florida)		0.01	3.55	20.2	0.20	0.17			0.11	0.05	10.1	18.8	
Venerupis decussata	Galician Rias (NW Iberian Peninsula)	0.012-0.108	0.029-0.463	1.08-7.03	7.37-19.30		0.037-0.115	2.18-5.04						Besada et al. (2007)
Venerupis philippi-narum	Galician Rias (NW Iberian Peninsula)	0.016-0.033	0.023-0.099	0.81-1.65	11.90-14.50		0.039-0.046	1.90-2.92						Besada et al. (2007)
Scrobicula-ria plana	Estuaries in SW England		14-1016*											Luoma et al. (1978) *Dry weight
S. plana	British estuaries	0.08-1.89**												Langston (1982) **different tissues
Spisula solida	Chile			4.9± 0.3-13.5± 0.4*	32± 2-208±8*		0.6± 0.2-13.0± 0.3*							De Gregori et al. (1996) *Dry weight
Venerupis corrugata	Galician Rias (NW Iberian Peninsula)	0.007-0.134	0.028-0.682	0.99-6.06	9.19-15.0		0.037-0.117	2.52-8.02						González-Quijano et al. (2005); Besada et al. (2007)

Table 2. Accumulation of organic pollutants and TBT in clams collected from field studies (values are ranges). For dry-weight conversion an average water content of 80% in fresh tissues can be used

Test species	Location	Organic pollutant ($ng\ g^{-1}$ wet weight except when otherwise stated)							Reference and notes
		Total aromatic hydrocarbons	PAHs	∑PCB	∑DDT	PCDDs	PCDFs	TBT	
Mercenaria mercenaria	Tampa Bay (Florida)		25.5	0.85*	1.8†				Nasci et al. (1999) *sum of 12 CBs; † pp', op'DDT+pp', op'DDE+ pp' op', DDD
	St. Petersburg (Florida)		2822.7	40.1*	33.7†				
Leukoma staminea	Gulf of Alaska	0-3500*							Fukuyama et al. (2000) *Dry weight
Rangia cuneata	Laguna de Pom (Mexico)	2900-147140*							Alvarez-Legorreta et al. (1994) *Dry weight
Venerupis decussata	Ebro delta (NE Iberian Peninsula)		20-120	1-18	0.6-7				Solé et al. (2000)
Venerupis philippinarum	Adriatic Sea			1.24-3.93*	0.7-0.8#	0.38-0.53 $\times10^{-3}$	0.38-1.38 $\times10^{-3}$		Bayarri et al. (2001) *sum of 7 CBs; #DDE only
Venerupis corrugata	Galician Rias (NW Iberian Peninsula)			1-5.8*	0.2-0.4†				González-Quijano et al. (2005); Besada et al. (2007) *sum of 10 CBs; †pp'DDT+pp'DDE+pp'DDD
Scrobicularia plana	Arcachon Bay (France)							28-165*	Ruiz et al. (1997) *Dry weight

FSW: filtered seawater.

As far as behavioural responses are concerned, the burrowing activity of many marine clam species (genus *Abra, Macoma, Mya, Protothaca, Ruditapes, Venerupis*) has provided the endpoint for standard sediment quality bioassays.

Interestingly, Kaschl and Carballeira (1999) report burrowing activity as a more sensitive response than shell closure in clams. Although other approaches such as recording number of buried individuals at a single time (Roper and Hickey, 1994; Roper et al., 1995; Nipper et al., 1998), or re-emergence after burial (Kaschl and Carballeira, 1999) have been reported, most studies quantify borrowing behaviour by periodic recording of number of emerging individuals. By counting the number of clams remaining on the sediment surface at short time intervals and adjusting the resulting ratios of buried animals to log-logistic or log-probit functions of time, a ET_{50} value (time needed for complete burial of a half of the population) that provides an objective estimate of burrowing speed can be obtained. The maximum duration of the test is often 24 h, but ET_{50} values in control sediments may be as low as a few minutes (see below). Clam reburial may be up to 20 times more sensitive as endpoint than short term mortality (Eisler, 1979; Hickey and Martin, 1995), although differences of only 4-6 folds between sublethal and lethal endpoints have also been reported (Eisler, 1979; Roper and Hickey, 1994; Ruiz et al., 1994). In addition, burrowing speed is an ecologically relevant endpoint, since delayed reburial leads to increased predation on these bivalves (Pearson et al., 1981). Concerning the clam burrowing bioassay methodology, several biotic and abiotic factors that interfere with the measured response must be taken into account. Table 3 shows typical control ET_{50} values described for different clam species. Disregarding species, clam size has a strong effect on burrowing speed, with smaller individuals showing much shorter ET_{50} values (Pfitzenmeyer and Drobeck, 1967; Stirling, 1975; Pearson et al., 1981; Phelps et al., 1983, 1985; Pariseau et al., 2007; St-Onge et al., 2007). Therefore standard methods normally advocate the use of juveniles (e.g. Phelps, 1989). In contrast clam density, tidal cycle, emersion period, salinity or pH showed moderate or null influence on burrowing activity (Phelps et al., 1985; Phelps, 1989; Riba et al., 2004; Pariseau et al., 2007). Other factors influencing clam reburial speed are species-specific, and must be standardised for the species in question prior to using the reburial test with monitoring purposes. Season seems to affect burrowing activity in *M. arenaria* (Pariseau et al., 2007). For short exposure times low temperatures tend to reduce burrowing speed (Pfitzenmeyer and Drobeck, 1967, and Phelps, 1989 for *M. arenaria*). Grain size showed also some effect on clam reburial, and each species may have its own substrate preferences. *M. arenaria* preferably burrows in fine sediment (Pfitzenmeyer and Drobeck, 1967; but see St-Onge et al., 2007) while *V. corrugata, V. decussata* and *V. philippinarum* tends to prefer sand (Kaschl and Carballeira, 1999; Shin et al., 2002). It is therefore plausible to use more than one species in order to test sediments with different textures from a single monitoring cruise.

Laboratory experiments (Phelps et al., 1983, 1985; Roper et al., 1995; Kaschl and Carballeira, 1999; Shin et al., 2002; Matozzo et al., 2004) proved that the clam reburial test was sensitive to copper and zinc spiking down to tens of ppm, to cadmium down to ppm units, and to nonylphenol down to tenths of ppm (but see also Roper and Hickey (1994) who did not find effects of chlordane on clam reburial). Pearson and coworkers (Pearson et al., 1981; Olla et al., 1983) have shown that artificially oiled sediments (1,000-10,000 ppm oil) inhibit the burrowing speed in the littleneck clam, *Leukoma staminea* (=*Protothaca staminea*), and the hard clam *M. mercenaria*. Ruiz et al. (1994) reported reduced reburial activity in juvenile *S. plana* at TBT concentrations down to 0.5 μg/L.

Table 3. Control ET_{50} values described for different clam species

Species	Size (mm)	Sediment	ET_{50} (min)*	Observations	References
Abra alba	10-15	sand, 1% OM	1.7 (1.1-2.8) 4.5 (2.8-7.2)	field experiments 17°C lab. exp.	Møhlenberg and Kiørboe (1983)
Macoma baltica	5-10	mud, 0.13% OM	10.2	10°C	McGreer (1979)
M. baltica	15-25	sand, 1% OM	3.7 (2.6-5.3) 14 (11-18)	field experiments 17°C lab. exp.	Møhlenberg and Kiørboe (1983)
Venerupis corrugata	10-14	sand, 0.3% OM	16 (14.3-16.9)	16°C	Kaschl and Carballeira (1999)
V. decussata	5-6	sand, 1.0% OM	5.3	20°C	J.C. Mariño-Balsa (pers. comm.)
V. decussata	10-14	sand	9.1-12.5	16°C	Kaschl and Carballeira (1999)
V. philippinarum	10	sand	0.6-2.4	20°C	DelValls et al. (2002); Riba et al. (2004)
V. philippinarum	30-40	coarse sand	50	20°C	Shin et al. (2002)
Leukoma staminea	15-25 25-35	coarse sand, 0.09%OC	15 (13.8-16.2) 19 (16.2-22.2)	13°C	Phelps et al. (1983)
L. staminea	26-35 36-45 46-55 56-65	sand	132 276 396 918	13°C	Pearson et al. (1981)
Mya arenaria	17-25	sand	16-35	4-26.5°C	Phelps (1989)
M. arenaria	35-50 50-65 65-75	sand	156 (134-177) 318 (280-367) 1092 (906-1314)	3.8-27.2°C	Pfitzenmeyer and Drobeck (1967)
Scrobicularia plana	2.3	<1mm	<5	15°C, 24 ppt	Ruiz et al. (1994)

Despite these promising results with spiked sediments, differences among natural sediments with a wide variation in metal content were not detected by the bioassay (Phelps et al., 1983; Roper et al., 1995; Shin et al., 2002), except for locations with exceptionally high levels of copper (above 1 g/Kg, Shin et al., 2002) or pesticides (above 100 mg/Kg of organophosphates, Møhlenberg and Kiørboe, 1983). Aging of the spiked sediment or addition of a chelating agent caused a remarkable reduction in the biological response (Phelps et al., 1983, 1985; Kaschl and Carballeira, 1999). Porewater copper concentration but not sediment copper concentration could explain the burrowing behaviour (Kaschl and Carballeira, 1999). The common conclusion is that the burrowing test detects the bioavailable fraction of the metal only and the results do not necessarily correlate with total metal concentrations.

In contrast, McGreer (1979) reported a correlation between the levels of mercury and cadmium in the sediments and decreased burrowing speed in the estuarine clam *M. balthica*, taking advantage of a field gradient of pollution caused by a sewage effluent.

The results of another group of studies using the clam reburial test are more difficult to compare because of lack of standardization in the endpoint selected to assess burrowing activity. Byrne and O'Halloran (1999) reported that burrowing of juvenile *V. phillipinarum* (=*Tapes semidecussatus*) was inhibited in Dublin Bay sediments which contained 58-198 Cu mg/kg dry weight, and advocate the use of this commercially available species for whole sediment toxicity testing in the Irish coast. They also found differences in the number of clams buried after 24 h in different natural sediments from the Irish coast, although they did not follow the burrowing time course and could not calculate ET_{50} values (Byrne and O'Halloran, 2000). However their study shows that the same sediments identified as polluted after 21 d exposure in lethal tests can be discriminated much faster using their 24h burial test. Nipper et al. (1998) recorded the number of juvenile *Macoma* buried after 10-min exposure to sediments, and found significant inhibition in the most polluted sediments in terms of trace metal/AVS ratio, rather than total metal concentrations, again supporting that the clam reburial test identifies bioavailable rather than total metal concentrations.

Standardization of a clam reburial test for a given species must include specification of a narrow interval of sizes, exposure temperature, and accurate descriptions of the quantitative relations between ET_{50} and grain size.

Future improvements of the test should include image analysis that helps in recording accurate results at short exposure times.

14.4. Ecotoxicological Bioassays with Early Life Stages of Clams

Anthropogenic impacts on the aquatic environment may be viewed from a physic-chemical (contamination) or biological (pollution) perspective. The biological effects of pollutants in the environment are more important than the mere presence of contaminants; with regard to environmental quality, the data from chemical analyses can only be interpreted before a biological background. It is therefore logical to integrate biological tools in the monitoring programs devoted to the assessment of environmental quality, traditionally based upon analytical chemistry determinations. Biological tools contribute to those monitoring nets with ecological relevance and cost-effectiveness, and they are frequently advocated as rapid screening tools for the detection of hot spots where more intensive investigation can later be focused on. In fact, ecotoxicological bioassays are essential in a monitoring program since they allow: i. detection of new pollutants a priori not expected, ii. assessment of the bioavailable fraction of the chemical only, and iii. identification of the combined effects of multiple chemicals.

The different ontogenetic stages in marine species differ in their sensitivity to pollutants, and embryo-larval development provides rapid and extremely sensitive responses susceptible of microscale testing. Thus, among the many methods employed in bioassays, those using meroplanktonic stages, such as sea urchin plutei or bivalve veligers, appear to be the most promising for coastal water quality assessment (e.g. Kobayashi, 1971). Many authors have proposed the use of early life stages of bivalves for marine toxicological studies because it is clearly established that they are more sensitive to toxic substances than the adults (e.g. Connor, 1972; reviewed by His et al., 1999). The concept of the "biological quality" of seawater itself was introduced by Wilson (1951), who found that pluteus larvae of the sea urchin *Echinus esculentus* were capable of developing in seawater obtained from the Celtic

Sea, but not in water from the vicinity of Plymouth. Subsequently Woelke (1961), Davis (1961) and Okubo and Okubo (1962) among many other scientists in America, Japan, Russia and Europe started using bivalve embryos as biological indicators of water quality impairment by chemical pollution (reviewed by His et al., 1999).

Concerning clams, those pioneer studies from the 60's and early 70's used two American species; the Veneridae hard clam *Mercenaria (=Venus) mercenaria*, and the Mactridae coot clam *Mulinia lateralis* (because of the very short generation time of ca. 60 days of the later). However, the bioassay has been later adapted to the European species of the Veneridae family *Polititapes virgineus (=Venerupis rhomboides), V. corrugata* and *V. decussata* (Mariño-Balsa et al., 2003; Beiras and Albentosa, 2004; Beiras and Saco-Álvarez, 2006). Early life stages of different bivalve species display similar sensitivity to pollutants.

For example, the median effective concentrations of the insecticide Sevin inhibiting normal embryogenesis in oysters, mussels and clams fall within a relatively narrow range of 2 to 4 mg/L, approximately (His et al., 1999). Beiras and Albentosa (2004) found similar sensitivity of *V. decussata* clam and *M. galloprovincialis* mussel embryos to different trace metals. Therefore the combination of different bivalve species with different spawning periods allows conducting the embryo-larval bioassay all year round. In the South Atlantic coast of Europe *V. corrugata* and *P. virgineus* are fertile during winter, *Mytilus* spp. during spring and autumn, and *V. decussata* during summer.

Clam embryo-larval bioassays can be conducted within a relatively short time period (24 to 48 hours after fertilization). Assessments of sublethal toxic effects of pollutants are based on the percentage of normal D-shaped larvae obtained at the end of embryogenesis (determination of the concentration that inhibits normal larval development in 50% of the fertilized eggs). The procedures for the European-Atlantic species of clams include the following.

A mature stock of adult clams is induced to spawn by placing ca. 2 kg of adults in warm seawater at 25ºC over a black bottom that allows identification of spawning individuals.

Once an individual starts spawning it is rapidly transferred into an individual vessel filled with u.v.-sterilized 0.45 µm-filtered seawater (FSW). The quality of the gametes is assessed under the microscope by taking a sample with a Pasteur pipette. Sperm must be mobile and eggs must be round-shaped. A dense suspension of suitable eggs are collected in a 50 mL measuring cylinder with FSW and fertilized with 2 mL of a suitable sperm suspension, under gentle stirring for a few minutes. Several aliquots of 20 µL are taken and number of eggs is counted in order to deliver the fertilized eggs into the testing vials at a density of 40 per mL. Delivery into the control and experimental vials must be finished within 45 min after fertilization. In order to control the quality of the procedures, 70 min after fertilization 100 individuals from the controls are observed in vivo to record the fertilization success, visible as the first cleavage of the embryo. For a test to be considered acceptable fertilization success must reach 80%, otherwise the test must be repeated with better biological material.

After 48 h incubation at 20ºC the vials are fixed with two drops of concentrated formalin and observed under an inverted microscope. Normal larval development is assessed in at least n=200 individuals per treatment (including all replicates of that treatment) by the perfect D-shape of the veliger (Figure 1a). Round-shaped embryos, trocophores, larvae with a protruding mantle (Figure 1b), convex-hinged larvae or larvae smaller than usual (Figure 1c) are all considered as abnormal. When results from different fertilizations are to be compared the response should be corrected by the respective control to take into account biological differences among stocks.

a)

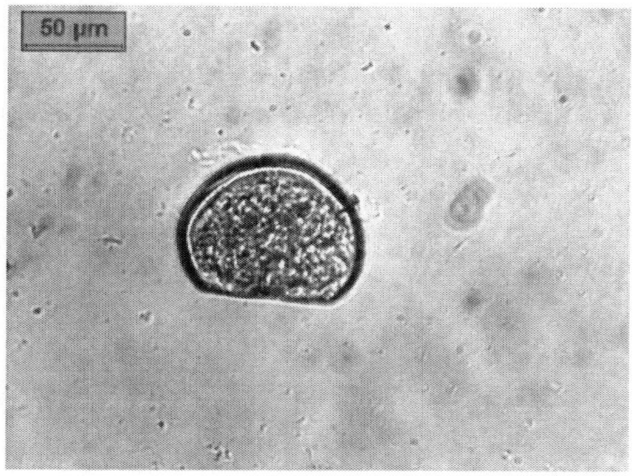

b)

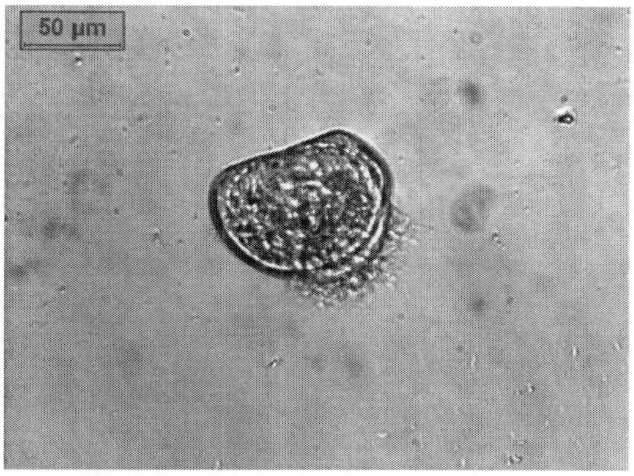

c)

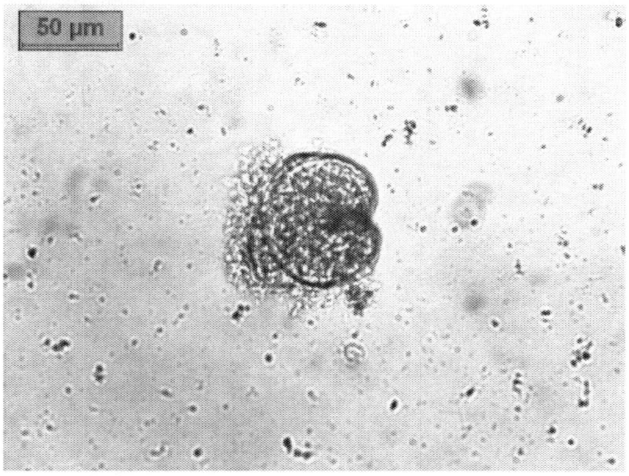

Figure 1. Normal (a) and abnormal (b,c) D-shaped larvae of *Venerupis corrugata* fixed with formalin after 48 h incubation of fertilized eggs at 20°C. Notice the abnormally protruding mantle and velum outside the prodissoconch in (b) and the abnormal convex hinge and smaller size in (c).

Tables 4 and 5 summarize the toxicity of trace metals on clam embryos and larvae respectively. Embryos are slightly more sensitive than larvae. Mercury, silver and copper are toxic at levels as low as ppb units, Pb and Zn at hundreds of ppb, and Ni at thousands of ppb. But the most toxic substance for clam early life stages so far tested is TBT that impairs *M. mercenaria* larval growth from 0.05 μg/L, inhibits normal embryogenesis at 1.1 μg/L, and causes 50% larval mortality at 1-1.6 μg/L (Roberts, 1987; Laughlin et al., 1989). Chlorine produced oxidants showed EC_{50} values for *M. mercenaria* and *M. lateralis* and *M. arenaria* embryos and larvae ranging from <5 to 500 μg/L (Roberts et al., 1975; Roberts, 1980; Roosenburg et al., 1980).

In a comprehensive study Davis and Hidu (1969) quantified the toxicity of pesticides using *M. mercenaria* embryos and larvae. The 48 h embryogenesis was a more sensitive response than the 10 d larval mortality.

Among the most toxic products we can highlight the Phygon (EC_{50}=14 μg/L), TCC (EC_{50}=32 μg/L), Delrad (EC_{50}=72 μg/L), and Omazene (EC_{50}=81 μg/L), and among the least toxic the Lindane (EC_{50}>10,000 μg/L)[1]. With intermediate toxicity they found the Aldrin (EC_{50}>10,000 μg/L but 10 d LC_{50}=410 μg/L), Guthion (EC_{50}=860 μg/L), Sevin (EC_{50}=3,820 μg/L) or Diuron (EC_{50}=2,530 μg/L).

A similarly comprehensive effort was made in the same laboratory using the same biological model to test the toxicity of surfactants, the active component of detergents (Hidu, 1965). Hidu's work covers "hard" detergents based on cationic, anionic and non-ionic surfactants.

Table 4. Toxicity of trace metals to clam embryos. EC_{50}: toxicant concentration causing 50% abnormal embryogenesis. Chlorides were used for Hg, Cu, Zn, Cd, nitrate for Ag and Pb, except when otherwise stated

Test species	Initial stage (time after fertilization)	Exposure conditions (time, temperature, salinity, density, seawater)	EC_{50} (μg metal ion l^{-1})						Reference and notes
			Hg	Ag	Cu	Zn	Pb	Cd	
Spisula solidissima	Egg Fertilized egg (1 h)	48 h, 20°C, 30‰, 30 ml^{-1}; 0.22 μm FSW uv sterilized		6.4-9.5 ca. 16					Eyster and Morse (1984); abn. lar. exc.
Munilia lateralis	Fertilized egg (2 h)	48 h, 21°C, 10,30‰, 75 ml^{-1}			18, 17				Morrison and Petrocelli (1990); sulfate.
Venerupis decussata	Fertilized egg	48 h, 20°C, 34‰, 20 ml^{-1}, ASW	4.2		9.1	129	156-312	424	Beiras and Albentosa (2004)

[1] However lower EC50 values for Lindane were more recently established using mussel embryos; see Beiras and Bellas (2008).

Table 5. Effects of trace metals on mortality, growth and metamorphosis of clam larvae. LC_{50}: toxicant concentration causing 50% mortality; otherwise EC_{50}: toxicant concentration causing 50% reduction in the end-point. Nitrate for Ag and chloride for the other metals were used

Test species.	Initial age/size	Exposure conditions (time, temperature., salinity, density, food, seawater)	End-point	LC_{50}/EC_{50} (μg metal ion l^{-1})					Reference and notes
				Hg	Ag	Cu	Zn	Ni	
Mercenaria mercenaria	2 d	8-10 d, 25°C, 10-12 ml^{-1}, mix algal food; 1μm FSW	mortality	14.7	32.4	16.4	195	5700	Calabrese et al. (1977a)
M. mercenaria	2 d	8-10 d, 25°C, 10-12 ml^{-1}; mix algal food; 1μm FSW	growth (length)	21.4	42.2	16.9	232	<1100	Calabrese et al. (1977a)
Munilia lateralis	9 d, 118 μm	72 h, 20-22°C, 34‰, 30 ml^{-1}, 1 μm FSW	Ca uptake	26.5		18.5	176		Ho and Zubkoff (1982)

FSW: filtered seawater.

Subsequent work by other authors focused on the biodegradable linear surfactants LAS (mainly dodecylbenzene sulphonate) and sodium dodecyl sulphate (SDS).

In agreement with the results of toxicity studies with heavy metals and biocides, embryos are again frequently affected earlier than larvae, and among the later growth is a more sensitive indicator of toxicity than mortality. The cationic surfactant lauryl pyridinium chloride is the most toxic compound in this group, with an embryogenesis EC_{50} of 8.5 to 90 μg L^{-1}. The new linear anionic surfactants are generally more toxic than the old nonlinear ones. The biodegradable surfactant LAS has an embryo EC_{50} ranging from 50 to 1700 (mean±sd: 383±493, n=10), but according to Cardwell et al. (1979) its degradation products are mostly non-toxic (EC_{50} from 1,270 to >10,000).

SDS, sometimes used as reference toxicant, has consistent effects on embryogenesis, with the exception of the study by Morrison and Petrocelli (1990), that report EC_{50} values an order of magnitude higher than the other studies. When these high values are excluded, average embryo EC_{50} is 741±233 (mean±sd, n=7). The toxicity of anionic surfactants is directly related to the length of the alkyl chain. The 14-C compounds are some 30 times more toxic than 10-C compounds (Cardwell et al., 1979).

Finally another set of studies with *M. mercenaria* and *M. lateralis* early life stages focused on the toxicity of crude and refined oils (Byrne and Calder, 1977; Gormly et al., 1996; Pelletier et al., 1997).

Despite the methodological heterogeneity, the results permit to draw some general conclusions. Refined oil is generally more toxic than crude. However, both are rarely toxic in environmentally realistic situations, i.e. at concentrations below 0.2 ppt. In contrast, the toxicity of used refined oil is higher (embryo EC_{50} of 40 μg/L, Byrne and Calder, 1977), probably due to the presence of highly toxic pyrogenic compounds such as polyaromatic hydrocarbons (PAH) and their hydroxylated derivatives. This emphasizes the necessity for adequate disposal of the residues from oil motor vehicles. Pelletier et al. (1997) have demonstrated the high toxicity of PAH's to early developmental stages of marine organisms, placing emphasis on the increase in toxicity caused by UV radiation, due to the formation of highly oxidizing intermediates. Pyrene under ultraviolet light had an EC_{50} as low as 0.23 μg/L for *M. lateralis* embryos.

Besides being used for basic research on specific (clearly identified and dosed) chemical compounds, bivalve embryo and larval bioassays are increasingly being used to monitor the quality of environmental samples of unknown composition.

The results of these investigations are rarely published in reviewed journals, as they often concern specific situations that are difficult to generalize with respect to other regions or seasons of the year. Frequently the results of this kind of studies show important small-scale variation that further complicates monitoring programs, and stresses the need for simple, rapid and affordable screening techniques.

Cryopreserved 48 h-old larvae of *V. philippinarum* have been also used in ecotoxicological bioassays to assess a pollution gradient in the German Bight (McFadzen, 1992). Larval survival steadily improved as the transect moved offshore, with higher mortalities occurring in the surface microlayer and sediment elutriates compared to the subsurface bulk waters.

In order to evaluate the toxicity of seawater and sediment exposed to the fuel spilled by the *Prestige* tanker, sank in November 2002 130 miles off cape Finisterre after 5 d trawling in zigzag trajectory, embryogenesis success of *V. corrugata* and *P. virgineus* was recorded after incubation in seawater and elutriates obtained from affected and unaffected areas. Fuel-polluted seawater showed a marked inhibition of embryogenesis, while sediment elutriates showed moderate to no toxicity (Mariño-Balsa et al., 2003; Beiras and Saco-Álvarez, 2006). The conclusions of these studies bear also implications for the strategy of monitoring in an oil spill event. Oil spill contingency plans focus on the follow-up and prediction of the trajectory of the superficial oil slick.

Along with the wind-driven drift of the oil slick on the surface, the dispersion of components of the hydrocarbon mixture accommodated into the liquid phase, whose trajectory is associated to subsurface hydrodynamics, must also be taken into account. The first is the cause of heavy mortalities on birds and mammals, but the later is the most dangerous for small size plankton and nekton organisms.

ACKNOWLEDGMENTS

I thank N. Trigo for her assistance in the preparation of this manuscript and all the personnel from the LEM-ECOTOX group for their one decade long efficient work developing biological tools in marine ecotoxicology. This study was partially funded by Research Project CTM2009-10908 from the Spanish *Ministerio de Ciencia e Innovación*.

REFERENCES

Abaychi, J. K. and Mustafa, Y. Z. (1988). The asiatic clam, *Corbicula fluminea*: an indicator of trace metal pollution in the Shatt al-Arab River, Iraq. *Environ. Poll.* 54, 109-122.

Absil, M. C. P., Kroon, J. J. and Wolterbeek, H. T. (1994). Availability of copper from phytoplankton and water for the bivalve *Macoma balthica*. II. Uptake and elimination from ^{64}Cu-labelled ditoms and water. *Mar. Biol.* 118, 129-135.

Alvarez-Legorreta, T., Gold-Bouchot, G. and Zapata.Pérez, O. (1994). Hydrocarbon Concentrations in Sediments and Clams (*Rangia cuneata*) in Laguna de Pom, Mexico. *Bull. Environ. Contam. Toxicol.* 52, 39-45.

Bayarri,S., Baldassarri, L. T., Lacovella, N., Ferrara, F., and Di Domenico, A. (2001). PCDDs, PCDFs, PCBs and DDE in edible marine species from the Adriatic Sea. *Chemosphere.* 43, 601-610.

Beiras, R. and Albentosa, M. (2004). Inhibition of embryo development of the commercial bivalves *Ruditapes decussatus* and *Mytilus galloprovincialis* by trace metals; implications for the implementation of seawater quality criteria. *Aquaculture.* 230, 205-213.

Beiras, R. and Saco-Álvarez, L. (2006). Toxicity of seawater and sand affected by the *Prestige* fuel-oil spill using bivalve and sea urchin embryogenesis bioassays. *Water Air Soil Pollut.* 177, 457-466.

Beiras, R. and Bellas, J. (2008). Inhibition of embryo development of the *Mytilus galloprovincialis* marine mussel by organic pollutants; assessment of risk for its extensive culture in the Galician Rias. *Aquaculture.* 277, 208-212.

Besada, V., Andrade, J. M., Schultze, F., and González, J. J. (2007). Levels of trace metals in coastal seafood. *Intnl. Symp. Mar. Sci.* Valencia, Spain.

Bilos, C., Colombo, J. C. and Rodriguez Presa, M. J. (1998). Trace metals in suspended particles, sediments and Asiatic clams (*Corbicula fluminea*) of the Río de la Plata Estuary, Argentina. *Environ. Poll.* 99, 1-11.

Bordin, G., McCourt, J., Cordeiro Raposo, F., and Rodriguez, A. (1996). Trace metals in the marine bivalve *Macoma balthica* in the Westerschelde estuary, The Netherlands. Part: 3 Variability of the role of cytosol in metal uptake by the clams. *Sci. Total Envir.* 180, 241-255.

Bourgoin, B. P., Risk, M. J. and Aitken, A. E. (1991). Factors Controlling Lead Availability to the Deposit-feeding Bivalve *Macoma balthica* in Sulphide-rich Oxic Sediments. *Estuar. Coast. Shelf Sci.* 32, 625-632.

Byrne, C. J. and Calder, J. A. (1977). Effect of water soluble fractions of crude, refined and waste oils on the embryonic and larval stages of the quahog clam *Mercenaria* sp. *Mar. Biol.* 40, 225-231.

Byrne, P. A. and O´Halloran, J. O. (1999). Aspects of Assaying Sediment Toxicity in Irish Estuarine Ecosystems. *Mar. Poll. Bull.* 39, 97-105.

Byrne, P. A. and O´Halloran, J. O. (2000). Acute and sublethal toxicity of estuarine sediments to the Manila Clam, *Tapes semidecussatus. Environ. Toxicol.* 15, 456-468.

Calabrese, A., Macinnes, J. R., Nelson, D. A., and Miller, J. E. (1977). Survival and growth of bivalve larvae under heavy-metal stress. *Mar. Biol.* 41, 179-184.

Cardwell, R. D., Woelke, C. E., Carr, M. I., and Sanborn, E. W. (1979). Toxic substance and water quality effects on larval marine organisms. *Washington Depart. Of Fisheries, Technical Report.* 45, 1-71.

Chong, K. and Wang, W.-X. (2001). Comparative studies on the biokinetics of Cd, Cr, and Zn in the green mussel *Perna viridis* and the Manila clam *Ruditapes philippinarum. Environ. Poll.* 115, 107-121.

Clement, L. E., Stekoll, S. M. and Shaw, D. G.(1980). Accumulation, Fractionation and Release of Oil by the Intertidal Clam *Macoma balthica. Mar. Environ.* 57, 41-50.

Connor, P. M. (1972). Acute toxicity of heavy metals to some marine larvae. *Mar. Pollut. Bull.* 3, 190-192.

Davis, H. C. (1961). Effects of some pesticides on eggs and larvae of oysters (*Crassostrea virginica*) and clams (*Venus mercenaria*). *Commer. Fish. Rev.* 23, 8-23.

Davis, H. C. and Hidu, H. (1969). Effects of pesticides on embryonic development of clams and oysters and on survival and growth of larvae. *US Fish Wld. S. Fish. B.* 67, 393-404.

De Gregori, I., Pinochet, H., Gras, N., and Muñoz, L. (1996). Variability of cadmium, cooper and zinc levels in molluscs and associated sediments from Chile. *Environ. Poll.* 92, 359-368.

Eisler, B. R. (1979). Behavioural responses of marine poikilotherms to pollutants. *Phil. Trans. R. Soc. Lond. B.* 286, 507-521.

Eyster, L. S. and Morse, M. P. (1984). Early shell formation during molluscan embryogenesis, with new studies on the surf clam, *Spisula solidissima. Am. Zool.* 24, 871-882.

Frouin, H., Pellerin, J., Fournier, M., Pelletier, E., Richard, P., Pichaud, N., Rouleau, C., Garnerot, F. (2007). Physiological effects of polycyclic aromatic hydrocarbons on soft-shell clam *Mya arenaria. Aquat. Toxicol.* 82, 120-134.

Fukuyama, A. K., Shigenaka, G. and Hoff, R. Z. (2000). Effects of residual Exxon Valdez oil on intertidal *Protothaca staminea*: mortality, growth, and bioaccumulation of hydrocarbons in transplanted clams. *Mar. Poll. Bull.* 40, 1042-1050.

Gagné, F., Blaise, C., Pellerin, J., Andre, C. (2007) Neuroendocrine disruption in Mya arenaria clams during gametogenesis at sites under pollution stress. *Mar. Environ. Res.* 64, 87-107.

González-Quijano, A, Besada, V., García, A., Schultze, F., and González, J. J. (2005). Contenido de compuestos organoclorados y metales pesados en almeja babosa (*Venerupis pullastra*) de las rías de Vigo y Pontevedra (NW de España).*6º Congreso Ibérico de Contaminación y Toxicología Ambiental.* Cádiz, Spain.

Gormly, N. M., Singer, S. and Genthener, F. J. (1996). Nontarget testing of microbial pest control agens using larvae of the coot clam, *Mulinia lateralis. Dis. Aquat. Organ.* 26, 229-235.

Hickey, C. W. and Martin, M. L. (1995). Relative sensitivity if five benthic invertebrate species to reference toxicants and resin-acid contaminated sediments. *Environ. Toxicol. Chem.* 8, 1401-1409.

Hidu, H. (1965). Effects of synthetic surfactants on the larvae of clams (*Mercenaria mercenaria*) and oysters (*Crassostrea virginica*). *J. Water Poll. Contr. Fed.* 37, 262-270.

His, E., Beiras, R. and Seaman, M. N. L. (1999). The assessment of marine pollution-bioassays with bivalve embryos and larvae. In: A. I. Southward, P. A. Tyler, C. M.

Young (Eds.), *Advances in Marine Biology, vol. 37*. pp. 1-178. Academic Press, London.

Ho, M. S. and Zubkoff, P. L. (1982). The effects of mercury, copper and zinc on calcium-uptake by larvae of the clam, *Mulinia-lateralis. Water Air Soil Poll.* 17, 409-414.

Jackim, E., Morrison, G. and Steele, R. (1977). Effects of environmental factors on radiocadmium uptake by four species of marine bivalves. *Mar. Biol.* 40, 303-308.

Kaschl, A. and Carballeira, A. (1999). Behavioural responses of *Venerupis decussata* (Linnaeus, 1758) and *Venerupis pullastra* (Montagu, 1803) to copper-spiked marine sediments. *Bol. Inst. Esp. Oceanogr.* 15, 1-4.

Kobayashi, N. (1971). Fertilised sea urchin eggs as an indicatory material for marine pollution bioassay, preliminary experiments. *Publs. Seto Mar. Biol. Lab.* 18, 379-406.

Langston, W. J. (1982). The distribution of mercury in British estuarine sediments and its availability to deposit-feeding bivalves. *J. Mar. Biol. Ass. U.K.* 62, 667-684.

Laughlin, R. B., Gustafson, R. and Pendoley, P. (1988). Chronic embryo-larval toxicity of tributylin (TBT) to the hard shell clam *Mercenaria mercenaria. Mar. Ecol. Prog. Ser.* 48, 29-36.

Laughlin, R. B., Gustafson, R. G. and Pendoley, P. (1989). Acute toxicity of tributyltin (TBT) to early life-history stages of the hard shell clam, *Mercenaria mercenaria. Bull. Environ. Contam. Toxicol.* 42, 352-358.

Lee, B.-G.,Wallace, W. G. and Luoma, S. N. (1998). Uptake and loss kinetics of Cd, Cr and Zn in the bivalves *Potamocorbula amurensis* and *Macoma balthica*: effects of sice and salinity. *Mar. Ecol. Prog. Ser.* 175, 177-189.

Luoma, S. N. and Bryan, G. W. (1978). Factors controlling the availability of sediment-bound lead to the estuarine bivalve *Scrobicularia plana. J. Mar. Biol. Ass. U.K.* 58, 793-802.

Mariño-Balsa, J. C., Pérez, P., Estévez-Blanco, P., Saco-Álvarez, L., Fernández, N., Beiras, R., (2003). Assessment of the toxicity of sediment and seawater polluted by the *Prestige* fuel spill using bioassays with clams (*Venerupis pullastra*, *Tapes decussatus*, *Venerupis rhomboideus*) and the microalga *Skeletonema costatum. Cienc. Mar.* 29, 115-122.

Matozzo, V., Ballarin, L. and Marin, M. G. (2004). Exposure of the clam *Tapes philippinarum* to 4-nonylphenol: changes in anti-oxidant enzyme activities and re-burrowing capability. *Mar. Poll. Bull.* 48, 563-571.

McFadzen, I. R. B. (1992). Growth and survival of cryopreserved oyster and clam larvae along a pollution gradient in the German Bigth. *Mar. Ecol. Prog. Ser.* 91, 215-220.

McGreer, E. R. (1979). Sublethal Effects of Heavy Metal Contaminated Sediments on the Bivalve *Macoma balthica* (L.) *Mar. Poll. Bull.* 10, 259-262.

Morcillo, Y., Ronis, M. J. J. and Porte, C. (1998). Effects of tributylin on the Phase Itestosterone metabolism and steroid tires of the clam *Ruditapes decussate. Aquat. Toxicol.* 42, 1-13.

Morrison, G. E. and Petrocelli, E. A. (1990). Suitability of *Mulinia lateralis* as a euryhaline toxicity test species. *Canadian Technical Report, Fisheries and Aquatic Sciences.* 1774, 337-340.

Møhlenberg, F. and Kiørboe, T. (1983). Burrowing and Avoidance Behaviour in Marine Organisms Exposed to Pesticide-contaminated Sediment. *Mar. Poll. Bull.* 14, 57-60.

Nasci, C., Da ros, L., Campesan, G., Van Vleet, E. S., Salizzatos, M., Sperni, L., and Pavoni, B. (1999). Clam transplantation and stress-related biomarkers as useful tools for assessing water quality in coastal environments. *Mar. Poll. Bull.* 39, 255-260.

Ng, T. Y.-T. and Wang, W.-X. (2004). Detoxificacion and effects of Ag, Cd, and Zn pre-exposure on mental uptake kinetics in the clam *Ruditapes philippinaum. Mar. Ecol. Prog. Ser.* 268, 161-172.

Nipper, M. G., Roper, D. S., Williams, E. K., Martin, M. L., Van Dam, F. L., and Mills, G. N. (1998). Sediment toxicity and benthic communities in mildly contaminated mudflats. *Environ. Toxicol. Chem.* 17, 502-510.

Okubo, K. and Okubo, T. (1962). Study on the bio-assay method for the evaluation of water pollution. II Use of the fertilized eggs of sea urchin and bivalves. *Bulletin of the Tokai Regional Fishery Research Laboratory.* 32, 131-140.

Olla, B. L., Bejda, A. J. and Pearson,W. H. (1983). Effects of Oiled Sediment on the Burrowing Behaviour of the Hard Clam, *Mercenaria mercenaria. Mar. Environ. Res.* 9, 183-193.

Pariseau, J., Myrans, B., Desrosiers, G., Chevarie, L., and Giguére, M. (2007). Influences of physical and biological variables on softshell clam (*Mya arenaria* Linneaus 1758) burial. *J. Shellfish Res.* 26, 391-400.

Pearson, W. H., Woodruff, D. L., Sugarman, P. C., and Olla, B. L. (1981). Effects of oiled sediment on predation on the Littleneck Clam, *Protothaca staminea*, by the Dungeness Crab, *Cancer magister. Estuar. Coast. Shelf Sci.* 13, 445-454.

Pelletier, M. C., Burgess, R. M., Ho, K. T., Kuhn, A., Mckinney, R. A., and Ryba, A. (1997). Phototoxicity of individual polycyclic aromatic hydrocarbons and petroleum to marine invertebrate larvae and juveniles. *Environ. Toxicol. Chem.* 16, 2190-2199.

Phelps, H. L., Hardy, J. T., Pearson, W. H., and Apts, C. W. (1983). Clam burrowing behaviour: inhibition by copper-enriched sediment. *Mar. Poll. Bull.* 14, 452-455.

Phelps, H. L., Pearson, W. H. and Hardy, J. T. (1985). Clam burrowing behaviour and mortality related to sediment copper. *Mar. Poll. Bull.* 16, 309-113.

Phelps, H. L. (1989). Clam burrowing bioassay for estuarine sediment. *Bull. Environ. Contam. Toxicol.* 43, 838-845.

Pfitzenmeyer, H. T. and Drobeck, K. G. (1967). Some factors influencing reburrowing activity of Soft-Shell Clam, *Mya arenaria. Ches. Scien.* 8 (3), 193-199.

Pierre Stecko, J. R. and Bendell-Young, L. I. (2000). Uptake of ^{109}Cd from sedimentsnby the bivalves *Macoma balthica* and *Protothaca staminea. Aquatic Toxicology* 47, 147-159.

Riba, I., Del Valls, T. A., Forja, J. M., and Gómez-Parra, A. (2003). The influence of pH and salinity on the toxicity of heavy metals in sediment to the estuarine clam *Ruditapes philippinarum. Environ. Toxicol. Chem.* 23, 1100-1107.

Riba, I., Del Valls. T. A., Forja, J. M., and Gomez-Parra, A. (2004). The influence of pH and salinity on the toxicity of heavy metals in sediment to the estuarine clam *Ruditapes philippinarum. Environ. Toxicol. Chem.* 23, 1100-1107.

Roberts, M. H. Jr., Diaz, R. J., Bender, M. E., and Huggett, R. J. (1975). Acute toxicity of chlorine to selected estuarine species. *J. Fisher. Res. Board Canada.* 32, 2525-2527.

Roberts, M. H. (1987). Acute toxicity of tributyltin choride to embryos and larvae of 2 bivalve molluscs, *Crassostrea virginica* and *Mercenaria mercenaria. Bull. Environ. Contam. Toxicol.* 39, 1012-1019.

Roosenburg, W. H., Rhoderick, J. C., Block, R. M., Kennedy, V. S., and Vreenegoor, S. M. (1980). Survival of *Mya arenaria* larvae (Mollusca: Bivalvia) exposed to chorine-produced oxidants. *Proc. Nat. Shellf. Ass.* 70, 105-111.

Roper, D. S. and Hickey, C. W. (1994). Behavioural responses of the marine bivalve *Macomona liliana* exposed to copper and chlordane-dosed sediments. *Mar. Biol.* 118, 673-680.

Roper, D. S., Nipper, M. G., Hickey, C. W., Martin, M. L., and Weatherhead, M. A. (1995). Burial, Crawling and Drifting Behaviour of the Bivalve *Macomona liliana* in Response to Common Sediment Contaminants. *Mar. Poll. Bull.* 31, 471-478.

Ruiz, J. M., Bryan, G. W. and Gibbs, P. E. (1994). Chromic toxicity of water tributyltin (TBT) and copper to spat of the bivalve *Scrobicularia plana* ecological implications. *Mar. Ecol. Prog. Ser.* 113, 105-117.

Ruiz, J. M., Szpunar, J. and Donard, O. F. X. (1997). Butylins in sediments and deposit-feeding bivalves Scrobicularia plana from Arcachon Bay, France. *Sci. Total Environ.* 198, 225-231.

Shin, P. K. S., Ng, A. W. M. and Cheung, R. Y. H. (2002). Burrowing responses of the short-neck clam *Ruditapes philippinarum* to sediment contaminants. *Mar. Poll. Bul.* 45, 133-139.

Sobral, P. and Widdows, J. (1997).Effects of Copper Exposure on the Scope for Growth of the Clam *Ruditapes decussates* from Southern Portugal. *Mar. Poll. Bull.*, 34 (12), 992-1000.

Solé, M., Porte, C., Barcelo, D., and Albaiges, J. (2000). Bivalves residue analysis for the assessment of coastal pollution in the Ebro Delta (NW Mediterranean). *Mar. Poll. Bull.* 40, 746-753.

St-Onge, P., Miron, G. and Moreau, G. (2007). Burrowing behaviour of the softshell clam (*Mya arenaria)* following erosion and transport. *J. Exp. Mar. Biol. Ecol.* 340, 103-111.

Stekoll, M. S., Clement, L. E. and Shaw, D. G. (1980). Sublethal effects of chronic oil exposure on the intertidal clam *Macoma balthica*. *Mar. Biol.* 57, 51-60.

Strandberg, J. D., Rosenfield, J., Berzins, I. K., and Reinisch, C. L. (1998). Specific localization of polychlorinated biphenyls in clams (*Mya arenaria*) from environmentally impacted sites. *Aquat. Toxicol.* 41, 343-354.

Strong, C. R. and Luoma, S. N. (1981) Variations in the correlation of body size with concentrations of Cu and Ag in the bivalve Macoma balthica. *Can. J. Fish. Aquat. Sci.* 38, 1059-1064.

Szefer, P., Wolowicz, M., Kusak, J., Deslous-Paoli, M., Czarnowski, W., Frelek, K., and Belzunce, J. (1999). Distribution of mercury and other trace metals in the cockle *Cerastoderma galucum* from the Mediterranean Lagoon Etang de Thau. *Arch. Environ. Contam. Toxicol.* 36, 56-63.

Thomas, C. A. and Bendell-Young, L. I. (1998). Linking the sediment geochemistry of an intertidal region to metal bioavailability in the deposit feeder *Macoma balthica*. *Mar. Ecol. Prog. Ser.* 173, 197-213.

Usero, J., Gónzalez-Regalado, E. and Gracia, I. (1996). Trace Metals in the Bivalve Mollusc *Chamelea gallina* from the Atlantic Coast of Southern Spain. *Mar. Poll. Bull.* 32, 305-310.

Wallace, W. G. and Lee, B.-G., Luoma, S. N. (2003).Subcellular compartmentalization of Cd and Zn in two bivalves. I. Significance of metal-sensitive fractions (MSF) and biologically detoxified metal (BDM). *Mar. Ecol. Prog. Ser.* 249, 183-197.

Wilson, D. P. (1951). A biological difference between natural waters. *J. Mar. Biol. Ass. U.K.* 30, 1-21.

Woelke C. E. (1961). Bioassay the bivalve larvae tool. *Proceedings of the Northwest Symposium on Water Pollution Research, US Dept. HEWPHS*, Portland, Oregon, pp. 113-123.

Wright, A. D. and Zamuda, C. D. (1987). Copper accumulation by two bivalve molluscs: Salinity effect is independent of cupric ion activity. *Mar. Environ. Res.* 23, 1-14.

INDEX

A

B

C

D

E

F

G

H

I

M

Q

R

S

T

U

V

W

Y

Z